METAMATERIALS

METAMATERIALS

Technology and Applications

Edited by Pankaj K. Choudhury

CRC Press
Taylor & Francis Group
Boca Raton London New York

CRC Press is an imprint of the
Taylor & Francis Group, an **informa** business

First edition published 2022
by CRC Press
2 Park Square, Milton Park, Abingdon, Oxon, OX14 4RN

and by CRC Press
6000 Broken Sound Parkway NW, Suite 300, Boca Raton, FL 33487-2742

British Library Cataloguing-in-Publication Data
A catalogue record for this book is available from the British Library

Library of Congress Cataloging-in-Publication Data
Names: Choudhury, P. K., editor.
Title: Metamaterials : technology and applications / edited by Pankaj K. Choudhury.
Other titles: Metamaterials (CRC Press)
Description: First edition. | Boca Raton : CRC Press, 2022. |
Includes bibliographical references and index. |
Summary: "This book is comprised of chapters on the topics pivoted to metamaterials, i.e., the artificially engineered mediums/structures, of micro- and nano-scale sizes. These will include novel applications of metamaterials in sensors, solar-cell designs, diffractive lenses, perfect absorbers, phase-shifters, THz imaging, antennas, etc" -- Provided by publisher.
Identifiers: LCCN 2021021776 (print) | LCCN 2021021777 (ebook) |
ISBN 9780367505080 (hardback) | ISBN 9780367505110 (paperback) |
ISBN 9781003050162 (ebook) Subjects: LCSH: Metamaterials. Classification: LCC TK7871.15.M48 M46 2022 (print) | LCC TK7871.15.M48 (ebook) | DDC 620.1/1--dc23
LC record available at https://lccn.loc.gov/2021021776
LC ebook record available at https://lccn.loc.gov/2021021777

ISBN: 978-0-367-50508-0 (hbk)
ISBN: 978-0-367-50511-0 (pbk)
ISBN: 978-1-003-05016-2 (ebk)

DOI: 10.1201/9781003050162

Typeset in Times
by MPS Limited, Dehradun

Dedication

Dedicated to
All lives to the COVID-19 pandemic
&
Those who strive for peace

Contents

Preface

Metamaterials have been in the research limelight for the past few years due to their exotic electromagnetic characteristics. These artificial structures are designed to achieve electromagnetic properties that do not occur in nature, and they are able to exhibit unique responses to incidence excitation. Their unusual features enable them to be used in a variety of technological applications, such as antennas, filters, absorbers, sensors, energy harvesters, cloaks and many others.

A meta-atom is the basic building block of a metasurface, which is usually a resonator comprising either plasmonic or dielectric materials. With this viewpoint, the role of engineered mediums remains greatly important due to the possibility of *on-demand* tailoring of the electromagnetic response. The propagation of surface electromagnetic waves at the interface of specially designed mediums has been widely investigated as many new forms of miniaturized devices can be used for novel applications.

This book emphasizes the fundamentals of metamaterials, describing the development of the field and the underlying theories, followed by the relevant advancements in the research arena. The authors, who are from different countries, contributed their recent research results, pivoted to metamaterial designs and experiments in fields ranging from optical materials, to antennas, to even microwave tubes. This illustrates the phenomenal growth of interest among R&D scientists focusing on engineered metamaterial technology-oriented applications. Both theoretical and experimental investigations are discussed, so the book can benefit expert scientists in universities and research laboratories as well as novice researchers, such as graduate students, to frame their own research topics/ideas and objectives.

This book comprises 13 chapters written by scientists from various countries. In Chapter 1, Subal Kar discusses a comprehensive roadmap of the progress of metamaterial and metasurface technology, as well as application viewpoints from its inception to recent times. The analytical treatment of metamaterials involves the homogenization process of a medium, which makes it essential to use effective medium theory. In Chapter 2, Guha and Basu present a review of effective medium theory and electromagnetic analysis of parameter retrieval techniques in the context of metamaterials. Their report incorporates simulation and experimental validations of retrieval techniques, as used in the case of metamaterial unit cells.

Achieving high-performance metamaterials at the device level requires fine adjustments of the constituting engineered meta-atoms at the micro/nano scale. This often results in unfeasible design processes owing to costs, complexity and time constraints. G. Oliveri *et al.,* in Chapter 3, review the recent advances in Material-by-Design technology, with specific attention to applicative scenarios emerging in communications and sensing. Chapter 4 focuses on tunable metamaterials. Ke Bi summarizes tuning methods for magnetically tunable, electrically tunable, thermally tunable and flexible metamaterials, emphasizing recent developments and technological potentials.

In Chapter 5, Kang *et al.* touch upon absorber applications of metamaterials in optical spans in the electromagnetic spectrum. In particular, the authors review theoretical background and strategies to achieve metamaterial-based near-perfect absorbers in the visible and infrared regimes, which can be harnessed for sensing, imaging, photo-detection and solar cell applications. In line with this, the theme of Chapter 6 is the use of metamaterials in designing low-frequency perfect absorbers. Khuyen *et al.* present a brief introduction of the operational progress of perfect absorbers as well as underlaying challenges and tendencies for the future development of low-frequency metamaterial-based perfect absorbers.

Tatjana Gric discusses the plasmonic properties of photonic metamaterials in Chapter 7 The author touches upon a few different forms of metamaterials comprising layered composites formed by dielectric mediums and graphene as well as their underlying physical mechanisms. In addition, the study incorporates a description of hyperbolic metamaterials, and the investigation demonstrates an application of studying cancer. In Chapter 8, Sreekanth and Singh report on tuning optical responses exhibited by specially designed hyperbolic metamaterials from the visible to terahertz frequencies by incorporating suitable functional materials within the multilayered structure. Also, the authors emphasize potential applications of active hyperbolic metamaterials in photonics and biomedical research.

Graphene has been greatly researched, and it remains extremely important in constructing tunable electronic and/or optical components. The amalgamation of graphene in designing metamaterials has led to a variety of technologically prudent nanoengineered mediums. Keeping this in mind, in Chapter 9, P.K. Choudhury explains the fundamentals of graphene and reviews the spectral features of a few types of graphene-supported metamaterial configurations that can be used for filters, absorbers and sensors. In Chapter 10, Nasri *et al.* discuss the use of specially designed asymmetric H-shaped metamaterials in sensing different estrogenic hormones, for example, 17β-estradiol (E2) – the most potent estrogen. The authors address the use of aptamers in the experimental study of various estrogenic molecules via metamaterials-based plasmonic sensing.

Cselyuszka *et al.*, in Chapter 11, report detailed insight into acoustic surface wave propagation at the boundary between fluid and hard-grooved surfaces. Within the context, the authors provide the theoretical background and analyses of spoof acoustic surface wave propagation as well as control and manipulation. Also, they touch upon the applications in sound trapping, collimation of sound, gas sensing and acoustic lensing.

In Chapter 12, Oleg Rybin talks about the principle of miniaturization of microwave patch antennas in achieving improved performance. The author presents the importance of substrate in enhancing the constitutive properties. The chapter reports that the use of larger number of layers yields better improvements with respect to power gain and efficiency of the antenna. The author shows that replacing a high dielectric substrate with metamaterial having the same values of effective relative permittivity/permeability enables one to create compact multi-band and multi-directional patch antennas.

Chapter 13 discusses the importance of metamaterials in constructing vacuum electron devices. Guha *et al.* review attempts made to improve the performance of conventional vacuum electron devices using various metamaterials. In the chapter, the authors analyze the effect of metamaterial assistance considering negative effective relative permittivity and permeability. Also, they model the constitutive parameters of metamaterials to study the vacuum electron devices through presenting analyses, simulations and experiments.

Overall, the included chapters are pivoted to novel applications of metamaterials in a wide spectral range. The themes are basically focused on the development of sensors, filters, absorbers, antennas and vacuum electronic devices. The editor of this book is thankful to all contributors for spending enough attention to summarize their research findings during the hard time of the ongoing pandemic. Indeed, it took a few more months than what was expected to realize this book, which essentially delayed the overall process. The editor highly appreciates the patience of all the contributors in this respect.

Finally, the editor takes this opportunity to acknowledge Marc Gutierrez of the CRC Press (Taylor & Francis Group) and the team for inviting him to take up this editorial task. Also, the editor extends sincere thanks to the Director of the Institute of Microengineering and Nanoelectronics (Universiti Kebangsaan Malaysia, Malaysia) for constant encouragement and help throughout. The endless support from his spouse, Swarnadurga, can in no way be forgotten.

Pankaj K. Choudhury

About the Editor

Pankaj K. Choudhury received a Ph.D. degree in physics in 1992. From 1992 to 1997, he was a Research Associate at the Department of Electronics Engineering, Institute of Technology, Banaras Hindu University (Varanasi, India). In 1997, he joined the Department of Physics, Goa University (Goa, India) as a Lecturer. In late 1999, he became a Researcher at the Center of Optics, Photonics and Lasers (COPL), Laval University (Quebec, Canada). From 2000 to 2003, he was with the Faculty of Engineering, Gunma University (Kiryu, Japan), as a Researcher. In May 2003, he received the position of Professor at the Faculty of Engineering, Multimedia University (Cyberjaya, Malaysia), where he was until late 2009. During that span, he also served the Telekom Research and Development (TMR&D, Malaysia) as a consultant for projects on optical devices. In late 2009, he became Professor at the Institute of Microengineering and Nanoelectronics (IMEN), Universiti Kebangsaan Malaysia (The National University of Malaysia, Malaysia). His research interests lie in the theory of optical waveguides, which include complex mediums, fiber optic devices and nanoengineered structures. He has published over 260 research papers, contributed chapters to 17 books, and edited and co-edited 7 research-level books. He is the reviewer for nearly four dozen research journals. He remains in the Editorial Board of *Optik – International Journal for Light and Electron Optics* (Elsevier, The Netherlands). Also, he is the Editor-in-Chief of the *Journal of Electromagnetic Waves and Applications* (Taylor & Francis, UK). He is a Fellow of IET and Senior Member of IEEE, OSA and SPIE.

1 Progress in Metamaterial and Metasurface Technology and Applications

Subal Kar M.Tech., Ph.D (Tech.)
Former Professor and Head, Institute of Radio Physics and Electronics, University of Calcutta, Kolkata, India

1.1 INTRODUCTION

Metamaterial, or phenomenologically left-handed material (LHM), is popularly known to make things "invisible." The "meta-material" is the combination of two terms: *meta* (whose meaning in Greek is "beyond") indicates that it exhibit properties not available in nature, and *material* means that it is constituted of permittivity and permeability. Metamaterial is capable of reversal of Snell's law of refraction and reversed Doppler Effect, and it produces reversal of Čerenkov radiation [1,2], see Fig. 1.1.

The history of artificial material exhibiting properties not available in nature can be traced back to the pioneering work of J.C. Bose on twisted jute pair polarizer at the millimeter-wave frequency in 1898 and K.F. Lindman's work on electromagnetic chirality at the microwave frequency in 1914, which may be considered to be the precursor of the present-day artificial chiral material. During 1940–1960, extensive work was also carried out on the so-called artificial dielectrics. However, the beginning of research on LHM (left-handed material), or metamaterial, can actually be reckoned from the seminal work of V.G. Veselago in 1968 [1]. Veselago examined the solutions to Maxwell's equations in hypothetical media having simultaneously negative isotropic permittivity and permeability, and he observed that such material would exhibit some counter-intuitive phenomena. He termed such materials, not found in nature, as "left-handed material". The term "metamaterial" for LHM was coined by R. Walser at a 1999 DARPA workshop on composite materials, where the prefix "meta" was chosen to convey that such composites transcend the properties of natural materials.

Technically speaking, metamaterial is artificially structured material (commonly metal-dielectric composite) having extrinsic inhomogeneity, but to an incident electromagnetic wave, it is effectively homogeneous. The structural properties, rather than the chemistry of material with which it is constituted, determine the

DOI: 10.1201/9781003050162-1

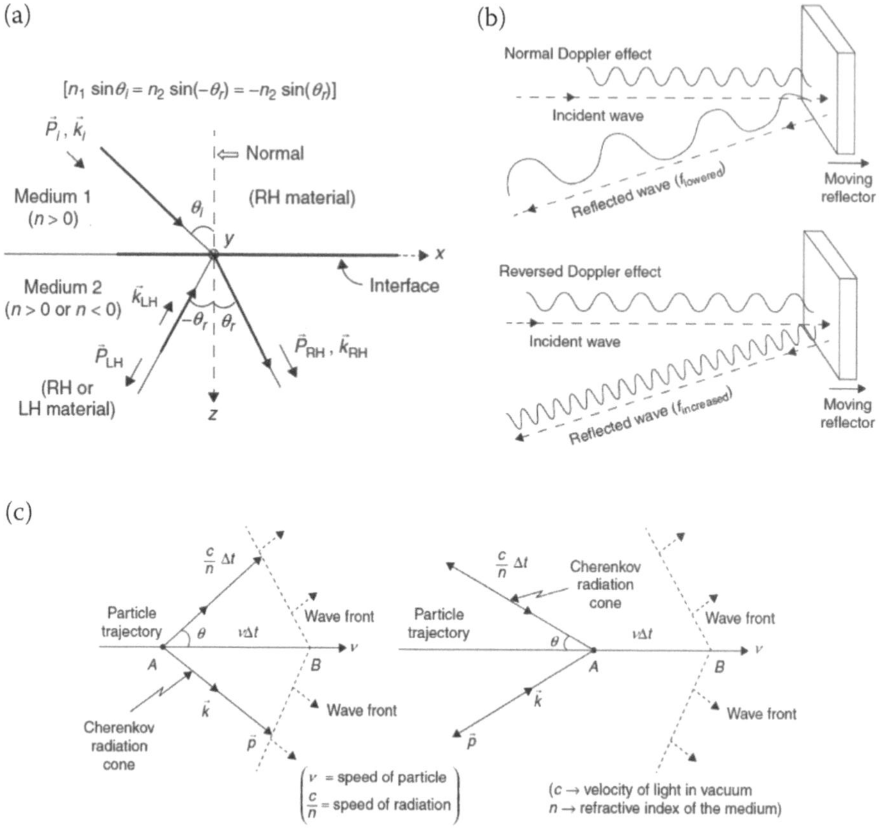

FIGURE 1.1 Counter-intuitive phenomena with metamaterial; (a) reversal of Snell's law of refraction, (b) reversed Doppler Effect and (c) reversal of Čerenkov radiation [2].

characteristics and functionalities of metamaterials. Such artificial material is realized with the unit cells in periodic structure having their dimensions commensurate with small-scale physics ($h << \lambda$, where h is the characteristic dimension of a unit cell and λ is the operating wavelength).

Permittivity and permeability are the constitutive parameters of any material. In usual/natural material, or right-handed material (RHM), both the permittivity and permeability are positive, whereas in LHM, or metamaterial, the permittivity and permeability are simultaneously negative. Unlike RHM, the permittivity and permeability are bulk properties of LHM rather than the elemental property, and thus, we talk of effective relative permittivity (ε_{reff}) and effective relative permeability (μ_{reff}) of LHM, or metamaterial.

Apart from being RHM or DPS (double positive), and LHM or DNG (double negative), the materials can be of other forms having either negative permittivity with positive permeability, or negative permeability with positive permittivity. Fig. 1.2 shows the plot of constitutive parameters (ε vs. μ) in the four quadrants. Negative permittivity below plasma frequency is possible in the ionosphere of

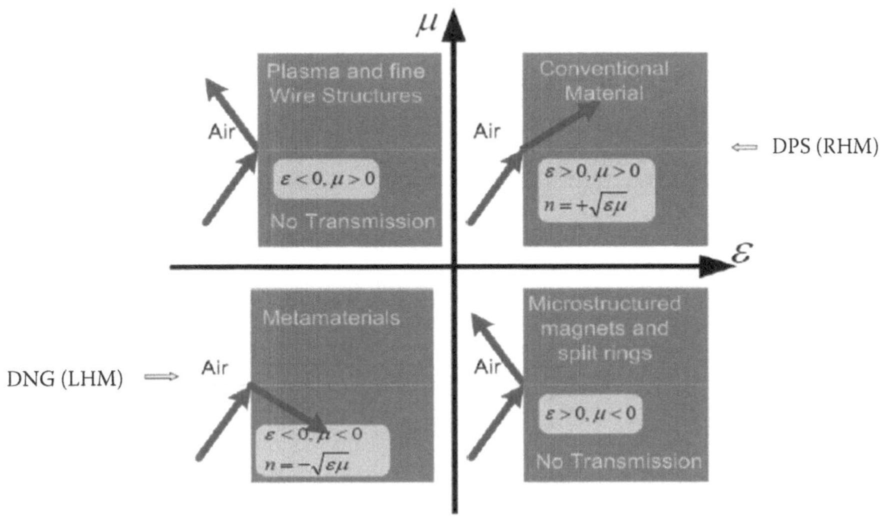

FIGURE 1.2 Plot of constitutive parameters of materials showing the possibility of different materials (Taken from the presentation of Subal Kar; 99th Indian Science Congress, 2012).

the Earth's atmosphere, which has been known for a long time. Negative permeability below plasma frequency is possible in gold and silver, but at the ultraviolet (UV) frequency.

John Pendry first showed the possibility for practically realizing the electric plasma at microwave frequency using an array of thin metallic wires in 1996 [3], and magnetic plasma using an array of split-ring resonators in 1999 [4], respectively, to realize the negative ε_{reff} and negative μ_{reff} below the corresponding plasma frequency. In actual design, the negative permittivity is realized with an array of metallic thin wires (TWs), see Fig. 1.3a, below its electric plasma frequency, and negative permeability with a matrix of C-shaped split-ring resonators (SRRs), see Fig. 1.3b, below its magnetic plasma frequency. Each unit cell in a periodic array of TWs or an SRR matrix, when irradiated with an electromagnetic signal, acts as an "electric atom" and "magnetic atom", respectively, mimicking the atomic arrangements as in the lattice

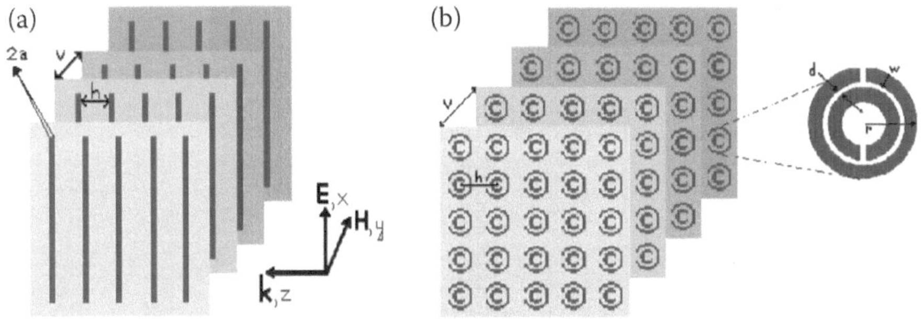

FIGURE 1.3 Plasmonic metamaterial (a) thin-wire (TW) array and (b) split-ring resonator (SRR) matrix [2].

TABLE 1.1
Dual of RHM gives the LHM [2]

Parameters for	β	Z_c	v_p	v_g	n
RHM 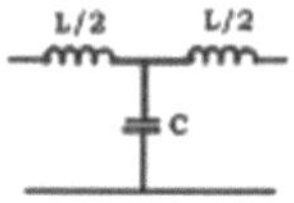	$\omega\sqrt{LC}$	$\sqrt{\frac{L}{C}}$	$\frac{1}{\sqrt{LC}}$	$\frac{1}{\sqrt{LC}}$	$\frac{\sqrt{LC}}{\sqrt{\mu_0\varepsilon_0}}$
LHM 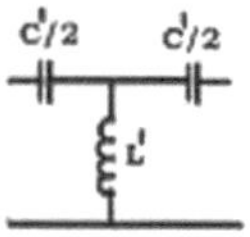	$-\frac{1}{\omega\sqrt{L'C'}}$	$\sqrt{\frac{L'}{C'}}$	$-\omega^2\sqrt{L'C'}$	$+\omega^2\sqrt{L'C'}$	$-\frac{1}{\omega^2\sqrt{L'C'\mu_0\varepsilon_0}}$

of natural material. The metamaterial realized with an array of TWs and a matrix of SRRs combined together constitutes a plasmonic metamaterial.

Table 1.1 shows the analogy between possible left-handed (LH) waves and the dual of the normal transmission line. Eleftheriades et al. [5] in 2002 proposed an alternative way to realize the LHM property using transmission lines.

The practical implementation of transmission line metamaterial is done by periodically loading a host transmission line with series capacitance and shunt inductance, as shown in Fig. 1.4. Effective metamaterial property is realizable only when the unit cell dimension (d) satisfies the condition: $d \ll \lambda$. Being non-resonant, the periodically loaded transmission line (PLTL) exhibits simultaneously low loss and broad bandwidth, and it is thus well suited for RF and microwave circuit applications.

It is interesting to note that Maxwell's wave equation can be used both for RHM and LHM, since the refractive index n appears as square power in the equation (see Eq. (1.1)) given below:

$$\nabla^2\Psi + n^2\frac{\omega^2}{c^2}\Psi = 0 \tag{1.1}$$

where $n^2 = \varepsilon_r\mu_r$ and Ψ represent the electric or magnetic field.

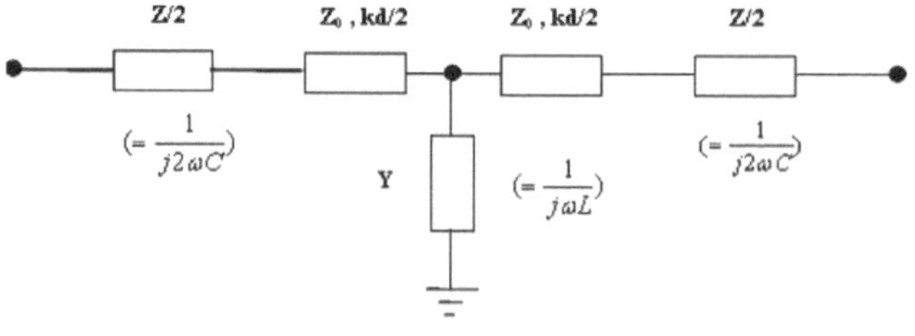

FIGURE 1.4 Schematic of periodically loaded transmission line (PLTL) to realize metamaterial property [2].

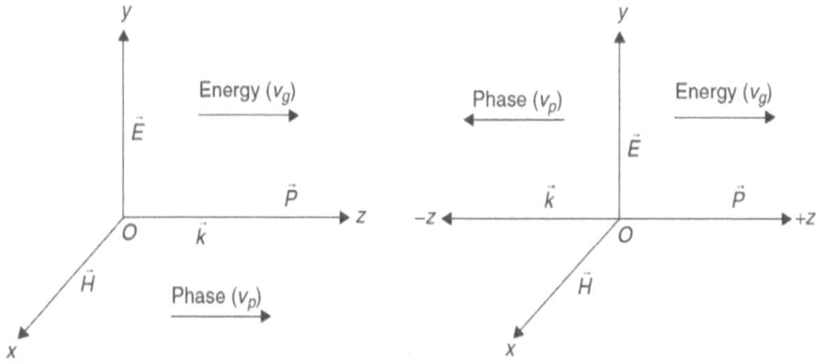

FIGURE 1.5 The triad formed by E, H and k vectors in the RHM and LHM, and the direction of Poynting vector P [2].

For RHM or double positive (DPS) medium, when both the ε_r and μ_r are positive, $n\,(=+\sqrt{\varepsilon_r\mu_r})$ is positive, whereas for LHM or DNG medium, when ε_r and μ_r are simultaneously negative, $n\,(=-\sqrt{\varepsilon_r\mu_r})$ is negative. However, Maxwell's wave equations are equally valid for signal propagation both in the case of RHM and also for LHM.

It may further be noted that E, H and k vectors form a left-handed triad in the case of LHM, unlike the RHM, in which E, H and k vectors form a right-handed triad, see Fig. 1.5. Hence, the LHM is said to support backward waves (as the k vector is in the $-z$-direction). However, the flow of electromagnetic energy, given by the Poynting vector, in LHM or metamaterial, like RHM, remains in the $+z$-direction (otherwise causality would have been violated).

1.2 DEVELOPMENT OF METAMATERIAL TECHNOLOGY – FROM INCEPTION TO RECENT TIMES

The first successful metamaterial was made at the UCSD (USA) in 2001 [6] using TW strips and SRRs to exhibit a negative refractive index at microwave frequency with both the TWs and SRRs printed as metal inclusions on a dielectric substrate, see Fig. 1.6a. It was a plasmonic metamaterial, with TWs exhibiting

(a) (b)

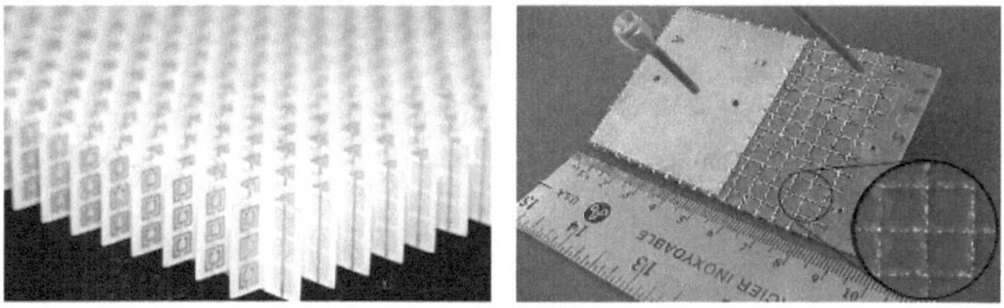

FIGURE 1.6 Two basic types of metamaterial; (a) Plasmonic metamaterial of TWs and SRRs, and (b) PLTL [2].

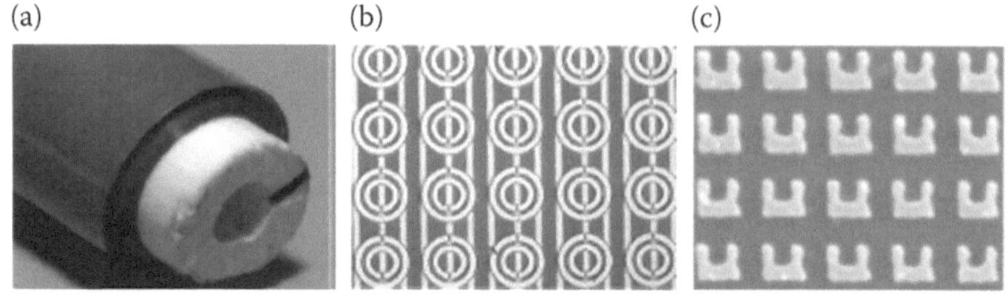

FIGURE 1.7 SRR structures at (a) 21 MHz, (b) 100 GHz and (c) 200 THz [7].

the effective negative permittivity below the electric plasma frequency of TWs, and SRRs exhibiting the effective negative permeability below the magnetic plasma frequency of SRRs. Since plasmonic metamaterial is inherently narrow-band due to the resonant nature of individual metamaterial inclusions, PLTL metamaterial was designed in 2002, see Fig. 1.6b [5], to realize the broadband metamaterial components, as the transmission line is non-resonant in nature, and thus, inherently broadband.

Since the proposal of realizing negative effective permeability with SRRs (more specifically with the spring role structure) by Pendry [4], various SRR structures have been developed from MHz to THz frequency range [7]; some of those structures are shown in Fig. 1.7.

1.2.1 Microwave to Infrared Metamaterial

Apart from different shaping of SRRs at different frequency ranges, as indicated in Fig. 1.7, variations of these magnetic inclusion structures have also been developed to realize different performance characteristics. The multiple split-ring resonator (MSRR) is used to increase the magnetization compared to the simple couplet-type SRR. However, both (SRR and MSRR) suffer from the problem of bi-anisotropy (that causes imbalance of current in the two rings due to asymmetric placement of splits in the two rings) [2]. The use of two splits in both rings and rotating one ring with respect to the other ring by 90° solves this problem and results in the so-called double split-ring resonator (DSRR) or labyrinth resonator (LR), as shown in Fig. 1.8 [8–10]. The LR is a fully symmetric metamaterial structure initially designed for high-frequency applications. It has two cuts in each ring, hence lowering the overall capacitance. Another variant of SRR is the spiral resonator (SR) [11]. Two more variants of SRRs are the two-turn split-ring resonator (TTSRR) and the non-bi-anisotropic split-ring resonator (NBSRR), which have advantages in specific applications [8,12]. In TTSRR, the two rings of SRR have a cut on the same side, and the two ends are cross-joined for connectivity; NBSRR is the symmetric version of TTSRR where bi-anisotropy is reduced [2]. A comparison of the performance characteristics of LR, MSRR and SR is given in Table 1.2 [2].

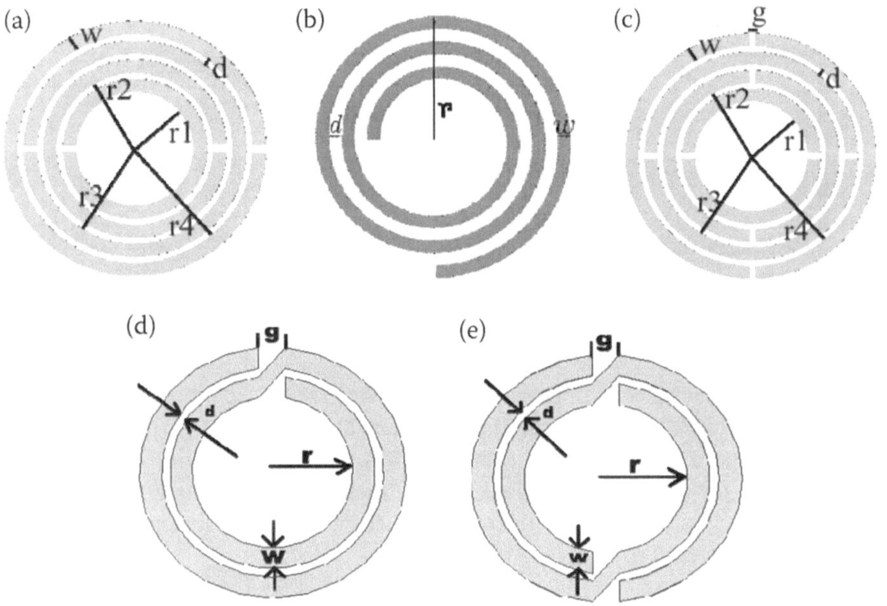

FIGURE 1.8 Variants of SRR; (a) MSRR, (b) SR, (c) LR, (d) TTSRR and (e) NBSRR [2].

TABLE 1.2
Comparative characteristics of LR, MSRR and SR

Structure type	f_r (GHz)	r (mm)	Δf (GHz)
LR	41.5	1.000	0.670
MSRR	21.0	0.648	0.278
SR	8.5	0.357	0.081

Table 1.2 shows that LR has the highest resonant frequency (f_r), provides larger bandwidth (Δf), and also has larger geometrical dimension (r) compared to the other two. This is because of the nature of capacitive loading of the structure concerned. Thus, LR will be a better magnetic inclusion structure at high frequency with larger bandwidth and better design tolerances added with the feature of not being bi-anisotropic like MSRR.

Other magnetic inclusion structures have also been discussed in literature; see Fig. 1.9 provide size-miniaturization at higher frequency of operation while designing metamaterial based couplers and filters. The U-shaped split-ring resonator (USRR) is a variant of SRR where two U-shaped strips are placed inverted on each other, which make it easy to fabricate at higher frequencies. The USRR structure was initially designed for infrared frequency applications [13]. In the broadside coupled split-ring resonator (BC-SRR), two split-rings of an SRR are on two sides of a substrate having a broadside coupling [14]; it is an alternative of an SRR to

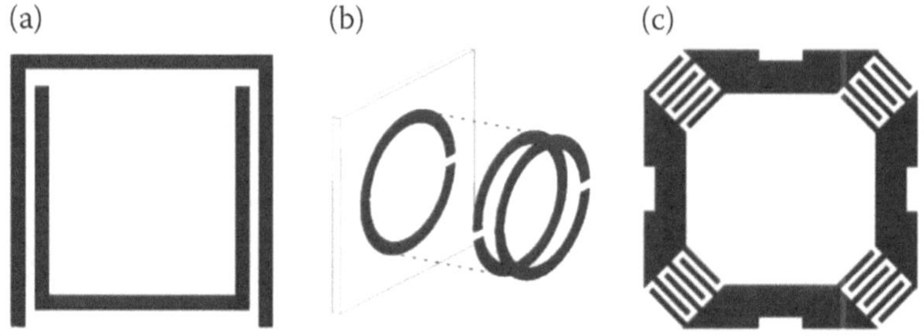

FIGURE 1.9 Other variants of SRR; (a) USRR, (b) BC-SRR and (c) ICRR [16].

avoid bi-anisotropic effects. In inter-digital capacitor loaded ring resonator (ICRR), a square ring is loaded with inter-digital capacitors at four corners [15]. ICRR is the only negative permeability structure that exhibits totally polarization-independent behavior with respect to the applied electric field due to its unique shape. A comparative study on the characteristics of all these magnetic inclusion structures is available in literature [16].

1.2.2 Optical Metamaterial

Design technology of metamaterial at optical frequency is quite different from those at microwave to THz frequencies. The ubiquitous split-ring metamolecule can be scaled down in size up to about 200 THz, but this scaling breaks down at higher frequencies as metal does not behave any more as a conductor, and it becomes transparent to the radiation for wavelengths shorter than 1.5 μm, i.e., beyond the 200 THz range. This scaling limit, combined with the fabrication difficulties of making nanometer-scale SRRs along with TWs (to form an SRR-TW combination), led to the development of alternative designs that are more suitable for optical regimes.

Cut-wire and fishnet are popular types of optical metamaterial structures that were developed in 2005 [17,18]. Cut-wires use pairs of metal nanostrips separated by a dielectric spacer, see Fig. 1.10a. Anti-parallel current flow in the pair results in magnetic resonance. Parallel current flow in the same strip causes electric resonance. However, in such structures, it is difficult to get overlapping

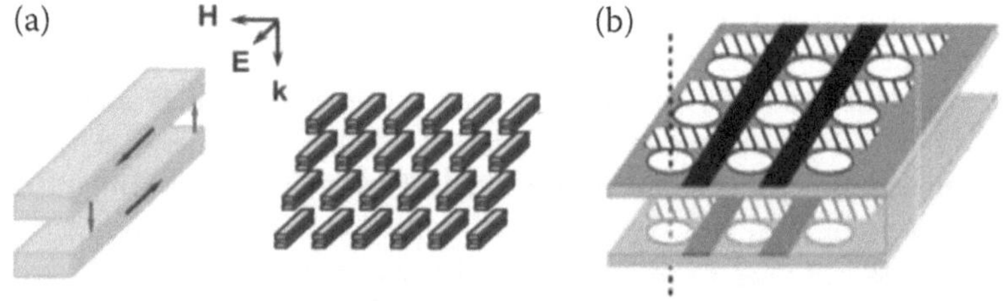

FIGURE 1.10 Optical metamaterial structures; (a) Cut-wire structure, and (b) fishnet structure [7].

$\varepsilon < 0$ and $\mu < 0$ zones. The particular design shown in the figure used a 2 mm × 2 mm array of nanorods imprinted on a glass substrate using electron-beam lithography. A negative refractive index of $n_{eff} \approx -0.3$ at the optical wavelength of 1.5 μm was reported [17].

Fishnet structure combines magnetic coupled strips (to provide $\mu < 0$) with continuous electric strips (to provide $\varepsilon < 0$), see Fig. 1.10b, over a broad spectrum. Hence, the overlapping frequency zone for simultaneously negative ε and μ is easily obtained at the optical frequency. In this design, a multilayer structure consisting of an Al_2O_3 dielectric layer between two gold films perforated with a square array of holes (838 nm pitch; 360 nm diameter) on a glass substrate was used [18]. The active regions for the electric (dark regions) and magnetic (hatched regions) responses are indicated. A minimum negative refractive index of $n_{eff} \approx -2$ was obtained around a 2 μm optical wavelength.

1.2.3 3D Metamaterial

Though it appears to be challenging, there is demand to fabricate three-dimensional (3D) metamaterials. Initial 3D metamaterials were made by creating multilayer structures using the challenging lift-off process and also by using a layer-by-layer technique that requires careful alignment. Complex 3D structures can be fabricated by electron-beam writing, focused-ion beam chemical vapor deposition, etc., but these methods are too complex and time consuming for mass production. Fabrication methods based on two-photon photopolymerization (TPP) are considered to be most promising for manufacturing large-area true 3D metamaterials. Direct single-beam laser writing and multiple-beam TPP techniques are the methods offering sub-diffraction resolution down to 100 nm. Nano-imprint lithography is also a successful method for fabricating 3D metamaterial.

1.2.4 All-dielectric Metamaterial

Unlike conventional metamaterials (that use noble metals like gold or silver on dielectric substrate), a class of new metamaterials, known as transparent metamaterials, are made entirely of dielectric materials or insulators and non-metals [19–21]. In fact, everything will be in a silicon platform, allowing integration of electronic and photonic devices on the same chip. The absence of metal in the metamaterial design will save light from getting unnecessarily lost as heat in the photonic device and interconnections [19]. Such metamaterials can make it possible for computer chips and interconnecting circuits to use photonics, instead of electronics, to process and transmit data, representing a potential leap in performance. In computers and consumer electronics, we still use copper wires between the different parts of the chip. But, if we can confine light to the same size as a nanoscale copper wire, we can have much faster clock speed, and hence, enormously fast data processing – the transparent metamaterial can make this possible [19,20] (Fig. 1.11).

The innovation in transparent metamaterial (TMTM) lies in modifying the phenomenon of total internal reflection (TIR) – the principle normally used to guide light in an optical fiber. Here the optical momentum of evanescent waves

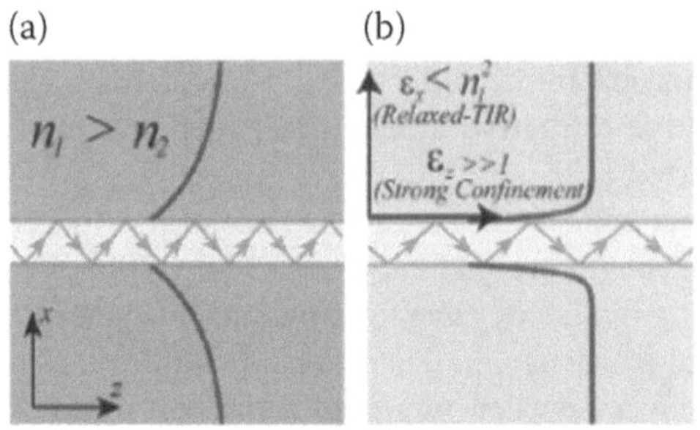

FIGURE 1.11 (a) Conventional TIR – the most power is in the cladding that decays slowly, and (b) relaxed-TIR in TMTM – light decays fast in cladding. [Reprinted/Adapted] with permission from Jahani and Jacob [19] ©The Optical society.

is controlled as opposed to conventional photonic devices, which manipulate propagating waves [20]. This dramatically reduces cross-talk (mutual coupling), the figure of merit of photonic integration, among close-by devices compared to any dielectric waveguide (slot, photonic crystal, etc.), making dense photonic integration possible [21].

The introduction of engineered anisotropy in permittivity tensor (ε_x = 4.8 and ε_z = 11.9) in the momentum of cladding (realizable with TMTM) brings about the phenomenon of relaxed-TIR [19].

Quantum information technologies, such as quantum cryptography, quantum information storage and optical quantum computing, demand effective stable sources of single photons and nanostructures to control the quantum dynamics (of these photons) [22]. Transparent metamaterial can aid in enhancing the single-photon radiation over a broad spectral range, the principle of which is discussed in Fig. 1.12. Here single-photon generation is based on coupling diamond nitrogen-vacancy (NV) centers (or silicon-vacancy centers) with a

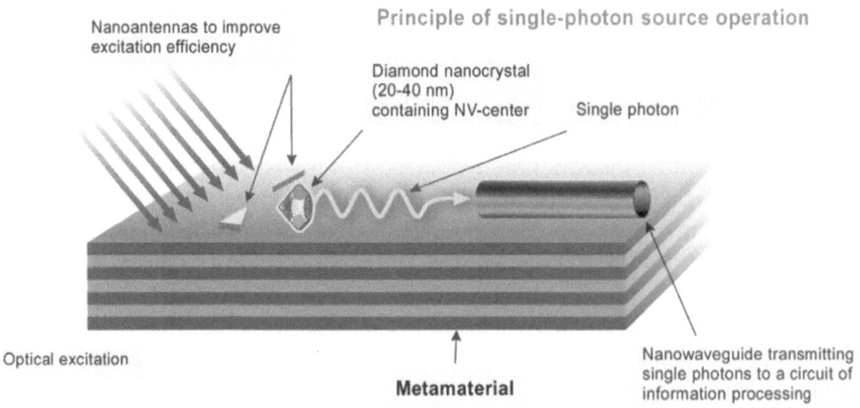

FIGURE 1.12 Principle of single-photon source with metamaterial cladding. Taken from the presentation of Subal Kar (104th Indian Science Congress, 2017).

metamaterial having hyperbolic dispersion characteristics in which the enhancement of single-photon emission is due to the presence of metamaterial (may be two orders of magnitude or so). Such single-photon sources may also have applications in nanochemistry to control chemical reactions at the level of individual molecules, biochemical analysis to determine the dynamics of molecular configuration, decoding DNA and so on [23].

1.3 APPLICATION SCENARIO OF METAMATERIALS

To unfold the application scenario of metamaterials, let us first mention some of their exotic application potentials – the "holy grails" of metamaterials. In 2000, John Pendry proposed the possibility of realizing *superlens* with metamaterial [24]. It is known that, at the image plane, the sub-wavelength object details are not obtainable when focusing is done by the natural or right-handed (RH) media due to the "diffraction limit". This happens because the evanescent waves, which contain the sub-wavelength details of the object, decay rapidly in an RH medium. However, the LH medium is found to be capable of growing (amplifying) the evanescent waves, possibly with surface plasmon coupling between the $z = 0$ and $z = d$ faces of the LHM slab (see Fig. 1.13) [25], thus overcoming the diffraction limit. The sub-wavelength details of the object are thus

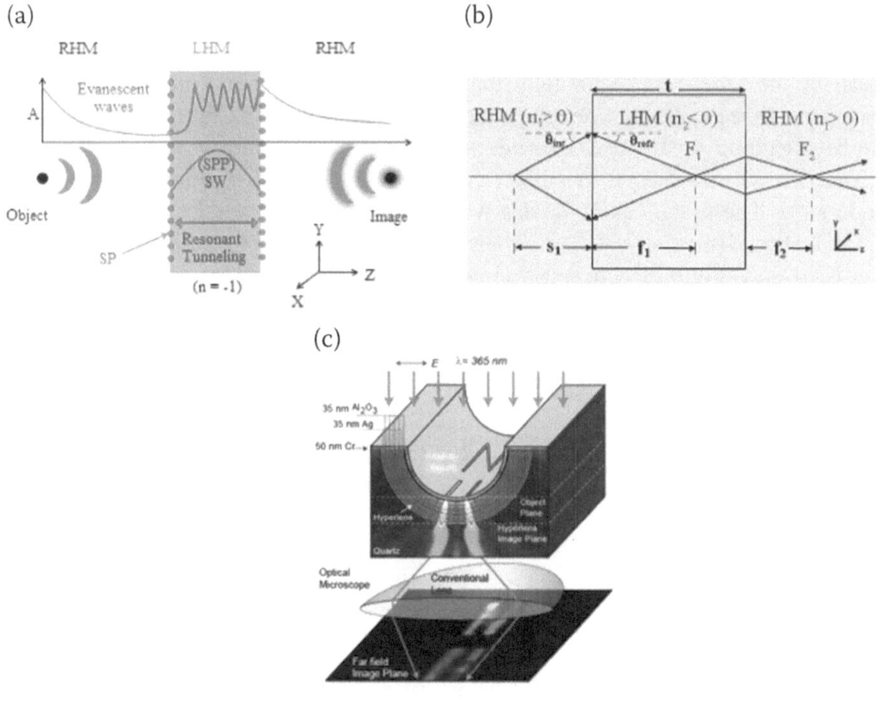

FIGURE 1.13 Concept of superlens. (a) Growth of evanescent waves via surface plasmon [25], (b) focusing by LHM plane slab [25], and (c) hyperlens based on the concept of superlens to magnify sub-diffraction limited objects [7].

obtainable at the image plane with LH medium focusing. The counter-intuitive LHM plane slab focusing is thus said to perform as a "superlens". This sub-wavelength imaging, possible with the superlens concept of metamaterial, is gaining enough enthusiasm that it might one day make it possible to image individual strands of DNA, thereby bringing about a revolution in medical research.

It is worth noting at this point that, unlike RHM, where focusing is done by curved surfaces (convex or concave lens), in LHM or metamaterial, focusing is realizable on flat interfaces between the positive- and negative-index media [5,25], see Fig. 1.13b. This is because the reversal of Snell's law (Fig. 1.13a) refracts the incident electromagnetic signal on the other side of normal compared to the natural material or RHM. Thus, the refracted wave within the slab makes a negative refracting angle (θ_{refr}), thereby converging at a focal point, F_1, while upon emergence from LHM slab, the ray again undergoes negative refraction and meets at another focal point, F_2. Hence, there exists a double-focusing effect [25].

Hyperlens, based on the concept of superlens, to magnify sub-diffraction limited objects and project the magnified images to the far field with conventional lens, has been demonstrated with resolution down to 125 nm at 365 nm working wavelength [26], see Fig. 1.13c. Hyperlens consists of a metamaterial formed by a curved periodic stack of Ag and Al_2O_3 deposited on a half-cylindrical cavity fabricated on a quartz substrate. Hyperlens might have possible applications in nanotechnology photolithography.

Among the many tropes found in science fiction and fantasy, few are more popular than the *cloaking device*. We are familiar with Harry Potter's invisibility cloak or the *Star Trek* technology that can make the whole Romulan warships disappear. Since 2006, the development of a metamaterial-based cloaking device has been gaining pace with extreme enthusiasm. However, it must be noted that the science-fiction movie type invisibility cloak is still a distant possibility, though not impossible. The first cloaking device was developed in 2006 by D. R. Smith et al. [27] of Duke University (USA). Their cloaking device at microwave frequency consisted of a group of concentric circles made of metamaterial (loops of copper wire stamped on fiber glass) with a cylindrical gap in the middle where the object (to be cloaked) was placed, see Fig. 1.14a. Their device could mask or make the

(a) (b)

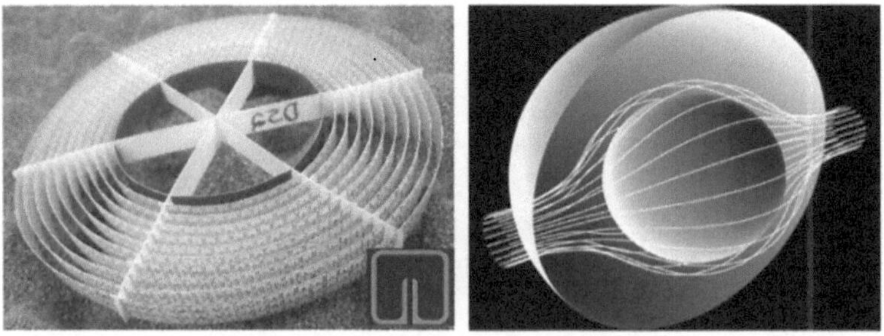

FIGURE 1.14 Demonstration of cloaking. (a) First cloaking device at microwave frequency, and (b) principle of cloaking with metamaterial [7].

object invisible from only one wavelength of the incident microwave signal. The device was not perfect, causing shadowing of microwaves, i.e., distortions.

The principle of cloaking by metamaterials depends on judicious control, i.e., graded variation of negative refractive index around the object to be cloaked. With reference to Fig. 1.14b, the object can be made invisible if there is no reflection from and also no transmission through or even no absorption of the incident electromagnetic signal in the object. The signal should just glide past the object. The trick is not a simple job as one has to make sure that waves from all angles are bending smoothly without scattering. In fact, when the electromagnetic signal is directed at the device, the wave would split, and it should be bent subtly around the device so that it is able to reform on the other side: the effect can be compared to river water flowing around a smooth rock, when no wakes are formed.

Other cloaking devices were also developed of which a few words about carpet cloaking may be mentioned. In April 2009, a team led by Xiang Zhang at UC Berkeley achieved "carpet cloaking" [28]. An object covered with a piece of cloth would normally be detectable based on its telltale bump, see Fig. 1.15, but with the new metamaterial, even the bump seems to vanish with such a cloaking device. They achieved the effect by drilling tiny nanoholes into the cloaking material, a silicon-based metamaterial. The cloaking system was operated near the infrared frequency and scalable to visible light. Carpet cloaking is capable of hiding microscopic objects. These may have potential use in optical computing; for example, such cloaks may be used to allow light to move more efficiently, by hiding the parts of a computer chip that get in the way of the beam. Also, expensive dielectric mirrors – special mirrors used to make printed circuits for electronics – can be ruined by tiny defects on their surfaces, which may be cloaked, making them look like perfect mirrors again.

We mentioned that the first experimental fabrication of metamaterial was to realize its basic property of the reversal of Snell's law and, hence, the negative

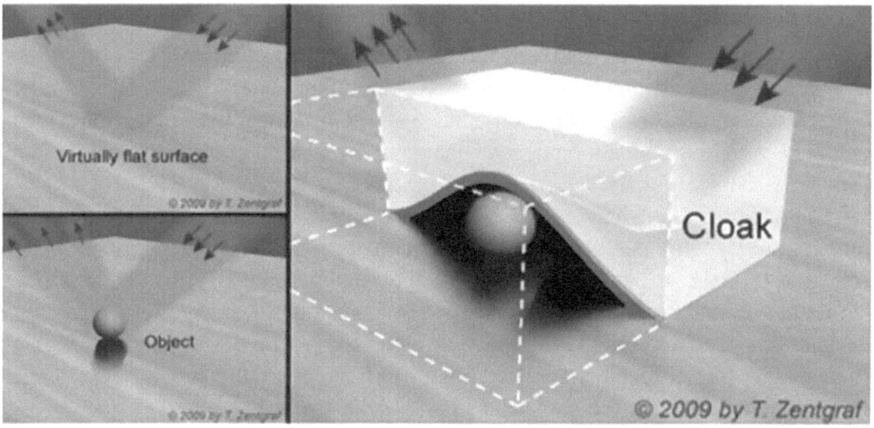

FIGURE 1.15 Concept of carpet cloaking [7].

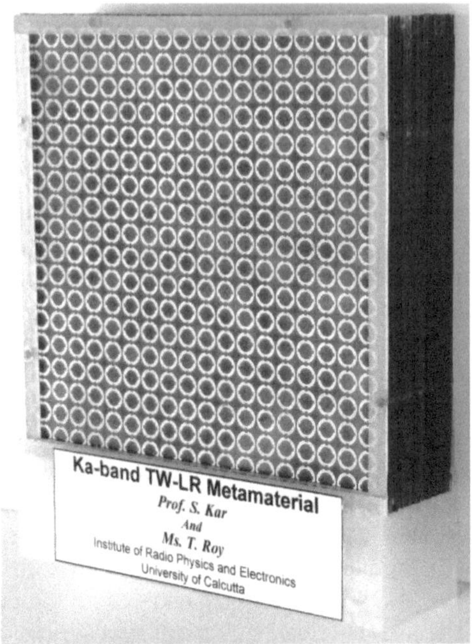

FIGURE 1.16 Plasmonic metamaterial designed with TW and LR with the performance characteristics: negative refractive index (n_{reff}) realized was: −1.84 at 31.25 GHz over a bandwidth of 3.5 GHz [2].

refractive index in 2001 at UCSD (USA). Following this, the applications of metamaterials have proliferated exponentially at different branches of electronics engineering, including those in antennae, filters, passive components, absorbers and so forth. Our group at Calcutta University (Kolkata, India), in collaboration with SAMEER (Kolkata, India) center and BARC (Mumbai, India), made the first successful plasmonic metamaterial in India [29] using the TW and Labyrinth Resonator (LR) combination, see Fig. 1.16.

We have also done successful design and development of metamaterial-inspired antenna and filters at microwave frequency. A complementary split-ring resonator (CSRR)-loaded microstrip patch antenna has shown 24% size reduction when compared with the conventional patch antenna with similar gain and bandwidth characteristics, see Fig. 1.17 [30].

Composite right/left-handed (CRLH) metamaterial filter has been designed and compared in size and performance with the conventional edge-coupled band-pass filter at 2.45 GHz. It has been found that the metamaterial-based filter has 67% size reduction and significant performance improvement, see Fig. 1.18 [31].

Many other researchers [32–35] have also worked on the development of highly efficient and miniaturized antennae, filters and directional couplers, etc.; some representative structures are shown in Fig. 1.19.

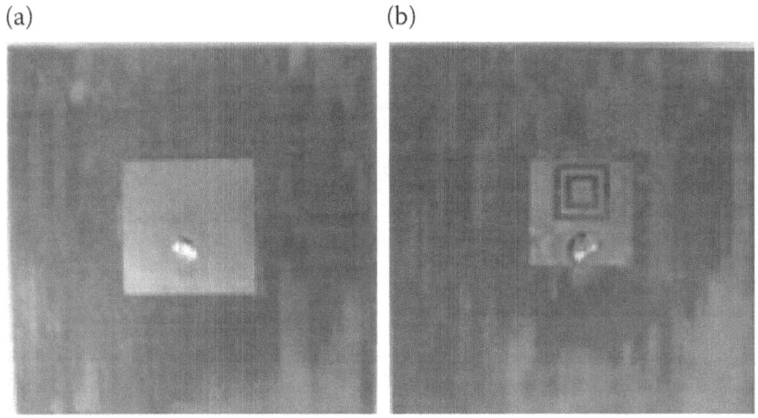

FIGURE 1.17 Microstrip patch antenna. (a) Conventional microstrip patch antenna, and (b) CSRR-loaded microstrip patch antenna [30].

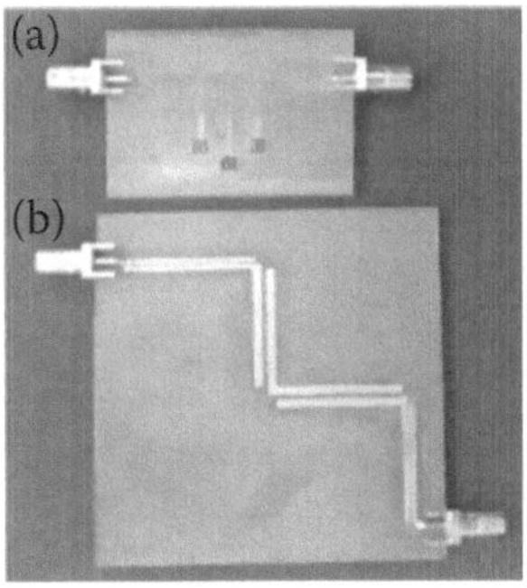

Property	CRLH BPF	Edge coupled BPF
Insertion loss	1.6 dB	2.3dB
Return loss	15 dB	18dB
Harmonics	Suppressed up to 10 GHz	No harmonic suppression
Size	4.3 cm x 3 cm	6.5 cm x 6 cm

FIGURE 1.18 Microwave band-pass filter (BPF); (a) CRLH-based metamaterial filter, and (b) conventional edge-coupled filter; also shown, filter performance comparison [31].

In addition to this, a waveguide-loaded SRR is also capable of exhibiting pass-band in the evanescent band and stop-band in the propagation band of waveguide, thereby providing the possibility of notch filter design, which is tailorable by SRR design dimensions [31], see Fig. 1.20.

Two more applications of metamaterial, namely absorbers and sensors, are also becoming very important these days. Absorbers at microwave frequency are required in bolometers, in anechoic chambers, for stealth purposes in defense applications and so on. In solar cells, such absorbers are also used to absorb solar energy. Conventional absorbers are carbon-foam or ferrite based, but are either thick or bulky in nature. Metamaterial absorbers are designed in such a way that they offer the input impedance to be equal to the impedance of free space, thereby limiting the reflections from the absorber structure. Since metamaterial absorbers are capable of offering ultra-thin thickness, conformal properties and compactness, they attract researchers and are practically the best-suited substitute for conventional absorbers.

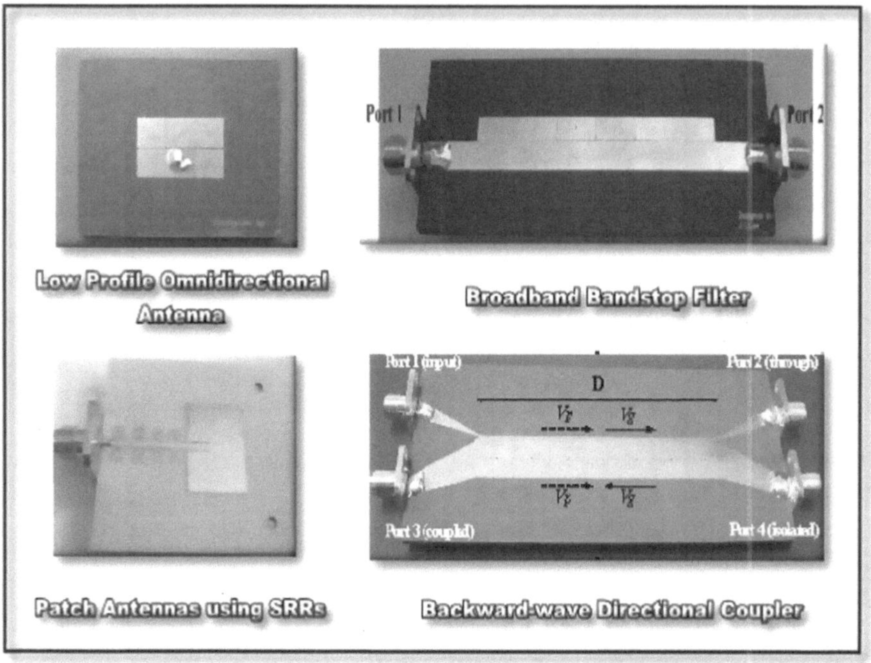

FIGURE 1.19 Metamaterial-based microwave passive components [7].

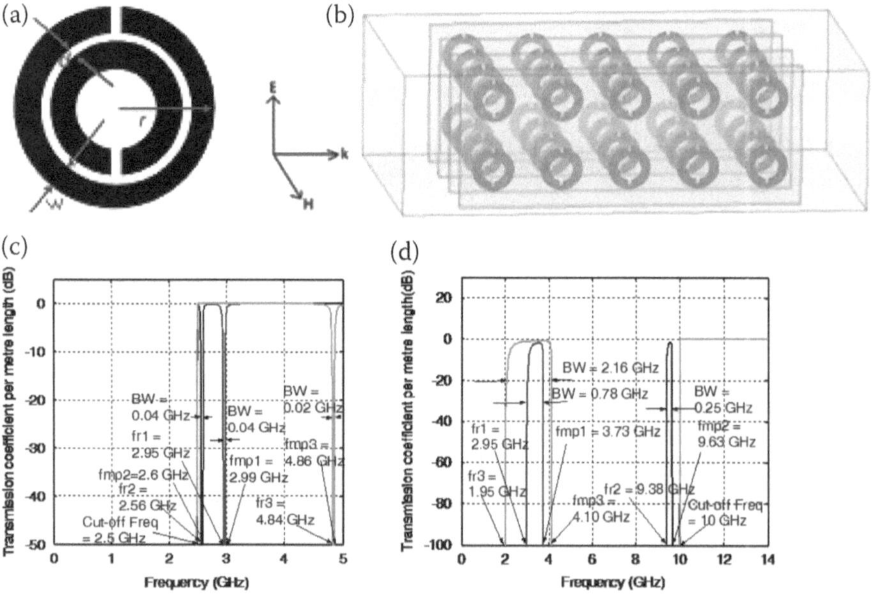

FIGURE 1.20 SRR-loaded waveguide and its characteristics: (a) SRR unit cell, (b) waveguide loaded with the sheets of SRR matrix, (c) stop-band in the pass-band of waveguide and (d) pass-band in the evanescent band of waveguide [31].

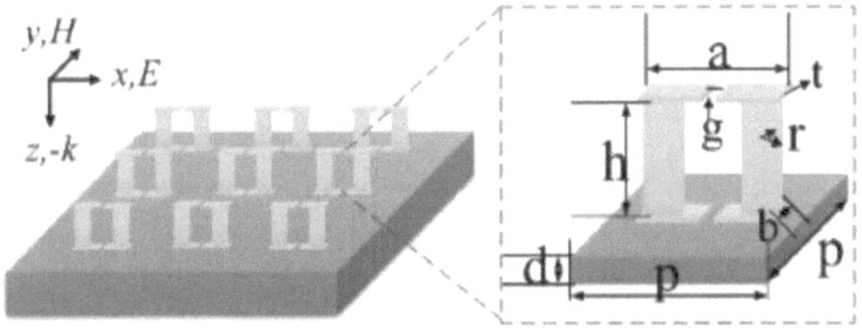

FIGURE 1.21 3D schematic drawing of the proposed metamaterial sensor and a single enlarged unit cell with its geometrical dimensions. [Reprinted/Adapted] with the permission from Wang et al. [36] ©The Optical society.

THz sensors are now using metamaterial assistance to significantly improve their sensor capabilities. Using a THz signal, the sensor determines the resonances of various molecular vibrations in a sample. As metamaterial magnetic inclusion structure (SRR/LR) is highly resonant, it aids in enhancing resonances and, hence, leads to better sensing with metamaterial assistance. The more the Q (quality factor) of metamaterial resonators is, the better the sensing capability. Effective sensing is evaluated by measuring the resonant response differences of metamaterial unit cells, thereby identifying and detecting minute differences in chemical and biochemical substances. In the structure shown in Fig. 1.21 [36], the Q-value of 30 has been realized with the maximum refractive index sensitivity of 788 GHz/RIU or 1.04 ×105 nm/RIU.

The use of a metallic structure in metamaterial design has inherent drawbacks; hence, the Q-factor gets compromised. If we use perfect absorbers designed with metamaterial, then we can get a higher Q-factor. Reaching high Q-factor resonance in these structures leads to enhanced sensor sensitivity to detect minute frequency shifts.

Bi-material sensors with metamaterial absorbers have been used for sensor design, see Fig. 1.22 [37]. This consists of a sensing element (absorber) that converts

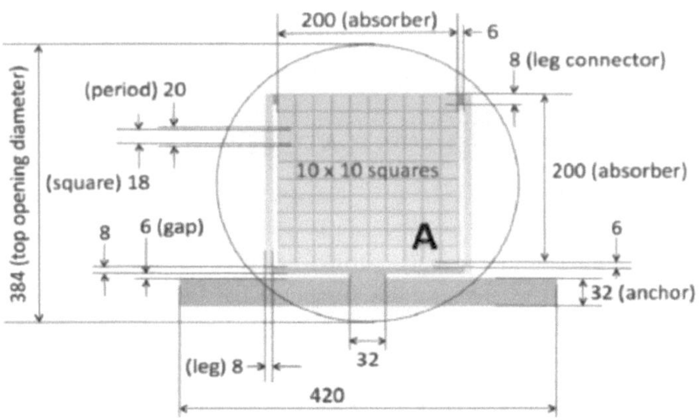

FIGURE 1.22 Structure of bi-material THz sensor using metamaterial absorber. [Reprinted/Adapted] with permission from Alves et al. [37] ©The Optical society.

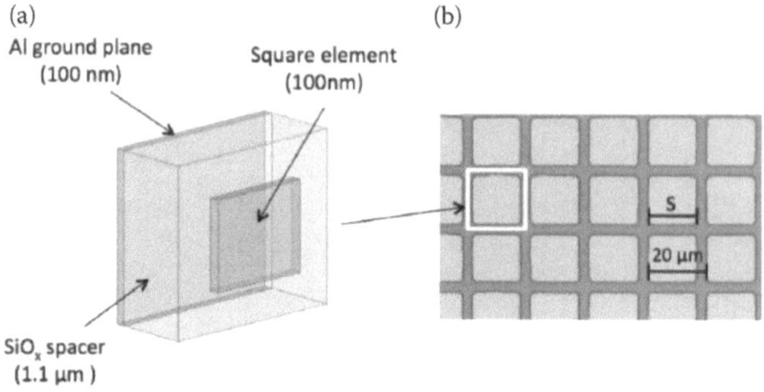

FIGURE 1.23 Metamaterial absorber; (a) unit cell, and (b) 1D periodic array. [Reprinted/Adapted] with permission from Alves et al. [37] ©The Optical society.

the incoming THz radiation to heat, which is eventually transmitted by conduction to two symmetrically located bi-material legs connected to the host substrate (heat sink) with supporting structures of lower thermal conductance (anchors). Temperature rise caused by the absorption of incident THz radiation results in deformation of bi-material legs. The deformation can be probed by different approaches, among which the optical readout is a simple technique that requires a reflective surface, which is embedded into the absorber. Such sensors have responsivity values as high as 1.2 deg/µW, have time constants as low as 200 ms, have minimum detectable power on the order of 10 nW, and can operate with low-power THz sources.

A "perfect" absorber can be constructed with the proper design of structural parameters [37]. The challenge remains in the design of a metamaterial film that is thin enough to provide low thermal capacitance, while providing structural strength, low stress and a flat reflective surface for an optical readout. Fig. 1.23 shows that a typical metamaterial absorber consists of a periodic array of Al square elements separated from an Al ground plane by a SiOx layer, a single unit cell being shown in Fig. 1.23a. Such a combination allows matching to the free-space impedance at specific frequencies, eliminating the reflection. Fig. 1.23b shows a periodic array of this metamaterial absorber [37].

1.4 METASURFACES AND APPLICATION POTENTIALS

Like metamaterial, similar intriguing properties and applications, including far complex wave-manipulation properties, can be realized using the 2D arrangements of engineered scatterers, now widely known as "metasurfaces." Metasurfaces or metafilms (as sometimes refereed in literature) are basically the 2D counterpart of metamaterials (which are 3D structures) consisting of an ultrathin planar arrangement of sub-wavelength-size building blocks of metallic patches or dielectric etchings. Such ultrathin structures have the unique ability to manipulate electromagnetic waves, with spatially arranged meta-atoms – fundamental building blocks of the metasurface – thereby blocking, absorbing, concentrating, dispersing or guiding the waves, from microwave through THz to optical frequencies.

Metamaterials are difficult to fabricate because a useful and practical metamaterial needs to be 3D, whereas metasurfaces being planer (2D) in structure can be easily fabricated using the planer fabrication tools. The planer fabrication process is cost effective, and being 2D, they can be easily integrated into other devices, which can make them a salient feature for nanophotonic circuits. Metasurface, which has followed as a derivative of metamaterial, is capable of tackling some of the critical challenges rooted in traditional metamaterials, such as high resistive loss from resonant plasmonic components and fabrication requirements for making 3D nanostructures. In the past few years, metasurfaces have achieved groundbreaking progress, providing unparalleled control of light, including the construction of arbitrary wave-fronts and realizing active and nonlinear optical effects.

Unlike metamaterials, a metasurface affects waves through modified boundary conditions instead of using bulk effective constitutive properties. To be more specific, the function of metamaterials is to realize artificial effective negative permittivity and permeability that changes the wave dynamics with the bulk constitutive properties, see Fig. 1.24a. But metasurfaces can arbitrarily control the wave-fronts, i.e., the amplitude, polarization, phase and frequency of the wave, as a function of position with the help of sub-wavelength scatterers in the planer structure of which they are made, see Fig. 1.24b.

Metasurfaces find applications in absorbers in which surface impedance can be varied and manipulated by patterning the 2D metasurface unit cells. When active and non-linear components are added to traditional metasurfaces, exceptional tunability and switching capability becomes possible.

Metasurfaces can be engineered with spatially varying boundary conditions to convert a given incident electromagnetic field into a desired scattered waveform, resulting in wave-front transformations. Using novel configurations of meta-atoms, the incoming plane waves can be deflected to preferable directions, manipulate their polarization state, and generate special beams. It is important to note that a metasurface can efficiently manipulate the wave-front of an incident electromagnetic wave through just the sub-wavelength propagation distance compared to the traditional 3D metamaterial that does the same at a distance far larger than the wavelength. Therefore, this can largely alleviate the propagation loss. The controllable

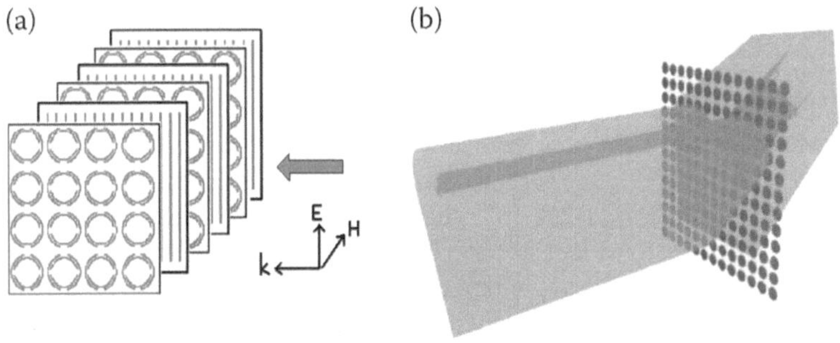

FIGURE 1.24 Metasurface vs. metamaterial. (a) periodic structure of metamaterial (b) metasurface showing wave-front transformation.

surface refractive index provided by metasurfaces can also be applied to lenses. In fact, they can be used to design 2D microwave/optical lenses like Luneburg and fish-eye lenses, which are applied in surface waveguides for antenna systems and planar microwave sources. Reconfigurable ultrathin surfaces provided by metasurface, resulting in wave-front transformations to the impinging waves to any imaginable degree, may form the basis for smart surfaces with a significant impact on nearly any electromagnetic and photonic application, from classical to quantum photonics, from radar to wireless technology.

Metasurfaces have already revolutionized the antenna design, especially the leaky-wave antennae, from microwave to THz and higher frequencies. Metasurface finds interesting applications in the design of plasmonic laser or lasing spaser, whose bandwidth can be controlled by tailoring the metamolecules of metasurface. Other applications of metasurfaces include cloaking and imaging. Since metasurfaces have high bio tissue sensitivity, they can be used in biosensors for inside-body examination and bio tissue discrimination, including cancer disease diagnosis.

Metasurfaces resemble frequency selective surfaces (FSSs) in many respects; they can replace FSSs in many applications due to the sub-wavelength nature of "meta-atoms" of which they are made of. An FSS is a periodic structure made of composite material (or sometimes with only dielectric material) and designed to be transparent in some frequency bands while reflecting, absorbing or redirecting the incident signal. However, in FSS, each individual element is resonant at the resonance frequency. FSS' typically have periodicity equal to half the wavelength of the resonant frequency. But, the periodicity of individual elements of metasurface is of sub-wavelength order, so is the individual elements of metasurface. Compared to FSSs, the sub-wavelength periodicity of the metasurface allows packing of a large number of unit cells in a constrained space, which is highly useful for radome design and in many other applications with limited space.

1.4.1 Applications of Metasurfaces

1.4.1.1 Absorbers

From microwave through optical frequencies, absorbers are required in many applications, such as anechoic chambers, solar cells, photodetectors and so forth. Conventional absorbers are usually composed of multilayer structures that are lossy and bulky, but the modern system requires miniaturized and compact absorbers. Metasurface-based absorbers are a good catch for their low profile, light weight and simplicity of construction with simple metallic structures. A passive absorber is shown in Fig. 1.25a [38] that consists of an array of gold material patches on MgF_2-substrate that are capable of absorbing energy independent of polarization and with a wide-angle range of up to ±80°. The proposed structure is designed to be polarization-independent in the *x*- and *y*-directions at normal incidence, thus yielding a polarization-insensitive property. Active metasurface absorbers have the advantage that they enable switchable absorption, tunable

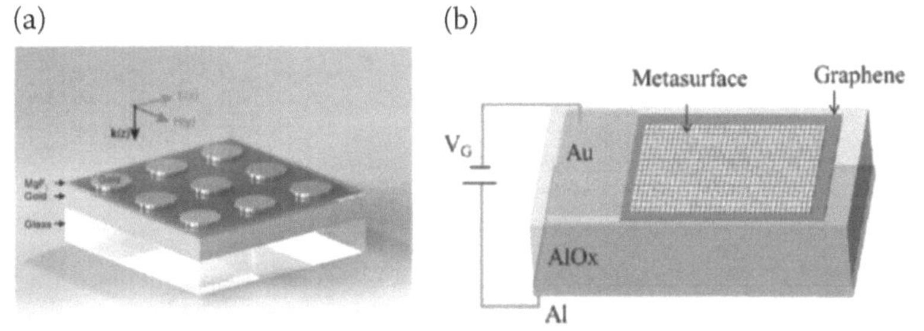

FIGURE 1.25 Metasurface absorbers; (a) passive absorber at infrared frequency, and (b) active tunable absorber at optical frequency. [Reprinted/Adapted] with permission from Li et al. [38] © Nanophotonics.

non-linear response and tunable resonant frequency. A tunable metasurface absorber, which can be used from tens of GHz to near infrared due to the broadband optical response of graphene, is shown in Fig. 1.25b.

1.4.1.2 Transformation Optics Applications

With the advent of metasurfaces, transformation optics has significantly benefitted. In fact, the introduction of metasurfaces for transformation optics has brought a new paradigm for efficiently controlling electromagnetic waves [39]. Transformation optics deals with the control of electromagnetic waves, leading to designer-demanded tailoring of its wave-front, thereby manipulating the wave dynamics at will. The wave-front shaping with conventional techniques like lens, hologram, etc., can be realized over a distance that is larger than the wavelength of operation. The same is the case with metamaterials or DNG materials in which the control of the wave-front of an electromagnetic signal is done by accumulating the phase through propagating over a distance larger than a wavelength. But, with metasurfaces, wave-front shaping and focusing of energy can be done over sub-wavelength distances, thus alleviating the propagation loss. This makes metasurfaces potentially superior for this purpose, and they are replacing bulky metamaterials in many applications. In fact, in metasurfaces, the electromagnetic signal undergoes a so-called "phase discontinuity" or "phase jump" caused by the interaction of the incident electromagnetic wave with the surface plasmon [40]. The surface plasmon in the metasurface originates due to an induced surface electromagnetic wave caused by impinging the electromagnetic wave that forces the charges present in the individual sub-wavelength elements of the metasurface to oscillate. Since the elements on a metasurface can be spatially varied, this variation can cause the currents on the surface to lead (or lag) depending on the individual resonant element. This localized phenomenon allows us to tailor the wave-fronts as they pass through a metasurface, and it leads to a variety of applications.

Realization of a polarization split in the visible region has been reported [41], see Fig. 1.26, that uses an all-dielectric gradient metasurface, composed of periodic arrangement of differently sized cross-shaped silicon nanoblocks

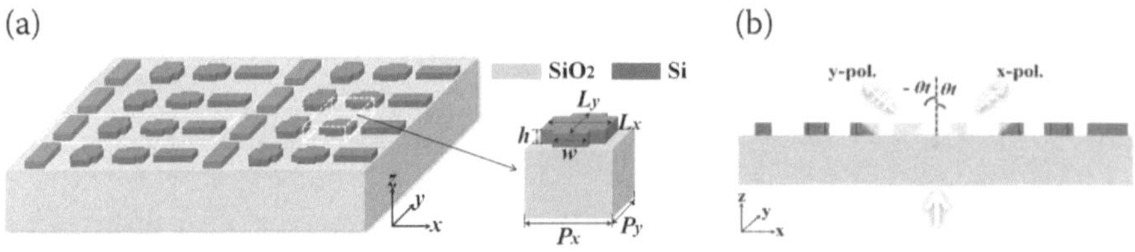

FIGURE 1.26 (a) The metasurface structure, and (b) polarization splitting. [Reprinted/Adapted] with permission from Li et al. [41] © Springer Nature.

resting on the fused silica substrate. The cross-shaped silicon block arrays can induce two opposite transmission-phase gradients along the x-direction for the linear x-polarization and along the y-direction for y-polarization. With proper design, the metasurface can separate the linearly polarized light into the x- and y-polarized ones, which propagate at the same angle along the left and right sides of the normal incidence in the x-z plane, as shown in Fig. 1.26b. The polarization beam splitter is expected to play an important role for future free-space optical devices.

Introduction of non-uniform impedance surfaces in lenses with the help of metasurface has resulted in very thin and ready-to-manufacture lenses. Fig. 1.27a shows one such metasurface lens in which the radii of the patches, i.e., the unit cells that make up the metasurface, are gradually decreased as one moves from the center of the lens [38,42]. On such a surface, the traveling electromagnetic wave encounters gradually varying surface impedance and the corresponding change in phase velocity. The impedance profile is obtained by combining the Luneburg lens design with TM surface wave dispersion relation. In another design, the metasurface elements have smoothly varying, i.e., asymmetric polygons, as shown in Fig. 1.27b. Both the methods are useful to design surfaces with spatially varying refractive indices.

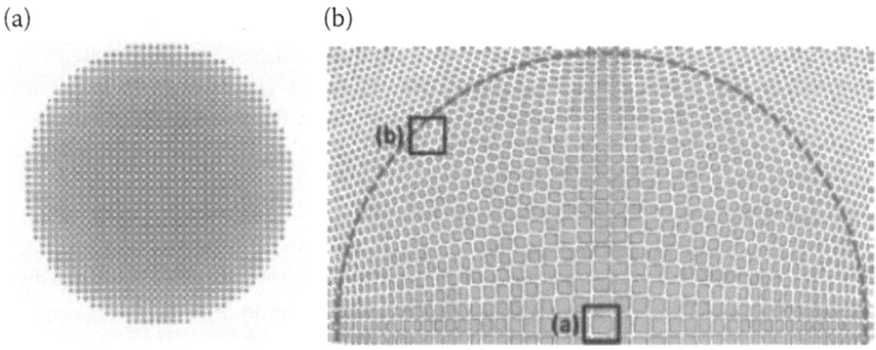

FIGURE 1.27 Luneburg lens; (a) with metasurface having gradually decreasing radii of unit cells from center to the rim, and (b) metasurface elements having smoothly varying, i.e., asymmetric polygons. [Reprinted/Adapted] with permission from Li et al. [38] © Nanophotonics.

1.4.1.3 Metasurface in Antenna Design

Metasurfaces have the unique ability and distinct advantage as media for radiating electromagnetic waves into free space apart from their wide applications in surface and free-space wave manipulation. Metasurface leaky-wave antennae have advantages in that they are low-profile and have simple feed structures. They also have frequency-dependent beam scanning properties (beam squint).

Fig. 1.28a [38,43] depicts a sinusoidally modulated graphene leaky-wave antenna with electronic beam-scanning capability. By applying bias voltages to the different grating pads beneath the graphene substrate, the graphene surface reactance can be modulated, resulting in a versatile beam-scanning capability at THz frequencies. Another 2D leaky-wave antenna operating at THz frequencies with tunable frequency and beam angle is shown in Fig. 1.28b [38,44]. It is based on tuning the conductivity of the graphene.

Metasurfaces have been successfully used to design and manufacture high-gain holographic antennae for which the surface wave is the major incident wave. The radiation for such antenna occurs when the phase matching between the forward and backward leaky waves occurs. The forward leaky wave arises from the grating, which has a larger periodicity, while the smaller period grating leads to a backward leaky wave. Based on this approach, circularly polarized leaky-wave antennae with a 26 dB gain have been realized [45]. The grating effect was produced by modulating the surface impedance of the metasurface. A significant advantage of this approach is that, instead of changing the antenna shape, in order to design a specific response, the metasurface modulation (i.e., surface impedance) is engineered [46].

Metasurfaces can also improve the performance of horn antennae. A metasurface designed using a genetic algorithm has been used as an inner surface for a conical horn, and the cross polar and side lobe levels have been improved over the entire Ku band [47]. A similar approach was applied to improve the performance of a hybrid mode square horn antenna using metasurfaces [48].

1.4.1.4 Lasing Spaser

The use of meta-molecules in developing "lasing spaser" is a dominant example of the application possibility of metasurfaces in photonics [49]. In spaser

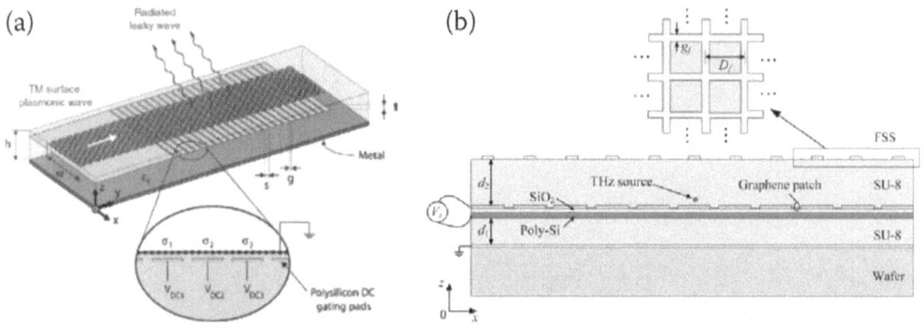

FIGURE 1.28 leaky-wave metasurface antennae. [Reprinted/Adapted] with permission from Li et al. [38] © Nanophotonics.

(Surface Plasmon Amplification by Stimulated Emission of Radiation), also known as plasmonic laser, the light-quanta-photons of the laser is being replaced with electronic excitations at the surface of metals called surface plasmons, which can have atomic-scale dimensions [50]. However, a spaser produces very little light, which is not collimated into a narrow beam.

But, if the emission can be fueled by plasmonic excitations in an array of coherently emitting meta-molecules (designed with magnetic inclusion structure, such as SRRs) supported by a gain medium (quantum-dot-doped dielectric), which can overcome the radiation losses and Joule losses in the metallic structure of meta-molecules, we have the "lasing spaser" [49,51], see Fig. 1.29. In contrast to conventional lasers that operate at wavelengths of suitable natural atomic or molecular transitions, the emission wavelength of lasing spaser can be controlled by meta-molecule design. Being the thinnest (~100 nm) laser, the lasing spaser promises new applications ranging from displays to high-speed communications.

The combination of artificial classical electromagnetic resonators (SRRs), forming the metasurface, plays the role of the active medium in the lasing spaser, just as an assembly of essentially quantum inversely populated atoms plays the same role in a conventional laser. These identical plasmonic resonators impose

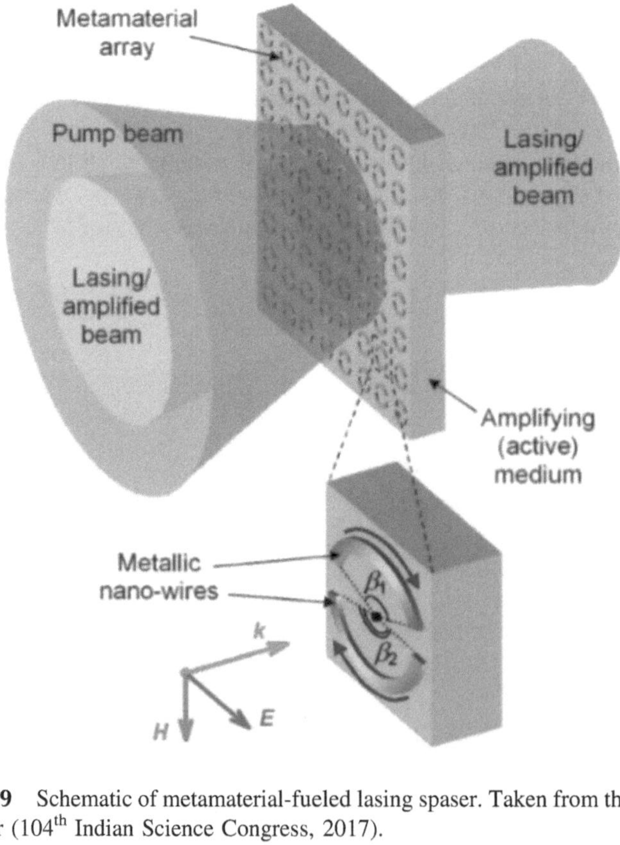

FIGURE 1.29 Schematic of metamaterial-fueled lasing spaser. Taken from the presentation of Subal Kar (104th Indian Science Congress, 2017).

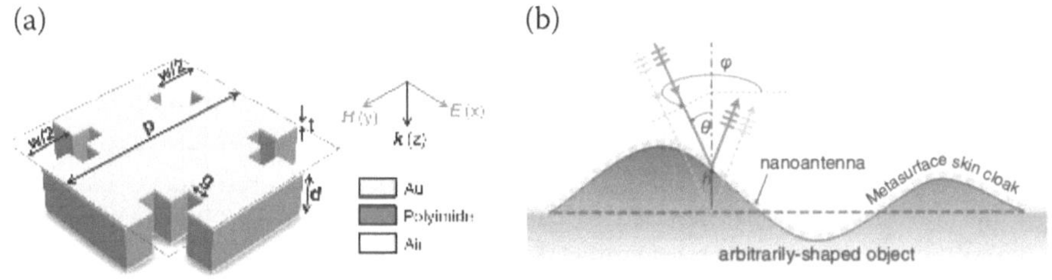

FIGURE 1.30 (a) Unit cell of the fishnet nanostructure used for the design of broadband metasurface filter. [Reprinted/Adapted] with the permission from Jiang et al. [52] © Springer Nature, and (b) schematic cross-section of metasurface skin cloak. [Reprinted/Adapted] from Xingjie et al. [53] under creative common license.

the frequency at which the device will lase, drawing energy from a supporting gain substrate. In a conventional laser, the direction of emission is dictated by the external resonator, and its coherence is underpinned by the stimulated emission of atoms in the gain medium. In a lasing spaser, the direction of emission is normal to the plane of array, where strong trapped-mode currents in the plasmonic resonators oscillate in phase.

1.4.1.5 Some More Applications of Metasurface

It has been shown that, by using extremely thin metasurfaces with deep subwavelength notches in a two-layered fishnet structure (a single unit cell of the fishnet nanostructure is shown in Fig. 1.30a), the dispersion characteristics can be engineered. This technique was then used to make a broadband metasurface filter [52].

Metasurface has also been used in the design of a very promising cloaking device at optical frequency. We have talked about carpet cloaking while discussing metamaterial applications, but such metamaterial-based optical cloaks use volumetric distribution of the material properties to gradually bend light and thereby obscure the cloaked region. Such a cloaking device is bulky and suffers from unnecessary phase shifts in the reflected light, making the cloaks detectable. However, metasurface-based skin cloaks [53], see Fig. 1.30b, operate at 730 nm wavelength and are able to conceal a 3D arbitrarily shaped object by complete restoration of the phase of the reflected light. Compared to the metamaterial-based cloak, the skin cloak based on metasurface is miniaturized in size (only 80 nm, one-ninth of the wavelength) and scalable for hiding macroscopic objects, too.

1.5 CONCLUSION

During the last two decades, metamaterials, together with their 2D counterpart, metasurfaces, have revolutionized the application scenario in the microwave through THz to optical frequencies. Advanced technological developments in these fields are opening up more efficient, miniaturized antennae, filters, couplers, sensors, wave-front manipulation and so forth. Exotic applications, such as cloaking, sub-wavelength imaging and transformation optics, are leading to innovations that could not have been dreamt of one or two decades ago. Metamaterial technology has developed from using simple printed circuit technology for fabrication of microwave metamaterials in the initial stage of its development, and have gone through various technological innovations for design and development of metamaterial at higher frequencies, including the use of nanowires for optical metamaterial. Totally dielectric-based transparent metamaterial is also a very important development for the design of photonic-integrated circuits. Though metamaterial applications started with the realization of a negative refractive index, its applications have exponentially proliferated into various common and exotic applications from efficient antenna design to cloaking devices and beyond. Metamaterials are bulky 3D devices; thus, metasurfaces were developed, which are 2D counterparts of metamaterials, the latter being easy to fabricate with planer technology and to integrate with other devices. With the advent of metasurfaces, wave-front manipulation leading to various applications in transformation optics within sub-wavelength extent has been possible. Recent advances with nanofabrication technology have made it possible to fabricate such ultrathin sub-wavelength planer structures with precision that was not possible even a decade ago. Thus, now we have laser spaser, skin cloaking and various other applications of metasurfaces that have become the order of the day, and many more innovative applications are emerging every morning.

In addition to all these developments, artificial intelligence, machine learning, and micro- and nano-technologies are expected to be the integral facets of future metamaterials/metasurfaces research. We can now confidently say that metamaterials and metasurfaces have evolved into a vibrant and multidisciplinary research field, with solid foundations, and have a bright future ahead of further growth and expansion. The field of research in metamaterials/metasurfaces is inherently multidisciplinary, combining the concepts from solid-state physics and materials science to microwave engineering and optics to plasmonics and nanotechnology. In the present time, metamaterial/metasurface research has arrived at a crossroad, at which multiple academic disciplines and industries must be brought together to initiate a mass transition from the proof-of-concept to practical applications in real-world problems in order to encourage their adoption as commercially viable alternatives to existing mainstream technologies.

REFERENCES

1. V. G. Veselago, "The electrodynamics of substances with simultaneously negative values of ε and μ," *Soviet Physics USPEKHI.*, vol. 10, no. 4, pp. 509–514, 1968.
2. Subal Kar, *Microwave Engineering: Fundamentals, Design and Applications*, ch. 15, New Delhi, India: Universities Press, 2016.
3. J. B. Pendry, J. Holden, W. J. Stewart, et al., "Extremely low frequency plasmons in metallic mesostructures," *Phys. Rev. Lett.*, vol. 76, no. 25, pp. 4773–4776, 1996.
4. J. B. Pendry, J. Holden, D. J. Robbins, et al., "Magnetism from conductors and enhanced nonlinear phenomena," *IEEE Trans. Microw. Theory Tech.*, vol. 47, no. 11, pp. 2075–2084, 1999.

5. G. V. Eleftheriades, A. K. Iyer, and P. C. Kremer, "Planar negative refractive index media using periodically L-C loaded transmission lines," *IEEE Trans. Microw. Theory Tech.*, vol. 50, no. 12, pp. 2702–2712, 2002.
6. R. A. Shelby, D. R. Smith, and S. Schultz, "Experimental verification of a negative index of refraction," *Science*, vol. 292, pp. 77–79, 2001.
7. Subal Kar and M. Ghosh, "Advances in metamaterial research," *Asian J. Phys.*, vol. 21, no. 1, pp. 1–16, 2012.
8. R. Marqués, J. Martel, F. Mesa, et al., "Left handed media simulation and transmission of EM waves in sub-wave-length SRR-loaded metallic waveguides," *Phys. Rev. Lett.*, vol. 89, pp. 183 901(1)–183 901(4), 2002.
9. I. Bulu, H. Caglayan, and E. Ozbay, "Experimental demonstration of labyrinth-based left-handed metamaterials," *Opt. Express.*, vol. 13, no. 25, pp. 10238–10247, 2005.
10. T. Roy, D. Banerjee, and Subal Kar, "Studies on multiple-inclusion magnetic structures useful for millimeter-wave left-handed metamaterial application," *IETE J. of Res.*, vol. 55, no. 2, 2009.
11. J. D. Baena, R. Marque´s, and F. Medina, "Artificial magnetic metamaterial design by using spiral resonators," *Phys. Rev. B.*, vol. 69, pp. 014402–014405, 2004.
12. J. D. Baena, J. Bonache, F. Martin, et al., "Equivalent-circuit model for split-ring resonators and complementary split-ring resonators coupled to planar transmission lines," *IEEE Trans. Microw. Theory Tech.*, vol. 53, no. 4, pp. 1451–1461, 2005.
13. J. Zhou, T. Koschny, and C. M. Soukoulis, "Magnetic and electric excitations in split ring resonators," *Opt. Express.*, vol. 15, pp. 17881–17890, 2007.
14. R. Marqués, F. Mesa, J. Martel, et al., "Comparative analysis of edge- and broadside-coupled split ring resonators for metamaterial design – Theory and experiments," *IEEE Trans. Antennas and Propagat.*, vol. 51, no. 10, pp. 2572–2581, 2003.
15. A. K. Iyer, and G. V. Eleftheriades, "Volumetric layered transmission-line metamaterial exhibiting a negative refractive index," *J. Opt, Soc. Am. B.*, vol. 23, no. 3, pp. 553–570, 2006.
16. A. Kumar, A. Majumder, S. Das, et al., "Simulation based characterization of negative permeability plasmonic structures at X band," *Proc. of Science and Information Conf. 2013*, pp. 671–679, October 7–9, 2013, London, UK.
17. V. M. Shalaev, W. Cai, U. K. Chettiar, et al., "Negative index of refraction in optical metamaterials," *Opt. Lett.*, vol. 30, no. 24, pp. 3356–3358, 2005.
18. S. Zhang, W. Fan, N. C. Panoiu, et al., "Experimental demonstration of near-infrared negative-index metamaterials," *Phys. Rev. Lett.*, vol. 95, pp. 137404–137407, 2005.
19. S. Jahani, and Z. Jacob, "Transparent sub-diffraction optics: Nanoscale light confinement without metal," *Optica.*, vol. 1, no. 2, pp. 96–100, 2014.
20. S. Jahani, and Z. Jacob, "Breakthroughs in photonics 2014: Relaxed total internal reflection," *IEEE Photon. J.*, vol. 7, no. 3, Article 0700505, 2005.
21. S. Jahani, and Z. Jacob, "All-dielectric metamaterials," *Nature Nanotechnol.*, vol. 11, pp. 23–26, 2016.
22. M. Y. Shalaginov, V. V. Vorobyov, J. Liu, et al., "Enhancement of single–photon emission from nitrogen-vacancy centres with N/(Al,Sc)N hyperbolic metamaterial," *Laser Photon. Rev.*, vol. 9, no. 1, pp. 120–127, 2015.
23. S. Takeuchi, "Recent progress in single-photon and entangled-photon generation and applications," *Jpn. J. Appl. Phys.*, vol. 53, no. 3, Article 030101, 2014.
24. J. B. Pendry, "Negative refraction makes a perfect lens," *Phys. Rev. Lett.*, vol. 85, no. 18, pp. 3966–3969, 2000.

25. M. Ghosh, and Subal Kar, "Metamaterial plane-slab focusing and sub-wavelength imaging – The concept, analysis and characterization," *Proc. of Science and Information Conf. 2013*, pp. 670–674, October 7–9, 2013, London, UK.
26. Z. Liu, H. Lee, Y. Xiong, et al., "Far-field optical hyperlens magnifying sub-diffraction-limited objects," *Science*, vol. 315, no. 5819, p. 1686, 2007.
27. D. Schurig, J. J. Mock, B. J. Justice, et al., "Metamaterial electromagnetic cloak at microwave frequencies," *Science*, vol. 314, no. 5801, pp. 977–980, 2006.
28. M. Gharghi, C. Gladden, T. Thomas Zentgraf, et al., "A carpet cloak for visible light," *Nano Lett.*, vol. 11, no. 7, pp. 2825–2828, 2011.
29. B. Das, "India joins metamaterial club," Nature (India), August 20, 2009. (http://www.nature.com/nindia/2009/090820/full/nindia.2009.273.html).
30. Subal Kar, M. Ghosh, A. Kumar, et al., "Complementary split-ring resonator-loaded microstrip patch antenna useful for microwave communication," *Int. J. Electron. Commun. Eng.*, vol. 10, no. 10, pp. 1321–1325, 2016.
31. Subal Kar, A. Kumar, A. Majumder, et al., "CRLH and SRR based microwave filter design useful for microwave communication," *Int. J. Computer, Electrical, Automation, Control and Information Eng.*, vol. 10, no. 4, pp. 629–633, 2016.
32. J. Yamauchi, K. Ninomiya, T. Ueda, et al., "Low-profile omnidirectional antennas based on pseudo-traveling-wave resonance using nonreciprocal metamaterials," *2017 IEEE Asia Pacific Microwave Conf. (APMC)*, November 13–16, 2017.
33. C. J. Lee, K. M. K. H. Leong, and T. Itoh, "Metamaterial transmission line based bandstop and bandpass filter designs using broadband phase cancellation," *2006 IEEE MTT-S Int. Microw. Symp. Dig.*, June 11–16, 2006.
34. B. Singh, H. Rana, A. Verma, et al., "SRR loaded microstrip patch antenna for bluetooth, HIPERLAN/WLAN and WIMAX," *2016 3rd Int. Conf. on Signal Processing and Integrated Networks (SPIN)*, February 11–12, 2016.
35. K. Yamagami, T. Ueda, and T. Itoh, "Enhanced bandwidth of asymmetric backward-wave directional couplers by using dispersion of nonreciprocal CRLH transmission lines," *2019 IEEE Asia-Pacific Microw. Conf. (APMC)*, December 10–13, 2019.
36. W. Wang, F. Yan, S. Tan, et al., "Ultrasensitive terahertz metamaterial sensor based on vertical split-ring resonators," *Photon. Res.*, vol. 5, no. 6, pp. 571–577, 2017.
37. F. Alves, F. Grbovic, B. Kearney, et al., "Bi-material terahertz sensors using metamaterial structures," *Opt. Express.*, vol. 21, no. 11, pp. 13256–13271, 2013.
38. A. Li, S. Singh, and D. Sievenpiper, "Metasurfaces and their applications," *Nanophotonics*, vol. 7, no. 6, pp. 989–1011, 2018. DOI: 10.1515/nanoph-2017-0120.
39. C. Sheng, H. Liu, and S. Zhu, "Transformation optics based on metasurfaces," *Science Bulletin*, vol. 64, pp. 793–796, 2019.
40. S. S. Bukhari, Y. Vardaxoglou, and W. Whittow, "A metasurfaces review: Definitions and applications," *Appl. Sci.*, vol. 9, Article 2727, 2019.
41. J. Li, C. Liu, T. Wu, et. al., "Efficient polarization beam splitter based on all-dielectric metasurface in visible region," *Nanoscale Res. Lett.*, vol. 14, Article 34, 2019.
42. M. Bosiljevac, M. Casaletti, F. Caminita, et al., "Nonuniform metasurface Luneburg lens antenna design," *IEEE Trans. Antennas Propag.*, vol. 60, pp. 4065–4073, 2012.
43. M. Esquius-Morote, J. S. Gomez-Dias, and J. Perruisseau-Carrier, "Sinusoidally modulated graphene leaky-wave antenna for electronic beam scanning at THz," *IEEE Trans. Terahertz Sci. Technol.*, vol. 4, pp. 116–122, 2014.
44. X. C. Wang, W. S. Zhao, J. Hu, et al., "Reconfigurable terahertz leaky-wave antenna using graphene-based high-impedance surface," *IEEE Trans. Nanotechnol.*, vol. 14, pp. 62–69, 2015.
45. B. Zhu, Y. Feng, J. Zhao, et al., "Switchable metamaterial reflector/absorber for different polarized electromagnetic waves," *Appl. Phys. Lett.*, vol. 97, Article 051906, 2010.

46. H. Wakatsuchi, S. Kim, J. Jeremiah, et al., "Waveform-dependent absorbing metasurfaces," *Phys. Rev. Lett.*, vol. 111, Article 245501, 2013.
47. M. Mavridou, K. Konstantinidis, and A. P. Feresidis, "Continuously tunable mm-wave high impedance surface," *IEEE Antennas Wireless Propag. Lett.*, vol. 15, pp. 1390–1393, 2016.
48. Y. Huang, L. S. Wu, M. Tang, et al., "Design of a beam reconfigurable THz antenna with graphene-based switchable high-impedance surface," *IEEE Trans. Nanotechnol.*, vol. 11, pp. 836–842, 2012.
49. N. I. Zheludev, S. L. Prosvirnin, N. Papasimakis, et al., "Lasing spaser," *Nature Photon.*, vol. 2, no. 6, pp. 351–354, 2008.
50. D. J. Bergman, and M. I. Stockman, "Surface plasmon amplification by stimulated emission of radiation: Quantum generation of coherent surface plasmons in nano-systems," *Phys. Rev. Lett.*, vol. 90, Article 027402, 2003.
51. V. A. Fedotov, M. Rose, S. L. Prosvirnin, et al., "Sharp trapped-mode resonances in planar metamaterials with a broken structural symmetry," *Phys. Rev. Lett.*, vol. 99, Article 147401, 2007.
52. Z. H. Jiang, S. Yun, L. Lin, et al., "Tailoring dispersion for broadband low-loss optical metamaterials using deep-subwavelength inclusions," *Sci. Repts.*, vol. 3, Article 1571, 2013.
53. N. Xingjie, Z. J. Zi Jing Wong, M. Mrejen, et al., "An ultrathin invisibility skin cloak for visible light," *Science*, vol. 349, no. 6254, pp. 1310–1314, 2017.

2 Review of Effective Medium Theory and Parametric Retrieval Techniques of Metamaterials

Raktim Guha Ph.D. scholar[1] *and*
B. N. Basu Professor (Adjunct)[2]
[1]Academy of Scientific and Innovative Research (AcSIR), Microwave Devices Area, CSIR-CEERI, Pilani, Rajasthan, India
[2]Sir J. C. Bose School of Engineering, Supreme Knowledge Foundation Group of Institutions, Mankundu, West Bengal, India

2.1 INTRODUCTION

The word *meta* means "beyond", and it is implied that *metamaterial* (MTM) refers to an artificially engineered composite material, the properties of which, when excited by incident electromagnetic waves, can be altered beyond what can be usually found in natural materials. The dimensions of an MTM unit cell are much smaller than the wavelength of the exciting electromagnetic wave, typically of the order of $\lambda/5$–$\lambda/10$, to satisfy the important "homogenization" criteria. This chapter reviews effective medium theory (EMT) and parametric retrieval techniques of MTMs. While retrieving the *effective* material parameters (the effective relative permittivity and the effective relative permeability), we usually consider the MTM as a homogeneous and isotropic material.

Approaches to EMT enable one to find the frequency-dependent constitutive parameters, namely the effective permeability μ and effective permittivity ε of MTM, treating the latter as a homogeneous material. One such approach, which does not require one to practically realize an MTM unit cell, can be used to determine, for a set of chosen material frequency parameters, whether the MTM would behave as μ (mu) negative (MNG), ε (epsilon) negative (ENG) or double negative (DNG), the latter signifying that both μ and ε are negative [1,2]. When the geometry and dimension of the unit cells making the medium are specified,

DOI: 10.1201/9781003050162-2

one can refer to another theoretical approach that is based on the simulated or measured refection/transmission coefficients [3–5]. In this second approach, by varying the dimensions of an MTM unit cell, one can achieve MNG, ENG or DNG response within a specific frequency band or in multiple frequency bands. In yet another approach, which is somewhat less popular, the equivalent resistance, inductance and capacitance are obtained from the dimensions of the MTM unit cell, and then the effective μ- and ε-values are found after doing electromagnetic analysis [6].

However, almost all MTMs are anisotropic in nature, warranting the representation of the retrieved MTM parameters in its tensor form. One can have questions about (i) the limiting condition of the homogenization procedure; (ii) the possibility of producing similar transmission and reflection coefficients of the actual MTM unit cell while replacing it with a homogeneous material having the same retrieved material parameters (of the actual MTM); and (iii) the parameter retrieval procedure of the MTM unit cell that has dimensions compared to the incident wavelength. An attempt has been made in this chapter to address the issues related to all these questions, including various constitutive parameter retrieval procedures of inhomogeneous [3,7,8], anisotropic [9–12], bi-anisotropic [13–17], uniaxial anisotropic [18,19], chiral [20–23] and hyperbolic [24–26] MTMs, excited with normal and oblique incident plane waves, thereby providing a prerequisite theoretical background to appreciate the technology and applications of the MTM dealt with in this book.

In Section 2.2, to follow, we discuss EMT of MTM. Section 2.3 covers the electromagnetic analysis of parameter retrieval techniques of MTM, encompassing the S-parameter retrieval techniques, the miscellaneous parameter retrieval techniques, the parameter retrieval of special MTMs and the simulation and experimental validation of the retrieval techniques. The chapter concludes with Section 2.4, a brief summary of the review.

2.2 EFFECTIVE MEDIUM THEORY

On a historical timeline, Hendrik Antoon Lorentz, a Dutch physicist in the late 19th century, derived the electromagnetic Lorentz force and Lorentz transformations. He also described, in classical terms, the interaction between atoms and electric fields by indicating the force between an electron and the nucleus of an atom as the spring-like force that follows Hooke's law. Therefore, an applied electric field creates an oscillating motion in the electron by stretching or compressing the spring-like force. This is called the Lorentz oscillator model [27].

The displacement of a mass attached to a spring is accurately described by the second-order linear differential equation with respect to time, considering the driving force and the linear damping force. Assuming that the spring obeys a linear relation, the forces responsible for motion of the mass along the y-direction are defined by Newton's second law and Hooke's law. Newton's second law, in terms of the displacement vector $\bar{y}$ of a particle of mass m, taking into account all the contributing forces, may be written as

$$m\frac{d^2\bar{y}}{dt^2} = \bar{F}_{\text{driving}} + \bar{F}_{\text{damping}} + \bar{F}_{\text{spring}} \tag{2.1}$$

where $\bar{F}_{\text{driving}}$, $\bar{F}_{\text{damping}}$ and $\bar{F}_{\text{spring}}$ represent the driving, damping and spring forces, respectively. Putting

$$\bar{F}_{\text{damping}} = -m\Gamma_e\frac{d\bar{y}}{dt} \text{ and } \bar{F}_{\text{spring}}\ (=-K\bar{y}) = -m\omega_{e0}^2\bar{y} \text{ (Hooke's law),}$$

where ω_{e0} is the electric plasma angular frequency, one can then rearrange Eq. (2.1) to write the driving force as

$$\bar{F}_{\text{driving}} = m\frac{d^2\bar{y}}{dt^2} + m\Gamma_e\frac{d\bar{y}}{dt} + m\omega_{e0}^2\bar{y}. \tag{2.2}$$

Here, K is the spring constant and Γ_e is the damping factor. Now, with reference to an electron taken as the particle of mass m and charge e, subject to the force due to a time-varying electric field $\bar{E}_y = \bar{E}_{0y}\exp(-j\omega t)$ of amplitude $\bar{E}_{0y}$ (supposedly directed along y) and angular frequency ω, the driving force can also be written as

$$\bar{F}_{\text{driving}} = -e\bar{E}_y. \tag{2.3}$$

Interpreting the time varying displacement of an electron as $\bar{y} = \bar{y}_0\exp(-j\omega t)$, with $\bar{y}_0$ as its amplitude, one can combine (2.2) and (2.3) to write

$$\bar{y}_0 = -\frac{e}{m}\frac{\bar{E}_{0y}}{(\omega_{e0}^2 - \omega^2) - j\Gamma_e\omega}. \tag{2.4}$$

"The dipole moment due to each electron is $\bar{p} = e\bar{y}_0$and the polarization, defined as the total dipole moment per unit volume $\bar{P}$, is given by the vector sum of all the dipoles in the unit volume. Assuming one dipole per molecule and an average number density of N molecules per unit volume, one obtains" an expression for $\bar{P}$ with its amplitude $\bar{P}_0$ given as [28]

$$\left.\begin{aligned} \bar{P}_0 &= N\bar{p}_0 = \frac{e^2}{m}\frac{N\bar{E}_{0y}}{(\omega_{e0}^2 - \omega^2) - j\Gamma_e\omega} = \varepsilon_0\chi_e(f)\bar{E}_{0y} \\ \bar{P}_0 &= \varepsilon_0\alpha N\bar{E}_{0,\text{eff}} \\ \alpha &= \frac{e^2}{\varepsilon_0 m}\frac{1}{\omega_{e0}^2 - (\omega^2 + \Gamma_e\omega)}. \end{aligned}\right\} \tag{2.5}$$

In (2.5), $\bar{p}_0$ is the amplitude of $\bar{p}$, the latter being defined following Eq. (2.4). In (2.5), the second expression in conjunction with the third expression is identical to the first expression, except with $\bar{E}_{0y}$ being interpreted as $\bar{E}_{0,\text{eff}}$. Here, $\chi_e(f)$ is the susceptibility of the medium in which the electronic motion is considered, the medium effective relative permittivity $\varepsilon_{r,\text{eff}}(f)$, depending on frequency $f\,(=\omega/2\pi)$, is given by

$$\varepsilon_{r,\text{eff}}(f) = 1 + \chi_e(f) = 1 + \frac{f_{ep}^2}{(f_{e0}^2 - f^2) - j\Gamma_e f/(2\pi)} \tag{2.6}$$

where $f_{ep} = ((1/(2\pi)^2)(Ne^2/\varepsilon_0 m))^{1/2}$ is the electric plasma frequency; $f_{e0}(=\omega_{e0}/2\pi)$ (corresponding to its angular frequency defined following Eq. (2.1)) is the electric resonant frequency; and Γ_e is the collision frequency representing the "electronic dissipation" in the medium, which also represents the damping factor, as mentioned following (2.2).

Further, according to the principle of electromagnetic duality, the frequency-dependent effective relative permeability $\mu_{r,\text{eff}}(f)$ of the medium can be defined as [28]

$$\mu_{r,\text{eff}}(f) = 1 + \chi_m(f) = 1 + \frac{f_{mp}^2}{(f_{m0}^2 - f^2) - j\Gamma_m f/(2\pi)} \tag{2.7}$$

where f_{mp} is the magnetic plasma frequency and f_{m0} is the magnetic resonant frequency of the medium. Γ_m is the collision frequency accounting for the magnetic losses in the medium. Further, a medium characterized by the frequency-dependent parameters $\varepsilon_{r,\text{eff}}(f)$ and $\mu_{r,\text{eff}}(f)$, obeying Eqs. (2.6) and (2.7), respectively, is also known as a dispersive medium.

The low-frequency limiting value ε_{rs} and the high-frequency limiting value $\varepsilon_{r\infty}$ of the effective relative permittivity of a medium may be written with the help of Eq. (2.6) as [27]

$$\left.\begin{aligned} \varepsilon_{rs} &= \varepsilon_{r,\text{eff}}(f \to 0) = 1 + \frac{f_{ep}^2}{f_{e0}^2} \\ \varepsilon_{r\infty} &= \varepsilon_{r,\text{eff}}(f \to \infty) = 1 \end{aligned}\right\}. \tag{2.8}$$

Hence, with the help of Eq. (2.8), one can write Eq. (2.6) in the following alternative form [27]:

$$\varepsilon_{r,\text{eff}}(f) = 1 + \frac{(\varepsilon_{rs} - \varepsilon_{r\infty})f_0^2}{(f_{e0}^2 - f^2) - j\Gamma_e f/(2\pi)}. \tag{2.9}$$

In a metallic medium, the electrons are not bound to the nuclei, and they can flow freely around their lattices. Hence, a significant change in the Lorentz oscillator model would take place as there is no restoring spring force, that is, $\bar{F}_{\text{spring}} = 0$, and consequently, $f_{e0} = 0$ as well, as can be appreciated from the expression for $\bar{F}_{\text{spring}}$ (Hooke's law) given following Eq. (2.1). Therefore, Eq. (2.6) reduces to

$$\varepsilon_{r,\text{eff}}(f) = 1 - \frac{f_{ep}^2}{f^2 + j\Gamma_e f/(2\pi)}. \tag{2.10}$$

Eq. (2.10) is known as the Drude-Lorentz-model, or simply the Drude-model expression, named after Paul Drude, for use in metals. An example of an actual MTM, realized using the Drude and Lorentz models using the curve-fitting approach, is given in Section 2.3.2.

However, the above discussion on the effective medium parameters strictly holds well only for a relatively less dense medium of polarizable objects. In a dense medium, the polarization at any point is affected by the fields, known as *local fields*, arising from nearby polarized objects. Therefore, the polarization of the medium is proportional to the effective field, which is the vector sum of the applied and local fields. Each polarizable object is considered within a small sphere surrounded by a uniformly polarized medium, instead of discrete dipoles at various locations [28]. Assuming the polarization $\vec{P}$ outside to be a constant, one can write the amplitude of the effective field $\bar{E}_{0,\text{eff}}$ in terms of the amplitude P_0 of $\vec{P}$ as [28]

$$\bar{E}_{0,\text{eff}} = \bar{E}_{0y} + \frac{\bar{P}_0}{3\varepsilon_0}. \tag{2.11}$$

With the help of the second expression in (2.5), one can obtain $\bar{E}_{0,\text{eff}} = P_0/\varepsilon_0 \alpha N$, and substituting $\bar{E}_{0y}$ from the first expression of (2.5) into Eq. (2.11), one can obtain $\bar{E}_{0,\text{eff}} = \bar{P}_0(1/(\varepsilon_0\chi_e(f)) + 1/(3\varepsilon_0))$. Comparing these two expressions for $\bar{E}_{0,\text{eff}}$, one can write

$$\frac{\bar{P}_0}{\varepsilon_0 \alpha N} = \bar{P}_0\left(\frac{1}{\varepsilon_0\chi_e(f)} + \frac{1}{3\varepsilon_0}\right)$$

which simplifies, after a little algebra, to

$$\chi_e(f) = \frac{N\alpha}{1 - N\alpha/3}. \tag{2.12}$$

Using Eq. (2.12), Ramakrishna and Grzegorczyk [28] obtained the effective relative permittivity of the medium $\varepsilon_{r,\text{eff}}(f)$ as

$$\varepsilon_{r,\text{eff}}(f) = \frac{1 + 2N\alpha/3}{1 - N\alpha/3} \tag{2.13}$$

which is known as the Clausius-Mossotti relation for dielectrics.

A composite medium contains small particles of radius $a<<\lambda$ and a dielectric of permittivity ε_i (say), randomly embedded in another bulk dielectric medium of dielectric permittivity ε_h(say). The Maxwell-Garnett formula for the effective relative permittivity of the composite medium, the derivation of which is kept outside the purview of this chapter, is given as [28–31]

$$\varepsilon_{r,\text{eff}}(f) = \varepsilon_h \frac{\varepsilon_i(1 + 2x) + 2\varepsilon_h(1 - x)}{\varepsilon_i(1 - x) + \varepsilon_h(2 + x)} \tag{2.14}$$

where $x = 4\pi a^3 L/3$ and L is the number density of the spheres. *"The Maxwell-Garnett approach incorporates the distortions due to the dipole field on an average and has been very successful in describing the properties of dilute random inhomogeneous materials"* [28].

2.3 ELECTROMAGNETIC ANALYSIS OF PARAMETER RETRIEVAL TECHNIQUES

In this section, we revisit the formulation of the most popular Nicolson-Ross-Weir (NRW) parameter retrieval technique [32–34], based on the measurement of the scattering coefficients (S_{11} for reflection and S_{21} for transmission) of the conventional materials. The researchers have improved the NRW method by removing the limitation of the electrical thickness of materials and the branch ambiguity, and they have evolved the method to extract the parameters of MTMs with very good accuracy. Besides the scattering (S) parameter extraction method [3–5,35–37], there are several methods of parameter retrieval of MTMs, such as the curve-fitting approach [38], the dispersion-equation method [39] and the field-averaging technique [40–44]. We review them briefly here.

Traditionally, the measurement of complex permittivity and permeability of linear materials is carried out at fixed frequencies in the frequency domain using the slotted-line and impedance-bridge techniques. In this context, Nicolson and Ross [32] proposed a method adopting a single time-domain measurement, based on the displays of the transmitted and reflected signals. In this technique, one needs to place an unknown annular sample under test in a microwave TEM-mode fixture and excite the sample with a sub-nanosecond baseband pulse (Fig. 2.1) [32,45,48]. With this time-domain measurement approach, a Fourier transformation is required to determine the permittivity and permeability from the measured transient response. However, with this method, the obtained constitutive values are frequency-band limited, depending on the time response of the pulse and its repetition frequency.

Weir [33] and Blakney and Weir [34] presented a computer-controlled automatic measurement system to measure the S_{11} and S_{21} parameters when a material sample was inserted in a waveguide or a TEM transmission line. They determined the complex permittivity and permeability of material directly in the microwave frequency domain.

There are several other methods to extract the complex constitutive parameters from the S_{11} and S_{21} measurements methods—such as NRW, Newton-Raphson, root-find algorithm, genetic algorithm and artificial neural network algorithm [45].

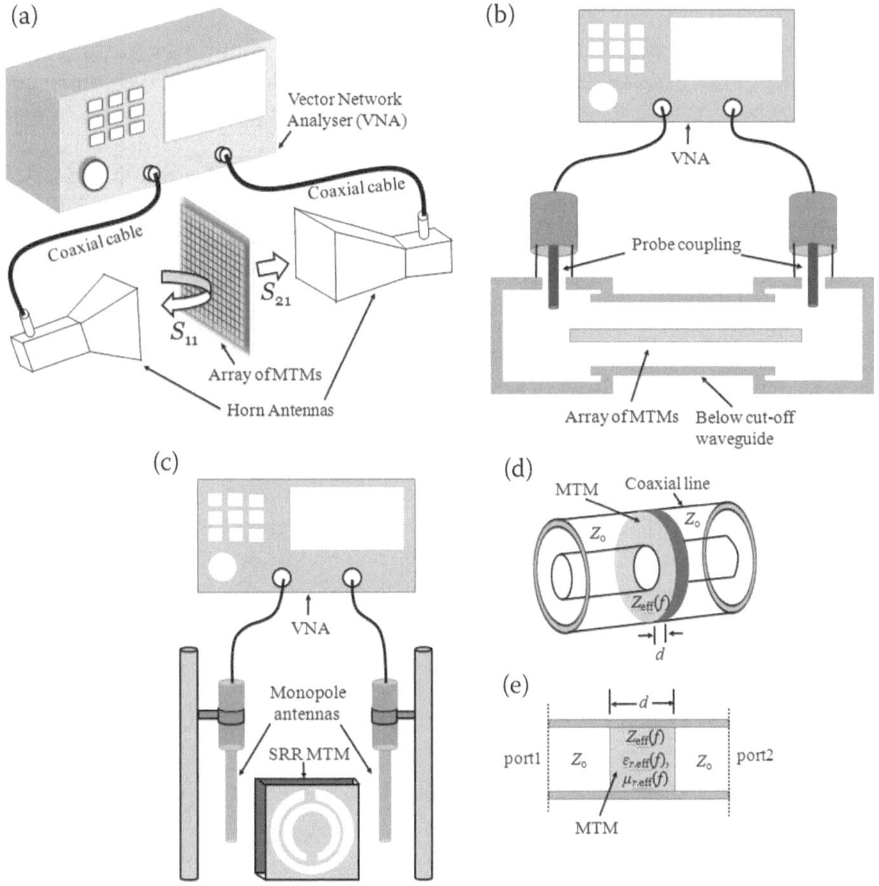

FIGURE 2.1 Schematics of the experimental setup to measure the material parameters from the reflection and transmission coefficients with reference to five cases of the setups based on the excitation of MUT with the material excited (a) in free-space medium [45], (b) in a below cutoff waveguide [46], (c) using two monopole antennas [47], (d) in a coaxial line [32] and (e) in a waveguide [48].

However, we restrict our discussion here to the NRW method for the characterization of only the conventional materials.

2.3.1 *S*-PARAMETER RETRIEVAL TECHNIQUES

The NRW method has been a standard technique for measuring the complex permittivity and permeability of homogeneous and isotropic materials for more than 50 years. One can categorize the S_{11} and S_{21} measurement methods with reference to the five cases based on the excitation of material under test (MUT). In the first of these cases, the MTM is excited in free space by a perpendicular- ($\perp$) or parallel- ($\parallel$) polarized plane wave incident at an angle θ_0 from the normal to the MTM assumed to be infinitely extended in the transverse directions, using the transmitting and receiving horn antennas in the measurement setup (Fig. 2.1a). In the second case, the array of MTMs is placed in a below cutoff waveguide (combined structure) to observe backward-wave propagation. In this case, the combined structure is excited with the help of cavities and probe coupling (Fig. 2.1b). In the third case, the characteristics (using S_{11} and S_{21}) of a magnetic MTM can be obtained with the help of two monopole antennas (Fig. 2.1c). In the fourth and fifth cases, the MTM is placed across the cross-sectional area of a coaxial transmission line (excited with TEM mode) and rectangular waveguide (excited with TE or TM mode), respectively (Fig. 2.1d and 2.1e). Three commonly used methods are compared in Table 2.1.

If d is infinite (Figs. 2.1d and 2.1e), the reflection coefficient Γ of a wave incident on the interface (boundary between the MUT and the free-space region) is simply given by [49]

$$\Gamma = (Z_{\text{eff}}(f) - Z_0)/(Z_{\text{eff}}(f) + Z_0) \tag{2.15}$$

where Z_0 is the characteristic impedance of free-space regions, and the same at the MUT region is given by $Z_{\text{eff}}(f)$, which may be a complex quantity.

For finite d, the transmission coefficient P along MUT may be written as

$$P = \exp(-jk_z d) \tag{2.16}$$

where k_z is the z-component of the propagation constant defined by

$$k_z^2 = k^2 - \kappa^2. \tag{2.17}$$

Here, $k = (\omega/c)(\varepsilon_{r,\text{eff}}(f)\mu_{r,\text{eff}}(f))^{1/2}$ is the propagation constant in the MUT region, where c is the speed of light. For the free-space system, $\kappa = k_0\sin\theta_0$, where $k_0 = (\omega/c)$; $\kappa = 0$ for the TEM-mode waveguide system; and $\kappa = \pi/a$ for, typically, TE_{10}-mode rectangular waveguide system [49]. For the method using the free-space region (Fig. 2.1a) [49]

TABLE 2.1
Comparison of Commonly Used Reflection-Transmission Measurement Methods [45]

Aspect	Free-space method	Coaxial line method	Waveguide method
Dielectric features	$\varepsilon_{r,\text{eff}}(f)$ and $\mu_{r,\text{eff}}(f)$	$\varepsilon_{r,\text{eff}}(f)$ and $\mu_{r,\text{eff}}(f)$	$\varepsilon_{r,\text{eff}}(f)$ and $\mu_{r,\text{eff}}(f)$
S-parameters	S_{11} and S_{21}	S_{11} and S_{21}	S_{11} and S_{21}
Frequency band	Wide	Wide	Discrete
MUT dimensions	Large	Small	Medium
Sample preparation	Easy	Easy	Difficult
Monitoring	Very easy	Easy	Difficult

$$Z_{\text{eff}}(f) = \begin{cases} \frac{k\eta_{\text{eff}}(f)}{k_z}, \perp\text{-pol} \\ \frac{k_z\eta_{\text{eff}}(f)}{k}, \|\text{-pol} \end{cases}; \quad Z_0 = \begin{cases} \frac{\eta_0}{\cos\theta_0}, \perp\text{-pol} \\ \eta_0\cos\theta_0, \|\text{-pol} \end{cases} \tag{2.18}$$

where the subscripts $\|$ – pol and $\perp$ – pol represent the parallel and perpendicular polarizations, respectively. Here, k_z can be read using Eq. (2.17). Further, for the TEM-mode waveguide system, $Z_{\text{eff}}(f) = \eta_{\text{eff}}(f)$ and $Z_0 = \eta_0$, where $\eta_{\text{eff}}(f) = \eta_0(\mu_{r,\text{eff}}(f)/\varepsilon_{r,\text{eff}}(f))^{1/2}$ and $\eta_0 = (\mu_0/\varepsilon_0)^{1/2}$. For the TE_{10}-mode rectangular waveguide system, $Z_{\text{eff}}(f) = \omega\mu_{r,\text{eff}}(f)/k_z$ and $Z_0 = \omega\mu_0/k_{z0}$, and $k_{z0}^2 = k_0^2 - \kappa^2$.

For all the five cases (Fig. 2.1), S_{11} (the transfer function between the reflected and incident signal ports), and S_{21} (the transfer function between the transmitted and incident signal ports (scattering matrix parameters) may be written as

$$S_{11} = \frac{(1 - P^2)\Gamma}{(1 - P^2\Gamma^2)}; \quad S_{21} = \frac{(1 - \Gamma^2)P}{(1 - P^2\Gamma^2)}. \tag{2.19}$$

Here, Γ and P can be read with the help of Eqs. (2.15), (2.18), (2.16) and (2.17), respectively.

In the original NRW method, there are two intermediate terms, which represent the sum and difference between S_{11} and S_{21}, respectively, and, using Eq. (2.19), can be written as

$$V_1 = S_{21} + S_{11} = \frac{P + \Gamma}{1 + \Gamma P}; \quad V_2 = S_{21} - S_{11} = \frac{P - \Gamma}{1 - \Gamma P}. \tag{2.20}$$

Now, one can obtain Γ and P in terms of measured values of S_{11} and S_{21} using Eq. (2.20). Hence, eliminating P, one may form the following quadratic equation in Γ:

$$\Gamma^2 - 2\Gamma X + 1 = 0 \tag{2.21}$$

where

$$X = \frac{1 - V_1V_2}{V_1 - V_2} = \frac{1 - S_{21}^2 + S_{11}^2}{2S_{11}}. \tag{2.22}$$

Thus, we obtain the expression for Γ, in terms of X given by (2.22), by solving Eq. (2.21), as follows:

$$\Gamma = X \pm (X^2 - 1)^{1/2}. \tag{2.23}$$

Regarding the sign ambiguity in the right-hand side of Eq. (2.23), Rothwell et al. [49] stated: “The sign ambiguity is usually resolved by assuming that the material is passive, since only one choice of sign satisfies the inequality $|\Gamma| \leq 1$.”

Therefore, now that Γ is known from Eq. (2.23), P can be obtained using Eq. (2.20) as [49]

$$P = \frac{V_1 - \Gamma}{1 - V_1 \Gamma} = |P| e^{j\phi} \tag{2.24}$$

where ϕ is the argument of the transmission term and bounded as $-\pi < \phi \leq \pi$. We may consider a dimensionless quantity $k'_z = k_z/k_0$ while equating Eqs. (2.16) and (2.24), taking natural logarithm in both sides of the equation

$$k'_z = \frac{m - \phi/2\pi}{d/\lambda_0} + j\frac{\ln|P|/2\pi}{d/\lambda_0} \tag{2.25}$$

where m is the branch index. Using Eqs. (2.17), (2.18), and (2.25), one can obtain $\varepsilon_{r,\mathrm{eff}}(f)$ and $\mu_{r,\mathrm{eff}}(f)$ considering three cases, as follows:

i. For a free-space system [49]):
   a. with perpendicular polarization

$$\varepsilon_{r,\mathrm{eff}}(f) = \frac{k'^2 F \cos\theta_0}{k'_z}; \; \mu_{r,\mathrm{eff}}(f) = \frac{k'_z}{F \cos\theta_0} \tag{2.26}$$

   b. with parallel polarization

$$\varepsilon_{r,\mathrm{eff}}(f) = \frac{k'_z F}{\cos\theta_0}; \; \mu_{r,\mathrm{eff}}(f) = \frac{k'^2 \cos\theta_0}{k'_z F} \tag{2.27}$$

where $F = (1 - \Gamma)/(1 + \Gamma)$, $k'^2 = k_z'^2 + \kappa'^2$, and $\kappa' = \kappa/k_0$.

ii. For a TEM-mode waveguide system [49]:

$$\varepsilon_{r,\mathrm{eff}}(f) = k'_z F; \; \mu_{r,\mathrm{eff}}(f) = \frac{k'_z}{F}. \tag{2.28}$$

where $k'_{z0} = k_{z0}/k_0$.

iii. Similarly, for a TE_{10}-mode rectangular waveguide system [49]:

$$\varepsilon_{r,\mathrm{eff}}(f) = k'^2 \frac{k'_{z0} F}{k'_z}; \; \mu_{r,\mathrm{eff}}(f) = \frac{k'_z}{k'_{z0} F} \tag{2.29}$$

The effective refractive index $n_{\mathrm{eff}}(f)$ of MUT can be calculated, with the help of Eq. (2.29), as

$$n_{\text{eff}}(f) = (\varepsilon_{r,\text{eff}}(f)\mu_{r,\text{eff}}(f))^{1/2}. \tag{2.30}$$

As a special case, in the free-space system, with normal incident ($\theta_0 = 0$), Eqs. (2.26) and (2.27) reduce to Eq. (2.28). Further, regarding Eq. (2.25), Rothwell et al. [49] stated, "*This equation reveals one of the known difficulties with the NRW method. Its value is not known a priori, but is related to the electrical thickness of the sample. Since the material parameters are unknown, n may be difficult to determine unless some test against known physics is employed, such as* $\varepsilon'' > 0$ *for passive materials. For engineered materials, however, this test may not be applicable*" (ε'' standing for the imaginary part of $\varepsilon_{r,\text{eff}}(f)$).

Improved NRW Methods for MTMs

The conventional parameter retrieval NRW method, even though it is extensively used, suffers from the ambiguities arising from the requirements of determining the branch index m (defined following Eq. (2.25)) and maintaining the MUT thickness $d < \lambda/2$. Yet, there is another problem with this method related to the periodicity of Eq. (2.16) for the wave propagating through the MUT, causing Eq. (2.25) to have an infinite number of roots [48]. However, there have been numerous studies done to overcome these ambiguities while maintaining very good accuracy in the experimental results of the extracted parameters. Here, we discuss some improved NRW methods that are used to extract the effective parameters of MTMs.

Ziolkowski [37] modified the NRW method to extract the constitutive parameters of MTMs by first-order approximating the exponential term appearing in Eq. (2.16) for the transmission coefficient, with the help of Taylor series expansion, as follows:

$$P \sim 1 - jk_z d. \tag{2.31}$$

However, the accuracy of the method that uses the first-order approximated expression is somewhat low when the measured materials have large sampling thicknesses [37].

Luukkonen et al. [48] proposed an improved NRW method, which did not involve any branch-seeking (because they measured the phase difference between the preceding measurement points rather than the phase delay at a given frequency point), thereby enabling the method to extract the materials parameters of an MUT, which has its thickness $d > \lambda/2$. In addition, this method was used to study the effect of thermal noise on the extracted material parameters.

The measured transmission coefficients of MUT at multiple frequency points ($\omega_0, \omega_1, \omega_2, \ldots, \omega_N$) will correspond to the arguments of Eq. (2.24) as $\phi_0, \phi_1, \phi_2, \ldots \phi_N$, respectively, for instance, ϕ_0 representing the phase at tse's first measurement point. Thus, the argument at the frequency point ω_N is given as [48]

$$\phi_N = \phi_0 + \sum_{i=1}^{N} \arg\left(\frac{P_{i-1}}{P_i}\right). \tag{2.32}$$

At this stage, with the help of Eq. (2.16), one can write

$$P_i = \exp(-jk_{z(i)}d), \quad P_{i-1} = \exp(-jk_{z(i-1)}d). \tag{2.33}$$

Now, one can put Eqs. (2.32) and (2.33) into Eq. (2.25) to obtain Eqs. (2.26)–(2.30) again with this improved method.

This extraction technique does not follow the branch seeking in the measured frequency band provided that $\Delta\phi = \phi_i - \phi_{i-1} < \pi$. Further, Luukkonen et al. [48] added: "*In order to unambiguously determine the values of the material parameters, the thickness d of the MUT should remain small at the lowest measurement frequency: d < λ/2, in which case m can be set to 0 at this frequency point. Otherwise, one needs to set the integer m corresponding to the electrical thickness of the sample at ω₀.*"

Electromagnetic MTMs are composite structures, formed either from periodic or random arrays of scattering elements, and should be considered as a continuous medium during the excitation of electromagnetic waves while maintaining the long wavelength limit [1,2]. However, using the simulation and experiment, it is well-known that some MTMs exhibit the scattering behavior with frequency-dependent $\varepsilon_{r,\text{eff}}(f)$ and $\mu_{r,\text{eff}}(f)$ [40,50,51]. In these MTMs, the conventional NRW method is not applicable. Therefore, a modified NRW method was proposed, assuming the medium to be homogeneous, where the *S*-parameters were inverted to obtain the complex impedance $Z_{\text{eff}}(f)$ and $n_{\text{eff}}(f)$ as follows [5]:

$$Z_{\text{eff}}(f) = \pm\left(\frac{(1+S_{11})^2 - S_{21}^2}{(1-S_{11})^2 - S_{21}^2}\right)^{1/2} \tag{2.34}$$

$$n_{\text{eff}}(f) = \frac{1}{k_0 d}\{[[\ln \exp(jk_0 n_{\text{eff}}(f)d)]'' + 2m\pi] - j[\ln \exp(jk_0 n_{\text{eff}}(f)d)]'\} \tag{2.35}$$

$$\varepsilon_{r,\text{eff}}(f) = n_{\text{eff}}(f)/Z_{\text{eff}}(f) \text{ and } \mu_{r,\text{eff}}(f) = n_{\text{eff}}(f)Z_{\text{eff}}(f). \tag{2.36}$$

Here and henceforth, the prime (′) and double-prime (″) denote the real and imaginary parts of a quantity, respectively.

However, Eqs. (2.35) and (2.36) are ambiguous while finding the correct branch index *m*. This is due to the inversion of logarithmic calculation [5]. The identification of the correct branch is very hard if the branches show the close proximity, which is influenced by the value of *d*. It was shown that the thinnest possible MUT gives the best results. If the MUT is passive, the condition $Z_{\text{eff}}'(f) \geq 0$ fixes the choice of sign in Eq. (2.34), and $n_{\text{eff}}''(f) \geq 0$ leads to the unambiguous result in Eq. (2.36). However, $n_{\text{eff}}'(f)$ is complicated by the branches of the logarithmic function [5]. Arslanagić et al. [52] argued over the assumption that it was unnecessary to find the correct branch. A large number of MTMs hold the passivity condition and, hence, also satisfy Eqs. (2.34)–(2.36).

Another extraction method based on the Kramers–Kronig (K-K) relation was developed to solve the branching problem caused by the complex logarithmic function [53–57]. Shi et al. [55] applied the first-order approximation of Taylor's series to the transmission term and combined it with the K-K method. The proposed method was not only to improve the accuracy of the extraction algorithm, but also to "*correctly solve the branch of the refractive index caused by complex exponential function and thus the proposed method is free from the limitation of the electrical thickness of the measured materials*." Further, Shi et al. [55] put Eq. (2.16) as

$$P = \left[\exp\left(-j\frac{k_z d + 2\pi m}{\nu}\right)\right]^{\nu} \quad (m = 0, \pm 1, \pm 2, \ldots) \tag{2.37}$$

where ν is an integer, which is typically set to a high value.

Applying the first-order approximation in the Taylor's series expansion [37] as in Eq. (2.31) to Eq. (2.37), one can obtain [55]

$$P = \left(1 - j\frac{k_z d + 2\pi m}{\nu}\right)^{\nu}. \tag{2.38}$$

By solving the νth root of Eq. (2.38), one can then write

$$P^{1/\nu} = 1 - j\frac{k_z d + 2\pi m}{\nu}. \tag{2.39}$$

The left-hand side of Eq. (2.39) has ν roots. Therefore, one can rewrite Eq. (2.39) as

$$P^{1/\nu} = |P|^{1/\nu} \exp\left(j\frac{\phi + 2(l-1)\pi}{\nu}\right) = 1 - j\frac{k_z d + 2\pi m}{\nu}. \quad (l = 1, \ldots, \nu). \tag{2.40}$$

For large value of ν, the right-hand side of Eq. (2.40) becomes approximately equal to 1. To maintain the roots of P closest to the real axis in the first and fourth quadrants of the complex plane, the value of integer l has to be controlled [55]. Further, we can write the complex refractive index as

$$n_{\text{eff}}(f) = n'_{\text{eff}}(f) + jn''_{\text{eff}}(f) = k_z/k_0. \tag{2.41}$$

Now, by rearranging Eq. (2.40), we can write

$$k_z = \nu\frac{j(\text{Re}(P^{1/\nu}) - 1) - (\text{Im}(P^{1/\nu}) + 2\pi m)}{d}. \tag{2.42}$$

Also, putting (2.42) into Eq. (2.41), we can get [55]

$$n'_{\text{eff}}(f) = -\nu\frac{\text{Im}(P^{1/\nu}) + 2\pi m}{k_0 d} \tag{2.43}$$

$$n''_{\text{eff}}(f) = \nu\frac{\text{Re}(P^{1/\nu}) - 1}{k_0 d}. \tag{2.44}$$

The branch ambiguity arises when $m \neq 0$ for MUTs with $d > \lambda/2$. Therefore, according to the K-K relation, one can obtain unambiguously the real part of $n_{\text{eff}}(f)$ as [55,57]

$$n_{\text{eff}}^{\prime\text{K-K}}(\omega) = 1 + \frac{2}{\pi}\int_0^{\infty}\frac{\omega' n''_{\text{eff}}(\omega)}{\omega^2 - \omega'^2}d\omega' \tag{2.45}$$

where ω' is the real part of ω, and the superscript refers to the K-K relation. Now, putting (2.44) into Eq. (2.45) and setting a high value to the upper limit (correspond to the highest frequency) of the integral, one can solve for the approximated real part of the complex refractive index. The value of m is determined while bringing n_{eff}' closest to $n_{\text{eff}}^{\prime\text{K-K}}(\omega)$. Using Eqs. (2.43)–(2.45) and Eq. (2.41), one can obtain $n_{\text{eff}}(f)$, and, with the help of Eqs. (2.41) and (2.34), one can obtain the MUT parameters from Eq. (2.36) [55].

In spite of wide usage with the advantages, such as the versatility and ease of implementation, the S-parameter extraction technique does not provide physical insight into MTM structures. Moreover, any mistake in branch selection in the technique may lead to false results. Also, the accuracy of algorithms in the technique is found to be highly dependent on the electrical thickness of MTMs.

2.3.2 Miscellaneous Parameter Retrieval Techniques

There are other parameter extraction techniques for MTMs, such as the curve-fitting approach [38], transfer matrix approach [3,58,59], field averaging approach [40,41], data-driven discontinuity detection approach [60], Snells's law approach [50,51,61], wave propagation approach [62,63], state-space approach [16,21], causality principle approach [22], quasi-mode theory approach [64], phase unwrapping approach [65] and asymmetrical strip line approach [66]. We will discuss some of these popular approaches in this subsection.

Curve-fitting Approach

Lubowski Schuman and Weiland [38] proposed a parameter fitting approach to the extraction of $\varepsilon_{r,\text{eff}}(f)$ and $\mu_{r,\text{eff}}(f)$ of MTMs. In this approach, $\varepsilon_{r,\text{eff}}(f)$ and $\mu_{r,\text{eff}}(f)$ of anisotropic and homogeneous slab are defined by the dispersive Drude and Lorentz models, respectively, which is an equivalent representation of actual MTM of the same thickness. The parameters of Drude and Lorentz models are obtained using the optimization algorithm. Basically, one has to obtain the same S-parameters (magnitude and phase of S_{11} and S_{21}) of the actual MTM, obtained using the simulation or measurement, by optimizing the Drude and Lorentz model parameters. Therefore, one can write $Z_{\text{eff}}(f)$ and $n_{\text{eff}}(f)$ as

$$Z_{\text{eff}}(f) = Z_0(\mu_{r,\text{eff}}(f)/\varepsilon_{r,\text{eff}}(f))^{1/2} \tag{2.46}$$

$$n_{\text{eff}}(f) = (\varepsilon_{r,\text{eff}}(f)\mu_{r,\text{eff}}(f))^{1/2}. \tag{2.47}$$

Here, $\varepsilon_{r,\text{eff}}(f)$ and $\mu_{r,\text{eff}}(f)$ are obtained using the dispersive Drude model in Eq. (2.10) and Lorentz model in Eq. (2.7), respectively. The signs of Eqs. (2.46) and (2.47) depend on the passivity condition ($Z'_{\text{eff}}(f) \geq 0$ and $n''_{\text{eff}}(f) \geq 0$) of the MTM. One can then write P from Eq. (2.16) as

$$P = \exp\left(-j\frac{\omega}{c}n_{\text{eff}}(f)d\right). \tag{2.48}$$

Now, substituting (2.15) and Eqs. (2.46)–(2.48) into Eq. (2.19), we can get S_{11} and S_{21}. The purpose of using the optimization algorithm for obtaining Drude and Lorentz model parameters is to provide the best fit between S_{11} and S_{21} of the actual MTM and the isotropic and homogeneous slab.

The merits of the curve-fitting approach are its straightforwardness, removal of branch ambiguities, etc. However, this approach is limited to non- and weak-bi-anisotropic MTM structures. The simple Drude and Lorentz models do not consider the magnetoelectric couplings of MTM.

Transfer Matrix Approach

One can observe the significant phase variation of transmitted and reflected waves across the unit cell at the operational frequency, in almost all MTMs, despite taking the unit cell dimensions smaller than the free-space wavelength, which invalidates the effective medium consideration. Therefore, when we apply the standard S-parameter extraction technique in inhomogeneous MTMs (discussed in Section 2.3.1) having resonant elements, the variety of artifacts in the retrieved material parameters arises due to the inhomogeneity (of MTMs). These artifacts are regarded as large fluctuation in $n_{\text{eff}}(f)$ and $Z_{\text{eff}}(f)$ of MTMs where the wavelength within MTMs can be smaller than the unit cell dimensions.

Hence, to overcome this situation, Smith et al. [3] proposed a modified S-parameter approach where the S-matrix can be found from the elements of the transfer matrix (T-matrix) [59]. The T-matrix relating the fields on one side of a planar slab to those on the other side can be defined as [3]

$$\bar{\text{F}}' = \text{T}\bar{\text{F}} \tag{2.49}$$

where

$$\bar{F} = \begin{pmatrix} E \\ H_{red} \end{pmatrix}. \tag{2.50}$$

Here, E and H_{red} are the complex electric and "reduced" magnetic field amplitudes located on the right- and left-hand faces of the slab. The primed and unprimed quantities refer to the quantities at the left- and right-hand faces of slab, respectively. Further, the effective medium consideration is invalid not only because the phase of transmitted and reflected waves varies significantly across the unit cell, even if it has a smaller thickness than the free-space wavelength, but also because the optical path length across the unit cell may not be small.

For periodic structures, this phase advancement will become still more significant. Therefore, the fields on one side of a unit cell corresponding to a periodic structure are related to those on the other side by a phase factor, as follows [3]

$$\bar{F}(x + d) = \exp(j\alpha d)\bar{F}(x) \tag{2.51}$$

where $\bar{F}$ is defined in Eq. (2.50) and α is the phase advance per unit cell. Eq. (2.51) is the Bloch condition and, with the help of Eq. (2.49), one can express it as

$$\bar{F}' = T\bar{F} = \exp(j\alpha d)\bar{F} \tag{2.52}$$

where the T-matrix is defined in Eq. (2.49) in conjunction with Eq. (2.50). Then, with the help of Eq. (2.52), one can obtain the dispersion relation of periodic structure by putting $|T - \exp(j\alpha d)I| = 0$, where I represents the identity matrix. Hence, taking $\det(T) = 1$ [3], the following simplified form of dispersion relation of periodic structure can be obtained:

$$2\cos(\alpha d) = T_{11} + T_{22} \text{ (dispersion relation)} \tag{2.53}$$

where T_{11} and T_{22} are the elements of the T-matrix [3].

Now, if the unit cell is not symmetric in the propagation direction, the standard retrieval procedure could not produce the unique solution of $n_{eff}(f)$. For instance, depending on the direction of propagation of the incident plane wave in the unit cell, standard S-parameter retrieval procedure gives [3]

$$\cos(n_{eff}(f)kd) = \frac{1}{2S_{21}}(1 - S_{11}^2 + S_{21}^2) \tag{2.54}$$

or

$$\cos(n_{eff}(f)kd) = \frac{1}{2S_{21}}(1 - S_{22}^2 + S_{21}^2). \tag{2.55}$$

Therefore, it is clear from Eqs. (2.54) and (2.55) that we cannot get a unique solution of $n_{eff}(f)$ for an asymmetric structure for which $S_{11} \neq S_{22}$. Now, comparing the forms of Eq. (2.53) with Eqs. (2.54) and (2.55), one can obtain the T-matrix elements in terms of the S-matrix elements, and again using Eq. (2.53), one can write

$$\cos(\alpha d) = \frac{1}{2S_{21}}(1 - S_{11}S_{22} + S_{21}^2). \tag{2.56}$$

$n_{eff}(f)$ can be recovered from the modified S-parameter retrieval method from Eq. (2.56) that utilizes all the elements of S-matrix. However, for both homogeneous and symmetric inhomogeneous unit cells, the retrieval process is the same because $S_{11} = S_{22}$.

In order to obtain $\varepsilon_{r,eff}(f)$ and $\mu_{r,eff}(f)$ using Eq. (2.36), one needs to know $n_{eff}(f)$ (the latter found from Eq. (2.56)) and Z_{red} (= E/H_{red}, for a continuous material), which can be obtained after carrying out a few mathematical steps while equating $T\bar{F} = \exp(j\alpha d)\bar{F}$ of Eq. (2.52). Further, the expression for $\exp(j\alpha d)$ can be obtained by solving $|T{-}\exp(j\alpha d)I| = 0$, and putting the expression for $\exp(j\alpha d)$ into that equation, one can obtain the final expression of Z_{red} as [3]

$$Z_{red} = \frac{(T_{22} - T_{11}) \mp ((T_{22} - T_{11})^2 + 4T_{12}T_{21})^{1/2}}{2T_{21}}. \tag{2.57}$$

The two roots of Eq. (2.57) indicate the two directions of wave propagation. Therefore, Z_{red} is not unique for inhomogeneous periodic structures as E/H_{red} will vary periodically with the structure [3].

Other Approaches

The above-mentioned parameter retrieval techniques are relatively popular. In addition to these techniques, there are several other parameter retrieval techniques that are extremely effective, such as the following:

i. The *field averaging approach* is used to determine macroscopic fields via averaging local fields obtained from the simulation or analysis, from which one can obtain the effective-medium parameters of these MTMs [40,41].
ii. The *data-driven (D-D) discontinuity detection approach* is based on detecting the discontinuity points in the real part of the refractive index and discerning the correct branch values at these discontinuities. This method is simple and does not require infinite frequency integration. It gives accurate values for MTM parameters at resonance and at the negative parameter region [60].
iii. The *Snells's law approach* is based on experimental scattering data at microwave frequencies on a structured MTM that exhibits a frequency band where $n_{eff}(f)$ is negative [50,51,61].
iv. The *wave propagation approach* is easy to implement and has no ambiguity in retrieving the material parameters of thick chiral MTMs [62,63].

v. The *state-space approach* is based on scattering data for a linearly polarized plane wave incident, normally on a homogeneous chiral slab, which is combined with the properties of a state transition matrix [16,21].
vi. The *causality principle approach,* based on the causality principle and K-K relations, removes the branch selecting problem, which has been proposed for extracting effective parameters of MTMs, and further extended to extract effective parameters of chiral MTMs as well [22].
vii. The *quasi-mode theory approach* calculates the self-energy, density of states, and mean-free paths for optical modes traveling inside a MTM and then determines the effective permittivity and permeability of the MTM by maximizing the density of states function [64].
viii. The *phase unwrapping approach* correctly chooses the branch by detecting successive phase jumps, using S_{11} and S_{21}, which exceed π at specific frequencies, especially at the resonant frequency region (Shi et al. 2016).
ix. The *asymmetrical strip line approach* is an experimental method, measuring the *S*-parameters of an asymmetrical strip-line partially filled with a sample, for extracting the electromagnetic parameters of MTMs between 10 MHz and 6 GHz, considering the two characteristics mentioned before: heterogeneity and anisotropy [66].

2.3.3 Parameter Retrieval of Some Special Type MTMs

In the *S*-parameter extraction techniques discussed in Section 2.3.1, it was assumed that the electromagnetic waves is incident normally ($\theta_0 = 0$) on the homogeneous and isotropic MTM slabs. Now, questions arise as to how to retrieve the MTM parameters if the incident wave is oblique (i.e., $\theta_0 \neq 0$) and what to do if the MTMs are other than homogeneous and isotropic in nature.

Inhomogeneous MTMs

In the *transfer matrix approach* for treating inhomogeneous MTMs, the conventional *S*-parameter retrieval methods are not suitable as $S_{11} \neq S_{22}$. Therefore, Smith et al. [3] proposed a modified *S*-parameter retrieval method that can be used for inhomogeneous structures.

Now, we consider multiple thin isotropic and homogeneous materials that are stacked or layered together to make an inhomogeneous MTM structure. In order to extract the effective relative parameters of this *m*-layer inhomogeneous structure, Shi et al. [8] proposed an electromagnetic field-based recursive method for solving the scattering parameters of the layered medium. A plane wave, which has an electric field parallel to the interface of two adjacent layers, is incident normally along the *z*-axis. Here, the electromagnetic fields in each region are expressed in terms of incident and reflected waves. Therefore, the *S*-parameters are recursively solved in each region for $(m - 1)$ times.

In the iterative solution procedure, $n_{eff}(f)$ and $Z_{eff}(f)$, the latter represented as $\eta_{eff}(f)$ in Shi et al. [8], are determined first for region 1 using the conventional NRW method. $n_{eff}(f)$ and $Z_{eff}(f)$ of region 1 are used to obtain the same for region 2, and so on. One can replace the layers of the actual MTM structure with the extracted $n_{eff}(f)$

and $Z_{\text{eff}}(f)$ of each region, and try to obtain the same S-parameters of the actual structure iteratively for $(m-1)$ times.

Uniaxial Anisotropic MTMs

Anisotropic materials comprise a class of complex materials, the properties of which are direction-dependent and described by tensors, instead of scalar structural parameters. In uniaxial MTMs, also called hyperbolic MTMs [24], one of the components of a diagonal tensor of $\varepsilon_{r,\text{eff}}(f)$ or $\mu_{r,\text{eff}}(f)$ is opposite in sign to the other two components [18]. For a uniaxial slab, since the optical axis is aligned with the z-axis (Fig. 2.2), the $\varepsilon_{r,\text{eff}}(f)$ and $\mu_{r,\text{eff}}(f)$ tensors are diagonal, as in Eq. (2.58). For uniaxial consideration, one can write $\varepsilon_{xx} = \varepsilon_{yy} = \varepsilon_o$ and $\varepsilon_{zz} = \varepsilon_e$; and $\mu_{xx} = \mu_{yy} = \mu_o$ and $\mu_{zz} = \mu_e$, where ε_o and μ_o are ordinary and ε_e and μ_e are extraordinary permittivity and permeability, respectively, of the uniaxial slab. The two most popular uniaxial MTMs are the arrays of metallic rods embedded in a host material and multilayer metal-dielectric stack (Fig. 2.2) [18].

$$\varepsilon_{r,\text{eff}}(f) = \begin{pmatrix} \varepsilon_{xx} & 0 & 0 \\ 0 & \varepsilon_{yy} & 0 \\ 0 & 0 & \varepsilon_{zz} \end{pmatrix}, \quad \mu_{r,\text{eff}}(f) = \begin{pmatrix} \mu_{xx} & 0 & 0 \\ 0 & \mu_{yy} & 0 \\ 0 & 0 & \mu_{zz} \end{pmatrix}. \tag{2.58}$$

In the presented retrieval approach [18], an inhomogeneous uniaxial MTM is replaced by a homogeneous uniaxial effective medium that is characterized by ε_o, μ_o, ε_e and μ_e – the approach referred to as the inverse-scattering problem. In the first step in this approach, the complex reflection and transmission coefficients of the actual system have been obtained for both the TE- and TM-polarizations, which are defined here as the forward problem (Fig. 2.3). While homogenization is valid, for normal incidence, the scattering wave parameter retrieval [67] yields ε_o and μ_o. For oblique incident scattering, the effective wave vector k_z for both the polarizations, together with the use of the dispersion relations in (2.59) and (2.60) to follow, obtained by solving Helmholtz's equation for a monochromatic plane wave, yields the extraordinary parameters ε_e, and μ_e. The retrieval steps of the approach are summarized in Fig. 2.3 [18].

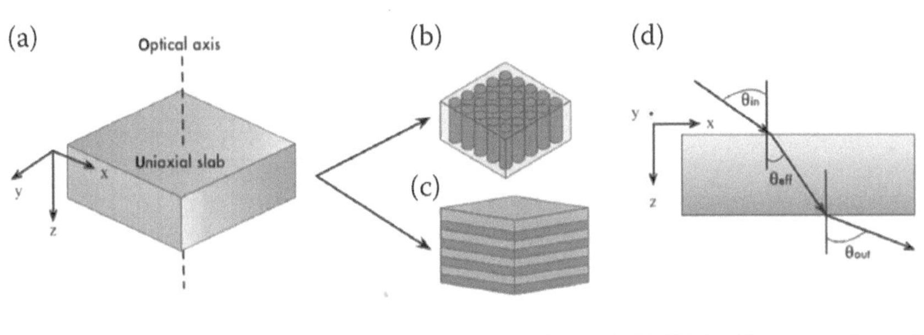

FIGURE 2.2 (a) Three-dimensional representation of uniaxial MTMs, (b) arrays of metallic rods embedded in a host material, (c) multilayer metal-dielectric stack, (d) projection of (a) onto the xz-plane, convention for the angle of incidence. [Source: Papadakis et al. [18]].

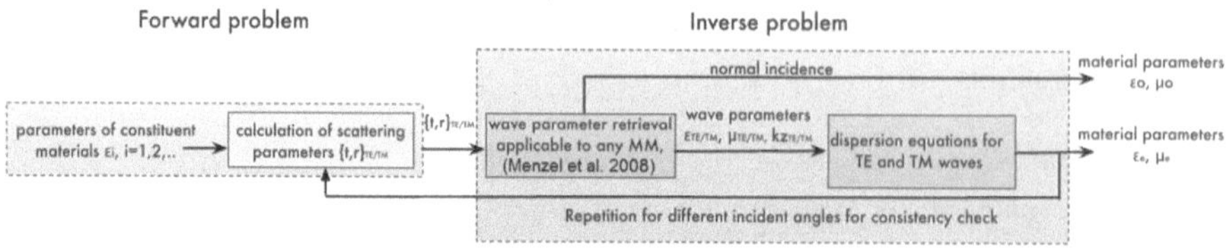

FIGURE 2.3 Retrieval steps. [Source: Papadakis et al. [18]].

For magnetically extraordinary wave: $\dfrac{k_x^2 + k_y^2}{\varepsilon_o \mu_e} + \dfrac{k_z^2}{\varepsilon_o \mu_o} = k_0^2$ (dispersion relation)

(2.59)

For electrically extraordinary wave: $\dfrac{k_x^2 + k_y^2}{\varepsilon_e \mu_o} + \dfrac{k_z^2}{\varepsilon_o \mu_o} = k_0^2$. (dispersion relation)

(2.60)

Bi-anisotropic MTMs with Oblique Incidence

Almost all embedded metallic resonant particles of MTMs show the cross-polarization effects, that is, an electric polarization as a response to an applied magnetic field, and vice versa. The bi-anisotropy is related to the existence of magnetoelectric coupling in artificial atoms of MTMs [68]. The SRR-based bi-anisotropic MTM is a pseudo chiral omega medium assumed to have diagonal $\varepsilon_{r,\text{eff}}(f)$ or $\mu_{r,\text{eff}}(f)$ [69]. For such MTMs, the *S*-parameter based extraction technique requires the measurement of six normally incident waves (three TE modes and three TM modes) along all the principle axes, in order to extract all seven unknown parameters of its tensors in Eq. (2.61) to follow [15]. If the medium is lossless, the retrieved constitutive parameters can be obtained analytically, and if the medium is lossy, the parameters can be obtained numerically [15].

A monochromatic plane wave, propagating from the top side of the bi-anisotropic slab of thickness *d*, is obliquely incident upon the slab at different angles (Fig. 2.4) [11]. Here, the incident electric fields are denoted by $\bar{E}_i$ (Fig. 2.4). The bi-anisotropic constitutive relations considered are [13]

$$\left.\begin{aligned} \bar{D} &= \varepsilon_{r,\text{eff}} \cdot \bar{E} + \bar{\bar{\xi}} \cdot \bar{H} \\ \bar{B} &= \mu_{r,\text{eff}} \cdot \bar{H} + \bar{\bar{\zeta}} \cdot \bar{E} \end{aligned}\right\},$$

where

$$\varepsilon_{r,\text{eff}}(f) = \begin{pmatrix} \varepsilon_{xx} & 0 & 0 \\ 0 & \varepsilon_{yy} & 0 \\ 0 & 0 & \varepsilon_{zz} \end{pmatrix}, \quad \mu_{r,\text{eff}}(f) = \begin{pmatrix} \mu_{xx} & 0 & 0 \\ 0 & \mu_{yy} & 0 \\ 0 & 0 & \mu_{zz} \end{pmatrix}, \qquad (2.61a)$$

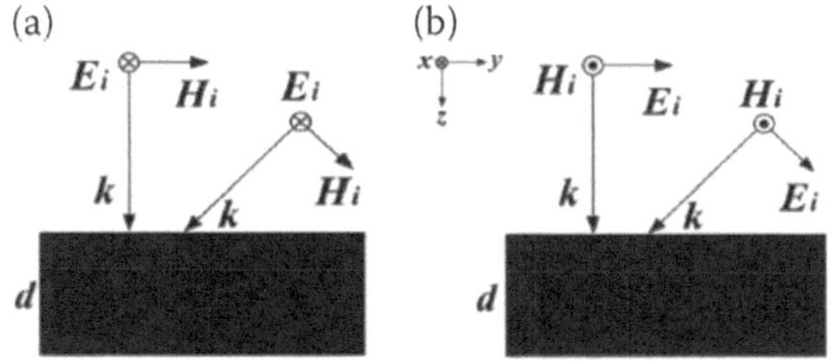

FIGURE 2.4 Normally and oblique incident of a plane waves in a MTM slab with (a) TE polarization and (b) TM polarization. [Source: Jiang et al. [11]].

$$\bar{\bar{\xi}} = \frac{1}{c}\begin{pmatrix} 0 & 0 & 0 \\ 0 & 0 & 0 \\ 0 & -j\xi_0 & 0 \end{pmatrix}, \quad \bar{\bar{\zeta}} = \frac{1}{c}\begin{pmatrix} 0 & 0 & 0 \\ 0 & 0 & j\xi_0 \\ 0 & 0 & 0 \end{pmatrix} \tag{2.61b}$$

where ξ_0 is the chirality parameter [13].

In order to obtain all the unknown parameters of Eq. (2.61), one needs to proceed, as instructed by Cohen and Shavit [13]: "......*four complex value measurements of the reflection S_{11} and four complex value measurements of the transmission S_{21}, at one oblique and one normal incidences for two orthogonal polarizations (TE and TM), are used to obtain 8 sets of equations in order to analytically extract 6 unknown parameters and 2 branch indices to obtain an explicit solution of the bianisotropic complex parameters. The remaining unknown chirality parameter is obtained from an additional oblique incidence measurement at both TE and TM polarizations using a numerical optimization procedure.*" [13].

Chiral MTMs

There are several methods to retrieve the material parameters of chiral MTMs, such as the wave propagation approach [62,63], the state-space approach [21], and the causality principle approach [22]. However, we will discuss briefly only the state-space approach here to keep the chapter concise.

In the state-space approach, the scattering data for a linearly polarized plane wave incident normally on a homogeneous, isotropic chiral slab are combined with the properties of a state transition matrix. The constitutive relations of the medium are given by [21]

$$\begin{pmatrix} \bar{\bar{D}} \\ \bar{\bar{B}} \end{pmatrix} = \begin{pmatrix} \varepsilon_0\varepsilon_r & -j\kappa/c \\ j\kappa/c & \mu_0\mu_r \end{pmatrix}\begin{pmatrix} \bar{\bar{E}} \\ \bar{\bar{H}} \end{pmatrix}$$

where κ is the so-called chirality parameter, which rotates the electric and magnetic fields. The state transition matrix $\bar{\Phi}(4 \times 4)$ is related to the transverse components of the electric and magnetic fields at the two boundaries of the chiral slab:

$$\begin{pmatrix} \bar{\bar{E}}_T(0) \\ \bar{\bar{H}}_T(0) \end{pmatrix} = \bar{\Phi} \begin{pmatrix} \bar{\bar{E}}_T(d) \\ \bar{\bar{H}}_T(d) \end{pmatrix}, \quad \bar{\Phi} = \exp(-\Gamma_\omega d),$$
$$\Gamma_\omega = \frac{\omega}{c}\Gamma = \begin{pmatrix} 0 & \frac{\omega}{c}\kappa & 0 & -j\omega\mu_0\mu_r \\ -\frac{\omega}{c}\kappa & 0 & j\omega\mu_0\mu_r & 0 \\ 0 & j\omega\varepsilon_0\varepsilon_r & 0 & \frac{\omega}{c}\kappa \\ -j\omega\varepsilon_0\varepsilon_r & 0 & -\frac{\omega}{c}\kappa & 0 \end{pmatrix} \tag{2.62}$$

where $\bar{\bar{E}}_T = (E_x, E_y)$ and $\bar{\bar{H}}_T = (H_x, H_y)$ are the matrix representations of the transverse components of the electric and magnetic fields, respectively. The state transition matrix $\bar{\Phi}$ of an isotropic chiral slab has various properties, some of which are useful for the parameter retrieval algorithm. The following two theorems are important for parameter retrieval [21]:

Theorem 1: The determinant of the state transition matrix of an isotropic chiral slab is equal to its unity.
Theorem 2: The state transition matrix of an isotropic chiral slab can be written as

$$\bar{\Phi} = \begin{pmatrix} \Phi_{11} & \Phi_{12} & \Phi_{13} & \Phi_{14} \\ -\Phi_{12} & \Phi_{11} & -\Phi_{14} & \Phi_{13} \\ \Phi_{31} & \Phi_{32} & \Phi_{11} & \Phi_{12} \\ -\Phi_{32} & \Phi_{31} & -\Phi_{12} & \Phi_{11} \end{pmatrix},$$

which has only six distinct elements: Φ_{11}, Φ_{12}, Φ_{13}, Φ_{14}, Φ_{31} and Φ_{32}. Once the state transition matrix $\bar{\Phi}$ is determined, one can write [21]

$$\Gamma = -\frac{c}{\omega d}\ln(\bar{\Phi}). \tag{2.63}$$

In order to compute the logarithm of a square matrix in Eq. (2.63), one can use the Cayley-Hamilton theorem [70,71]. One can retrieve the medium parameters by comparing the Γ matrix obtained in terms of the constitutive parameters in Eq. (2.62) and obtained from the scattering coefficients in Eq. (2.63) [21].

Using this method, one can obtain the scattering parameters relying on the direct computation of the transfer matrix of the slab, which is different from the conventional scattering parameter retrieval method, which involves acquiring n_{eff} and Z_{eff}. In addition, "*the proposed approach allows avoiding nonlinearity of the problem but requires getting enough equations to fulfil the task which was provided by considering some properties of the state transition matrix,*" as stated by Zarifi et al. [21].

2.3.4 Simulation and Experimental Validation of the Retrieval Techniques

In this subsection, we discuss some of the retrieval techniques applied on some specific MTM structures, giving a comparison between the irrelative accuracies of the techniques. We also review some of the popular experimental setups to measure S-parameters.

Comparison between Parameter Retrieval Techniques

We know (Sections 2.3.1 and 2.3.2) that there are several techniques, to retrieve MTM parameters, which have their relative pros and cons. Therefore, it is difficult to pick the best retrieval technique from among them. Here, we consider the K-K method (popular and extensively used) and D-D discontinuity detection approach (comparatively new among other techniques) to retrieve the parameters of metal strip and square SRR-based composite MTMs of two unit cells [60]. Using the K-K method, one can obtain $n_{\text{eff}}(f)$ with the help of Eqs. (2.43)–(2.45) and Eq. (2.41) (Fig. 2.5a), and both $\varepsilon_{r,\text{eff}}(f)$ and $\mu_{r,\text{eff}}(f)$ can be obtained with the help of Eq. (2.36) using Eqs. (2.41) and (2.34), where a plane wave propagating along the direction of periodicity of the two unit cells and electric field is oriented parallel to the length of the metal strip while the magnetic field is oriented parallel to the center axis of the SRRs [60]. However, the K-K method has a limitation due to the truncation in the upper limit of the K-K integral (Eq. 2.45). Therefore, to find the proper values of m at high frequencies, one needs to extend the frequency band of the K-K integral, which is practically unmanageable, and in spite of the extension, the method saturates at a maximum k_0d value [60].

In the D-D method (defined as the proposed method in Fig. 2.5), unlike the K-K method, one can select the correct values of m in $n'_{\text{eff}}(f)$ directly without taking the help of $n''_{\text{eff}}(f)$. Being independent of d and the frequency region, the D-D method can retrieve the parameters of MTMs unambiguously. However, the K-K method gives incorrect $n'_{\text{eff}}(f)$ (Fig. 2.5a) due to the selection of incorrect values of m

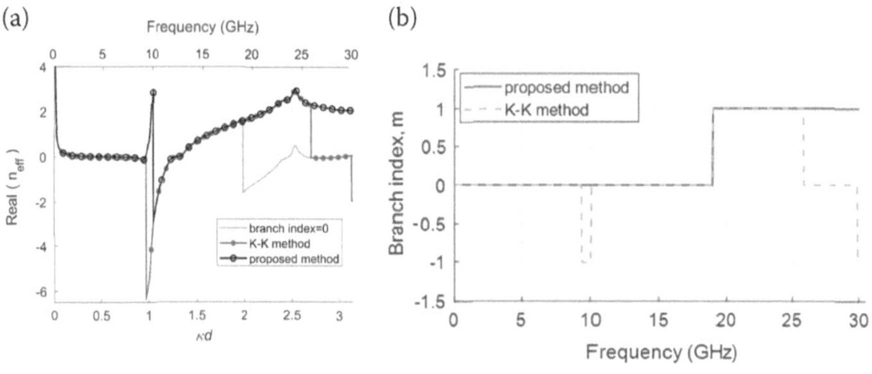

FIGURE 2.5 (a) Extracted refractive index of; (b) branch index of two unit cells of metal strip and square SRRs based composite MTMs. [Source: Aladadi and Alkanhal [60]].

(Fig. 2.5b) in the regions between 9.24 GHz< f < 9.99 GHz (0.967 < k_0d < 1.04) and f > 25.8 GHz (k_0d > 2.7). With a further increase of d, the performance frequency region of the K-K method reduces while the D-D method remains unaffected.

Experimental Setup for Parameter Retrieval of MTMs

It is essential to validate the analytical parameter retrieval methods of MTMs against some extensively used and popular experimental measurements with due consideration to their constraints, as follows.

Negative refraction is one of the unique properties of left-handed MTMs, exhibited in the microwave frequency region, which was experimentally verified by Shelby et al. [50], Parazzoli et al. [51] and Ran et al. [72,73]. For this purpose, a parallelogram-shaped 2D MTM slab, made of SRRs and metal strips for the DNG response, is placed in a parallel-plate waveguide (PPW) with absorbers placed on the sides. A microwave detector can move in a straight path along the broad wall of PPW. Correspondingly, in the Shelby et al. [50] experimental setup of a prism-shaped MTM slab, the detector can be placed to follow a circular path, which is placed at one end of the PPW to detect the power transmitted through the MTM slab at frequencies in which only the first mode (equivalent to a plane wave) can propagate.

Parazzoli et al. [51] presented an experimental setup (Fig. 2.6) for the Snell's law experiment on a negative index of refraction material (NIM), which is different from that in Shelby et al. [50], in which (i) a free-space environment is used as opposed to a waveguide measurement system; (ii) data are taken at two different distances from the sample to waveguide detector; and (iii) numerical simulations showed excellent agreement with the experimental observations. In this case, Parazzoli et al. [51] considered a 1D NIM structure, defined as a structure where the patterns reside on one set of parallel planes only. The detector of the electric field was placed at 33 cm and 66 cm, respectively, from the front face of the NIM wedge. A lens was placed at the exit of the horn antenna to provide both the focused and collimated-mode electromagnetic wave to the NIM sample; the sample was placed in the focal region of the lens. The detector was moved on a circular arc centered on the backside of the wedge. Parazzoli et al.

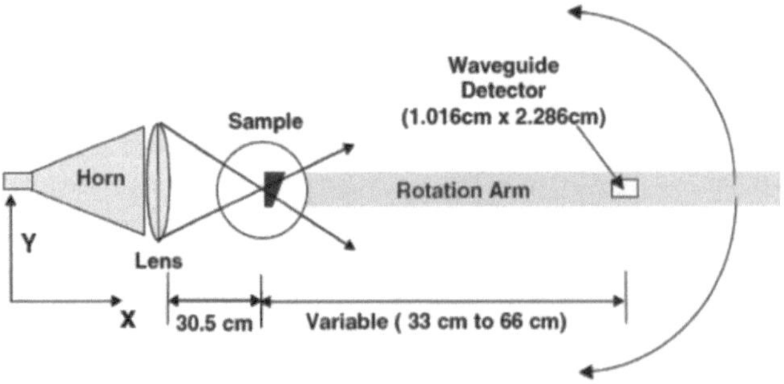

FIGURE 2.6 Schematic of the setup used in the Snell's law experiment showing the conical horn, lens, sample and waveguide detector. [Source: Parazzoli et al. [51]].

observed that the phase velocity reverses its sign in the NIM wedge, but the Poynting vector is always directed toward the wave propagation, as expected [51].

In order to determine the magnetic resonant frequency, Aydin et al. [47] measured the transmission through a single unit cell of SRR using two monopole antennas as the transmitter and the receiver. A single SRR unit cell is placed between the monopole antennas (Fig. 2.1c). The length of the monopole antennas is each $\lambda/2$, considered to work at the frequency range covering the value of f_{m0} of the SRRs. The incident wave propagates from one antenna to another, with $\bar{E}$ oriented perpendicular and $\bar{H}$ oriented parallel to the center axis of the MTM. The monopole antennas are then connected to a vector network analyzer (VNA) to measure the transmission coefficients. A dip in the transmission spectrum of SRR structure can be attributed to the resonant nature of the SRR. The same technique has been used by Bilotti et al. [6] as well.

In order to study the wave propagation in a waveguide well below the cutoff frequency of the waveguide, which can provide backward-wave propagation, one can use an array of SRRs, which will exhibit negative permeability, and an empty waveguide supporting below cutoff waveguide mode, which can exhibit negative permittivity. For the sake of experimental appreciation, a periodic array of SRRs was placed on the plane of symmetry of a square waveguide supporting the TE_{10} mode [46]. The first and the last cells of the array were placed partially out of the waveguide to strengthen its excitation. Both the ends of the combined structure were connected with a coaxial-to-waveguide adapter to measure the S_2 parameter in the VNA (Fig. 2.1b). A measured transmission region well below the cutoff frequency validates the theoretical findings [46].

Some experimental constraints may arise during the measurement of S-parameters of MTMs. It may be hard to achieve a pure plane wave during measurement, contrary to what is predicted by simulation. However, this may be achieved with PPW by reducing the interference from the external environment. Furthermore, it is also hard to achieve the negative permittivity with the array of rods for which the rods should touch both the top and bottom plates of waveguide [74].

2.4 SUMMARY AND CONCLUSION

In this chapter, we briefly reviewed and discussed EMT of MTMs and also discussed the merits and drawbacks of several well-known parameter retrieval techniques of MTMs. It is simple to represent an MTM by assuming it to be homogeneous and isotropic in nature. The effective relative permittivity and the effective permeability of an MTM are represented by the Drude model and the Lorentz model, respectively. This representation gives enormous analytical control to the researchers in their study of MTM properties and responses, such as MNG, ENG and DNG, and negative refractive index, without calling for the actual MTMs. However, such a study does not give one any physical insight into the actual MTM structure.

The NRW method has now become the most popular in the extraction of complex permittivity and permeability of homogeneous and isotropic conventional materials. However, from the standpoint of its practical applicability for parameter retrieval, the NRW method is not suitable since most of the MTMs are inhomogeneous and anisotropic in nature. Among the alternative methods globally sought for, the transmission and reflection of incident wave, or S-parameter retrieval method, has subsequently proved to be appropriate and, hence, has been used extensively. However, this method mainly suffers from branch ambiguity and the inherent assumption of extremely small sample thickness of MTMs. These lacunas of the method were overcome by introducing the first-order approximation of Taylor's series to the transmission term, and combining the method with the K-K method.

There are other approaches to the parameter extraction technique for MTMs, such as the curve-fitting approach, transfer matrix approach, field averaging approach, data-driven discontinuity detection approach, Snells' law approach, wave propagation approach, state-space approach, causality principle approach, quasi-mode theory approach, phase unwrapping approach and asymmetrical strip line approach. These approaches to retrieval techniques following continuous research have now come out to be more powerful to investigate into different ranges of MTMs encompassing inhomogeneous, anisotropic, bi-anisotropic, uniaxial anisotropic, chiral and hyperbolic MTMs, excited with normally and obliquely incident plane waves.

Further, the parameter extraction techniques have been validated against commercial simulation software and experiments. For experimental appreciation, an MTM under test is placed in free space, keeping it equidistant from the transmitting and receiving horn antennas or monopole antennas. The unit cell is excited with normally or obliquely incident plane waves. The MTM under test can also be placed inside a rectangular waveguide or a coaxial transmission line to measure the transmission and reflection coefficients from which to calculate the effective relative permittivity and permeability.

It is believed that the EMT and parametric retrieval techniques of MTM reviewed in this chapter will prove to be useful in designing MTM-assisted microwave components, such as phase shifter, directional coupler, branch-line coupler, power divider, filter, cavity, super lens/hyper-lens, antenna, radome, absorber and sensor. At the same time, they should be useful in designing MTM-assisted vacuum electron devices as well.

REFERENCES

1. J. B. Pendry, A. J. Holden, D. J. Robbins, and W. J. Stewart, "Magnetism from conductors and enhanced nonlinear phenomena," *IEEE Trans. Microw. Theo. Tech.*, vol. 47, no. 11, pp. 2075–2084, 1999.
2. J. B. Pendry, A. J. Holden, W. J. Stewart, and I. Youngs, "Extremely low frequency plasmons in metallic mesostructures," *Phys. Rev. Lett.*, vol. 76, no. 25, pp. 4773–4776, 1996.

3. D. R. Smith, D. C. Vier, T. H. Koschny, and C. M. Soukoulis, "Electromagnetic parameter retrieval from inhomogeneous metamaterials," *Phys. Rev. E.*, vol. 71, no. 3, p. 036617, 2005.
4. D. R. Smith, S. Schultz, P. Markoš, and C. M. Soukoulis, "Determination of effective permittivity and permeability from reflection and transmission coefficients," *Phys. Rev. B.*, vol. 65, no. 19, p. 195104, 2002.
5. X. Chen, T. M. Grzegorzyk, B.-I. Wu, J. Pacheco, Jr., and J. A. Kong, "Robust method to retrieve the constitutive effective parameters of metamaterials," *Phys. Rev. E*, vol. 70, no. 1, p. 016608, 2004.
6. F. Bilotti, A. Toscano, L. Vegni, K. Aydin, K. B. Alici, and E. Ozbay, "Equivalent-circuit models for the design of metamaterials based on artificial magnetic inclusions," *IEEE Trans. Microw. Theo. Tech.*, vol. 55, no. 12, pp. 2865–2873, 2007.
7. L. L. Hou, J. Y. Chin, X. M. Yang, X. Q. Lin, R. Liu, F. Y. Xu, and T. J. Cui, "Advanced parameter retrievals for metamaterial slabs using an inhomogeneous model," *J. Appl. Phys.*, vol. 103, no. 6, p. 064904, 2008.
8. Y. Shi, Z. Li, K. Li, L. Li and C. Liang, "A retrieval method of effective electromagnetic parameters for inhomogeneous metamaterials," *IEEE Trans. Microw. Theo. Tech.*, vol. 65, no. 4, pp. 1160–1178, 2017.
9. T. M. Grzegorczyk, M. Nikku, X. Chen, B.-I. Wu, and J. A. Kong, "Refraction laws for anisotropic media and their application to left-handed metamaterials," *IEEE Trans. Micro. Theo. Tech.*, vol. 53, no. 4, pp. 1443–1450, 2005a.
10. A. Castanié, J.-F. Mercier, S. Félix, and A. Maurel, "Generalized method for retrieving effective parameters of anisotropic metamaterials," *Opt. Express.*, vol. 22, no. 24, pp. 29937–29953, 2014.
11. Z. H. Jiang, J. A. Bossard, X. Wang, and D. H. Werner, "Synthesizing metamaterials with angularly independent effective medium properties based on an anisotropic parameter retrieval technique coupled with a genetic algorithm," *J. Appl. Phys.*, vol. 109, no. 1, p. 013515, 2011.
12. Y. T. Aladadi and M. A. S. Alkanhal, "Extraction of tensor parameters of general biaxial anisotropic materials," *AIP Advances*, vol. 10, no. 2, p. 025113, 2020.
13. D. Cohen and R. Shavit, "Bi-anisotropic metamaterials effective constitutive parameters extraction using oblique incidence S-parameters method," *IEEE Trans. Ante. Prop.*, vol. 63, no. 5, pp. 2071–2078, 2015.
14. U. C. Hasar, G. Buldu, Y. Kaya, and G. Ozturk, "Determination of effective constitutive parameters of inhomogeneous metamaterials with bianisotropy," *IEEE Trans. Microw. Theo. Tech.*, vol. 66, no. 8, pp. 3734–3744, 2018.
15. X. Chen, B.-I. Wu, J. A. Kong, and T. M. Grzegorczyk, "Retrieval of the effective constitutive parameters of bianisotropic metamaterials," *Phys. Rev. E.*, vol. 71, no. 4, p. 046610, 2005.
16. D. Zarifi, M. Soleimani, and A. Abdolali, "Electromagnetic characterization of biaxial bianisotropic media using the state space approach," *IEEE Trans. on Anten. Prop.*, vol. 62, no. 3, pp. 1538–1542, 2014.
17. F.-J. Hsieh and W.-C. Wang, "Full extraction methods to retrieve effective refractive index and parameters of a bianisotropic metamaterial based on material dispersion models," *J. Appl. Phys.*, vol. 112, no. 6, p. 064907, 2012.
18. G. T. Papadakis, P. Yeh, and H. A. Atwater, "Retrieval of material parameters for uniaxial metamaterials," *Phys. Rev. B*, vol. 91, no. 15, p. 155406, 2015.
19. A. F. Mota, A. Martins, J. Weiner, F. L. Teixeira, and B.-H. V. Borges, "Constitutive parameter retrieval for uniaxial metamaterials with spatial dispersion," *Phys. Rev. B*, vol. 94, no. 11, p. 115410, 2016.

20. C. Menzel, C. Rockstuhl, T. Paul, and F. Lederer, "Retrieving effective parameters for quasiplanar chiral metamaterials," *Appl. Phys. Lett.*, vol. 93, no. 23, p. 233106, 2008a.
21. D. Zarifi, M. Soleimani, and A. Abdolali, "Parameter retrieval of chiral metamaterials based on the state-space approach," *Phys. Rev. E*, vol. 88, no. 2, p. 023204, 2013a.
22. D. Zarifi, M. Soleimani, and V. Nayyeri, "Parameter retrieval of chiral metamaterials based on the causality principle," *Int. J. RF Microw. Comput. Eng.*, vol. 23, no. 5, pp. 610–618, 2013b.
23. R. Zhao, T. Koschny, and C. M. Soukoulis, "Chiral metamaterials: Retrieval of the effective parameters with and without substrate," *Opt. Expr.*, vol. 18, no. 14, pp. 14553–14567, 2010.
24. V. P. Drachev, V. A. Podolskiy, and A. V. Kildishev, "Hyperbolic metamaterials: New physics behind a classical problem," *Opt. Expr.*, vol. 21, no. 12, pp. 15048–15064, 2013.
25. L. Ferrari, C. Wu, D. Lepage, X. Zhang, and Z. Liu, "Hyperbolic metamaterials and their applications," *Prog. Quant. Electro.*, vol. 40, pp. 1–40, 2015.
26. Z. Guo, H. Jiang, and H. Chena, "Hyperbolic metamaterials: From dispersion manipulation to applications," *J. Appl. Phys.*, vol. 127, no. 7, p. 071101, 2020.
27. I. F. Almog, M. S..Bradley, and V. Bulović, *The Lorentz Oscillator and Its Applications*, pp. 1–34, MIT lecture note, 2011.
28. S. A. Ramakrishna and T. M. Grzegorczyk, *Physics and Applications of Negative Refractive Index Materials*, Bellingham, Washington, USA: CRC Press, 2008.
29. T. C. Choy, *Effective Medium Theory: Principles and Applications*, Oxford, UK: Oxford University Press, 2015.
30. Vadim A. Markel, "Introduction to the Maxwell Garnett approximation: Tutorial," *J. Opt. Soc. Am. A.*, vol. 33, no. 7, pp. 1244–1256, 2016a.
31. Vadim A. Markel, "Maxwell Garnett approximation (advanced topics): Tutorial," *J. Opt. Soc. Am. A.*, vol. 33, no. 11, pp. 2237–2255, 2016b.
32. A. M. Nicolson and G. F. Ross, "Measurement of the intrinsic properties of materials by time-domain techniques," *IEEE Trans. Instru. Meas.*, vol. IM-19, no. 4, pp. 377–382, 1970.
33. W. B. Weir, "Automatic measurement of complex dielectric constant and permeability at microwave frequencies," *Proc. IEEE.*, vol. 62, no. 1, pp. 33–36, 1974.
34. T. L. Blakney, and W. B. Weir, "Comments on 'automatic measurement of complex dielectric constant and permeability at microwave frequencies,'" *Proc. IEEE.*, vol. 63, no. 1, pp. 203–205, 1975.
35. S. O'Brien and J. B. Pendry, "Magnetic activity at infrared frequencies in structured metallic photonic crystals," *J. Phys.: Cond. Matt.*, vol. 14, no. 25, pp. 6383–6394, 2002.
36. P. Markoš and C. M. Soukoulis, "Transmission properties and effective electromagnetic parameters of double negative metamaterials," *Opt. Expr.*, vol. 11, no. 7, pp. 649–661, 2003.
37. R. W. Ziolkowski, "Design, fabrication, and testing of double negative metamaterials," *IEEE Trans. on Anten. Prop.*, vol. 51, no. 7, pp. 1516–1528, 2003.
38. G. Lubowski, R. Schuman, and T. Weiland, "Extraction of effective metamaterial parameters by parameter fitting of dispersive models," *Microw.Opt. Tech. Lett.*, vol. 42, no. 2, pp. 285–288, 2007.
39. R. A. Shore and A. Yaghjian, "Traveling waves on two and three dimensional periodic arrays of lossless scatterers," *Radio Sci.*, vol. 42, no. 6, p. RS6S21, 2007.

40. D. R. Smith, D. C. Vier, N. Kroll, and S. Schultz, "Direct calculation of permeability and permittivity for a left-handed metamaterial," *Appl. Phys. Lett.*, vol. 77, no. 14, pp. 2246–2248, 2000.
41. D. R. Smith and J. B. Pendry, "Homogenization of metamaterials by field averaging," *J. Opt. Soc. of America B.*, vol. 23, no. 3, pp. 391–403, 2006.
42. J. Lerat, N. Malléjac, and O. Acher, "Determination of the effective parameters of a metamaterial by field summation method," *J. Appl. Phys.*, vol. 100, no. 8, p. 084908, 2006.
43. A. Pors, I. Tsukerman, and S. I. Bozhevolnyi, "Effective constitutive parameters of plasmonic metamaterials: Homogenization by dual field interpolation," *Phys. Rev. E*, vol. 84, no. 1, p. 084908, 2006.
44. I. Tsukerman, "Nonlocal homogenization of metamaterials by dual interpolation of fields," *J. Opt. Soc. of America B*, vol. 28, no. 12, pp. 2956–2965, 2006.
45. T. Ozturk and M. Guneser, "Measurement methods and extraction techniques to obtain the dielectric properties of materials," *Electrical and Electronic Properties of Materials*, IntechOpen, pp. 83–108, 2018.
46. P. Castro, J. Barroso and J. Neto, "Experimental study on split-ring resonators with different slit widths," *J. Elect. Ana. Appl.*, vol. 5, no. 9, pp. 366–370, 2013.
47. K. Aydin, I. Bulu, K. Guven, M. Kafesaki, C. M. Soukoulis and E. Ozbay, "Investigation of magnetic resonances for different split-ring resonator parameters and designs," *New J. Phys.*, vol. 7, no. 168, pp. 1–15, 2005.
48. O. Luukkonen, S. I. Maslovski, and S. A. Tretyakov, "A stepwise Nicolson–Ross–Weir-based material parameter extraction method," *IEEE Anten.Wire. Prop. Lett.*, vol. 10, pp. 1295–1298, 2011.
49. E. J. Rothwell, J. L. Frasch, S. M. Ellison, P. Chahal, and R. O. Ouedraogo, "Analysis of the Nicolson-Ross-Weir method for characterizing the electromagnetic properties of engineered materials," *Prog. Elect. Res.*, vol. 157, pp. 31–47, 2016.
50. R. A. Shelby, D. R. Smith, and S. Schultz, "Experimental verification of a negative index of refraction," *Science*, vol. 292, no. 5514, pp. 77–79, 2001.
51. C. G. Parazzoli, R. B. Greegor, K. Li, B. E. C. Koltenbah, and M. Tanielian, "Experimental verification and simulation of negative index of refraction using Snell's law," *Phys. Rev. Lett.*, vol. 90, no. 10, p. 107401, 2003.
52. S. Arslanagić, T. V. Hansen, N. A. Mortensen, A. H. Gregersen, O. Sigmund, R. W. Ziolkowski, and O. Breinbjerg, "A review of the scattering-parameter extraction method with clarification of ambiguity issues in relation to metamaterial homogenization," *IEEE Anten.Prop. Mag.*, vol. 55, no. 2, pp. 91–106, 2013.
53. K.-E. Peiponen, V. Lucarini, E. M. Vartiainen, and J. J. Saarinen, "Kramers-Kronig relations and sum rules of negative refractive index media," *Eur. Phys. J. B.*, vol. 41, pp. 61–65, 2004.
54. Z. Szabó, G. Park, R. Hedge, and E. Li, "A unique extraction of metamaterial parameters based on Kramers–Kronig relationship," *IEEE Trans. on Microw. Theo. Tech.*, vol. 58, no. 10, pp. 2646–2653, 2010.
55. Y. Shi, T. Hao, L. Li, and C.-H. Liang, "An improved NRW method to extract electromagnetic parameters of metamaterials," *Microw. Opt. Tech. Lett.*, vol. 58, no. 3, pp. 647–652, 2016.
56. Z. Szabó, "Closed form Kramers–Kronig relations to extract the refractive index of metamaterials," *IEEE Trans. on Microw. Theo. Tech.*, vol. 65, no. 4, pp. 1150–1159, 2017.
57. V. Lucarini, J. J. Saarinen, K. E. Peiponen, and E. M. Variainen, *Kramers-Kronig Relations in Optical Materials Research*, Berlin: Springer, 2005.

58. D. R. Smith, "Analytic expressions for the constitutive parameters of magnetoelectric metamaterials," *Phys. Rev. E*, vol. 81, no. 3, p. 036605, 2010.
59. D. M. Pozar, *Microwave Engineering*. Hoboken, NJ, USA: Wiley, 2011.
60. Y. T. Aladadi and M. A. S. Alkanhal, "Extraction of metamaterial constitutive parameters based on data-driven discontinuity detection," *Opt. Mater. Express*, vol. 9, no. 9, pp. 3765–3780, 2019.
61. M. Navarro-Cía, M. Beruete, F. Falcone, M. Sorolla, and I. Campillo, "Polarization-tunable negative or positive refraction in self-complementariness-based extraordinary transmission prism," *Prog. Elect. Res.*, vol. 103, pp. 101–114, 2010.
62. A. Andryieuski, R. Malureanu, and A. V. Lavrinenko, "Wave propagation retrieval method for metamaterials: Unambiguous restoration of effective parameters," *Phys. Rev. B.*, vol. 80, no. 19, p. 193101, 2009.
63. A. Andryieuski, R. Malureanu, and A. V. Lavrinenko, "Wave propagation retrieval method for chiral metamaterials," *Opt. Expr.*, vol. 18, no. 15, pp. 15498–15503, 2010.
64. S. Sun, S. T. Chui, and L. Zhou, "Effective-medium properties of metamaterials: A quasimode theory," *Phys. Rev. E.*, vol. 79, no. 6, p. 066604, 2009.
65. Y. Shi, Z.-Y. Li, L. Li, and C.-H. Liang, "An electromagnetic parameters extraction method for metamaterials based on phase unwrapping technique," *Wave Random Complex*, vol. 26, no. 4, pp. 417–433, 2016b.
66. S. Gómez, P. Quéffélec, A. Chevalier, A. C. Tarot, and A. Sharaiha, "Asymmetrical stripline based method for retrieving the electromagnetic properties of metamaterials," *J. Appl. Phys.*, vol. 113, no. 2, p. 024912, 2013.
67. C. Menzel, C. Rockstuhl, T. Paul, F. Lederer, and T. Pertsch, "Retrieving effective parameters for metamaterials at oblique incidence," *Phys. Rev. B.*, vol. 77, no. 19, p. 195328, 2008.
68. R. Marqués, F. Medina, and R. Rafii-El-Idrissi, "Role of bianisotropy in negative permeability and left-handed metamaterials," *Phys. Rev. B.*, vol. 65, no. 14, p. 144440, 2002.
69. A. Serdyukov, I. Semchenko, S. Tretyakov, and A. Sihvola, *Electromagnetics of Bi-anisotropic Materials: Theory and Applications*. Amsterdam, Netherland: Gordon and Breach, 2001.
70. J. Hao and L. Zhou, "Electromagnetic wave scatterings by anisotropic metamaterials: Generalized 4 × 4 transfer-matrix method," *Phys. Rev. B.*, vol. 77, no. 9, p. 094201, 2008.
71. M. Kafesaki, I. Tsiapa, N. Katsarakis, T. H. Koschny, C. M. Soukoulis, and E. N. Economou, "Left-handed metamaterials: The fishnet structure and its variations," *Phys. Rev. B*, vol. 75, no. 23, p. 235114, 2007.
72. L. Ran, J. Huangfu, H. Chen, X. Zhang, and K. Chen, "Beam shifting experiment for the characterization of left-handed properties," *J. Appl. Phys.*, vol. 95, no. 5, pp. 2238–2241, 2004.
73. S. I. Tarapov, S. Y. Polevoy, and N. N. Beletski, 'Gyrotropic metamaterials and polarization experiment in the millimeter waveband', in Oleksiy Shulika and Igor Sukhoivanov (Eds.), *Contemporary Optoelectronics: Materials, Metamaterials and Device Applications*, pp. 115–129, London, UK: Springer, 2015.
74. Nader Engheta and Richard W. Ziolkowski (Eds.), *Metamaterials: Physics and Engineering Explorations*, New York, USA: John Wiley & Sons, 2006.

3 Engineered Metamaterials through the Material-by-Design Approach

G. Oliveri[1], M. Salucci[1], M. A. Hannan[1],
A. Monti[2], S. Vellucci[3], F. Bilotti[3],
A. Toscano[3], and A. Massa[1,4,5]
[1]CNIT - "University of Trento" Research Unit, Trento – Italy
[2]Niccolò Cusano University, Rome – Italy
[3]Department of Engineering, ROMA TRE University, Rome – Italy
[4]ELEDIA Research Center (ELEDIA@UESTC - UESTC) School of Electronic Engineering, Chengdu, China
[5]ELEDIA Research Center (ELEDIA@TSINGHUA – Tsinghua University), Beijing, China

3.1 INTRODUCTION AND RATIONALE

In the last two decades, metamaterial (MTM) technology has been consistently among the most active research areas in electromagnetic (EM) engineering and science [1–7]. The academic and industrial interest for the synthesis of materials with desired properties is motivated by their unprecedented field manipulation capabilities that enabled their effective adoption in a wide set of practical applications, ranging from to MHz to THz frequencies and beyond [1–4,6,7]. Furthermore, the wave manipulation features of MTM-based devices have been generalized and transferred to other heterogeneous scientific domains, including mechanics and acoustics [8].

In such a framework, the most widely employed paradigms for the synthesis and the optimization of MTM-based devices are based on the decomposition of the problem at hand into two separate sub-problems: (i) the design of "black-box" materials with equivalent physical properties (e.g., surface impedance, refractive index, permittivity, permeability) suitable for yielding the required device performance/behavior [6,7], and (ii) the synthesis of arrangements of standard materials matching those unconventional physical properties, which

DOI: 10.1201/9781003050162-3

are often obtained with suitable homogenization methodologies applied to 1D/2D/3D periodic arrangements of a reference unit cell [9–11]. The arising approaches proved to be very effective in a wide variety of cases, including MTM lenses and cloaking devices [12–17], metasurfaces [18] and MTM-enhanced antennas [19–24].

On the other hand, it is worth highlighting that those strategies may not fully exploit the whole set of degrees-of-freedom (DoFs) available in a complex MTM-based device. As a matter of fact, the actual target of an EM synthesis is rarely that of achieving a desired bulk physical property, whereas a macro-scale performance index or "task" (e.g., the antenna gain or the sidelobe level, the lens focusing capability, the transmission/reflection coefficient of a metasurface, the radar cross-section of a cloaked object) is usually of interest. Starting from this observation, the possibility of improving the achievable performance by considering the macro-scale system task as the design objective (rather than the local matching of a specific equivalent material property) has been recently investigated in MTM engineering and science [25–27]. More specifically, the Material-by-Design (MbD) framework has emerged as a powerful, flexible and modular paradigm to "*address the task-oriented design of advanced systems comprising artificial materials whose constituent properties are driven by the device functional requirements*" [25]. The fundamental idea of MbD, which is an instance of the System-by-Design (SbD) paradigm, defined in short as "How to deal with complexity" [26,28–31], is that any device comprising artificial materials can be synthesized from a fundamentally new perspective considering (i) its macro-scale functionality/task as the design objective and (ii) the micro-scale properties of the standard materials composing the device as the DoFs of the synthesis problem at hand [25].

The MbD paradigm is indeed motivated by several features of MTM engineering and science problems:

- A task-oriented design avoids focusing on a *local* matching objective, but rather treats the whole system and its complexity during the synthesis process; thus, it potentially promotes a more effective exploration of the functional landscape.
- The implementation of an MbD-based approach does not necessarily imply the exploitation of an homogenization step, thus avoiding the complexity of such a task, especially when dealing with complex structures that possibly feature anisotropic effects [9–11].
- Thanks to the modularity of the arising implementation scheme, which is inherited by the SbD framework [26,28–35], customization to every problem of the resulting design procedure (in terms of considered functional blocks and methodological choices) can be easily implemented.
- The MbD concept naturally and seamlessly enables single- and multi-objective designs as well as dealing with complex constraints through a suitable customization of the associated logical blocks.

- Since the MbD goal is not the local matching of a specific EM property, but rather of a set of global performance indicators, the arising formulation seamlessly supports the definition of multiphysics objectives as well as performance metrics.

Thanks to such features, several MbD customizations have already been implemented and assessed, for instance, the design of (i) task-oriented wide-angle impedance matching (WAIM) metasurfaces, (ii) inhomogeneous lenses for array performance enhancement, and (iii) wave polarization meta-layers. Of course, although very effective and customizable, MbD implementation also yields to non-trivial challenges from theoretical and practical viewpoints. These challenges are mainly related to the multi-scale nature of the arising design problems that intrinsically corresponds to a considerable computational complexity as a consequence of the curse of dimensionality [25,26,28]. A careful combination of macro-scale modeling and micro-scale DoF control procedures as well as design algorithms are usually required to properly address realistic large-scale problems. Accordingly, the formulation of the MTM-enhanced design problem within the MbD framework is discussed in the following section by presenting the theoretical motivation, the concept, the applicative guidelines and the features of state-of-the-art MbD strategies for engineered MTM design. A discussion of the current trends and future developments in MbD research is also illustrated.

This chapter is organized as follows. After an introduction on the concept of MbD and its fundamental features (Section 3.2), the customization of the MbD to selected applicative state-of-the-art examples is detailed (Section 3.3). Some conclusions and final remarks follow (Section 3.4).

3.2 THE MbD PARADIGM: CONCEPT AND FEATURES

3.2.1 The MbD Synthesis Loop

The fundamental driving concept in MbD is the possibility to design advanced systems comprising artificial materials in a "task-oriented" fashion. Accordingly, MbD can be seen as a meta-paradigm in which (i) the generic functional logic complies with the prescribed task-oriented principle, while (ii) the specific implementation and resulting design details depend on the target scenario, application and constraints.

Generally speaking, the MbD paradigm can be described through the meta-loop in Fig. 3.1. Such a flowchart features four fundamental functional blocks:

- The *EM/multiphysics micro-modeling* block (Fig. 3.1). Such a block is responsible for the computation of the EM micro-level response (or its multiphysics counterpart depending on the specific problem) starting from the descriptors of the designed device at the micro-scale level. The output of such a function does not describe the overall device performance index, but rather a local property (e.g., the scattering matrix expansion [26] in the case of periodic structures) that requires further processing for being linked to the measure of the effectiveness of the complete device.

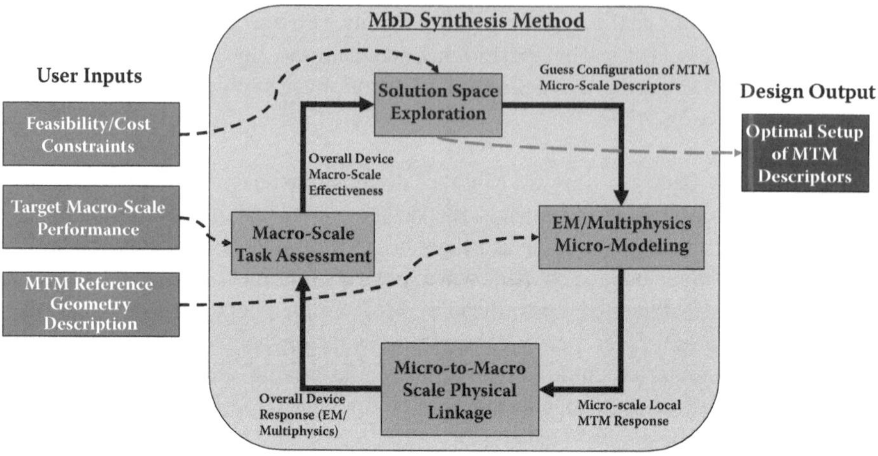

FIGURE 3.1 MbD paradigm – Functional flowchart of a generic MbD process.

- The *micro- to macro-scale physical linkage* block (Fig. 3.1). The functionality implemented here is the estimation of the overall EM or multiphysics response (e.g., scattering cross-section) of the device starting from its micro-scale simulated performance.
- The *macro-scale task assessment* block (Fig. 3.1). This block is devoted to the evaluation of the effectiveness of the device at the macro-scale level according to the user-defined guidelines and the device-simulated performance at the macro-scale level.
- The *solution space exploration* block (Fig. 3.1). This block implements, usually carried out through an iterative process, the functional landscape analysis of the design objective and the identification of the best setup for the device micro-scale descriptors.

These four functional blocks are usually combined in the so-called *MbD synthesis loop* that embodies the fundamental representation of the MbD paradigm from a logical viewpoint (Fig. 3.1). It is worthwhile to point out that not all blocks are required in every MbD customization. Sometimes the functional role of one block can be split into a cascade of operations to guarantee the maximum scalability and efficiency of the design process (see Section 3.3.3). On the other hand, several methods have been conceived and proposed in the literature to address each of the above functionalities, and they can be used for effective system integration within the MbD synthesis loop. The following section provides a detailed description of each functional block and illustrates selected benchmark implementation choices.

3.2.2 The MbD Functional Blocks – Objectives and Examples

With reference to the macro-blocks mentioned in Section 3.2.1, the fundamental objective of the *EM/multiphysics micro-modeling* block (Fig. 3.1) is to handle

(i) as inputs, the micro-scale descriptors of the designed artificial material [e.g., the geometry and the material setup of the unit cell that is replicated on a regular grid to implement the desired artificial material (Section 3.3.2, Section 3.3.4) or the contour of the designed lens (Section 3.3.3)], and (ii) as outputs, the EM or multiphysics micro-level response of the artificial material [e.g., the Floquet coefficients (Section 3.3.2) or the the transmission/reflection matrices expansion (Section 3.3.4)]. Because of such a wide variety of formulations and objectives, several approaches have been proposed to handle this functionality depending on the applicative context [26,36,37]. As for planar single- or multi-layer EM artificial materials, periodic boundary conditions and perfect matching layers as well as generalized scattering matrices and resulting cascade operations have been considered in EM-formulated MbD problems [26]. Alternative strategies based on the computation of equivalent transmission/reflection matrices or equivalent surface impedances have been used as well [36,37]. Anyway, regardless of the approach and unless canonical geometries [38–40], full-wave numerical simulations and/or semi-analytical techniques are often required [26], they usually involve non-trivial computational complexities; thus, the micro-modeling block can result in a bottleneck when handling realistic geometries. To address such a potential MbD limitation, the use of Learning-by-Examples (LBEs) methods has been recently proposed [31,41]. More specifically, the micro-scale EM response of the artificial material is actually "predicted" by exploiting suitable statistical learning algorithms after an initial training phase [31,41].

The *micro- to macro-scale physical linkage* block (Fig. 3.1) is often very challenging from both the theoretical and the practical viewpoint. As a matter of fact, the computation of the relationship between the micro-scale response of a structure and its macro-scale features is a common and key item across many EM problems, including the design of transmit arrays and reflect arrays [31]. The main objective of this functional block is to give a faithful estimation of the large-scale system property (e.g., the array gain, the device EM signature, the active element impedance mismatch, the polarization state of the transmitted or reflected wave), starting from its micro-scale response in a numerically efficient manner. Since the full-wave modeling of the entire device and artificial material structure is often a computationally unfeasible option for the design phase [26,42], the exploitation of the superposition principle is usually an effective choice for large-scale planar periodic and quasi-periodic layouts [26,31]. Alternatively, the use of the generalized scattering matrix method or of the *generalized sheet transition conditions* based on surface impedance or susceptibility tensors has been proposed [42]. Nevertheless, owing to the wide range of MbD applicative scenarios, a general-purpose solution for the micro- to macro-scale physical linkage problem has not been proposed yet, and the implementation of such a phase requires suitable customization to the scenario at hand.

The *macro-scale task assessment* block (Fig. 3.1) is aimed at establishing the connection between the physical response of the system coming from previous functional phases and the design algorithm. Mathematically, this consists of defining the problem *cost function,* which quantifies the mismatch of the synthesized solution from the user-defined task. It is worth remarking that the typology and the features of the implementation of the *macro-scale task assessment* (i.e., convexity of the cost function, degree of non-linearity, presence of multiple minima) have a considerable impact on the choice of the subsequent functional block [43]. From a logical viewpoint, both "matching" [27] (i.e., to match a reference such as a power pattern shape) and "minimization/maximization" definitions [26] (i.e., the minimization/maximization of a quality index) are commonly adopted, whereas the dimensionality (i.e., single- or multi-objective in nature) is a direct consequence of the user-defined task of the device [26,27].

The aim of the *solution space exploration* block (Fig. 3.1) is that of exploring the cost function landscape to identify the descriptor setup (at the micro-scale) that guarantees the best macro-scale performance as quantified by the *macro-scale task assessment* block. Unless elementary MbD problems have closed-form solutions available, guess solution sets are iteratively generated with a wide variety of available local/global optimizers [43–46]. Nevertheless, the choice of the most proper class of optimization strategies among the available ones and its configuration is a non-trivial step yielding an MbD implementation that guarantees competitive search capabilities. As a matter of fact, according to the "no-free lunch" optimization theorems [43], there is not an "optimal" tool for any optimization problem [44,45]. Owing to the vast literature on the topic, the relevance of such aspects within the more general SbD framework [28] and the focus of this chapter, the interested reader can refer to the guidelines and discussion in the cited references [26,28–31] for the best choice of *solution space exploration* techniques.

3.3 MbD AT WORK IN METAMATERIAL-BASED SENSING AND COMMUNICATION APPLICATIONS

3.3.1 Customization of MbD in Applicative Scenarios

Starting from the meta-loop in Section 3.2, the application of the MbD paradigm in a specific scenario/context essentially requires the user to specify (i) the definition and mathematical coding of the macro-scale task that the device comprising the artificial material will have to address according to the applicative scenario and user-defined objectives, (ii) the description of the DoFs of the artificial material/device at hand (e.g., geometric and EM descriptors of its unit cell), (iii) the identification of the functional blocks for the customization of the MbD paradigm to the framework at hand, and (iv) the choice of the implementation strategies to be adopted/integrated within each functional block. Within this framework, the modularity of the MbD approach actually enables the user to revisit the definition of the functional blocks and modify/substitute their implementation without the need to change the entire design loop, hence allowing the exploration of different tradeoffs and objectives depending on the obtained performance and design constraints. The following sections present the customization of the MbD approach to three representative synthesis problems concerned with (i) wide-angle impedance matching layers for next-generation phased arrays (Section 3.3.2), (ii) MTM-enhanced compact phased arrays (Section 3.3.3) and (iii) wave manipulation MTM devices (Section 3.3.4).

3.3.2 MbD-designed Metamaterials for Wide-angle Impedance Matching Layers

The design of MTM-based WAIMs for enhancing the field-of-view (FoV) of next-generation phased arrays has been addressed with the MbD paradigm in Oliveri et al. [26]. The fundamental objective of WAIMs [47,48] is to stabilize the fluctuations of the active reflection coefficient caused by the coupling effects among the single array radiators when scanning [26,30,49], to have stable performance on a wide FoV and to avoid the issues caused by impedance oscillations such as blind spots [26,30,49]. Toward this goal, the micro-scale field manipulation features of the artificial MTM constituting the WAIM have been exploited to yield a macro-scale active reflection coefficient stabilization [26,30,49].

We explore the setup investigated in Oliveri et al. [26] that consists of a regular array of truncated waveguides [Fig. 3.2(a)] displaced on a regular lattice with vectors d_1 and d_2 in the $x - y$ plane [Fig. 3.2(a)] and covered by a WAIM layer composed of a multiscale artificial MTM structure [Fig. 3.2(b)].

To deal with such a design problem within the MbD framework, first the mathematical formulation of the macro-scale task has been coded into the minimization of the following cost function [26]

$$\Phi^{WAIM}(g) \triangleq \frac{\int_{f_{\min}}^{f_{\max}} \left\{ \int_0^{\frac{\pi}{2}} \int_0^{2\pi} |\Gamma(\theta, \varphi, f;\, g)|^2 d\varphi d\theta \right\} df}{\pi^2 (f_{\max} - f_{\min})} \tag{3.1}$$

that is the *integral power reflection* given by the ratio between the antenna reflected and input powers over the whole scanning range and bandwidth (i.e., $[f_{\min}, f_{\max}]$) [26], where g is the vector of the solution descriptors, $|\cdot|$ is the magnitude operator and $\Gamma\ (\theta, \varphi, f;\, g)$ is the active reflection coefficient at the steering angle (θ, φ) and frequency f [50].

As for the description of the DoFs, different choices have been adopted in the literature depending on the scenario and problem constraints [26,30,49]. For instance, the entries of g are the slab permittivities and thicknesses when homogeneous dielectric slabs with controllable dielectric constants were considered in Oliveri et al. [30,49]. Otherwise, the L entries of

$$g \triangleq \{g_l,\ l = 1, \ldots, L\} \tag{3.2}$$

are the micro-scale geometrical descriptors [Fig. 3.2(b)] of identical microstrip unit cells periodically printed on a dielectric substrate with relative dielectric permittivity ε_{sub}, dielectric loss tangent $\tan \delta_{sub}$ and thickness t_{sub} to build a WAIM layer. Such a WAIM implementation has been chosen because of the reduced costs, the fabrication simplicity, the low profile and the robustness of such a printing technology if compared to alternative artificial material implementations [26].

As for the identification of the functional blocks of the MbD process, the customization presented in Oliveri et al. [26] (Fig. 3.3) included (i) an EM

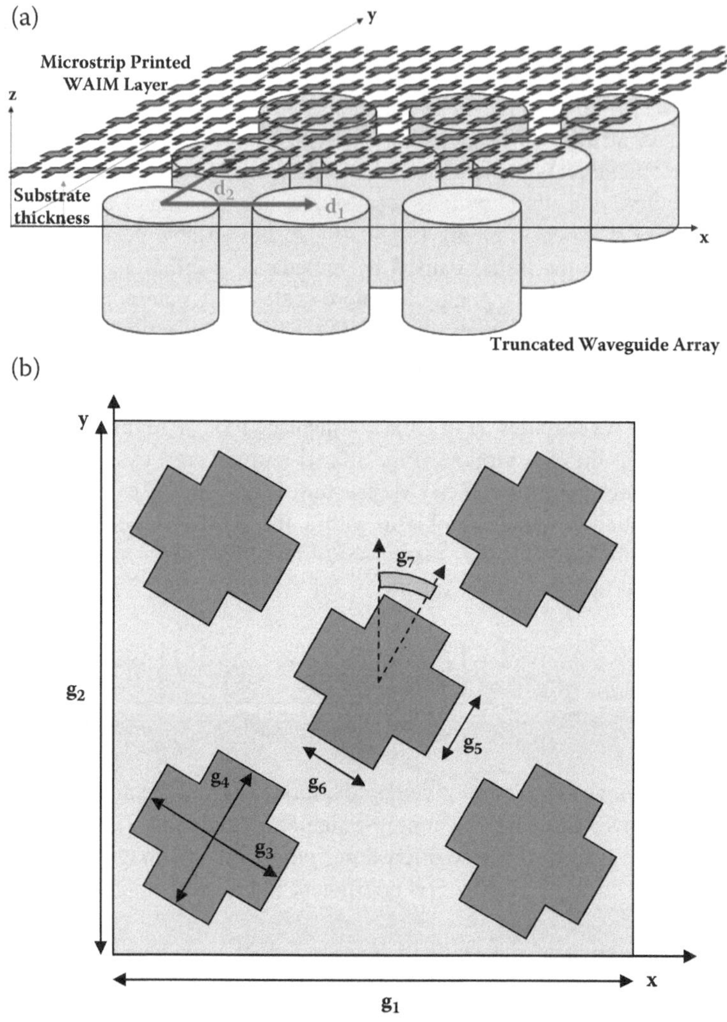

FIGURE 3.2 MbD customization to WAIM design – Geometry of (a) the WAIM-coated macro-scale phased array configuration and (b) the WAIM micro-scale unit cell [26].

micro-modeling block, corresponding to the *WAIM unit cell scattering computation* functional phase, (ii) a micro- to macro-scale physical linkage block, implemented through a combination of a *WAIM homogenization* phase and a *macro-scale voltage reflection coefficient computation* one, collectively resulting in the computation of $\Gamma(\theta, \varphi, f; \mathrm{g})$, (iii) a macro-scale task assessment block corresponding to the *cost function computation* phase, and finally (iv) the *solution space exploration* block. It is worth noticing that all the standard phases for the MbD paradigm are required for customization of the WAIM design problem (Fig. 3.3 vs. Fig. 3.1), even though the micro- to macro-scale physical linkage functionality has been split into two sub-phases for efficiency and flexibility reasons.

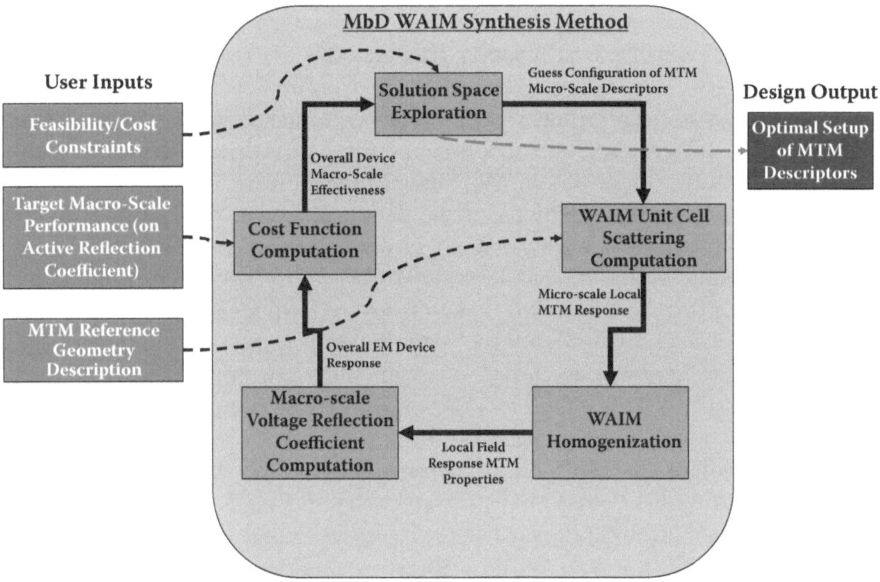

FIGURE 3.3 MbD customization to WAIM design – Functional flowchart of the MbD-based WAIM design process [26].

The choice of implementation strategies has been subsequently carried out [26]. More specifically (Fig. 3.3), see the following:

- The *WAIM unit cell scattering computation* was performed with a full-wave numerical simulator based on the Method-of-Moments with periodic boundary conditions, which was applied to a single-cell (micro-scale) WAIM model [26]. This choice was made because of the flexibility and the efficiency of the resulting computation, owing to the limited size of the discretized region [26].
- The *WAIM homogenization* was carried out with the technique illustrated in Smith et al. [51], owing to its numerical efficiency, flexibility and accuracy when dealing with sub-wavelength WAIM lattices.
- The *macro-scale voltage reflection coefficient computation* was implemented with the modal analysis method [26,30], which determines the antenna active admittance through the truncated Floquet series [26,30], hence avoiding expensive full-wave numerical calculations.
- The *cost function computation* was yielded by suitably discretizing frequency and angular domains [26].
- The *solution space exploration* block was implemented, starting from the analysis of the features and properties of the cost function at hand (in terms of non-linearity and presence of multiple minima) and following the guidelines in the cited references [44–46]. Accordingly, a global search procedure, inspired by the inertial-weight version of the particle swarm optimization

technique, was iteratively repeated until a convergence condition, based on the stagnation of the cost function, was met [26].

The effectiveness of the resulting MbD process was evaluated in Oliveri et al. [26] by considering different aperture shapes and lattice geometries. Despite its technological/realization simplicity, an MbD-based printed WAIM layer significantly mitigated the undesired oscillations of the active reflection coefficient in large phased arrays, both in narrow-band and in wideband working conditions. Moreover, the MbD technique synthesized multi-scale WAIM layers with performance very close to that of the corresponding state-of-the-art [30] ideal layouts featuring homogeneous layers with arbitrary permittivity/permeability tensors [26]. Furthermore, the comparisons carried out with full-wave numerical simulations have also demonstrated the robustness of the MbD-based designs, even when neglecting any homogenization process [26].

To give proof of the MbD performance, a single-layer MbD-based WAIM (Fig. 3.4) featuring the "cross" unit cell geometry in Fig. 3.2(b) and coating a circular waveguide array based on a hexagonal lattice with $\mathbf{d}_1 = 1.067 \times 10^{-2}\hat{x}$ [m] and $\mathbf{d}_1 = 5.33 \times 10^{-3}\hat{x} + 9.23 \times 10^{-3}\hat{y}$ [m] as obtained in Oliveri et al. [26] is discussed hereinafter. The plots of the active reflection coefficient as a function of the scan angle in the $\varphi = 90$ [deg] plane are compared in Fig. 3.4(a) [26] when dealing with (i) the uncoated phased array ("No WAIM") or its WAIM-coated counterparts assuming (ii) an ideal multi-layer structure with arbitrary permittivity values as discussed in Oliveri et al. [30] ("[Oliveri 2015]"), (iii) an MbD-designed arrangement assuming perfectly transparent substrate ("[Oliveri 2017], $\varepsilon_{\text{sub}} = 1.0$") or (iii) with Polytetrafluoroethylene (PTFE) substrate ("[Oliveri 2017], $\varepsilon_{\text{sub}} = 2.08$"). The results of the HFSS modeling [Fig. 3.4(b)] of this latter configuration are reported in Fig. 3.4(a) as well, for the sake of completeness ("[Oliveri 2017], $\varepsilon_{\text{sub}} = 2.08$, HFSS"). Such results show that the array equipped

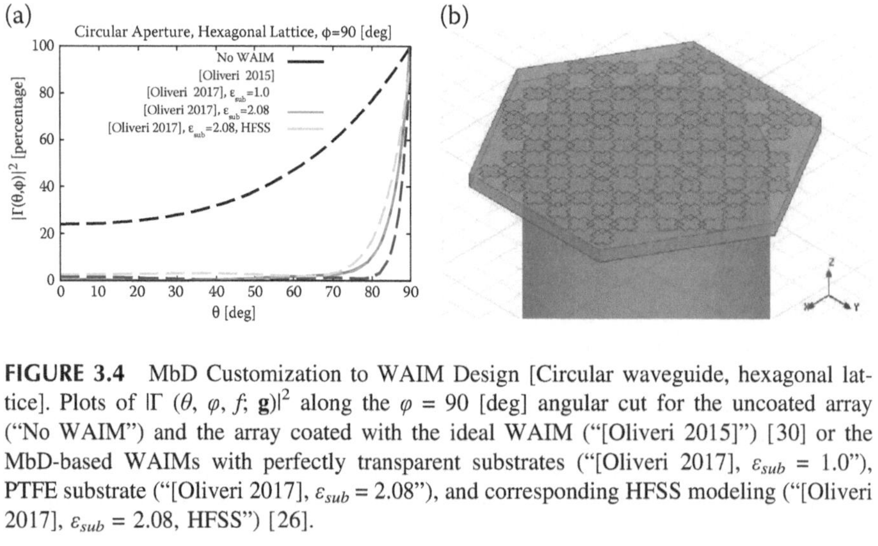

FIGURE 3.4 MbD Customization to WAIM Design [Circular waveguide, hexagonal lattice]. Plots of $|\Gamma\ (\theta,\ \varphi,\ f;\ \mathbf{g})|^2$ along the $\varphi = 90$ [deg] angular cut for the uncoated array ("No WAIM") and the array coated with the ideal WAIM ("[Oliveri 2015]") [30] or the MbD-based WAIMs with perfectly transparent substrates ("[Oliveri 2017], $\varepsilon_{sub} = 1.0$"), PTFE substrate ("[Oliveri 2017], $\varepsilon_{sub} = 2.08$"), and corresponding HFSS modeling ("[Oliveri 2017], $\varepsilon_{sub} = 2.08$, HFSS") [26].

with an MbD-based realistic WAIM affords very low integral power reflection values for the overall structure $[\Phi^{WAIM}(\mathbf{g})|_{[Oliveri\ 2017]} \approx 1.4\%$ – Fig. 3.4(a)], close to the ideal WAIM setup $[\Phi^{WAIM}(\mathbf{g})|_{[Oliveri\ 2015]} \approx 1.1\%$ – Fig. 3.4(a)] and significantly better than the "No WAIM" configuration, which affords a 28.8% total integral power reflection [Fig. 3.4(a)]. Furthermore, the results reported in Oliveri et al. [26] also positively compared to those obtained with ideal WAIM configurations [30] in terms of computational efficiency, since a complete synthesis process took less than 2 hours on a standard laptop [26]. Such accuracy and efficiency are a direct consequence of the modularity and the scalability of the MbD as well as of the implementation choices of each functional block within the MbD flowchart (Fig. 3.3).

3.3.3 Phased Array Enhancement through Metamaterial Lenses and MbD

We consider a completely different scenario next to further assess the flexibility and the customizability of the MbD paradigm. It concerns enhancing the performance of phased arrays through the exploitation of field-manipulation microwave lenses coating the structure [27]. Toward this, the MbD was used to simultaneously yield the following, without affecting/modifying the radiation features of the arrangement with respect to its original architecture [27]:

- Reduction of the aperture and the number of array elements (i.e., a simplification of the structure).
- Modification of the coating lens geometry to comply with the aerodynamics of the hosting structure (e.g., for avionics applications).

To reach the desired macro-scale "focusing" and beam-shaping capabilities, Salucci et al. exploited the micro-scale field manipulation capabilities of the inhomogeneous MTM that coats the phased array [27]. With reference to the time-harmonic setup discussed in Salucci et al. [27], let us consider a reference linear phased array of N elements with n-th radiator ($n = 1,\ldots, N$) is excited with a complex excitation J_n, an it is located at

$$\mathbf{r}_n = \left(x_n = -\frac{L}{2} + (n-1)\,d,\ y_n = h\right) \tag{3.3}$$

where d is the inter-element spacing, h is the distance from the ground plane and $L = (N - 1) \times d$ is the array aperture [Fig. 3.5(a)]. The MbD process was aimed at synthesizing a coating lens with extension Ω', external contour $\partial\Omega'$ and permittivity/permeability distributions $\underline{\underline{\varepsilon}}'(\mathbf{r}')/\underline{\underline{\mu}}'(\mathbf{r}')$, respectively, to miniaturize the antenna array so that it fit (i) a smaller aperture $\tilde{L} < L$ while complying with (ii) a non-regular groundplane profile τ_l and external aerodynamic profile τ_u [Fig. 3.5(b)] without losing its radiation performance [27].

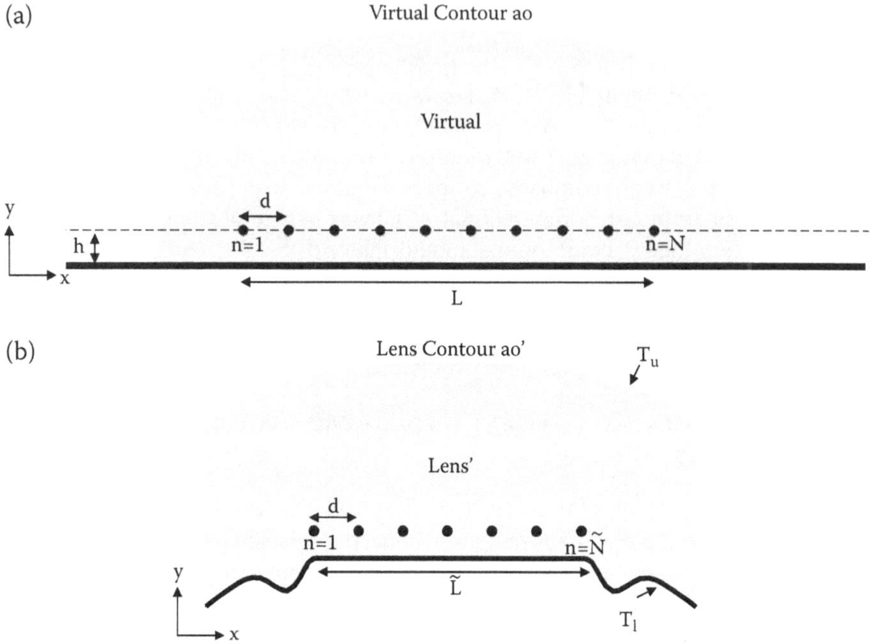

FIGURE 3.5 MbD Customization to Array Miniaturization – Geometry of (a) the reference layout inside the "virtual" contour $\partial\Omega$, (b) the final MbD arrangement inside the conformal/aerodynamic lens contour $\partial\Omega'$ [27].

The objective of the MbD process is computing $\underline{\underline{\varepsilon}}'(\mathbf{r})$, $\underline{\underline{\mu}}'(\mathbf{r})$ and the positions $\tilde{\mathbf{r}}'_n$ and excitations J'_n, $n = 1,\ldots, \tilde{N}$ $(\tilde{N} < N)$ of the miniaturized array that minimizes the following cost function [27]

$$\Phi^{Lens}(\mathbf{g}) \triangleq \int_{\Psi} \left| \tilde{\mathbf{E}}(\mathbf{r}'; \mathbf{g}) - \mathbf{E}(\mathbf{r})|_{\mathbf{r}=\mathbf{r}'} \right|^2 d\mathbf{r}' \tag{3.4}$$

Ψ being the matching area (complying with $\Psi \cap \Omega' = \varnothing$), while $\mathbf{E}(\mathbf{r})$ and $\tilde{\mathbf{E}}(\mathbf{r}'; \mathbf{g})$ are the field radiated by the reference array and the miniaturized one, respectively [27].

As for the DoFs, they could be straightforward identified as the collection of $\tilde{\mathbf{r}}'_n$, J'_n, $n = 1,\ldots, \tilde{N}$, and the spatially discretized versions of $\underline{\underline{\varepsilon}}'(\mathbf{r})$, $\underline{\underline{\mu}}'(\mathbf{r})$ [52]. However, such a choice would result in a huge solution space to be handled [27]. A more efficient DoF definition was followed in Salucci et al. [27] by leveraging the quasi-conformal transformation electromagnetics source-inversion (QCTO-SI) framework [50]. Thanks to QCTO-SI properties [50], Salucci et al. were able to fully describe the solution space in terms of the designed lens contour $\partial\Omega'$ [Fig. 3.5(b)] and the reference array "virtual" contour $\partial\Omega$ [Fig. 3.5(a)] [27]. Accordingly, $\mathbf{g}$ was defined as the collection of the curve descriptors of $\partial\Omega'$ and $\partial\Omega$.

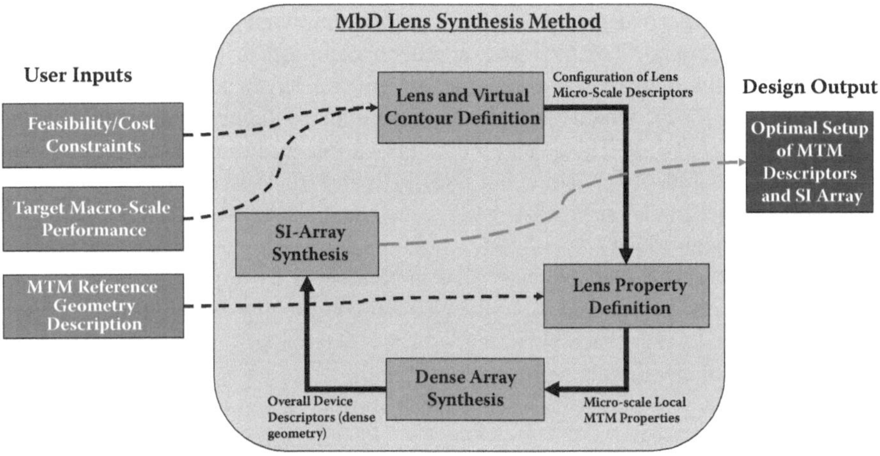

FIGURE 3.6 MbD Customization to Array Miniaturization – Functional flowchart of the MbD-based lens design process [27].

As for the MbD functional blocks, a completely different MbD flowchart was obtained with respect to the example in Section 3.3.2. More specifically, the customization presented in Salucci et al. [27] (Fig. 3.6) included (i) a solution space exploration functionality corresponding to the *lens and virtual contour definitions* block, (ii) the EM micro-modeling functionality corresponding to the *lens property definition* block, and (iii) a micro- to macro-scale physical linkage block implemented as the combination of a *dense array synthesis* block and an *SI-array synthesis* block. The following unique features of the MbD approach as applied to this test case can then be pointed out:

- The implemented QCTO-based approach [27] did not feature an iterative design process, since it is a "one-shot" process devoted to minimizing (3.4) regardless of the user-defined DoF setup. Such an observation has suggested potential extensions of the approach to include additional objectives beyond (3.4).
- The computation of the macro-scale objective (3.4) is not required within the MbD process, since it is implicitly included within both the *lens property definition* block and the *SI-array synthesis* block, as detailed in the following.

Consequently, none of the standard phases for the MbD paradigm have been implemented nor has their order been kept unaltered (Fig. 3.6 vs. Fig. 3.1).

As for the implementation strategies, the following choices have been adopted (Fig. 3.6):

- The *lens and virtual contour definition* has been based on a hybrid encoding featuring both linear-piecewise and spline segments [27] given the nature of the specified contours, which can be efficiently described on a per-segment basis (Fig. 3.5).

- The *lens property definition* has been carried out by means of the generalized QCTO method [50,53,54]. The approach computes a smooth geometrical mapping function between $\partial\Omega'$ and $\partial\Omega$ (Fig. 3.5), to be used for defining $\underline{\underline{\varepsilon}}'(\mathbf{r})$ and $\underline{\underline{\mu}}'(\mathbf{r})$, which theoretically guarantee to fit the optimal condition [i.e., $\Phi^{Lens}(\underline{\underline{g}}) = 0$]. This is an optimal choice because of the efficiency of the QCTO method, its capability to exactly match (3.4) without requiring any iterative optimization processes [27], and the stability and robustness of the arising design [55–57];
- The *dense array synthesis* has been implemented by applying the mapping function between $\partial\Omega'$ and $\partial\Omega$ to the element coordinates of the reference array $\{\mathbf{r}_n;\ n = 1,\ldots, N\}$, hence obtaining a "dense" layout still with N elements but with a smaller inter-element spacing [27,50].
- The *SI-array synthesis* has been yielded through a source inversion process [50], which computes the positions $\tilde{\mathbf{r}}'_n$ and the excitations J'_n, $n = 1,\ldots, \tilde{N}$ ($\tilde{N} < N$) of the miniaturized array, still complying by definition with (3.4). It is worthwhile to point out that such an inversion has required only the pseudo-inversion of the radiation kernel matrix [27].

The potentialities of the resulting MbD synthesis were evaluated in Salucci et al. [27] for different contour shapes and reference arrays. The general outcome was that a non-trivial reduction of the antenna aperture and number of elements can be obtained without major degradations in the radiated field. Moreover, Salucci et al. also numerically proved that there is a trade-off between reference array miniaturization and artificial material complexity (i.e., in terms of permittivity/permeability ranges and degree of anisotropy) [27].

As a representative example, we look at the performance of the MbD-based synthesis when dealing with a $N' = 17$ conformal layout matching the radiation performance of a uniform reference linear arrangement featuring $N = 20$, $d = \frac{\lambda}{2}$, $J_n = 1.0$ ($n = 1, \ldots, N$), and $h = \frac{\lambda}{4}$ [27]. The dominant component of the relative electric permittivity (i.e., the "zz" one [27]) of the MbD-based lens shows, as expected, spatially larger variations in the lower region of the lens, which is subject to a more significant deformation to match the conformal support profile τ_l [Fig. 3.7(a)]. However, the permittivity values are always bounded to ranges compatible with recent MTM design and fabrication processes [27]. The comparison of the far-field normalized radiation patterns of the reference, the dense, and the final MbD arrangement [Fig. 3.7(b)] indicates that (i) the pattern radiated by the lens-coated miniaturized array faithfully matches the reference one as quantitatively confirmed by the values of the far-field integral matching error ξ [i.e., $\xi_{MbD} = 1.77\%$ – Fig. 3.7(b)], (ii) a significant worsening of the matching occurs if the MbD lens is removed, regardless of the array layout [e.g., $\xi_{dense\ no\ lens} = 12.2\%$ – Fig. 3.7(b)]. The results in Salucci et al. [27] also demonstrated the reliability of the MbD-based strategy to handle the transformation of reference layouts with different dimensions, excitations, sidelobe levels or pattern shapes, as well as smaller/larger values of the inter-element spacing d.

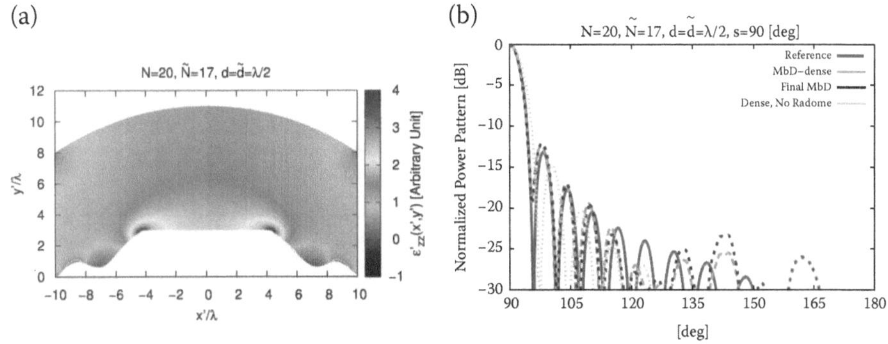

FIGURE 3.7 MbD Customization to Array Miniaturization [$N = 20$, $\tilde{N} = 17$, $d = \tilde{d} = \frac{\lambda}{2}$] – Plots of (a) the zz-component of the relative permittivity distribution of the MbD conformal radome and (b) the normalized power patterns radiated by the reference array, the dense array (with and without radome), and the final MbD layout [27].

3.3.4 Wave Manipulation MTM Devices Based on MbD

The last example of the MbD-based synthesis is concerned with the design of wave manipulation devices that modify the polarization state of transmitted fields [36,37]. In this framework, the MbD paradigm was applied to simultaneously yield, for all incidence angles and operative frequencies, (i) a low reflection from the MTM screen (i.e., the minimization of back reflection by the metasurface) and (ii) a user-defined polarization for the transmitted field [36,37]. More specifically, the micro-scale field manipulation qualities of the artificial MTM, which composes the wave polarizer, were exploited to enforce desired properties of the macro-scale transmission coefficient [36,37].

In Oliveri et al. [37], the structure at hand consists of a stack of N identical planar MTMs with inter-layer spacing d_n, $n = 1,\ldots, N-1$ (Fig. 3.8). Each layer of a the multi-layer metasurface (Fig. 3.8) is composed of a regular combination of microstrip printed unit cells, each with an arbitrary geometry, arranged in a rectangular lattice with periodicity p_x, p_y along $\hat{x}$ and $\hat{y}$, respectively.

Concerning the mathematical formulation of the macro-scale task, the wave polarizer objective was defined as [36,37]

$$\Phi^{POL}(\mathbf{g}) \triangleq \frac{\int_{f_{\min}}^{f_{\max}} \left\{ \int_{\theta_{\min}}^{\theta_{\max}} \int_{\varphi_{\min}}^{\varphi_{\max}} \left(\alpha_1 |\Gamma^T - \Gamma(\theta, \varphi, f; \mathbf{g})|^2 + \alpha_2 |\mathcal{E}^T - \mathcal{E}(\theta, \varphi, f; \mathbf{g})|^2 \right) d\varphi d\theta \right\} df}{(\theta_{\max} - \theta_{\min}) \times (\varphi_{\max} - \varphi_{\min}) \times (f_{\max} - f_{\min})} \quad (3.5)$$

where $\Gamma(\theta, \varphi, f; \mathbf{g})$ is the macro-scale field reflection coefficient when a time-harmonic $\hat{x}$-polarized plane wave at frequency f impinges on the metasurface with incidence angle (θ, φ) (Γ^T being its corresponding target value), $\mathcal{E}(\theta, \varphi, f; \mathbf{g})$ is the

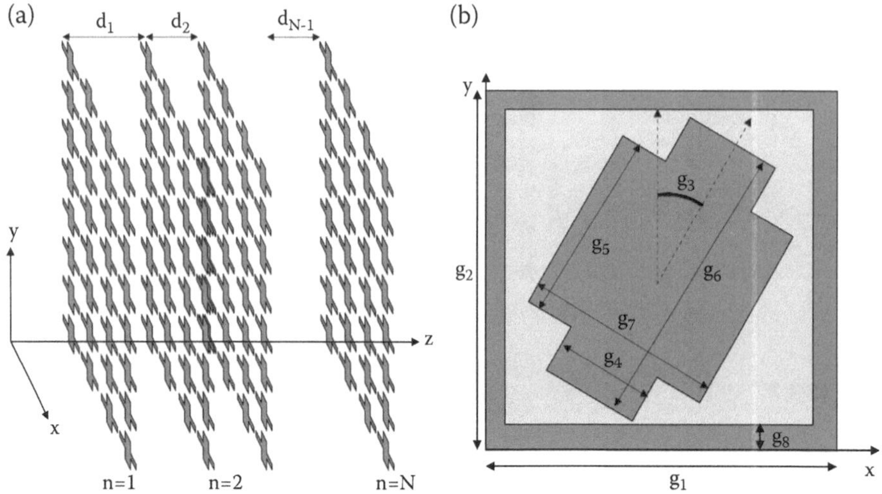

FIGURE 3.8 MbD Customization to Wave Polarizer Design – Geometry of (a) the multi-layer macro-scale wave polarizer and (b) the micro-scale unit cell [36,37].

macro-scale ellipticity angle of the transmitted wave in the same conditions ($\mathcal{E}^T$ being the corresponding target value), $[\theta_{min}, \theta_{max}]$, $[\varphi_{min}, \varphi_{max}]$, and $[f_{min}, f_{max}]$ being the elevation angular range, the azimuth angular range and the bandwidth, respectively.

As for the DoFs, owing to the adopted multiscale formulation and thanks to the MbD capability to handle the arising complexity [36,37], the g vector entries are (i) the micro-scale geometrical descriptors of the microstrip printed unit cell (e.g., the length and thickness of the arms of the cross, the thickness of the surrounding ring, the angular tilt of the cross – Fig. 3.8), and (ii) the inter-layer spacing parameters d_n, $n = 1, \ldots, N - 1$.

The MbD functional blocks are [36,37] (Fig. 3.9) (i) an EM micro-modeling block, denoted as *EM simulation* block [36,37], (ii) a micro- to macro-scale physical linkage functionality implemented through the *EM polarizer homogenization* block, (iii) a macro-scale task assessment block concerned with the *cost function evaluation* phase and (iv) the *solution space exploration* block. For each block, the following computational strategies were implemented [36,37]:

- Both full-wave models, including periodic boundary conditions [37] as well as a forward solver based on a fast modal strategy [36] featuring a spectral MoM approach [58], were chosen for the *EM simulation* to deduce the EM scattering matrices of the structure. The choice of the most proper strategy to compute the micro-level response of the unit cell was driven by the efficiency and the accuracy of the computation process as well as by the flexibility of the addressable layouts [36,37].
- The *EM polarizer homogenization* was implemented to deduce the equivalent layer properties (i.e., ellipticity and reflection coefficient of the overall

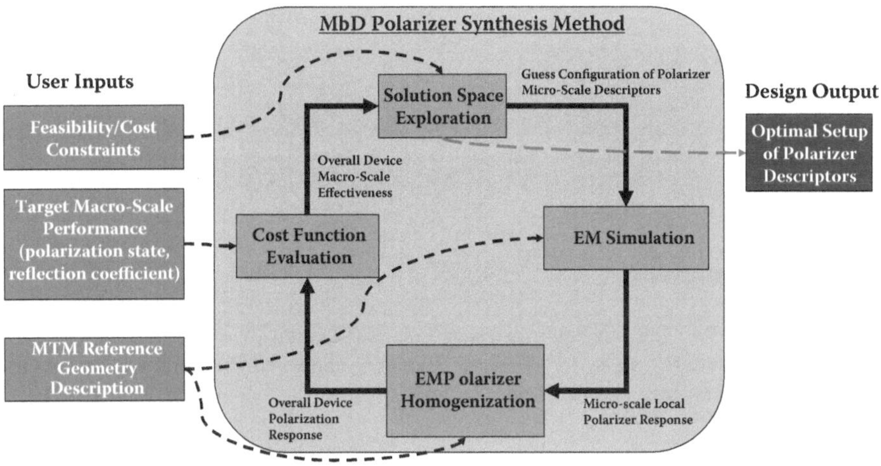

FIGURE 3.9 MbD customization to wave polarizer design – Functional flowchart of the MbD-based wave polarizer design process [36,37].

polarizer) starting from the micro-scale electromagnetic response computed by the simulation block. Toward this end, the combination of the incident wave with the equivalent periodic EM scattering matrices was employed [36,37], thanks to the marginal impact of edge effects when large regular arrangements are at hand.

- The *cost function evaluation* was straightforwardly implemented by suitably discretizing (3.5) in the frequency and the angular domains [36,37].
- The *solution space exploration* block was implemented by customizing a particle swarm optimization-based global search procedure [36,37] to the functional features and properties of (3.5).

The effectiveness and potentialities of the arising MbD design process were evaluated in Bekele et al. and Oliveri et al. [36,37] when dealing with different polarizer geometries, bandwidth and target performance. The numerical analysis presented in Bekele et al. and Oliveri et al. [36,37] demonstrated the effectiveness and efficiency of the MbD paradigm in dealing with multiple design criteria and objectives concerning wave polarization and the reflection coefficient. To give some insight to the interested reader, the MbD synthesis of a linear to circular polarization converter with $N = 3$ layers [37] is discussed next. The design process was carried out by assuming $\Gamma^T = 0$ (no reflection), $\mathcal{E}^T = 45$ [deg] (left circular polarization as a target), $[\theta_{min}, \theta_{max}] = [0, 10]$ [deg], $\varphi_{min} = \varphi_{max} = 0$ [deg], and $[f_{min}, f_{max}] = [8.5, 10.5]$ [GHz] [37]. The plots of $\Gamma\ (\theta, \varphi, f;\ \mathbf{g})$ [Fig. 3.10(a)] and $\mathcal{E}\ (\theta, \varphi, f;\ \mathbf{g})$ [Fig. 3.10(b)] for the arrangement synthesized in Oliveri et al. [37] confirm the transmission of an almost perfectly polarized wave [i.e., $\mathcal{E}\ (\theta, \varphi, f;\ \mathbf{g}) > 37$ [deg] – Fig. 3.10(a)] with stable reflection properties in all working frequencies and angles of incidence [i.e., $|\Gamma\ (\theta, \varphi, f;\ \mathbf{g})| < -11$ [dB] - Fig. 3.10(b)] [37]. More specifically, the transmission performance turned out

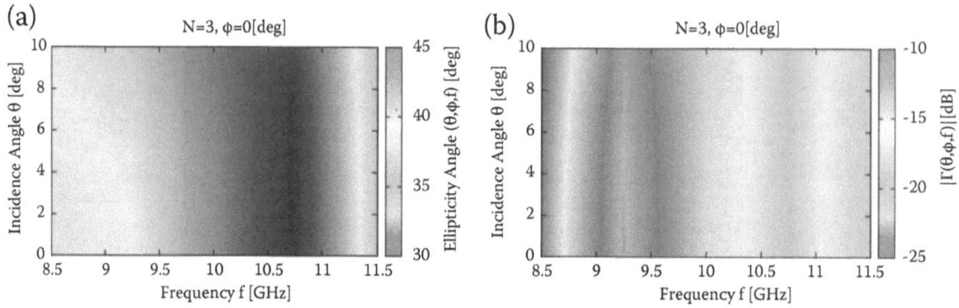

FIGURE 3.10 MbD Customization to Wave Polarizer Design [N = 3 Linear to Circular Polarization Converter] – Behavior of (a) the ellipticity angle and (b) the reflection coefficient versus the frequency and the incidence angle [37].

excellent at the higher end of the spectrum [i.e., $|\Gamma\ (\theta, \varphi, f; \mathbf{g})| < -15$ [dB] when $f > 8.7$ [GHz] - Fig. 3.10(b)] and only marginally affected by the incidence angle (Fig. 3.10) [37].

3.4 FINAL REMARKS, CURRENT TRENDS AND FUTURE PERSPECTIVES

The MdB paradigm has recently emerged as a powerful, flexible and modular framework to *"address the task-oriented design of advanced systems comprising artificial materials whose constituent properties are driven by the user-defined functional requirements of the system"* [25]. The fundamental concept behind the MbD is that the intrinsic DoFs of any MTM-based system/device can be exploited in a more efficient and effective manner if (i) the design process considers the macro-scale functionality/task as the MTM synthesis objective (rather than its micro-scale EM response) and (ii) the synthesis descriptors are actually defined as the micro-scale properties of the standard materials employed in the device building (rather than its "equivalent" representation) [25].

Accordingly, a review of the most recent advances in the application of MbD to complex EM systems/devices was presented by focusing on (i) the definition of the fundamental MbD features, (ii) the customization process to a set of specific applicative scenarios, (iii) a demonstration of the flexibility and the potentialities of the arising MbD-based synthesis processes and (iv) MbD performance in recently synthesized MTM-enhanced wave manipulation systems. Within such a framework, the following key aspects were illustrated:

- The adoption and customization of the MbD paradigm allows one to formulate MTM synthesis in a task-oriented fashion, hence avoiding focus on micro-scale matching objectives (usually irrelevant from the applicative viewpoint), but rather potentially enabling an enhanced exploration of the functional landscape of the design from the macro-scale perspective.
- The modularity of the MbD concept, which is inherited by the SbD framework, enables full customization of the resulting design procedure in terms of

functional blocks and implementation choices, as proved in the MbD synthesis of WAIM-enhanced phased arrays (Section 3.3.2), microwave lenses for array miniaturization (Section 3.3.3) and wave polarizers (Section 3.3.4).
- Single- and multi-objective syntheses as well as complex design constraints are naturally and easily included in the MbD process.
- The MbD paradigm may seamlessly support multiphysics objectives as well as various performance constraints thanks to its task-oriented nature.

Despite such successful results and outcomes, many open challenges still exist in the application and customization of MbD within MTM engineering. Current and future efforts within this research framework are expected to deal with (i) the integration of MbD techniques with advanced artificial intelligence approaches, (ii) the numerical and experimental validation of multiphysics MbD-designed MTMs, and (iii) the extension and generalization of MbD-based techniques to deal with the real-time control of reconfigurable and smart EM MTMs.

ACKNOWLEDGMENTS

This work benefited from the networking activities carried out within the Project "CLOAKING METASURFACES FOR A NEW GENERATION OF INTELLIGENT ANTENNA SYSTEMS (MANTLES)" funded by the Italian Ministry of Education, University and Research within the PRIN2017 Program under grant 2017BHFZKH (CUP: E64I19000560001), within the Project "SMARTOUR – Piattaforma Intelligente per il Turismo" (Grant no. SCN_00166) funded by the Italian Ministry of Education, University and Research within the Program "Smart cities and communities and Social Innovation," and within the Grant No. 61721001 funded by the National Natural Science Foundation of China.

REFERENCES

1. N. Engheta and R. Ziolkowski, Eds., *Metamaterials: Physics and Engineering Explorations*, New York, USA: Wiley, 2006.
2. J. B. Pendry, D. Schurig, and D. R. Smith, "Controlling electromagnetic fields," *Science*, vol. 312, pp. 1780–1782, June 2006.
3. T. J. Cui, D. Smith, and R. Liu, Eds., *Metamaterials: Theory, Design, and Applications*, New York, USA: Springer, 2010.
4. D.-H. Kwon and D. H. Werner, "Transformation electromagnetics: An overview of the theory and applications," *IEEE Antennas Propag. Mag.*, vol. 52, no. 1, pp. 24–46, Feb. 2010.
5. D. H. Werner and D.-H. Kwon, Eds., *Transformation Electromagnetics and Metamaterials: Fundamental Principles, and Applications*, New York, USA: Springer, 2014.
6. F. Yang and Y. Rahmat-Samii, *Surface Electromagnetics with Applications in Antenna, Microwave, and Optical Engineering*, Cambridge, UK: Cambridge University Press, 2019.
7. G. Oliveri, D. H. Werner, and A. Massa, "Reconfigurable electromagnetics through meta-materials – A review," *Proc. IEEE*, vol. 103, no. 7, pp. 1034–1056, July 2015.

8. T.-C. Lim, *Mechanics of Metamaterials with Negative Parameters*. Singapore: Springer, 2020.
9. M. G. Silveirinha and C. A. Fernandes, "Homogenization of 3-D-connected and nonconnected wire metamaterials," *IEEE Trans. Microw. Theory Tech.*, vol. 53, no. 4, pp. 1418–1430, April 2005.
10. O. Ouchetto, H. Ouchetto, S. Zouhdi, and A. Sekkaki, "Homogenization of Maxwell's equations in lossy biperiodic metamaterials," *IEEE Trans. Antennas Propag.*, vol. 61, no. 8, pp. 4214–4219, August 2013.
11. E. Martini, G. M. Sardi, and S. Maci, "Homogenization processes and retrieval of equivalent constitutive parameters for multisurface-metamaterials," *IEEE Trans. Antennas Propag.*, vol. 62, no. 4, pp. 2081–2092, April 2014.
12. C. Mateo-Segura, A. Dyke, H. Dyke, S. Haq, and Y. Hao, "Flat Lunenburg lens via transformation optics for directive antenna applications," *IEEE Trans. Antennas Propag.*, vol. 62, no. 4, pp. 1945–1953, April 2014.
13. J. C. Soric, R. Fleury, A. Monti, A. Toscano, F. Bilotti, and A. Alù, "Controlling scattering and absorption with metamaterial covers," *IEEE Trans. Antennas Propag.*, vol. 62, no. 8, pp. 4220–4229, August 2014.
14. D. Bao, R. C. Mitchell-Thomas, K. Z. Rajab, and Y. Hao, "Quantitative study of two experimental demonstrations of a carpet cloak," *IEEE Antennas Wireless Propag. Lett.*, vol. 12, pp. 206–209, 2013.
15. F. Bilotti, S. Tricarico, and L. Vegni, "Plasmonic metamaterial cloaking at optical frequencies," *IEEE Trans. Nanotechn.*, vol. 9, no. 1, pp. 55–61, January 2010.
16. A. Monti, J. C. Soric, A. Alù, A. Toscano, and F. Bilotti, "Anisotropic mantle cloaks for TM and TE scattering reduction," *IEEE Trans. Antennas Propag.*, vol. 63, no. 4, pp. 1775–1788, April 2015.
17. J. C. Soric, A. Monti, A. Toscano, F. Bilotti, and A. Alù, "Multiband and wideband bilayer mantle cloaks," *IEEE Trans. Antennas Propag.*, vol. 63, no. 7, pp. 3235–3240, July 2015.
18. E. Martini, M. Mencagli, D. Gonzalez-Ovejero, and S. Maci, "Flat optics for surface waves," *IEEE Trans. Antennas Propag.*, vol. 64, no. 1, pp. 155–166, January 2016.
19. Y. Luo, J. Zhang, L. Ran, H. Chen, and J. A. Kong, "New concept conformal antennas utilizing metamaterial and transformation optics," *IEEE Antennas Wireless Propag. Lett.*, vol. 7, pp. 509–512, 2008.
20. A. Monti, J. Soric, A. Alu, F. Bilotti, A. Toscano, and L. Vegni, "Overcoming mutual blockage between neighboring dipole antennas using a low-profile patterned metasurface," *IEEE Antennas Wireless Propag. Lett.*, vol. 11, pp. 1414–1417, 2012.
21. J. Lei, J. Yang, X. Chen, Z. Zhang, G. Fu, and Y. Hao, "Experimental demonstration of conformal phased array antenna via transformation optics," *Sci. Rep.*, vol. 8, no. 1, id. 3807, pp. 1–12, 2018.
22. A. Monti, M. Barbuto, A. Toscano, and F. Bilotti, "Nonlinear mantle cloaking devices for power-dependent antenna arrays," *IEEE Antennas Wireless Propag. Lett.*, vol. 16, pp. 1727–1730, 2017.
23. D. Ramaccia, D. L. Sounas, A. Alù, F. Bilotti, and A. Toscano, "Nonreciprocity in antenna radiation induced by space-time varying metamaterial cloaks," *IEEE Antennas Wireless Propag. Lett.*, vol. 17, no. 11, pp. 1968–1972, November 2018.
24. S. Vellucci, A. Monti, M. Barbuto, A. Toscano, and F. Bilotti, "Waveform-selective mantle cloaks for intelligent antennas," *IEEE Trans. Antennas Propag.*, vol. 68, no. 3, pp. 1717–1725, March 2020.
25. A. Massa and G. Oliveri, "Metamaterial-by-design: Theory, methods, and applications to communications and sensing – Editorial," *EPJ Applied Metamaterials*, vol. 3, no. E1, pp. 1–3, 2016.

26. G. Oliveri, M. Salucci, N. Anselmi, and A. Massa, "Multiscale system-by-design synthesis of printed WAIMs for waveguide array enhancement," *IEEE J. Multiscale Multiphysics Computat. Techn.*, vol. 2, pp. 84–96, 2017.
27. M. Salucci, G. Oliveri, N. Anselmi, and A. Massa, "Material-by-design synthesis of con- formal miniaturized linear phased arrays," *IEEE Access*, vol. 6, pp. 26367–26382, 2018.
28. A. Massa, G. Oliveri, P. Rocca, and F. Viani, "System-by-Design: a new paradigm for handling design complexity," in *8th European Conf. Antennas Propag.*, The Hague, The Netherlands, 2014, pp. 1180–1183.
29. G. Oliveri, L. Tenuti, E. Bekele, M. Carlin, and A. Massa, "An SbD-QCTO approach to the synthesis of isotropic metamaterial lenses," *IEEE Antennas Wireless Propag. Lett.*, vol. 13, pp. 1783–1786, 2014.
30. G. Oliveri, F. Viani, N. Anselmi, and A. Massa, "Synthesis of multi-layer WAIM coatings for planar phased arrays within the system-by-design framework," *IEEE Trans. Antennas Propag.*, vol. 63, no. 6, pp. 2482–2496, June 2015.
31. G. Oliveri, A. Gelmini, A. Polo, N. Anselmi, and A. Massa, "System-by-design multi-scale synthesis of task-oriented reflectarrays," *IEEE Trans. Antennas Propag.*, vol. 68, no. 4, pp. 2867–2882, April 2020.
32. G. Oliveri, F. Bilotti, A. Toscano, and A. Massa, "System-by-design paradigm as applied to the synthesis of innovative field manipulation devices including task-oriented meta-materials," *8th International Congress on Advanced Electromagnetic Materials in Mi- crowaves and Optics* (Metamaterials 2014), Copenhagen, Denmark, pp. 229–231, August 25–30, 2014.
33. L. Tenuti, G. Oliveri, F. Viani, F. Bilotti, A. Toscano, and A. Massa, "A system-by-design approach for the synthesis of multi-layer mantle cloaks," *2015 IEEE AP-S International Symposium and USNC-URSI Radio Science Meeting*, Vancouver, BC, Canada, pp. 59–60, July 19–25, 2015.
34. L. Tenuti, G. Oliveri, A. Monti, F. Bilotti, A. Toscano, and A. Massa, "Design of mantle cloaks through a system-by-design approach," *10th European Conference on Antennas and Propagation (EuCAP 2016)*, Davos, Switzerland, pp. 1–3, April 11–15, 2016.
35. G. Oliveri, E. Bekele, M. Salucci, L. Tenuti, G. Gottardi, T. Moriyama, T. Takenaka, F. Bilotti, A. Toscano, and A. Massa, "A system-by-design approach to the synthesis of mantle cloaks for large dielectric cylinders," *PIERS 2016 in Shanghai*, Shanghai, China, August 8–11, pp. 3144–3145, 2016.
36. E. T. Bekele, G. Oliveri, E. Martini, S. Maci, and A. Massa, "A Material-by-design strategy for the design and optimization of multisurface-metamaterial polarizers," *8th European Conference on Antennas and Propagation (EuCAP 2014)*, The Hague, 2014, pp. 1965–1968.
37. G. Oliveri, F. Apolloni, A. Gelmini, E. T. Bekele, S. Maci, and A. Massa, "Numerical homogenization and synthesis of wave polarizers through the material-by-design paradigm," *9th European Conference on Antennas and Propagation* (EuCAP 2015), Lisbon, 2015, pp. 1–4.
38. S. A. Tretyakov, S. Maslovski, and P. A. Belov, "An analytical model of metamaterials based on loaded wire dipoles," *IEEE Trans. Antennas Propag.*, vol. 51, no. 10, pp. 2652–2658, October 2003.
39. D. L. Sounas, T. Kodera, and C. Caloz, "Electromagnetic modeling of a magnetless non-reciprocal gyrotropic metasurface," *IEEE Trans. Antennas Propag.*, vol. 61, no. 1, pp. 221–231, January 2013.
40. I. Yoo and D. R. Smith, "Analytic model of coax-fed printed metasurfaces and analysis of antenna parameters," *IEEE Trans. Antennas Propag.*, vol. 68, no. 4, pp. 2950–2964, April 2020.

41. M. Salucci, L. Tenuti, G. Oliveri, and A. Massa, "Efficient prediction of the EM response of reflectarray antenna elements by an advanced statistical learning method," *IEEE Trans. Antennas Propag.*, vol. 66, no. 8, pp. 3995–4007, August 2018.
42. Y. Vahabzadeh, N. Chamanara, K. Achouri, and C. Caloz, "Computational analysis of metasurfaces," *IEEE J. Multiscale Multiphysics Comput. Techn.*, vol. 3, pp. 37–49, 2018.
43. D. H. Wolpert and W. G. Macready, "No free lunch theorems for optimization," *IEEE Trans. Evol. Comput.*, vol. 1, no. 1, pp. 67–82, April 1997.
44. P. Rocca, M. Benedetti, M. Donelli, D. Franceschini, and A. Massa, "Evolutionary optimization as applied to inverse scattering problems," *Inverse Problems*, vol. 25, no. 12, pp. 1–41, December 2009.
45. P. Rocca, G. Oliveri, and A. Massa, "Differential evolution as applied to electromagnetics," *IEEE Antennas Propag. Magazine*, vol. 53, no. 1, pp. 38–49, February 2011.
46. J. Robinson and Y. Rahmat-Samii, "Particle swarm optimization in electromagnetics," *IEEE Trans. Antennas Propag.*, vol. 52, no. 2, pp. 397–407, February 2004.
47. S. Sajuyigbe, M. Ross, P. Geren, S. A. Cummer, M. H. Tanielian, and D. R. Smith, "Wide angle impedance matching metamaterials for waveguide-fed phased-array antennas," *IET Microw. Antennas Propag.*, vol. 4, no. 8, pp. 1063–1072, August 2010.
48. E. Magill and H. A. Wheeler, "Wide-angle impedance matching of a planar array antenna by a dielectric sheet," *IEEE Trans. Antennas Propag.*, vol. 14, no. 1, pp. 49–53, January 1966.
49. G. Oliveri, A. Polo, M. Salucci, G. Gottardi, and A. Massa, "*SbD*-based synthesis of low-profile *WAIM* superstrates for printed patch arrays," *IEEE Trans. Antennas Propag.*, in press. doi:10.1109/TAP.2020.3044686
50. G. Oliveri, E. T. Bekele, M. Salucci, and A. Massa, "Array miniaturization through QCTO-SI metamaterial radomes," *IEEE Trans. Antennas Propag.*, vol. 63, no. 8, August 2015.
51. D. R. Smith, S. Schultz, P. Markos, and C. M. Soukoulis, "Determination of the effective permittivity and permeability of metamaterials from reflection and transmission coefficients," *Phys. Rev. B, Condens. Matter*, vol. 65, no. 19, May 2002, Art. no. 195104.
52. D. E. Brocker, J. P. Turpin, P. L. Werner, and D. H. Werner, "Optimization of gradient index lenses using quasi-conformal contour transformations," *IEEE Antennas Wireless Propag. Lett.*, vol. 13, pp. 1787–1791, 2014.
53. G. Oliveri, E. T. Bekele, D. H. Werner, J. P. Turpin, and A. Massa, "Generalized QCTO for metamaterial-lens-coated conformal arrays," *IEEE Trans. Antennas Propag.*, vol. 62, no. 8, August 2014.
54. D.-H. Kwon, "Quasi-conformal transformation optics lenses for conformal arrays," *IEEE Antennas Wireless Propag. Lett.*, vol. 11, pp. 1125–1128, September 2012.
55. J. Gao, C. Wang, K. Zhang, Y. Hao, and Q. Wu, "Beam steering performance of compressed Luneburg lens based on transformation optics," *Results in Physics*, vol. 9, pp. 570–575, June 2018.
56. R. Yang, W. Tang, and Y. Hao, "Wideband beam-steerable flat reflectors via transformation optics," *IEEE Antennas Wireless Propag. Lett.*, vol. 10, pp. 1290–1294, 2011.
57. G. Oliveri, E. T. Bekele, M. Salucci, and A. Massa, "Transformation electromagnetics miniaturization of sectoral and conical metamaterial-enhanced horn antennas," *IEEE Trans. Antennas Propag.*, vol. 64, no. 4, pp. 1510–1513, April 2016.
58. E. Martini, G. M. Sardi, F. Caminita, and S. Maci, "Linear-to-circular planar multisurface- metamaterial polarizer designed by anisotropic effective constitutive parameters," *IEEE International Symposium on Antennas Propag. (APS/URSI 2013)*, Orlando, Florida, USA, July 7–12, 2013.

4 Tunable Metamaterials

Ke Bi and Jianchun Xu
State Key Laboratory of Information Photonics and Optical Communications, School of Science, Beijing University of Posts and Telecommunications, Beijing, China

4.1 INTRODUCTION

Metamaterial is a kind of artificial structure or material with exotic physical properties that natural materials do not possess. The electromagnetic parameters can be controlled by artificial structures, which provide a way to control the electromagnetic waves [1–5]. Metamaterial design has a large degree of freedom. It can work in different frequency ranges by properly designing the structural parameters. Therefore, metamaterials have a wide application prospect in the fields of radio frequency (including absorbing material, antenna, etc.), terahertz (THz; including THz sensor, detector, etc.) and optics (including perfect lens, invisible cloak, super-resolution imaging, etc.). The electromagnetic properties of metamaterials are realized by the electromagnetic responses of specific structures. However, the fixed structure of metamaterial results in a specific range of operation frequency. Beyond this range, the exotic electromagnetic properties will be reduced or even disappear. This phenomenon means that once the operating frequency is changed, it is necessary to redesign the metamaterial structure to achieve the same property, thus limiting its practicability. Obviously, the practicality of metamaterials will be greatly increased if the properties of metamaterials can be controlled by changing the external field without changing the structure.

The electromagnetic parameters of dielectric metamaterials consisting of some dielectric units are highly sensitive to the external fields, such as magnetic field, electric field, temperature and strain. The electromagnetic characteristics of metamaterials can be dynamically controlled by a reasonable introduction of the external field. These tunable metamaterials have wide application prospects in modern communication and radar systems, such as absorbing materials, microwave devices and smart antennas. In order to achieve the tunable properties of dielectric metamaterials, various methods have been extensively studied, including magnetically tunable [6–8], electrically tunable [9–11] and thermally tunable methods [12–14]. As a typical tunable metamaterial, the permeability and permittivity of magnetically tunable metamaterials can be adjusted by the external magnetic field. Compared with other tunable dielectric metamaterials, magnetically tunable dielectric metamaterials have the advantages of wider tuning range and faster response speed [15–17]. The electrically tunable dielectric metamaterial properties can be achieved

DOI: 10.1201/9781003050162-4

by changing the material characteristics of the unit cell with bias voltage [18–20]. In addition, thermally tunable dielectric metamaterials are mainly based on the dependence of dielectric constant on temperature [21–23]. Besides, the electromagnetic property of flexible metamaterials can be tuned by the strain or stress deformation, which is a benefit for the future wearable systems [24,25]. This chapter classifies the control methods of tunable dielectric metamaterials from the point of external field control and reviews the main research progress, tuning mechanism and future application prospects of magnetically, electrically and thermally tunable dielectric and flexible metamaterials.

4.2 MAGNETICALLY TUNABLE DIELECTRIC METAMATERIALS

As a gyromagnetic medium, ferrite can exhibit ferromagnetic resonance under the external bias magnetic field, thus achieving negative magnetic permeability. The frequency ranges with negative permeability can be tuned by adjusting the external magnetic field. Therefore, ferrite materials have been widely applied in tunable devices, such as negative refraction metamaterials, metamaterial antennas and metamaterial band-stop filters [26–33].

4.2.1 Ferrite/Wire Composite Structure

To achieve the magnetically tunable negative refraction property, Zhao et al. [26] proposed a ferrite-based metamaterial by using yttrium iron garnet (YIG) ferrite rods and metal wires. The structure consists of YIG rods and Cu wires, and the external magnetic field is along the long axis direction of ferrite (see Fig. 4.1a). Fig. 4.1b presents the measured transmission of the metamaterial sample under different magnetic fields. As the external magnetic field increases from 1,600 Oe to 2,300 Oe, the center frequency increases from 8.2 GHz to 10.7 GHz, and the response speed reaches 3.5 GHz/kOe. This structure achieves a dynamic, continuous and reversible magnetic tunable behavior in a wide band range, which indicates that the operating frequency can be easily adjusted by changing the external magnetic field. The dual-band property [27] is then achieved by further introducing different sizes of ferrite into the above structure. The YIG rods with different sizes are placed on both sides of the printed circuit board (PCB), shown in Fig. 4.1a. As the external magnetic field increases from 1,900 Oe to 2,300 Oe, the first pass-band frequency increases from 9.1 GHz to 10.2 GHz, and the second pass-band frequency increases from 9.5 GHz to 10.6 GHz. These results demonstrate that these structures have a magnetically tunable property.

By interacting with the magnetic field of an electromagnetic wave, ferromagnetic resonance can arise in ferrite under an applied magnetic field. The effective permeability can be expressed as [28]

$$\mu_{eff}(\omega) = 1 - \frac{F\omega_{mp}^2}{\omega^2 - \omega_{mp}^2 - i\Gamma(\omega)\omega} \tag{4.1}$$

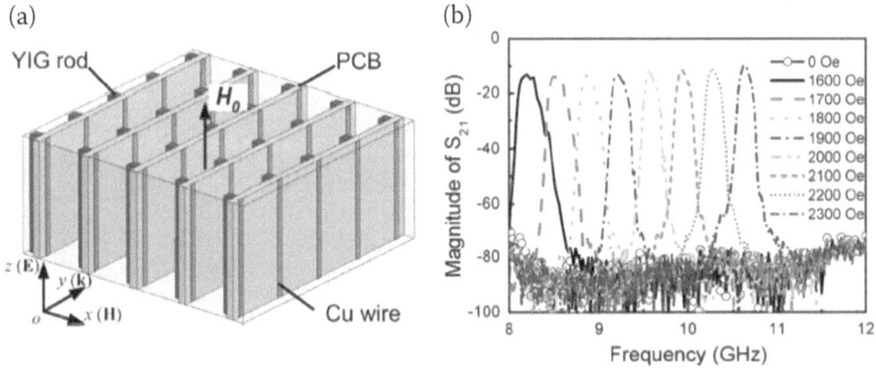

FIGURE 4.1 (a) Schematic of tunable left-handed material (LHM) consisting of YIG rods and copper wires, (b) measured transmission for the metamaterial under a series of applied magnetic fields [26].

$$\Gamma(\omega) = \left(\frac{\omega^2}{\omega_r + \omega_m} + \omega_r + \omega_m\right)\alpha \tag{4.2}$$

$$\omega_{mp} = \sqrt{\omega_r(\omega_r + \omega_m)} \tag{4.3}$$

where α is the damping coefficient of ferromagnetic precession, γ is the gyromagnetic ratio, $F = \omega_m/\omega_r$, $\omega_m = 4\pi M_s\gamma$ is the characteristic frequency of ferrite, and M_s is the saturation magnetization. The ferromagnetic resonance frequency can be expressed as

$$\omega_r = \gamma\sqrt{[H_0 + (N_x - N_z)4\pi M_s][H_0 + (N_y - N_z)4\pi M_s]} \tag{4.4}$$

where H_0 is the external magnetic field, and N_x, N_y and N_z are the demagnetization factors in the x-, y- and z-directions, respectively. From Eqs. (4.1) to (4.4), one can infer that the magnetic permeability of ferrite strongly depends on the resonance frequency. Due to the two different sizes of ferrite rods, two negative permeability regions will appear near the two ferromagnetic resonance frequencies. Also, two left-handed pass-bands will be formed when the metal wire array provides the negative permittivity.

The ferrite/metal wire structure is a typical structure of the left-handed material. This structure not only achieves the low-loss left-handed magnetically tunable characteristics, but also obtains simple construction and certain advantages in device miniaturization.

4.2.2 Structure of Ferrite Metamaterial Filter

Our group proposed a band-pass filter structure consisting of two YIG ferrite arrays with different saturation magnetizations [30]. The schematic diagrams of a

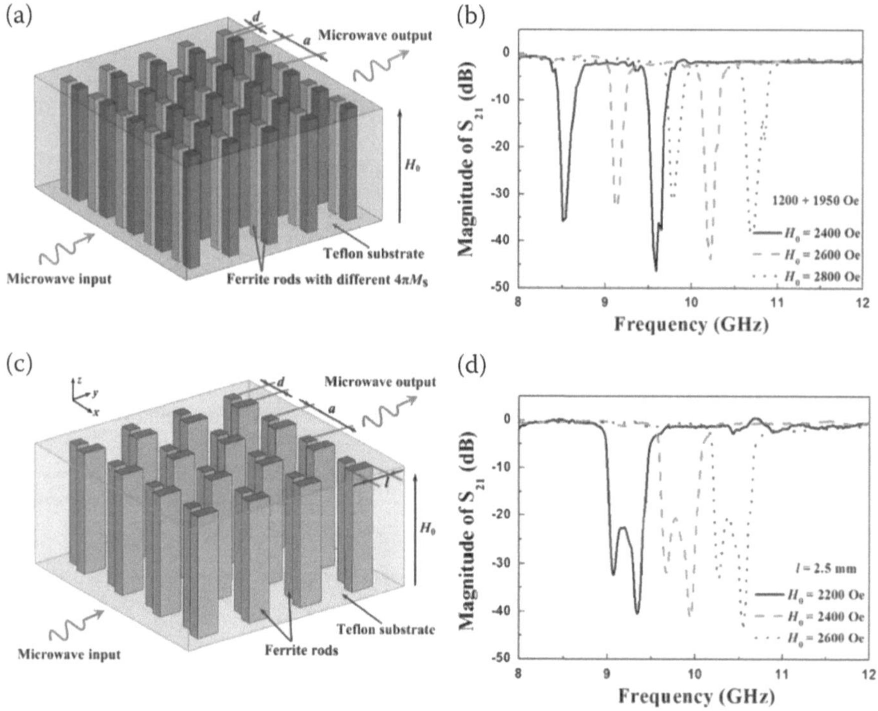

FIGURE 4.2 (a) Schematic diagram of the magnetically tunable microwave band-pass filter using ferrite-based metamaterial structure and (b) measured transmission spectra for the magnetically tunable microwave band-pass filter with a series of H_0 [30]. (c) Schematic diagram of the tunable microwave band-stop filter using ferrite-based metamaterial structure, (d) measured transmission spectra for the magnetically tunable microwave band-stop filter with a series of H_0 [31].

magnetically tunable microwave band-pass filter using ferrite-based metamaterial structure and transmission properties are shown in Figs. 4.2a and 4.2b, respectively. The saturation magnetizations of the two YIG rods are 1,200 Oe and 1,950 Oe, while other parameters are exactly the same. Eq. (4.4) shows that the ferrite ferromagnetic resonance frequency is not only affected by the external magnetic field, but also by the saturation magnetization. It increases as the saturation magnetization increases. Therefore, due to the different saturation magnetizations of the two kinds of ferrite rods, their ferromagnetic resonance frequencies are different and correspond to the two separate stop-bands in Fig. 4.2b, respectively. A pass-band with an insertion loss of about –2 dB is formed between the two stop-bands. The central frequency of this pass-band can be dynamically controlled by an external magnetic field.

Negative permeability can be generated by ferromagnetic resonance. Our group [31] also proposed a band-stop ferrite-based metamaterial filter using YIG rods with different sizes. As indicated schematically in Fig. 4.2c, this structure consists of two kinds of ferrite arrays that are only different in size. The measured transient response of the proposed device, as illustrated in Fig. 4.2d, shows that the structure has 500 MHz bandwidth (–3 dB) and –1.5 dB insertion loss. The central frequency of the stop-band increases as the external magnetic field increases, which indicates that the structure has magnetically tunable characteristics.

Ferrite-based metamaterial microwave band-stop and band-pass filters have a simple structure and excellent performance, which provide a new way to design microwave filters.

4.2.3 Ferrite/Dielectric Composite Structure

According to Bethe theory, electromagnetic waves cannot transmit through metal plates with sub-wavelength holes. If a structure can be designed to enable electromagnetic waves to transmit through the metal plates with sub-wavelength holes, it is expected to be applied in the microwave and optical devices. Based on the ferrite/dielectric composite structure, our group designed a dual-band magnetically tunable extraordinary transmission structure [32]. Fig. 4.3a shows the schematic diagram of the magnetically tunable filter based on extraordinary transmission. The transmission spectra are shown in Fig. 4.3b. Two pairs of

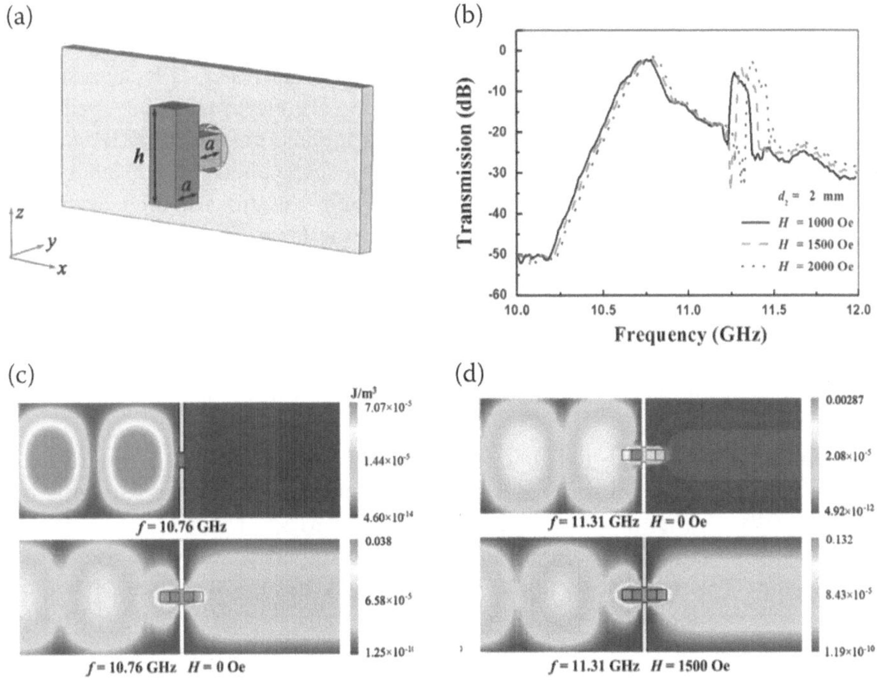

FIGURE 4.3 (a) Schematic diagram of the dual-band magnetically tunable filter based on extraordinary transmission, (b) measured transmission spectra of the proposed filter with a series of H, (c) simulated electric energy density distribution in the xy-plane at 10.76 GHz and (d) simulated electric energy density distributions in the xy-plane at 11.31 GHz [32].

dielectric cubes and ferrite rods are placed symmetrically on both sides of the sub-wavelength hole. The first pass-band is caused by the Mie resonance of dielectric cubes, and the second pass-band is caused by the ferromagnetic resonance of ferrite rods. The pass-band generated by the Mie resonance cannot be changed by the external magnetic field, while the pass-band generated by the ferromagnetic resonance can be dynamically adjusted by the external magnetic field. When the external magnetic field reaches 2,000 Oe, the peak values of the two transmission bands reach −1.3 dB and −2 dB.

Figs. 4.3c and 4.3d show the process of electromagnetic waves transmitted through the sub-wavelength hole. Without the ferrite and dielectric cubes, the electromagnetic waves cannot pass through the metal plate. When ferrite rods and dielectric cubes are placed on both sides of the sub-wavelength hole and the external magnetic field is 0 Oe, the electromagnetic wave with a frequency of 10.76 GHz can transmit through the metal plate due to the Mie resonance caused by dielectric cube, which achieves extraordinary transmission. When the external magnetic field is 1,500 Oe, ferromagnetic resonance takes place in ferrite, and the extraordinary transmission is also achieved at the ferromagnetic resonance frequency.

In addition, Wang et al. [33] realized a nonreciprocal Fano-resonant metamolecule structure with magnetically biased ferrite and dielectric cubes. A dielectric cube is placed above the center of ferrite cuboid. In the case of a single ferrite cuboid, there is no obvious resonance dip in the electromagnetic spectrum of ferrite. For a single dielectric cube, strong resonance is generated at 14.17 GHz, corresponding to the second-order Mie resonance mode. The electromagnetic spectrum of the proposed metamolecule shows an asymmetric Fano dip at 13.66 GHz and a weaker mode at 14.05 GHz. The Fano-type spectral shape almost disappears in the S_{12} curve of metamolecule, indicating that this system has strong nonreciprocity at the Fano dip. Besides, the resonance frequency of the Fano dip shows a blue shift with the increase of the external magnetic field, which reflects the magnetically tunable characteristics.

The ferrite-dielectric-based tunable metamaterials can achieve dual-band characteristics by utilizing ferromagnetic resonance of ferrites and Mie resonance of dielectrics. This design idea is hopeful to be applied in modulators, isolators and other devices.

Huang et al. [34] reported that ferrite can be integrated into a traditional passive metamaterial absorber (MA) as a substrate or superstrate. The authors systematically analyzed two ferrite-based tunable MAs. The metal resonance ring array is etched on one side of the FR4 printed circuit board, while the metal ground plane covers another side. In the first MA design, the ferrite layer is inserted between the FR4 layer and ground plane, and the ferrite and FR4 layer together act as a substrate. In the second MA design, the ferrite is covered on the FR4 layer, and another FR4 spacer is placed between the ferrite and the passive MA to avoid the interaction between the ferrite and metal resonators. The experimental and simulated results show that these two MAs exhibit a distinct absorption peak when the magnetic field is 0 Oe. As the magnetic field strength increases gradually, the absorption peaks of both MAs move toward the high-frequency region, and the frequency shift rates of the two MAs are 0.36 MHz/Oe and 0.18 MHz/Oe, respectively. These results confirm that the proposed ferrite-based MAs have magnetically tunable properties.

The design of a metal/ferrite/metal structure-based MA is also studied in Li et al. [35]. Unlike ordinary metal/dielectric/metal structures, this structure obtains an adjustable absorption peak within an ultra-wide frequency band due to the introduction of ferrite. The simulated results show that the corresponding absorption frequency increases from 0.2 GHz to 7.6 GHz when the external magnetic field increases from 10 Oe to 2,600 Oe, and the absorption peaks are above 0.9. The experimental results further validate the absorbing performance of the proposed structure. As the external magnetic field increases from 600 Oe to 1,000 Oe, the frequency of absorption peak increases from 2.2 GHz to 3.2 GHz.

Recently, a broadband ferrite-based tunable MA was reported [36], with eight U-shaped ferrite particles and a metal ground plate. There are four absorption peaks in the range of 8.55–10 GHz due to the magnetic and dielectric losses. These four peaks result in 1.5 GHz absorption bandwidth. The simulated and measured results both demonstrate that the resonance frequency of the proposed metamaterial changes from 8 GHz to 10 GHz when the applied magnetic field increases from 2,500 Oe to 3,000 Oe, which illustrates the magnetically tunable property of the proposed MA. Our group also designed a magnetically tunable ferrite-based metamaterial perfect absorber structure [37]. This structure is composed of a ferrite rod array and a metal plate. Both simulations and experiments confirm that the absorption peak generated by ferromagnetic resonance occurs in the frequency band of 8–12 GHz when a magnetic field bias is applied. When the magnetic field is 2,000 Oe, the absorptivity reaches 99.2%. As the magnetic field increases from 2,000 Oe to 2,400 Oe, the absorption peak shifts from 8.97 GHz to 10.02 GHz, showing a magnetically tunable behavior.

4.3 ELECTRICALLY TUNABLE DIELECTRIC METAMATERIALS

Due to the compatibility with electronic information technology, electrically tunable metamaterials have attracted extensive concern in the fields of materials and electronics. According to the different materials, this section mainly introduces three kinds of electrically tunable metamaterials: graphene-based, varactor diode-based, and liquid crystal-based metamaterials.

4.3.1 GRAPHENE

Recently, graphene has attracted the attention of researchers because of its excellent properties in optics, electricity and mechanics. An electrically tunable polarizer based on the periodic array of graphene ribbons was proposed by Zhu et al. [38]. The simulation results indicate that the TM-polarization absorption is only 0.75%, and the TE-polarization absorption rate can reach 99.86% under normal incidence excitation. The absorption of TM-polarization is insensitive to Fermi energy. On the contrary, the TE-polarization has a strong dependence on Fermi energy. With the increase of Fermi energy, the absorption frequency increases rapidly.

Another variation of the electrically tunable metasurface with graphene-based optical antennas was proposed by Yao et al. [39], in which the metasurface structure can be incorporated into a sub-wavelength-thick optical cavity to produce an electrically tunable perfect absorber. The absorber can switch the critical coupling state through the gate voltage applied on the graphene to realize the modulation depth of up to 100%. Moreover, the ultrathin (thickness < $\lambda_0/10$) high-speed (up to 20 GHz) optical modulators within a wide wavelength range (5–7 μm) are realized. As shown in Fig. 4.4a, the proposed structure is composed of a metal film, a dielectric layer and a graphene layer with plasmonic structure. Figs. 4.4b and 4.4c show the simulated and measured reflection spectra. According to the electric transmission characteristics of graphene samples, the charge neutral point $V_{CNP} = 0$ V (where the concentration of electrons and holes in graphene are the same) and carrier mobility that is used to obtain the simulated reflectance can be determined. With the gate voltage tuned away from the charge neutral point, the charge carrier concentration in the graphene sheet increases, and

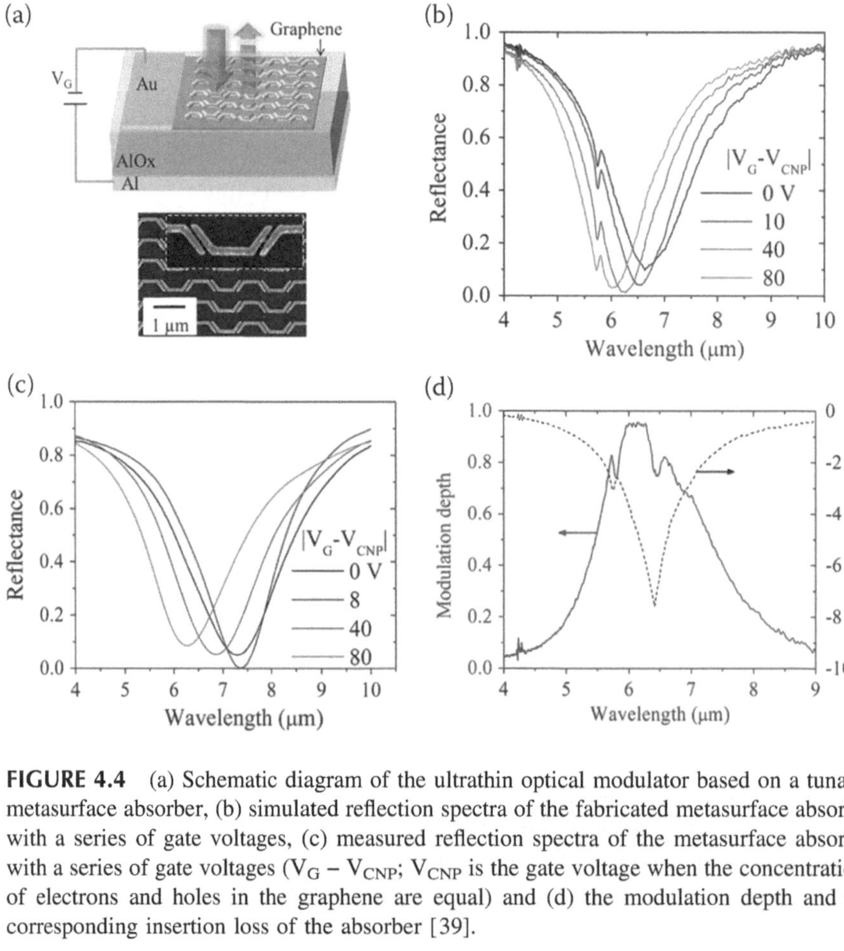

FIGURE 4.4 (a) Schematic diagram of the ultrathin optical modulator based on a tunable metasurface absorber, (b) simulated reflection spectra of the fabricated metasurface absorber with a series of gate voltages, (c) measured reflection spectra of the metasurface absorber with a series of gate voltages ($V_G - V_{CNP}$; V_{CNP} is the gate voltage when the concentrations of electrons and holes in the graphene are equal) and (d) the modulation depth and the corresponding insertion loss of the absorber [39].

the resonance frequency of metasurface shows a blue shift, as shown in Fig. 4.4c. When the gate voltage increases to 40 V, the measured minimum reflection value appears near 6.3 μm and then increases with the further increase of the grid voltage. Fig. 4.4d shows that the modulation depth is more than 95% at 6 μm and more than 50% within the wide wavelength range of 5.4 μm to 7.3 μm.

To design a dynamically electrically tunable broadband metasurface absorber, the graphene analog of electromagnetically induced transparency (EIT) can be employed [40]. This proposed structure is composed of graphene plasmonic analogs, a metal plate and an SiO_2 isolation layer. The simulated results show that the Fermi energy level of graphene can be changed by adjusting the electrostatic gating, thus dynamically tuning the operating bandwidth between broadband absorption and narrow-band absorption. Wang et al. [41] also proposed a low-cost plasmonic metasurface integrated with a single graphene layer for dynamic modulation of mid-infrared light. The plasmonic metasurface consists of a split magnetic resonator (MR) array. The extraordinary optical transmission (EOT) was observed by exciting the magnetic plasmas in the split MRs. Also, the introduction of the split can provide enhanced fields. Changing the Fermi energy of graphene can tune the EOT. Ulteriorly, the metasurface in Zhang et al. [42] can realize an independent dual-tunable property in the microwave band. The geometry and dimensions of the metasurface are shown in Fig. 4.5a. With the help of varactor diodes and graphene-based structure, the dynamically independent regulation of the magnitude and resonance frequency of the absorption peak is achieved. Fig. 4.5b shows the simulated

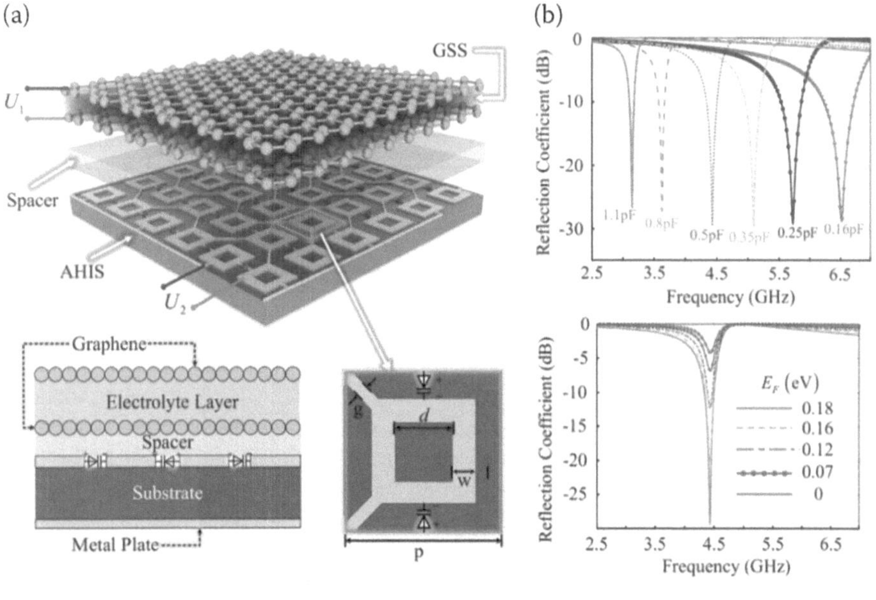

FIGURE 4.5 (a) Schematic diagram of the proposed graphene-based tunable metasurface, and (b) simulated reflection spectra of the proposed tunable metasurface with a series of capacitances and Fermi energies [42].

reflectance of metasurface with various capacitances and Fermi energies. The Fermi energy of graphene and the capacitance of varactor diodes can be controlled by biases voltage, thus realizing the working frequency change from 3.41 GHz to 4.55 GHz and the reflection amplitude variation from –3 dB to –30 dB.

The graphene-based electronic tunable metamaterials have extremely high modulation depth and width, which are expected to be employed in optical modulation and optical switch.

4.3.2 Varactor Diode

Varactor diode is a semiconductor device based on the principle of variable capacitance between PN junctions. An electrically tunable metamaterial is presented in [43]. In this design, varactor diodes are loaded on the split of square loops. The proposed absorber has a compact planar structure with a simple feeding network. It has been experimentally demonstrated that this absorber obtains flexible frequency tunability in the range of 5–6 GHz. As the voltage increases from 0 V to 12 V, the corresponding resonance frequency shifts from 5.18 GHz to 5.68 GHz. When the voltage is higher than 10 V, the change of reflection spectra is slight, because the capacitance of varactor diode is almost unchanged with the further increase of applied voltage. Besides, the reflection coefficient becomes smaller (from –8.90 dB to –19.57 dB) when the capacitance is reduced by raising the applied voltage, which means a high-quality factor of the absorbers is realized.

Fu et al. [44] theoretically and experimentally studied the electrically tunable Fano-type resonance phenomenon of asymmetric metal wire pairs loaded with varactor diodes. The Fano-type transmission spectrum with high-quality factor Q is caused by the interference between the dipole and quadrupole modes. The ohmic loss of series resistance in varactor diodes plays a major role in absorption. At the Fano-type resonance frequency, the two wires synchronously show the strongest electric resonance, and Fano-type resonance exhibits a large group delay. When the polarization voltage changes from 0 V to 8 V, the Fano-type resonance frequency shifts from 3.11 GHz to 3.27 GHz, showing a blue shift of 0.16 GHz. Meanwhile, the lower and upper resonance frequencies, respectively, show a blue shift of 0.19 GHz and 0.23 GHz.

To realize multi-frequency electromagnetic modulation, an electrically tunable metamaterial was proposed and demonstrated in [45]. This structure includes a single-warped wire and a paired-warped wire, in which a PIN diode is loaded in the gap of the paired wires. The EIT-like spectrum is manipulated by controlling the electronic mode of the single wire and the magnetic mode of the paired wires. The electromagnetic coupling is realized by adjusting the switch state and the resonance intensity of magnetic mode. The experimental results illustrate the electromagnetic modulation in three narrow bands on the EIT-like spectrum, and the modulation contrast of up to 31 dB is realized by using metamaterials.

The metasurface in [46] consists of an asymmetric ABA geometry structure with two different PIN diodes, and it can achieve a dynamic switch between perfect

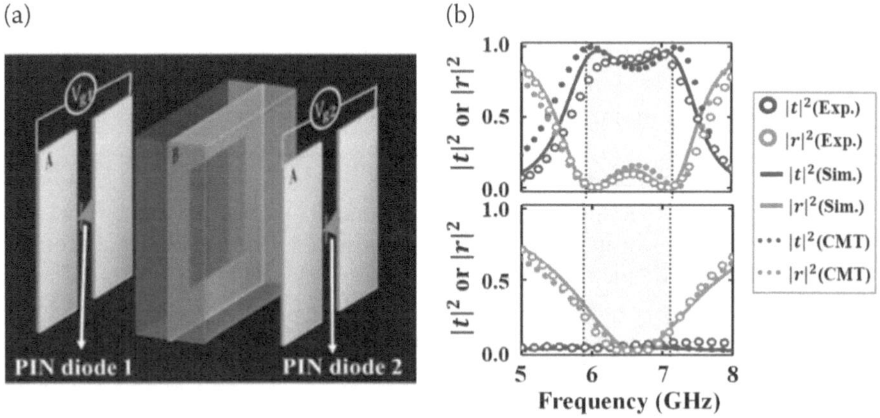

FIGURE 4.6 (a) Schematic diagram of the ABA geometry structure, and (b) transmittance (blue) and reflectance (red) with different bias voltages [46].

transparency and perfect absorption (as indicated in Fig. 4.6a). As illustrated in Fig. 4.6b, this structure can realize two perfect transmission peaks in a range of 5.8–7.4 GHz when the bias voltages of PIN diodes are set to 0 V. By employing the different bias voltages of the two PIN diodes, the intrinsic connections between transmission and reflection of the proposed structure are broken, thus resulting in perfect absorption with the transparency window. Based on the coupled-mode theory (CMT) and phase diagram, Li et al. proposed a dynamically tunable metasurface. All the results demonstrate that the transmission and absorption properties of the trilayer metasurface can be dynamically controlled by the asymmetrical bias voltages of the loaded PIN diodes.

The introduction of varactor diodes in the design of metamaterials realizes the supernormal electric field modulation of metamaterials. This design idea provides for the practical application of MAs and metamaterial modulators in microwave circuits.

4.3.3 Liquid Crystal

Liquid crystal is a kind of interstate substance between the crystalline and liquid states. When liquid crystal is powered on, its arrangement becomes orderly; when it is powered off, its arrangement is disordered, thus showing special physical properties. In 2007, Zhao et al. [47] brought a new vision to the problem of electrical tunability by introducing the nematic liquid crystal (NLC) into metamaterials. In this design, the split-ring resonator (SRR) array is immersed in NLC. With the increase of electric field strength, the arrangement of NLC tends to be orderly, and the frequency ranges with negative permeability move to low frequency.

Recently, by electronically controlling the NLC orientation, Kowerdziej also achieved the tunability of rod-split-square resonator-based metamaterial at THz frequency [48]. The non-polarized THz electromagnetic wave, acting as the probe light, is usually incident from the SRR array side to the metamaterial device. To align the liquid crystal molecules, an electric field with a modulation frequency of 1 kHz is applied. The experimental results show that the electromagnetic properties of metamaterial devices can be effectively adjusted (the transmittance changes up to 19%) by tuning the amplitude of AC (alternating current) bias voltage from 0 V to 300 V.

According to Isić et al. [49], an electrically tunable THz MA based on NLC is also proposed, as shown in Fig. 4.7a. The proposed structure includes a ground electrode carrying a liquid crystal cell capped by a polymer film. This film is used to support the top electrode and package the liquid crystal in the device. The gold layer deposited on the polymer film is etched into a triangular lattice pattern of circular patches connected by thin wires, as shown in Fig. 4.7b. Without the applied bias voltage, the NLC molecules are along with the horizontal orientation of thin polymer deposited on the top and bottom electrodes. Therefore, the permittivity along the z-axis is equal to ε_0. When the voltage is increased, the liquid crystal molecules become aligned with the z-axis, and the permittivity increases to ε_e, resulting in a red shift of resonance frequency, as shown in Fig. 4.7c. Fig. 4.7d shows that the reflectance at the target frequency f_0 increases from below 0.4% (when $U = 0$ V) to above 93% (when $U = 7$ V), corresponding to a modulation depth of more than 23 dB. The reflection spectra of incident waves with p-polarization and s-polarization (electric field orientates along x-axis and y-axis, respectively) are almost the same, which indicates that the device is insensitive to polarization.

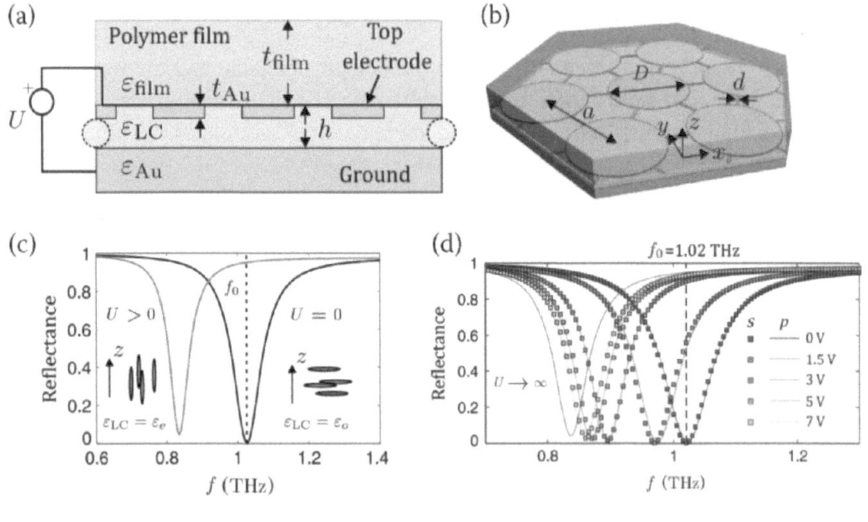

FIGURE 4.7 (a) Schematic diagram of the proposed tunable absorber (cross-section), (b) perspective view of the hexagonal unit cell, (c) modulation of the reflectance spectra by U, calculated assuming $\varepsilon_{LC} = \varepsilon_0, \varepsilon_e$ for $U = 0$ and $U > 0$, respectively, and (d) reflectance spectra with a series of bias U for s- and p-polarized incident waves [49].

Another variation of tunable THz MA with anisotropic liquid crystal was proposed by Hokmabadi et al. [50], in which the complementary split-ring resonator (CSRR) was used as a unit cell. The liquid crystal was filled between the CSRR and the Cu ground, so etching and a supportive dielectric layer were unnecessary. The results showed that, as the applied bias voltage increased from 0 V to 5 V, the resonance frequency obtained a frequency shift of 5 GHz, while the full-width at half-maximum (FWHM) bandwidth and the absorption rate remained at 0.025 THz and 90%, respectively. By applying the uniaxial model to simulate the absorption spectrum of liquid crystal simulated with various bias voltages, the effective refractive index of liquid crystal changed between 1.5 and 1.7. The design of a new type of heterojunction with zero-index metamaterial (ZIM) and liquid crystal was also studied in [51]. Its refractive index depended on the applied electric field intensity. By controlling the direct current (DC) voltage applied to the anisotropic dielectrics, electrically tuning perfect transmission and nearly perfect reflection were achieved. The structure can be used as an electrically adjustable switch in a waveguide system, which has good adjustable performance and high sensitivity.

In addition to the above three types of electrically tunable metamaterials, Li et al. [52] designed an electrically tunable superconducting niobium nitride (NbN) metamaterial device and analyzed its optical transmission characteristics with a hybrid coupling model. The maximum transmission coefficient at 0.507 THz was 0.98 when the temperature was 4.5 K. Upon increasing the applied voltage to 0.9 V, the maximum transmission coefficient decreased to 0.19. The relative transmission change of 80.6% made the device an effective narrow-band THz switch. Moreover, the peak frequency changed from 0.507 THz to 0.425 THz, which meant that this device could be applied in frequency selection. To realize fast modulation, the previous structure was optimized, and an electrically tunable superconducting metamaterial, which could dynamically modulate THz waves, was proposed in [53]. The metamaterial unit cell consisted of a square ring resonator and an SRR. The row of SRRs was connected to the electrodes on both sides of the chip through a continuous NbN line. The maximum modulation depth in the transmission window of the device was 79.8%. Controlled by the external sinusoidal signal, the device could achieve a modulation speed of about 1 MHz at 0.345 THz.

4.4 THERMALLY TUNABLE DIELECTRIC METAMATERIALS

In addition to magnetic field and electric field, temperature field is also one of the external fields commonly used to adjust the performance of materials and devices – a promising way to realize tunable metamaterials.

4.4.1 Vanadium Dioxide

Vanadium dioxide (VO_2) is a metal oxide with phase transition properties. Its low transformation temperature (341 K) makes it widely applied in optical, electronic and photoelectric devices. Wang et al. [54] proposed a thermally tunable infrared metamaterial based on VO_2. Numerical simulations show that, when VO_2 is in a metallic state, there is a wide absorption peak at the wavelength of 10.9 μm, but when it assumes a dielectric phase at the phase transition temperature of 341 K, the wavelength changes to 15.1 μm, illustrating a large tunability (38.5%) of the resonance wavelength.

The physical mechanism of magnetic resonance is associated with the plasma in the metallic phase VO_2 and the optical phonons in the dielectric phase VO_2. In [55], a hybrid MA was obtained by using the VO_2 film. The proposed structure included an electric split resonator ring (eSRR) array. In each eSRR unit cell, one inner resonator was embedded in the outer resonator pairs. The VO_2 film was only placed between the inner resonators and substrate. The measured phase transition temperature of the VO_2 film was about 340 K. In the room temperature condition, the absorption peaks appeared at 9.36 GHz and 18.6 GHz, and the maximum absorption rates were 87.0% and 93.0%, respectively. The absorption characteristics changed significantly when the device temperature increased from room temperature to 345 K. As for low-frequency response, the peak absorption rate slightly decreased from 87.0% at 9.36 GHz to 71.7% at 9.98 GHz. At high frequencies, the maximum absorption rate was significantly reduced from 93.0% at 18.6 GHz to 39.4% at 19.1 GHz. Therefore, in the VO_2-based microwave MA, a relative amplitude modulation of about 57.6% of the microwave absorption was achieved. Considering the frequency shift, the modulation depth of this MA could reach 63.3% at 18.6 GHz.

To design a thermally tunable MA, a VO_2 ground plate can be employed [56]. The unit cell of metamaterial consists of a three-layer structure: Au layer at the top, ZnS dielectric layer in the middle and VO_2 ground plate at the bottom. When the VO_2 ground plate is in a low-temperature dielectric phase, the composite structure forms a metal/dielectric/dielectric three-layer resonator. In this case, the metamaterial acts as a frequency selective surface. The VO_2 will turn to the metal phase when the ground is heated to 341 K. In this case, the metamaterial forms a typical MA structure. The gold layer on the top and the VO_2 layer on the bottom can simultaneously maintain electrical resonance mode and magnetic resonance mode. At 22.5 THz, the metamaterial switches from a high-reflective state at a low temperature (below 341 K) to a low reflection state at high temperature (above 341 K), while VO_2 changes from the insulating phase to the metallic phase. The reflectivity of metamaterial structure is about 35% at 313 K, which is higher than the reflectivity of around 8% at 353 K. At 34 THz, it switches from a low reflectance state at low temperature ($R \approx 15\%$) to a high reflectance state at high temperature ($R \approx 55\%$). Obviously, the combination of VO_2 and metamaterials has realized the thermally tunable characteristics of metamaterials. This research provides a new idea for the dynamic thermal radiation regulation of new electronic, optical and thermal devices.

For broadband property, Lei et al. [57] proposed a thermally tunable metasurface absorber consisting of chromium (Cr) top caps, VO_2 spacers and a Cr film substrate. As shown in Fig. 4.8a, each unit cell in the metasurface has a different size. The losses as a function of frequency at different temperatures are presented in Fig. 4.8b. When the VO_2 is under an insulating state, a relative bandwidth with 90% absorption is 97% (from 1627–4696 nm). The unit cells with different sizes can generate multiple resonant modes. The combination of these modes realizes the

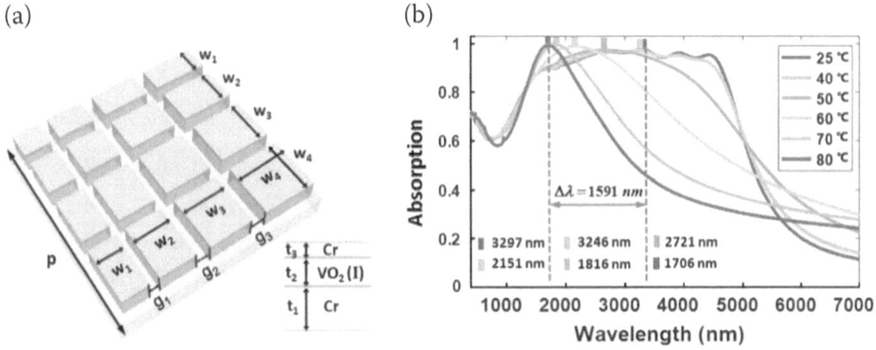

FIGURE 4.8 (a) Schematic diagram of the metamaterial ultra-broadband absorber with VO_2 spacer, and (b) measured absorption curves of the VO_2-based MA with a series of device temperatures [57].

broadband property of the absorber at room temperatures from 298 K to 313 K. When VO_2 is under a metallic state with the temperature increase, the frequency of absorption peak shifts from 3,297 nm to 1,706 nm, and its 90% absorption bandwidth decreases below 20%. Except for the traditional applications in thermally tunable devices, this design has the potential to be applied in smart nanostructures or devices.

Recently, utilizing "hybrid structures" integrated with phase-change materials, our group completed two designs of thermally tunable silicon-based all-dielectric metasurfaces (SAMs) [58]. In these designs, a $BaTiO_3$ film or a VO_2 film was deposited on the fabricated SAM. According to the results, it can be conjectured that the permittivity of $BaTiO_3$ and VO_2 films increases with external temperature, thus resulting in red shifts of the resonance frequency. Besides, the degeneration of resonance strength is caused by the increase of dielectric loss, which is related to the imaginary part of the permittivity. To prepare large-scale, high-precision and flexible all-dielectric metamaterials, we proposed a micro-template-assisted self-assembly method, in which heat-shrinkable substrate is applied to adjust the distance between microsphere resonators [59]. Under the heat treatment, these resonators will be closely packed, and a broadband reflection of the proposed metamaterial is then achieved. Obviously, the electromagnetic properties of the proposed metamaterial are affected by the distance between these resonators. Thus, this method has a potential to design various high-efficiency THz devices with thermally tunable characteristics.

4.4.2 Indium Antimonide

Zhu et al. [60] proposed a thermally tunable metamaterial that consisted of a metal SRR array with semiconductor indium antimonide (InSb). Because the permittivity of InSb has a strong dependence on temperature, the resonance frequency of metamaterials can be continuously modulated in the THz region by increasing temperature. The resonance frequency has a large blue shift of about 65%.

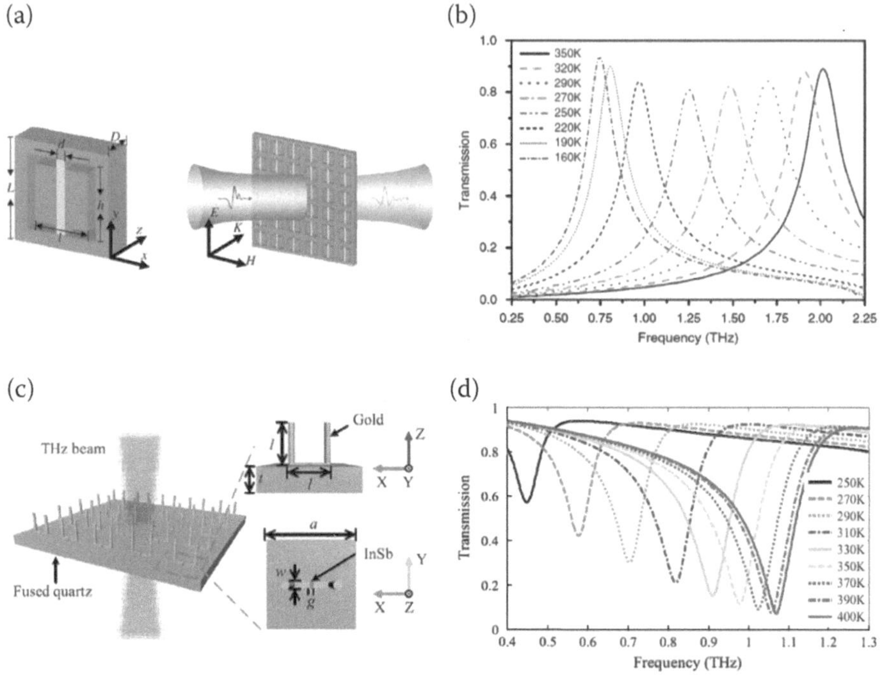

FIGURE 4.9 (a) Schematic diagram of the device, (b) simulated transmission spectra of the device at various temperatures with the FDTD method [61], (c) schematic illustration of the proposed metamaterials and (d) THz wave transmission spectra at various temperatures [62].

With the InSb semiconductor strips embedded in the sub-wavelength metallic holes array, a thermally tunable metamaterial filter is reported [61]. Its structure and dimensions are listed clearly in Fig. 4.9a. According to the simulated absorption spectra shown in Fig. 4.9b, it can be inferred that the resonance frequency of the filter can be adjusted by controlling the temperature of InSb. At 160 K, InSb shows typical dielectric features. The maximum transmission peak can reach 91.0%, and the FWHM is about 235 GHz at 0.74 THz. When the temperature is 290 K, InSb exhibits typical metallic features. The transmission peak value shifts to 1.71 THz, and the maximum value decreases by 84.5%. As the temperature further increases to 350 K, the transmission peak value shifts sharply to 2.02 THz, with a maximum value of 89.1%.

Li et al. [62] theoretically investigated the broadband thermally tunable metamaterials with a negative refractive index in the THz region. Two vertical L-shaped metallic structures, as depicted in Fig. 4.9c, are fused on the quartz substrate, and InSb is inserted in the bottom gap of the two L-shaped metals. The THz absorption spectra of metamaterials are shown in Fig. 4.9d at the corresponding temperatures. As the temperature changes, the capacitance of the gap changes with the InSb properties. When the temperature increases from 250 K to 400 K, the resonance frequency changes from 0.45 THz to 1.07 THz, and the transmission amplitude at the resonance frequency reduces from 57% to 7%.

Besides, when the temperature rises to 400 K, the metamaterials show negative refraction in the ranges of 0.4–0.9 THz and 1.06–1.15 THz.

In the THz band, the complex permittivity of InSb can be approximately described by the simple Drude model when the temperature changes from 300 K to 350 K [63]:

$$\varepsilon(\omega) = \varepsilon_\infty - \omega_p^2/(\omega^2 + i\gamma\omega) \tag{4.5}$$

where ω is the angular frequency, ε_∞ represents the high-frequency dielectric constant, γ is the damping constant, and the plasma frequency $\omega_P = (Ne^2/\varepsilon_0 m^*)^{1/2}$, where N is the carrier density, m^* is the effective mass, e is the electronic charge, and ε_0 is the vacuum permittivity. The damping constant γ ($\gamma = e/m^*\mu$) of InSb is inversely proportional to the electron mobility μ that is affected by the temperature. Therefore, γ will change with temperature, thus affecting the permittivity of InSb. However, the electron mobility changes barely within the frequency range of 0.1–1.5 THz and the temperature range of 300–350 K. Therefore, it is reasonable to neglect the influence of temperature on γ. In addition, the plasma frequency ω_P of InSb strongly depends on the temperature T. When the temperature is between 300 K and 350 K, the energy gap of InSb changes barely with the temperature, and the carrier density N (cm^{-3}) in InSb can be expressed as [64]:

$$N = 5.76 \times 10^{14} T^{3/2} \exp(-0.26/2\kappa_B T) \tag{4.6}$$

where K_B is the Boltzmann constant. The temperature will change the carrier concentration N and then change the ω_P. Therefore, in the far-infrared part of the THz region, the $\varepsilon(\omega)$ of InSb is very sensitive to temperature.

4.4.3 Strontium Titanate

In 2008, Zhao et al. [23] proposed a thermally tunable dielectric metamaterial based on the temperature sensitivity of dielectric ceramic material barium strontium titanate ($Ba_{1-x}Sr_xTiO_3$) doped with MgO. As the temperature increases from 258 K to 308 K, the frequency of Mie resonance shifts from 13.65 GHz to 19.28 GHz. Similar to barium strontium titanate, strontium titanate ($SrTiO_3$, STO) obtains typical perovskite structure with high permittivity and low loss. It is a kind of electronic functional ceramic material with wide applications, and its temperature sensitivity can be improved by employing a reasonable doping process.

In recent years, Luo et al. [65] designed a thermally tunable dual-band THz MA based on STO. The unit cell of metamaterial is composed of two nested closed square ring resonators, an STO dielectric substrate and a metal layer. The results show that the absorber has two obvious absorption peaks at 0.096 THz and 0.137 THz, with the peak values of 97% and 75%, respectively. Because the permittivity of the STO substrate changes with temperature, the resonance frequency shifts about 25% and 27% as the absorber is cooled from 400 K to 250 K. The tunability is more than 53% when the temperature decreases to 150 K.

Further, Luo et al. [66] cut off the four corners of the outer rings of the two nested square rings, realizing the tri-band absorption. The results show that the absorber has three unique absorption peaks in the frequency range of 0.05–0.35 THz at the frequencies of 0.129 THz, 0.198 THz and 0.316 THz with the peak values of 99.3%, 99.1% and 94.6%, respectively. By increasing the temperature, the resonance frequency of the absorber can be continuously adjusted in the THz region. When the temperature decreases from 400 K to 200 K, the frequency tuning depth at room temperature can reach 67.3%.

Zhao et al. [67] proposed a Si/STO composite all-dielectric metamaterial and experimentally demonstrated the thermally tunable properties of the electromagnetic parameters. The schematic diagram and a fabricated sample are shown in Figs. 4.10a and 4.10c. As depicted in Fig. 4.10b, the simulation results show that the first and second resonance frequencies of the Si/STO all-dielectric metamaterial are lower than the one without STO (red arrow mark). As the dielectric constant of STO decreases with an increase in temperature, the first resonance frequency changes

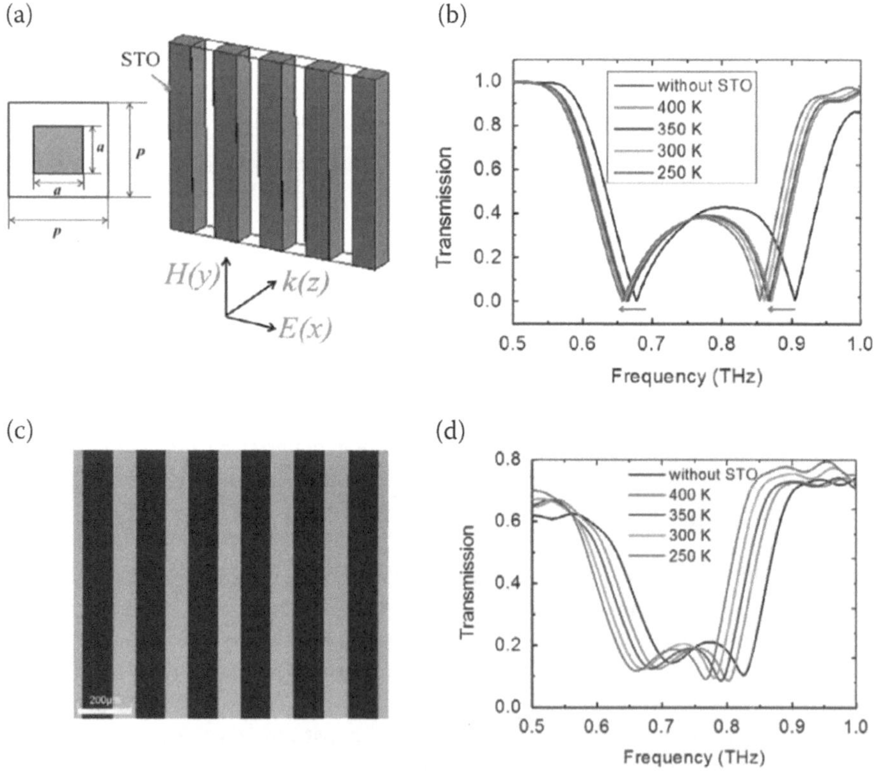

FIGURE 4.10 (a) Schematic diagram of tunable silicon/STO all-dielectric metamaterial, (b) transmission spectra of the tunable silicon/STO all-dielectric metamaterial with a series of temperatures, (c) photograph of the fabricated SAMs and (d) measured transmission spectrum of the tunable silicon/STO all-dielectric metamaterial with a series of temperatures [67].

from 0.657 THz to 0.665 THz, and the second resonance frequency changes from 0.769 THz to 0.801 THz when the temperature increases from 250 K to 400 K. The experimental results illustrated in Fig. 4.10d show that, when the temperature increases from 250 K to 400 K, the first resonance frequency changes from 0.662 THz to 0.695 THz, and the second resonance frequency changes from 0.769 THz to 0.801 THz. The experimental results are consistent with the simulation results. The proposed metamaterial structure has a thermally tunable performance.

Dielectric metamaterials based on the dielectric sensitivity of STO exhibit excellent performance in THz absorber designs. In addition, due to its good thermal tunability and low loss, it is expected to be applied in tunable THz detecting and sensing devices.

4.5 FLEXIBLE METAMATERIALS

Nowadays, there is an increasing interest in the development of flexible devices due to their potential applications in displays, personal electronics, artificial skins, as well as in medical treatment [68–72]. During the past two decades, various flexible devices have been designed and fabricated on different kinds of substrates [73–79].

4.5.1 Polyimide

Compared with metamaterials fabricated on silicon, metamaterials with polyimides substrate have excellent advantages of high sensitivity, low cost, conformable adhesion and flexibility. Therefore, polyimides are usually employed in the designs of tunable metamaterials. In 2009, Melik et al. proposed a flexible metamaterial that consisted of an SRR array layer, a dielectric layer, an Au layer and a polyimide substrate for wireless strain sensing with great sensitivity and low nonlinear error [80]. Its structure and fabricated sample are depicted clearly in Fig. 4.11a. Due to the low elastic modulus and great mechanical deformation of the polyimide substrate, the presented metamaterial has high sensitivity that is beneficial for responding to the small frequency shifts in the transmission spectra. As illustrated in Fig. 4.11b, the flexible metamaterial sensor exhibits a 0.292 MHz/kgf sensitivity and an associated 3% nonlinearity-error, when the applied load F changes from 27 kgf to 252 kgf. Compared with the silicon-based sensor, the proposed flexible metamaterial sensor has an enormous promotion in sensitivity and nonlinearity errors.

Recently, dynamically tunable THz metamaterials have garnered considerable interest due to their exotic electromagnetic characteristics. By transferring an SRR array with patterned GaAs patches to a thin polyimide layer, Fan et al. [81] realized the design of optically tunable THz metamaterials. The electromagnetic response of the presented metamaterial is affected by the optical excitation of GaAs. According to the differential transmission of metamaterial, the modulation depth at the LC resonance frequency of 0.98 THz varies from 20% to 60% as the pump power increases from 0.25 mW to 8mW. Moreover, it can achieve a similar modulation depth in a broad range of 1.1–1.8 THz. This design has great potential in nonlinear and multifunctional applications of sensors, modulators and perfect absorbers.

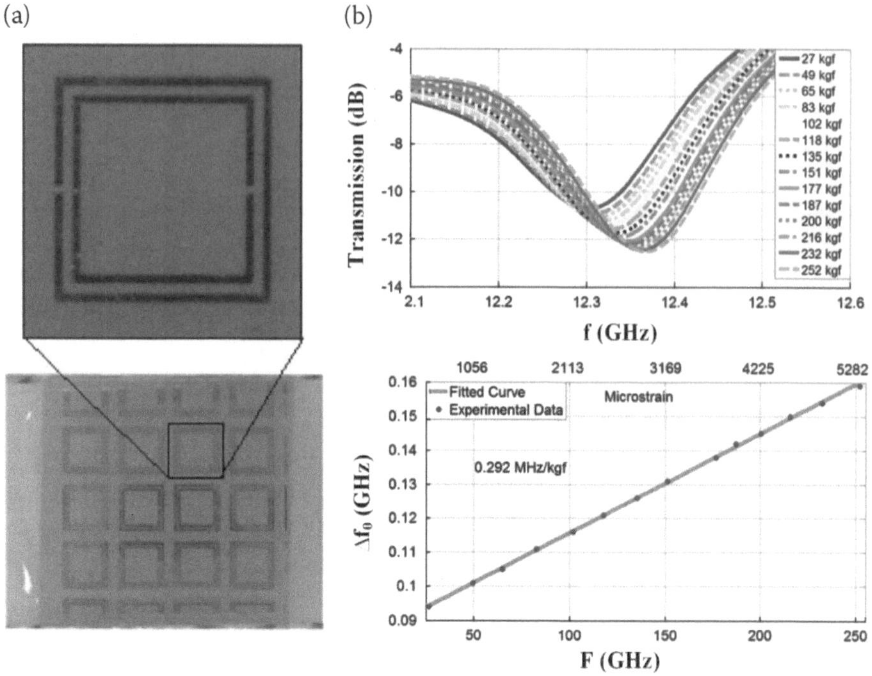

FIGURE 4.11 (a) Fabricated structure of the proposed sensor, and (b) transmission spectra of the flexible metamaterial sensor with different external force and the corresponding nonlinearity errors [80].

4.5.2 Ecoflex

Ecoflex is a highly stretchable silicone elastomer and exhibits good performance in skin safe, tear strength and elongation break. Therefore, Ecoflex is suitable for the design of tunable metamaterials. Using liquid metal encased in Ecoflex, Liu et al. [82] proposed a tunable meta-atom in the X-band frequency range to provide a structural basis for reconfigurable metamaterials. The meta-atom includes a liquid metal SRR and the external flexible elastomer (Ecoflex). Figs. 4.12a and 4.12b illustrate the measured and simulated transmission spectra of the meta-atom with different stretching ratios. With normal incidence, the resonance frequency of the meta-atom shifts from 10.54 GHz to 7.67 GHz, as the stretch ratio alters from 17% to 72%. In the case of another measurement configuration (see the inset of Fig. 4.12b), the resonance frequencies also exhibit obvious red shifts when the stretch level change from 17% to 72%. Three kinds of measurement configurations can excite resonance. Similar tunable characteristics can be observed in the last measurement configuration. During the changing of surrounding elastomer, the liquid metal remains continuous, and the tunable range of resonance frequency can cover 70% of the whole X-band. These features can provide a simple and effective structure for tunable metamaterials and obtain potential applications in wearable and cloaking devices.

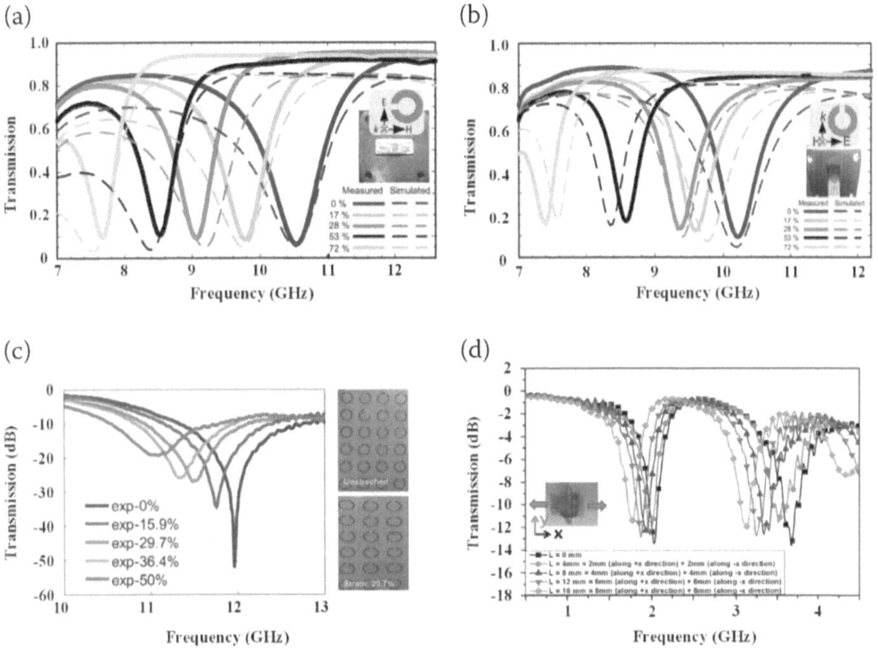

FIGURE 4.12 Measured and simulated transmission spectra of the tunable liquid metal meta-atom at various stretching ratios with (a) normal incidence, and (b) grazing incidence [82]. (c) Simulated transmission spectra of the meta-skin with various stretch ratios [83]. (d) Measured transmission spectra of the proposed sensor with different stretched lengths [84].

Based on the same structure, Yang et al. [83] further presented a wearable microwave meta-skin with tunable and cloaking effects. The above meta-atoms are used as unit cells of the meta-skin. These meta-skin consists of six metamaterial layers with 15 × 15 unit cells. Like the second measurement configuration illustrated in Fig. 4.12b, the direction of magnetic field **H** is vertical to the SRR. Thus, the circulating current is stimulated in the SRR, and the magnetic resonance is generated. Obviously, the resonance frequency of meta-skin, as illustrated in Fig. 4.12c, changes with the stretch ratios. Besides, the resonance frequency also is modulated by changing the spacing between two neighboring layers, which extends the frequency tuning range of the meta-skin. Then, a dielectric nylon rod is wrapped by a single layer metamaterial to demonstrate tthe cloaking effect. The measured results verified that the proposed metamaterial can significantly reduce the scattering gain of the covered object in the range of 8–12 GHz, which is beneficial for the cloaking effect.

The SRR, as mentioned above, is a well-known structure in sensor designs. Eom et al. [84] employed this structure to design a stretchable sensor for detecting strain direction and intensity. This sensor uses coplanar waveguide (CPW) technology, and two complementary SRRs are etched on the ground plane. Utilizing the microfluidic channels, the liquid metal is injected into the high-stretchable Ecoflex elastomer to form patterned resonant structures. According to the equivalent circuit model of SRR, the resonance frequency can be determined by

$$f = \frac{1}{2\pi\sqrt{LC}} \tag{4.7}$$

Two SRRs with different sizes, thus, introduce two resonance peaks at 2.03 GHz and 3.6 GHz. The dual band is a distinguishing feature of this work. One resonance peak is mainly affected by the deformation of one SRR, which can provide an indicator to identify the direction of the applied strain.

Fig. 4.12d depicts the measured transmission spectra when the stretched length is the same in two sides of the sensor. From this picture, both of the two resonance frequencies have obvious red shifts with the increase of stretched length. The relationships between the frequency shift and stretching length are linear and fit well with the linear equations, providing precise detection of strain direction and intensity.

4.5.3 Polydimethylsiloxane (PDMS)

Due to the merits of nontoxic, optically clear and remarkable elasticity, polydimethylsiloxane (PDMS) is often used as a flexible substrate in THz and photonic devices. In 2009, Olcum et al. [85] designed an elastomeric metal grating to control the resonance condition of surface plasmon polaritons by tuning the periods. As the strain increases, the elastomeric grating is stretched, its period is increased, and the surface plasmon resonance wavelength has red shift. To realize broadband tunable metamaterial in the infrared band, Pryce et al. [86] adhered the U-shaped Au SRRs array on a PDMS substrate. In this design, the variation of period was negligible, and the deformation of metamaterial mainly resulted in changing the coupling between the unit cells, which affected the resonance frequency shift. Four kinds of metamaterials with different arrays were investigated: SRR-bar, asymmetric coupled SRR, square SRR and double SRR arrays. Not only were the stretched situations discussed, but the relaxed samples were also considered. The simulated and measured results both demonstrated that the resonance frequencies of the proposed metamaterials were tuned by the deformation.

With a complex Au nanorod array fabricated on a stretchable PDMS substrate, a reconfigurable metasurface was reported in [87]. As shown in Fig. 4.13a, the

(a)

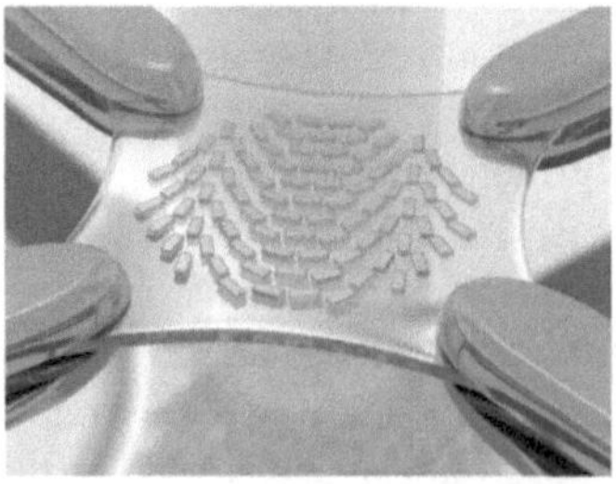

(b)

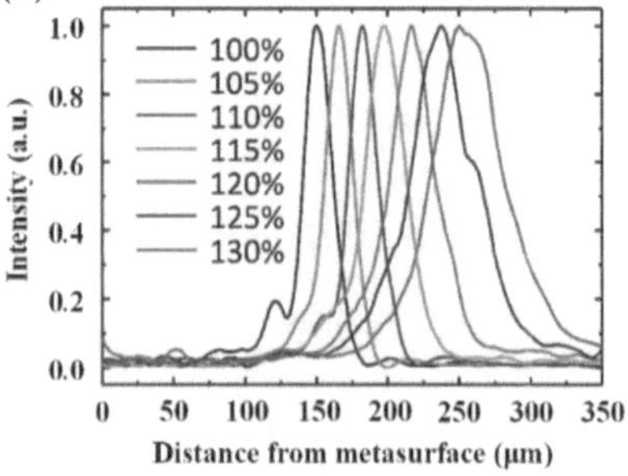

FIGURE 4.13 (a) Schematic illustration of the proposed metasurface. (b) Measured intensity distributions with various stretch ratios along the optical axis [87].

relative position of unit cells can be changed by stretching the flexible PDMS, thus continuously tuning the position-dependent phase discontinuity. This tuning mechanism was applied in the design of a reconfigurable optical device to adjust the anomalous refraction angle of the visible light. To demonstrate the practical application effect of the proposed metasurface in wavefront control, a flat zoom lens was fabricated on PDMS. Fig. 4.13b illustrates the experimental intensity distributions along the optical axis. As the PDMS is stretched from 100% to 130%, the focal length of lens can be continuously changed from 150 μm to 250 μm. Taking advantage of the shrink effect of PDMS, a tunable metamaterial with an adjustable narrowing gap from 140 nm to sub-10-nm was proposed by Liu ta al [88]. The gold disk arrays were fabricated on the PDMS substrate by interference lithography and gold deposition technology. With the precise control of gap distance between adjacent nanoparticles, the transmission spectra of metal metamaterial exhibit remarkable red shift. Moreover, the relationship between the gap distance and applied strain shows nearly perfect linearity. It is worthy to note that, due to the shrink effect, the gap distance is reduced with increasing applied strain.

Besides, Lee et al. [89] proposed two single-layer flexible metasurfaces that can be applied as a THz half-waveplate (HWP) and a quarter-waveplate (QWP). In this design, a rectangular ring resonator array was fabricated on a stretchable PDMS substrate to form the tunable flexible metasurface, as illustrated in Fig. 4.14a. Owing to the strong coupling between the adjacent resonators, the proposed metasurface obtains extraordinarily high anisotropic effective permittivity. The large phase retardation (180°) and resonance frequency of the incident beam can be achieved by deformation. As shown in Fig. 4.14b, the metasurface with small gap distance acts as an HWP, and the metasurface with large gap distance works as a QWP. In addition, these metasurfaces have spectral tunability.

Similar to the case of normal incidence discussed in 2015 [82], Liu et al. (2019) [90] presented an SRR-based meta-atom to realize biaxial tunability. The square metal SRRs composed of silver nanowires were transferred onto PDMS by using the patterned filtration method. To demonstrate the biaxial tunability, the meta-atom was placed inside a waveguide to introduce magnetic resonance. The simulated and measured transmission spectra of meta-atoms with different strain along the x- and y-directions are depicted in Figs. 4.14c and 4.14d, respectively. With the increase of applied strain along the x-direction, the resonance frequency gradually shifts from 11.927 GHz to 10.804 GHz. As the applied strain changes from 0% to 100%, the resonance frequency has red shift, which can cover 90% of the X-band. This design extended the tuning method and developed a novel preparation method for the fabrication of meta-atoms, and these meta-atom-based metamaterials benefit electromagnetic applications of wearable devices, sensors and cloaking.

Recently, all-dielectric metamaterial has attracted much interest due to its low intrinsic loss, high transmission, and high diffraction efficiency. For example, a flexible THz all-dielectric metamaterial reported in our research was designed by using the elastomeric medium, as indicated in Fig. 4.15a and 4.15b [91]. The proposed metamaterial is composed of a single-layer ceramic microspheres embedded in PDMS polymer. As the direction of applied strain is changed, the

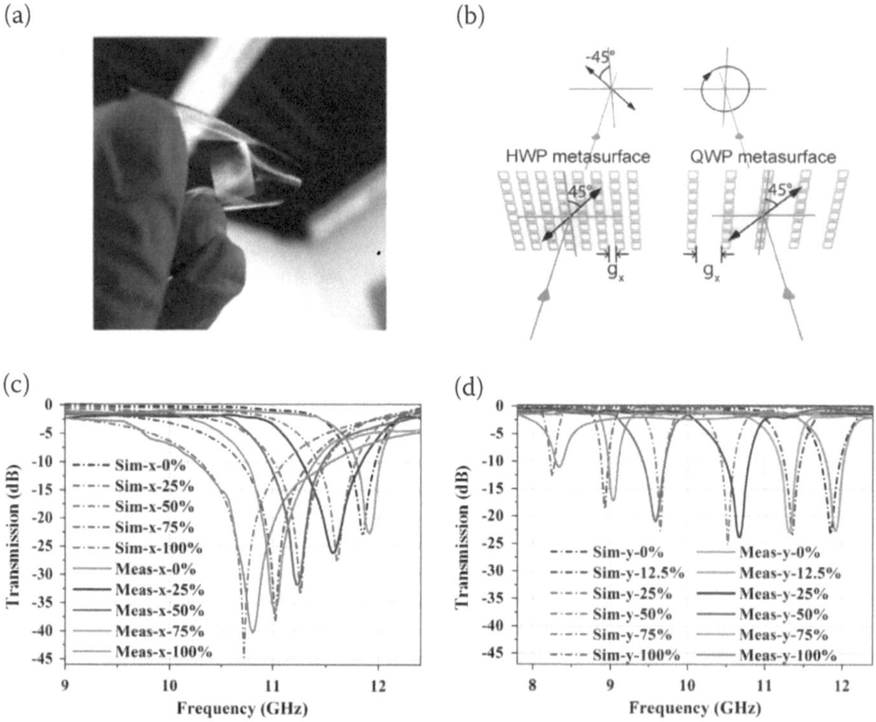

FIGURE 4.14 (a) Digital image of the proposed stretchable metasurface, (b) the metasurfaces works as an HWP and a QWP [89]. (c) Measured and simulated transmission spectra of the tunable meta-atom in the strain of 0–50% along the *x*-direction and (d) *y*-direction [90].

transverse and longitudinal coupling effects will be enhanced or reduced, which will make the magnetic resonance shift to higher or lower frequency. Particularly, a red shift of magnetic resonance will be achieved when the applied strain is along the electronic field, and a blue shift will emerge when the applied strain is along the magnetic field. This phenomenon can be explained with the dipole-dipole coupling theory. All the simulated and measured results demonstrate the tunability of the proposed metamaterial.

To solve the problem of polarization-dependence, a tunable all-dielectric metasurface with polarization insensitivity was presented by Zhang et al. [92]. As illustrated in Fig. 4.15c, the square TiO_2 array was embedded in an elastomeric PDMS substrate. In fact, the cross-section of square TiO_2 is an inverse trapezoid with a corner around 72°. In the full visible range of 450–650 nm, the TiO_2 metasurface stimulated by *x*-direction polarization beam or *y*-direction polarization beam reflects different colors when the strain applied in the same direction changes from 0% to 40%. As shown in Fig. 4.15d, the *x*-direction polarization and *y*-direction polarization both can achieve the same effect, which verifies the polarization-insensitive tunability of the proposed

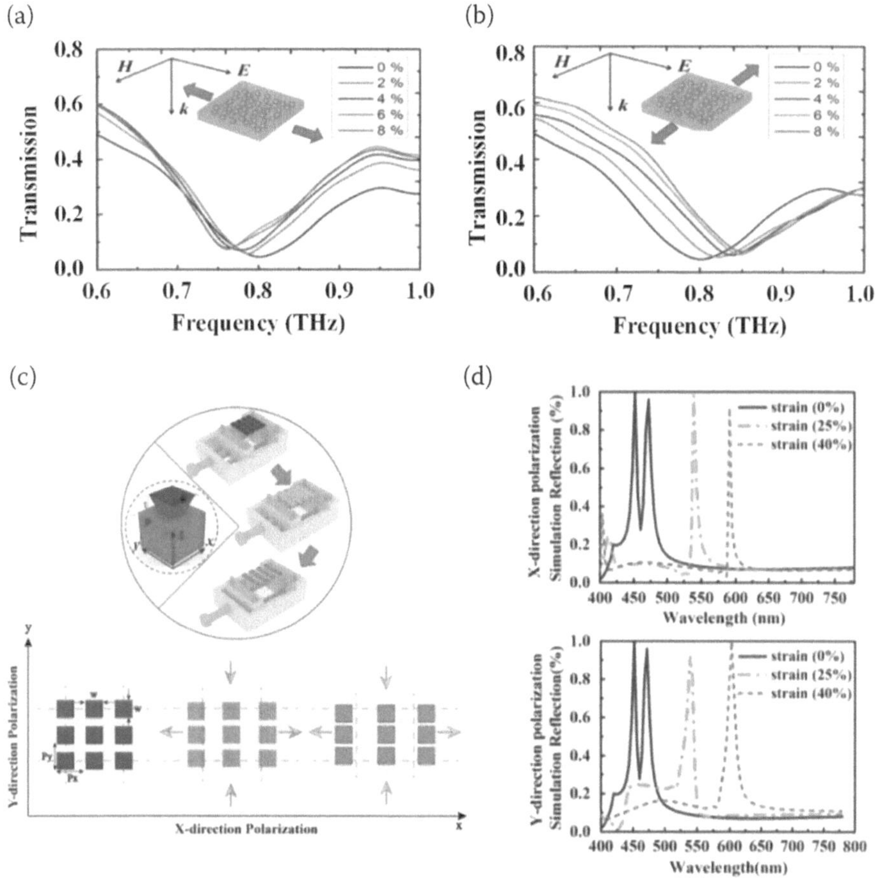

FIGURE 4.15 Transmission spectra of the proposed flexible metamaterial with (a) strain direction along the electric field, and (b) strain direction along the magnetic field [91]. (c) Schematic diagram of a tunable TiO_2-based metasurface, and (d) the simulated reflection spectra with different strain and polarizations [92].

stretchable TiO_2 metasurface. This design paves a way to polarization-insensitive full-spectrum color printing, and it has great application potential in functional color display, anticounterfeiting, security bank notes and point-of-care devices.

4.5.4 Paper

Recently, paper-based substrates have created an exciting avenue in the area of flexible devices' fabrication, as these are thin, light and inexpensive. Compared to other flexible substrates, the most attractive feature of paper-based substrates is their ability to be easily rolled or folded into 3D configurations to meet various design considerations for different applications.

In 2019, Zhang et al. from Shandong University reported an efficient pen-on-paper approach to construct metasurfaces on paper with silver units for electromagnetic shielding [93,94]. By using a roller-ball pen filled with conductive ink, Zhang et al. directly wrote metallic resonators on a paper substrate (Fig. 4.16a). Compared to metal-based shields, which usually have been limited by their high density, poor flexibility and limited tuning of shielding effectiveness, the paper-based metasurfaces not only exhibited remarkable electromagnetic shielding properties with a small thickness (thickness reduced more than 90% compared with that of traditional shielding materials), but also had a high mechanical bearing capacity, as well as lightweight characteristics, which made them ideal candidates for specific frequency electromagnetic filters.

It is worth pointing out that the paper used as substrate comes from waste bins. In Zhang's work, their paper-based metasurfaces can be fabricated even on a crumpled wastepaper, without hugely affecting the EM properties. The transmission coefficient (S_{21}) was measured and compared with that of its original uncrumpled shape at various frequencies, as shown in Fig. 4.16b. Although the absorption peak shows a slight shift toward higher frequency due to the

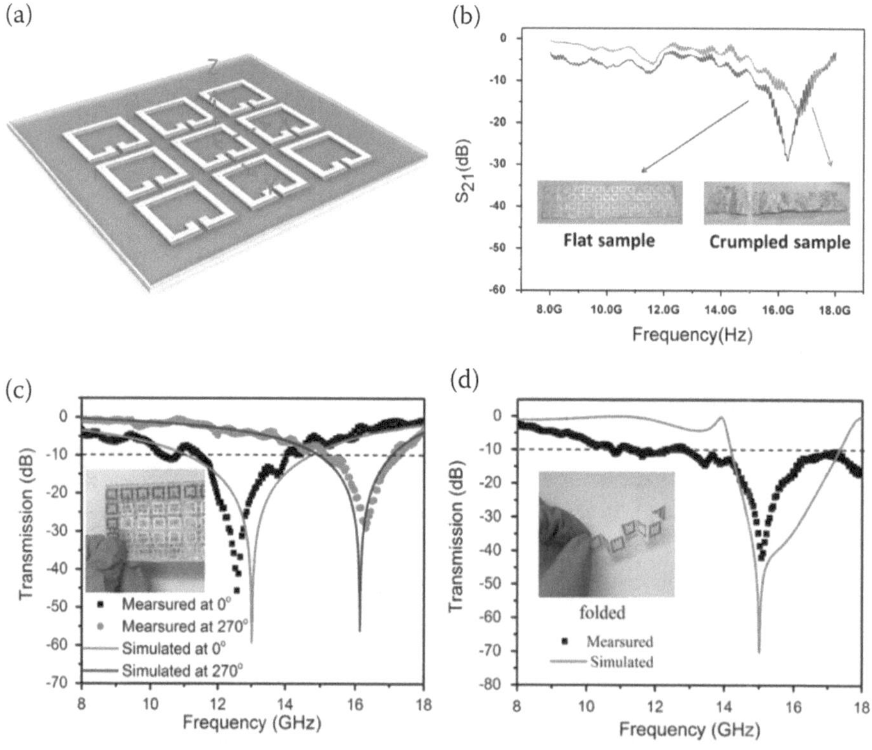

FIGURE 4.16 (a) Schematic diagram of paper-based metasurface. (b) Comparison of S_{21} between flat and crumpled printing paper-based metasurfaces. The transmission spectra of paper-based metasurfaces. (c) Unfolded sample, and (d) folded samples [93].

deformation of SRRs in a crumpled sample, the minimum S_{21} value of the absorption peak still remains at –20 dB, which indicates a good shielding performance even in the wrinkled state.

Moreover, both the shielding effectiveness and the shielding frequency range can be designed specifically by controlling the geometric parameters. The interesting part is that, due to the flexibility of substrate, the paper-based metasurface can be folded or piled up to form 3D multilayer stacking, which provides tunable EM properties for a wide variety of applications (Figs. 4.16c and 4.16d).

According to a recent report, households generate over 4 kg of paper waste every week, let alone the tremendous amount of wastepaper generated by businesses. Thus, the utilization of wastepaper as the substrate material for metasurface not only would be an effective solution to the problem regarding the disposal of paper waste, but also would create new opportunities for turning wastepaper into a solution for eliminating electromagnetic pollution.

Based on the above discussion, the environmentally friendly nature and pliability of paper, as well as the simple fabrication process, make paper-based metasurfaces promising candidates for future "green" electronics.

4.6 CONCLUSIONS

One of the main factors that restricts the application of electromagnetic metamaterials is bandwidth problems. The appearance of dielectric metamaterials not only broadens the scope of metamaterial designs and reduces the loss, but it also makes the physical properties of metamaterials tunable with the external field. The emergence of dielectric metamaterials solves the problem of narrow bandwidth of metamaterials to a certain extent, and it gives metamaterials new opportunities for application. Based on the existing literature, this chapter introduces the main tuning methods of tunable dielectric metamaterials and flexible metamaterials, focusing on the four different tuning mechanisms, namely magnetically, electrically and thermally tunable dielectric and flexible metamaterials. Besides, there are also several tuning methods, such as geometrically tunable and rotation angle tunable. However, compared to the magnetically, electrically and thermally tunable metamaterials, these tunable metamaterials are difficult to integrate with microwave devices. Therefore, we do not introduce these articles in detail. The design idea of tunable metamaterials, which is combined with the characteristics of natural materials, not only provides a new idea for the design of metamaterials, but also is expected to be widely used in microwave communication devices, such as filters, absorbers, modulators, and so on.

ACKNOWLEDGMENTS

This work was supported by the National Natural Science Foundation of China (Grant Nos. 51972033, 61774020, 61905021, and 51788104), Beijing Youth Top-Notch Talent Support Program, Science and Technology Plan of Shenzhen City (Grant No. JCYJ20180306173235924, JCYJ20180305164708625).

REFERENCES

1. M. Kadic, T. Buckmann, R. Schittny, and M. Wegener, "Metamaterials beyond electromagnetism," *Rep. Prog. Phys.*, vol. 76, p. 126501, 2013.
2. A. Poddubny, I. Iorsh, P. Belov, and Y. Kivshar, "Hyperbolic metamaterials," *Nat. Photonics.*, vol. 7, pp. 948–957, 2013.
3. G. Oliveri, D. H. Werner, and A. Massa, "Reconfigurable electromagnetics through metamaterials—A review," *P. IEEE*, vol. 103, pp. 1034–1056, 2015.
4. V. G. Veselago, "The electrodynamics of substances with simultaneously negative values of ε and μ," *Sov. Phys. Usp.*, vol. 10, pp. 509–514, 1968.
5. D. R. Smith, W. J. Padilla, D. C. Vier, S. C. Nemat-Nasser, and S. Schultz, "Composite medium with simultaneously negative permeability and permittivity," *Phys. Rev. Lett.*, vol. 84, pp. 4184–4187, 2000.
6. K. Bi, W. J. Liu, Y. S. Guo, G. Y. Dong, and M. Lei, "Magnetically tunable broadband transmission through a single small aperture," *Sci. Rep.*, vol. 5, p. 12489, 2015.
7. Y. X. Wang, Y. Qin, Z. Z. Sun, and P. Xu, "Magnetically controlled zero-index metamaterials based on ferrite at microwave frequencies," *J. Phys. D Appl. Phys.*, vol. 49, p. 405106, 2016.
8. X. Liu, L. Y. Xiong, X. Yu, S. L. He, B. Zhang, and J. L. Shen, "Magnetically controlled terahertz modulator based on Fe3O4 nanoparticle ferrofluids," *J. Phys. D Appl. Phys.*, vol. 51, p. 105003, 2018.
9. C. Y. Ye, Z. H. Zhu, W. Xu, X. D. Yuan, and S. Q. Qin, "Electrically tunable absorber based on nonstructured graphene," *J. Optics.*, vol. 17, p. 125009, 2015.
10. A. Komar, Z. Fang, J. Bohn, J. Sautter, M. Decker, A. Miroshnichenko, T. Pertsch, I. Brener, Y. S. Kivshar, and I. Staude, "Electrically tunable all-dielectric optical metasurfaces based on liquid crystals," *Appl. Phys. Lett.*, vol. 110, p. 071109, 2017.
11. O. Balci, N. Kakenov, E. Karademir, S. Balci, S. Cakmakyapan, E. O. Polat, H. Caglayan, E. Özbay, and C. Kocabas, "Electrically switchable metadevices via graphene," *Sci. Adv.*, vol. 4, p. eaao1749, 2018.
12. W. Lewandowski, M. Fruhnert, J. Mieczkowski, C. Rockstuhl, and E. Górecka, "Dynamically self-assembled silver nanoparticles as a thermally tunable metamaterial," *Nat. Commun.*, vol. 6, p. 6590, 2015.
13. R. Kowerdziej, M. Olifierczuk, and J. Parka, "Thermally induced tunability of a terahertz metamaterial by using a specially designed nematic liquid crystal mixture," *Opt. Express.*, vol. 26, pp. 2443–2452, 2018.
14. Y. S. Guo, H. Liang, X. J. Hou, X. L. Lv, L. F. Li, J. S. Li, K. Bi, M. Lei, and J. Zhou, "Thermally tunable enhanced transmission of microwaves through a subwavelength aperture by a dielectric metamaterial resonator," *Appl. Phys. Lett.*, vol. 108, p. 051906, 2016.
15. K. Bi, L. Y. Zeng, K. J. Chai, Z. X. Fan, L. J. R. Liu, Q. M. Wang, and M. Lei, "Magnetically tunable microwave bandpass filter structure composed of ferrite rods and metallic slits," *Appl. Phys. Lett.*, vol. 107, p. 064103, 2015.
16. H. J. Yang, T. L. Yu, Q. M. Wang, and M. Lei, "Wave manipulation with magnetically tunable metasurfaces," *Sci. Rep.*, vol. 7, p. 5441, 2017.
17. B. Du, Z. Xu, J. Wang, and S. Xia, "Magnetically tunable ferrite-dielectric left-handed metamaterial," *Prog. Electromagn. Res.*, vol. 66, pp. 21–28, 2016.
18. Y. Bai, K. J. Chen, T. Bu, and S. L. Zhuang, "An electrically tunable terahertz metamaterial modulator with two independent channels," *J. Appl. Phys.*, vol. 119, p. 124505, 2016.
19. W. Z. Xu, F. F. Ren, J. D. Ye, H. Lu, L. J. Liang, X. M. Huang, M. K. Liu, I. V. Shadrivov, D. A. Powell, and G. Yu, "Electrically tunable terahertz metamaterials

with embedded large-area transparent thin-film transistor arrays," *Sci. Rep.*, vol. 6, p. 23486, 2016.
20. D. J. Park, J. H. Shin, K. H. Park, and H. C. Ryu, "Electrically controllable THz asymmetric split-loop resonator with an outer square loop based on VO_2," *Opt. Express.*, vol. 26, pp. 17397–17406, 2018.
21. K. Bi, G. Y. Dong, X. J. Fu, and J. Zhou, "Ferrite based metamaterials with thermo-tunable negative refractive index," *Appl. Phys. Lett.*, vol. 103, p. 131915, 2013.
22. X. J. Fu, X. Q. Xi, K. Bi, and J. Zhou, "Temperature-dependent terahertz magnetic dipole radiation from antiferromagnetic $GdFeO_3$ ceramics," *Appl. Phys. Lett.*, vol. 103, p. 211108, 2013.
23. Q. Zhao, B. Du, L. Kang, H. J. Zhao, Q. Xie, B. Li, X. Zhang, J. Zhou, L. T. Li, and Y. G. Meng, "Tunable negative permeability in an isotropic dielectric composite," *Appl. Phys. Lett.*, vol. 92, p. 051106, 2008.
24. L. Q. Cong, N. N. Xu, J. Q. Gu, R. Singh, J. G. Han, and W. L. Zhang, "Highly flexible broadband terahertz metamaterial quarter-wave plate," *Laser Photonics Rev.*, vol. 8, pp. 626–632, 2014.
25. K. Bertoldi, V. Vitelli, J. Christensen, and M. Van Hecke, "Flexible mechanical metamaterials," *Nat. Rev. Mater.*, vol. 2, p. 1706, 2017.
26. H. J. Zhao, J. Zhou, L. Kang, and Q. Zhao, "Tunable two-dimensional left-handed material consisting of ferrite rods and metallic wires," *Opt. Express.*, vol. 17, pp. 13373–13380, 2009.
27. K. Bi, J. Zhou, H. J. Zhao, X. M. Liu, and C. W. Lan, "Tunable dual-band negative refractive index in ferrite-based metamaterials," *Opt. Express.*, vol. 21, pp. 10746–10752, 2013.
28. H. J. Zhao, J. Zhou, Q. Zhao, B. Li, L. Kang, and Y. Bai, "Magnetotunable left-handed material consisting of yttrium iron garnet slab and metallic wires," *Appl. Phys. Lett.*, vol. 91, p. 131107, 2007.
29. K. Bi, J. Zhou, X. M. Liu, C. W. Lan, and H. J. Zhao, "Multi-band negative refractive index in ferrite-based metamaterials," *Prog. Electromagn. Res.*, vol. 140, pp. 457–469, 2013.
30. K. Bi, W. T. Zhu, M. Lei, and J. Zhou, "Magnetically tunable wideband microwave filter using ferrite-based metamaterials," *Appl. Phys. Lett.*, vol. 106, p. 173507, 2015.
31. Q. M. Wang, L. Y. Zeng, M. Lei, and K. Bi, "Tunable metamaterial bandstop filter based on ferromagnetic resonance," *AIP Adv.*, vol. 5, p. 077145, 2015.
32. Q. M. Wang, K. Bi, and M. Lei, "Magnetically tunable dual-band transmission through a single subwavelength aperture," *Appl. Phys. Lett.*, vol. 106, p. 194102, 2015.
33. X. B. Wang, G. Q. Zhang, H. H. Li, and J. Zhou, "Magnetically tunable Fano resonance with enhanced nonreciprocity in a ferrite-dielectric metamolecule," *Appl. Phys. Lett.*, vol. 112, p. 174103, 2018.
34. Y. J. Huang, G. J. Wen, W. R. Zhu, J. Li, L. M. Si, and M. Premaratne, "Experimental demonstration of a magnetically tunable ferrite based metamaterial absorber," *Opt. Express.*, vol. 22, pp. 16408–16417, 2014.
35. W. Li, J. Wei, W. Wang, D. W. Hu, Y. K. Li, and J. G. Guan, "Ferrite-based metamaterial microwave absorber with absorption frequency magnetically tunable in a wide range," *Mater. Design.*, vol. 110, pp. 27–34, 2016.
36. W. J. Wang, C. L. Xu, M. B. Yan, A. Wang, J. Wang, M. D. Feng, J. F. Wang, and S. B. Qu, "Broadband tunable metamaterial absorber based on U-shaped ferrite structure," *IEEE Access*, vol. 7, pp. 150969–150975, 2019.
37. M. Lei, N. Y. Feng, Q. M. Wang, Y. N. Hao, S. G. Huang, and K. Bi, "Magnetically tunable metamaterial perfect absorber," *J. Appl. Phys.*, vol. 119, p. 244504, 2016.

38. Z. H. Zhu, C. C. Guo, K. Liu, J. F. Zhang, W. M. Ye, X. D. Yuan, and S. Q. Qin, "Electrically tunable polarizer based on anisotropic absorption of graphene ribbons," *Appl. Phys. A.*, vol. 114, pp. 1017–1021, 2014.
39. Y. Yao, R. Shankar, M. A. Kats, Y. Song, J. Kong, M. Loncar, and F. Capasso, "Electrically tunable metasurface perfect absorbers for ultrathin mid-infrared optical modulators," *Nano Lett.*, vol. 14, pp. 6526–6532, 2014.
40. G. Yao, F. R. Ling, J. Yue, C. Y. Luo, Q. Luo, and J. Q. Yao, "Dynamically electrically tunable broadband absorber based on graphene analog of electromagnetically induced transparency," *IEEE Photonics J.*, vol. 8, p. 7800808, 2016.
41. Z. P. Wang, Y. Deng, and L. F. Sun, "An electrically tunable metasurface integrated with graphene for mid-infrared light modulation," *Chinese Phys. B.*, vol. 26, p. 114101, 2017.
42. J. Zhang, X. Z. Wei, I. D. Rukhlenko, H. T. Chen, and W. R. Zhu, "Electrically tunable metasurface with independent frequency and amplitude modulations," *ACS Photonics.*, 7, pp. 265–271, 2019.
43. J. F. Zhu, D. L. Li, S. Yan, Y. J. Cai, Q. H. Liu, and T. Lin, "Tunable microwave metamaterial absorbers using varactor-loaded split loops," *Europhys. Lett.*, vol. 112, p. 54002, 2015.
44. Q. H. Fu, F. L. Zhang, Y. C. Fan, X. He, T. Qiao, and B. T. Kong, "Electrically tunable Fano-type resonance of an asymmetric metal wire pair," *Opt. Express.*, vol. 24, pp. 11708–11715, 2016.
45. Y. C. Fan, T. Qiao, F. L. Zhang, Q. H. Fu, J. J. Dong, B. T. Kong, and H. Q. Li, "An electromagnetic modulator based on electrically controllable metamaterial analogue to electromagnetically induced transparency," *Sci. Rep.*, vol. 7, p. 40441, 2017.
46. Y. Li, J. Lin, H. J. Guo, W. J. Sun, S. Y. Xiao, and L. Zhou, "A tunable metasurface with switchable functionalities: From perfect transparency to perfect absorption," *Adv. Opt. Mater.*, vol. 8, p. 1901548, 2020.
47. Q. Zhao, L. Kang, B. Du, B. Li, J. Zhou, H. Tang, X. Liang, and B. Z. Zhang, "Electrically tunable negative permeability metamaterials based on nematic liquid crystals," *Appl. Phys. Lett.*, vol. 90, p. 011112, 2007.
48. R. Kowerdziej, M. Olifierczuk, J. Parka, and J. Wrobel, "Terahertz characterization of tunable metamaterial based on electrically controlled nematic liquid crystal," *Appl. Phys. Lett.*, vol. 105, p. 022908, 2014.
49. G. Isić, B. Vasić, D. C. Zografopoulos, R. Beccherelli, and R. Gajić, "Electrically tunable critically coupled terahertz metamaterial absorber based on nematic liquid crystals," *Phys. Rev. Appl.*, vol. 3, p. 064007, 2015.
50. M. P. Hokmabadi, A. Tareki, E. Rivera, P. Kung, R. G. Lindquist, and S. M. Kim, "Investigation of tunable terahertz metamaterial perfect absorber with anisotropic dielectric liquid crystal," *AIP Adv.*, vol. 7, p. 015102, 2017.
51. Y. Y. Cao, Q. Q. Meng, and Y. D. Xu, "Electrically tunable electromagnetic switches based on zero-index metamaterials," *J. Optics.*, vol. 20, p. 025103, 2018.
52. C. Li, C. H. Zhang, G. L. Hu, G. C. Zhou, S. L. Jiang, C. T. Jiang, G. H. Zhu, B. B. Jin, L. Kang, and W. W. Xu, "Electrically tunable superconducting terahertz metamaterial with low insertion loss and high switchable ratios," *Appl. Phys. Lett.*, vol. 109, p. 022601, 2016.
53. C. Li, J. B. Wu, S. L. Jiang, R. F. Su, C. H. Zhang, C. T. Jiang, G. C. Zhou, B. B. Jin, L. Kang, and W. W. Xu, "Electrical dynamic modulation of THz radiation based on superconducting metamaterials," *Appl. Phys. Lett.*, vol. 111, p. 092601, 2017.
54. H. Wang, Y. Yang, and L. P. Wang, "Wavelength-tunable infrared metamaterial by tailoring magnetic resonance condition with VO2 phase transition," *J. Appl. Phys.*, vol. 116, p. 123503, 2014.

55. Q. Y. Wen, H. W. Zhang, Q. H. Yang, Z. Chen, Y. Long, Y. L. Jing, Y. Lin, and P. X. Zhang, "A tunable hybrid metamaterial absorber based on vanadium oxide films," *J. Phys. D Appl. Phys.*, vol. 45, p. 235106, 2012.
56. R. Naorem, G. Dayal, S. A. Ramakrishna, B. Rajeswaran, and A. Umarji, "Thermally switchable metamaterial absorber with a VO_2 ground plane," *Opt. Commun.*, vol. 346, pp. 154–157, 2015.
57. L. Lei, F. Lou, K. Y. Tao, H. X. Huang, X. Cheng, and P. Xu, "Tunable and scalable broadband metamaterial absorber involving VO_2-based phase transition," *Photonics Res.*, vol. 7, pp. 734–741, 2019.
58. C. W. Lan, H. Ma, M. T. Wang, Z. H. Gao, K. Liu, K. Bi, J. Zhou, and X. J. Xin, "Highly efficient active all-dielectric metasurfaces based on hybrid structures integrated with phase-change materials: From terahertz to optical ranges," *ACS Appl. Mater. Inter.*, vol. 11, pp. 14229–14238, 2019.
59. K. Bi, D. Q. Yang, J. Chen, Q. M. Wang, H. Y. Wu, C. W. Lan, and Y. P. Yang, "Experimental demonstration of ultra-large-scale terahertz all-dielectric metamaterials," *Photonics Res.*, vol. 7, pp. 457–463, 2019.
60. J. Zhu, J. G. Han, Z. Tian, J. Q. Gu, Z. Y. Chen, and W. L. Zhang, "Thermal broadband tunable terahertz metamaterials," *Opt. Commun.*, vol. 284, pp. 3129–3133, 2011.
61. W. Li, D. F. Kuang, F. Fan, S. J. Chang, and L. Lin, "Subwavelength B-shaped metallic hole array terahertz filter with InSb bar as thermally tunable structure," *Appl. Optics.*, vol. 51, pp. 7098–7102, 2012.
62. W. L. Li, Q. L. Meng, R. S. Huang, Z. Q. Zhong, and B. Zhang, "Thermally tunable broadband terahertz metamaterials with negative refractive index," *Opt. Commun.*, vol. 412, pp. 85–89, 2018.
63. S. T. Bui, X. K. Bui, T. T. Nguyen, P. Lievens, Y. P. Lee, and D. L. Vu, "Thermally tunable magnetic metamaterials at THz frequencies," *J. Optics.*, vol. 15, p. 075101, 2013.
64. J. G. Han, and A. Lakhtakia, "Semiconductor split-ring resonators for thermally tunable terahertz metamaterials," *J. Mod. Optic.*, vol. 56, pp. 554–557, 2009.
65. C. Y. Luo, Z. Z. Li, Z. H. Guo, J. Yue, Q. Luo, G. Yao, J. Ji, Y. K. Rao, R. K. Li, and D. Li, "Tunable metamaterial dual-band terahertz absorber," *Solid State Commun.*, vol. 222, pp. 32–36, 2015.
66. C. Y. Luo, D. Li, Q. Luo, J. Yue, P. Gao, J. Q. Yao, and F. R. Ling, "Design of a tunable multiband terahertz waves absorber," *J. Alloy. Compd.*, vol. 652, pp. 18–24, 2015.
67. Y. J. Zhao, B. W. Li, C. W. Lan, K. Bi, and Z. W. Qu, "Tunable silicon-based all-dielectric metamaterials with strontium titanate thin film in terahertz range," *Opt. Express.*, vol. 25, pp. 22158–22163, 2017.
68. Q. L. Li, J. Y. Zhang, Q. H. Li, G. H. Li, X. Y. Tian, Z. W. Luo, F. Qiao, X. Wu, and J. Zhang, "Review of printed electrodes for flexible devices," *Front. Mater.*, vol. 5, p. 77, 2019.
69. K. Wang, H. P. Wu, Y. N. Meng, Y. J. Zhang, and Z. X. Wei, "Integrated energy storage and electrochromic function in one flexible device: An energy storage smart window," *Energ. Environ. Sci.*, vol. 5, pp. 8384–8389, 2012.
70. H. L. Lv, Z. H. Yang, P. L. Wang, G. B. Ji, J. Z. Song, L. R. Zheng, H. B. Zeng, and Z. J. Xu, "A voltage-boosting strategy enabling a low-frequency, flexible electromagnetic wave absorption device," *Adv. Mater.*, vol. 30, p. 1706343, 2018.
71. L. Gao, "Flexible device applications of 2D semiconductors," *Small.*, vol. 13, p. 1603994, 2017.
72. K. Song, J. Kim, S. Cho, N. Kim, D. Jung, H. Choo, and J. Lee, "Flexible-device injector with a microflap array for subcutaneously implanting flexible medical electronics," *Adv. Healthc. Mater.*, vol. 7, p. 1800419, 2018.

73. L. Lin, Y. F. Hu, C. Xu, Y. Zhang, R. Zhang, X. N. Wen, and Z. L. Wang, "Transparent flexible nanogenerator as self-powered sensor for transportation monitoring," *Nano Energy*, vol. 2, pp. 75–81, 2013.
74. P. K. Singh, K. A. Korolev, M. N. Afsar, and S. Sonkusale, "Single and dual band 77/95/110 GHz metamaterial absorbers on flexible polyimide substrate," *Appl. Phys. Lett.*, vol. 99, p. 264101, 2011.
75. X. L. Xu, B. Peng, D. H. Li, J. Zhang, L. M. Wong, Q. Zhang, S. J. Wang, and Q. H. Xiong, "Flexible visible-infrared metamaterials and their applications in highly sensitive chemical and biological sensing," *Nano Lett.*, vol. 11, pp. 3232–3238, 2011.
76. T. Schmaltz, A. Y. Amin, A. Khassanov, T. Meyer-Friedrichsen, H. G. Steinrück, A. Magerl, J. J. Segura, K. Voitchovsky, F. Stellacci, and M. Halik, "Low-voltage self-assembled monolayer field-effect transistors on flexible substrates," *Adv. Mater.*, vol. 25, pp. 4511–4514, 2013.
77. K. Nomura, H. Ohta, A. Takagi, T. Kamiya, M. Hirano, and H. Hosono, "Room-temperature fabrication of transparent flexible thin-film transistors using amorphous oxide semiconductors," *Nature*, vol. 432, pp. 488–492, 2004.
78. J. Zhou, Y. D. Gu, P. Fei, W. J. Mai, Y. F. Gao, R. S. Yang, G. Bao, and Z. L. Wang, "Flexible piezotronic strain sensor," *Nano Lett.*, vol. 8, pp. 3035–3040, 2008.
79. Y. Zhang, X. Q. Yan, Y. Yang, Y. H. Huang, Q. L. Liao, and J. J. Qi, "Scanning probe study on the piezotronic effect in ZnO nanomaterials and nanodevices," *Adv. Mater.*, vol. 24, pp. 4647–4655, 2012.
80. R. Melik, E. Unal, N. K. Perkgoz, C. Puttlitz, and H. V. Demir, "Flexible metamaterials for wireless strain sensing," *Appl. Phys. Lett.*, vol. 95, p. 181105, 2009.
81. K. B. Fan, X. G. Zhao, J. D. Zhang, K. Geng, G. R. Keiser, H. R. Seren, G. D. Metcalfe, M. Wraback, X. Zhang, and R. D. Averitt, "Optically tunable terahertz metamaterials on highly flexible substrates," *IEEE Trans. Terahertz. Sci. Technol.*, vol. 3, pp. 702–708, 2013.
82. P. Liu, S. M. Yang, A. Jain, Q. G. Wang, H. W. Jiang, J. M. Song, T. Koschny, C. M. Soukoulis, and L. Dong, "Tunable meta-atom using liquid metal embedded in stretchable polymer," *J. Appl. Phys.*, vol. 118, p. 014504, 2015.
83. S. M. Yang, P. Liu, M. D. Yang, Q. G. Wang, J. M. Song, and L. Dong, "From flexible and stretchable meta-atom to metamaterial: A wearable microwave meta-skin with tunable frequency selective and cloaking effects," *Sci. Rep.*, vol. 6, p. 21921, 2016.
84. S. Eom and S. Lim, "Stretchable complementary split ring resonator (CSRR)-based radio frequency (RF) sensor for strain direction and level detection," *Sensors*, vol. 16, p. 1667, 2016.
85. S. Olcum, A. Kocabas, G. Ertas, A. Atalar, and A. Aydinli, "Tunable surface plasmon resonance on an elastomeric substrate," *Opt. Express.*, vol. 17, pp. 8542–8547, 2009.
86. I. M. Pryce, K. Aydin, Y. A. Kelaita, R. M. Briggs, and H. A. Atwater, "Highly strained compliant optical metamaterials with large frequency tunability," *Nano Lett.*, vol. 10, pp. 4222–4227, 2010.
87. H. S. Ee and R. Agarwal, "Tunable metasurface and flat optical zoom lens on a stretchable substrate," *Nano Lett*, vol. 16, pp. 2818–2823, 2016.
88. W. J. Liu, Y. Shen, G. H. Xiao, X. Y. She, J. F. Wang, and C. J. Jin, "Mechanically tunable sub-10-nm metal gap by stretching PDMS substrate," *Nanotechnology*, vol. 28, p. 075301, 2017.
89. S. Lee, W. T. Kim, J. H. Kang, B. J. Kang, F. Rotermund, and Q. H. Park, "Single-layer metasurfaces as spectrally tunable terahertz half-and quarter-waveplates," *ACS Appl. Mater. Interfaces.*, vol. 11, pp. 7655–7660, 2019.

90. C. Liu, J. Cai, X. H. Li, W. Q. Zhang, and D. Y. Zhang, "Flexible and tunable electromagnetic meta-atom based on silver nanowire networks," *Mater. Design.*, vol. 181, p. 107982, 2019.
91. C. W. Lan, K. Bi, B. W. Li, Y. J. Zhao, and Z. W. Qu, "Flexible all-dielectric metamaterials in terahertz range based on ceramic microsphere/PDMS composite," *Opt. Express.*, vol. 25, pp. 29155–29160, 2017.
92. C. Zhang, J. X. Jing, Y. K. Wu, Y. B. Fan, W. H. Yang, S. Wang, Q. H. Song, and S. M. Xiao, "Stretchable all-dielectric metasurfaces with polarization-insensitive and full-spectrum response," *ACS Nano.*, vol. 14, pp. 1418–1426, 2020.
93. Z. Y. Wang, X. Y. Fu, Z. D. Zhang, Y. L. Jiang, M. Waqar, P. T. Xie, K. Bi, Y. Liu, X. W. Yin, and R. H. Fan, "Based metasurface: Turning waste-paper into a solution for electromagnetic pollution," *J. Clean Prod.*, vol. 234, pp. 588–596, 2019.
94. X. Y. Fu, Z. D. Zhang, Y. L. Jiang, W. J. Zhang, K. Bi, K. Sun, and R. H. Fan, "Flexible 2.5 D metamaterial with high mechanical bearing capacity for electromagnetic interference filters at microwave frequency," *Adv. Eng. Mater.*, vol. 22, p. 1901126, 2019.

5 Metamaterials-based Near-perfect Absorbers in the Visible and Infrared Range

Kyungnam Kang, Seongmin Im, and Donghyun Kim
School of Electrical and Electronic Engineering, Yonsei University, Seoul, Republic of Korea

5.1 INTRODUCTION

Metamaterials (MTMs) are state-of-the-art artificial media that usually consist of metal and/or dielectric structures on a scale of sub-wavelength. MTMs demonstrate interesting electrodynamic properties and effects that are not found in natural materials, including negative refractive index [1,2], electromagnetic (EM) wave cloaking [3–6], inverse Doppler effect [7], superlensing [8], energy harvesting [9–11], plasmonic detection [12] and perfect absorption [13,14]. With the development of nanotechnology, MTMs can be fabricated to fit the design of different applications based on particular sizes and shapes, thereby allowing the possibility of implementing artificial optical resonance characteristics that would not have been found in nature.

In particular, near-perfect absorbers using MTMs can be utilized over diverse frequency bands, such as microwave [15,16], terahertz, near-infrared [17–23], and visible range [24,25]. Landy et al. first proposed a metamaterial perfect absorber (MMPA), which is composed of two split-ring resonators that are connected by an inductive ring parallel to the split-wire; it is highly absorptive over a narrow GHz frequency [14]. The work has been expanded to MMPA research in the diverse frequency regime, and various applications, such as solar cell [13,25], biological sensing [26,27], thermal imaging [28], photodetection [29,30] and absorption filtering [31]. Recently, advances in plasmonics have been combined with MTMs, which led to many novel MMPA structures operating at the visible and infrared wavelength regimes. Techniques based on the excitation of surface plasmon (SP) have been traditionally used for biosensing, often by way of SP polaritons (SPPs) and localized SP (LSP) [32–37]. Localization of SP was performed to produce super-localized light volume for biomolecular detection [38–40]. The use of SP

DOI: 10.1201/9781003050162-5

excitation may be extended for MTMs to produce enhanced effective absorption characteristics with a small target volume while ensuring design flexibility.

MMPAs can be classified into narrow-band, multiple-band, and broadband absorbers depending on the bandwidth of absorption characteristics. Because MMPAs consist basically of metal-dielectric-metal structures that may cause plasmonic resonance, which is responsible for the narrow response of absorption, MMPAs tend to be well customized for single bands [31,41,42]. Since applications, including solar cells and photodetection, often need broad absorption characteristics, MMPAs have often been developed as broadband absorbers, such as dual band [43,44], triple band [45] and more [46,47]. In the infrared and visible frequency range, however, the research on broadband perfect absorbers remains immature because the SPPs and LSPs generated with sub-wavelength metallic structures produce narrow-band characteristics. Although single narrow-band is designed rather easily and can be fabricated for perfect absorption, broadband perfect absorption requires a more complicated MTM structure and would, therefore, be difficult to design and fabricate.

Broadband MMPAs have been studied mainly with four techniques. The first is planar arrangement tailoring the size of metallic patterns on the top layer used to develop multiple resonances, as shown in Fig. 5.1a [48]. The second is vertical arrangement stacking multiple metal-dielectric top layers. This achieves broadband absorption, as shown in Fig. 5.1b [49]. The third is to add lumped circuit elements, such as resistors, capacitors and diodes, as shown in Fig. 5.1c [50]. The welded form accompanies various resonant modes connected together and reduces the quality factor. Fig. 5.1d shows the metal and dielectric nanocomposites that trap EM waves to be absorbed by metallic nanoparticles for excitation of LSPs [51].

In this chapter, we discuss theoretical backgrounds of MMPAs in Section 5.2. Performance of various MMPA structures designed for visible and infrared frequency regime is explored in Section 5.3. Narrow-band and broadband near-perfect absorbers can be applicable to various optical devices, such as sensors, solar cells, light-emitting diodes and spectroscopic elements: these possibilities are described in Section 5.4. Concluding remarks are provided in Section 5.5.

5.2 PRINCIPLES OF NEAR-PERFECT ABSORPTION USING MTMS

5.2.1 Theoretical Backgrounds

Conventional periodic structures, such as plasmonic crystal and grating, have been employed as perfect absorbers [52–56]. However, MMPAs, which consist of three components, i.e., metallic periodic arrays, a dielectric layer and a continuous metallic film, have been widely studied as promising structures of MTMs due to their high absorption, polarization insensitivity and design flexibility. First, metallic periodic arrays play an important role in MMPA design because structural parameters, such as periodicity, size, shape and thickness, can be adjusted to satisfy the impedance-matching condition under which zero reflection can be obtained at resonant frequencies. In contrast, a dielectric layer acts as a spacer where an incident EM wave can stay and be dissipated. The third continuous metallic film completely blocks all EM waves accompanying no transmission.

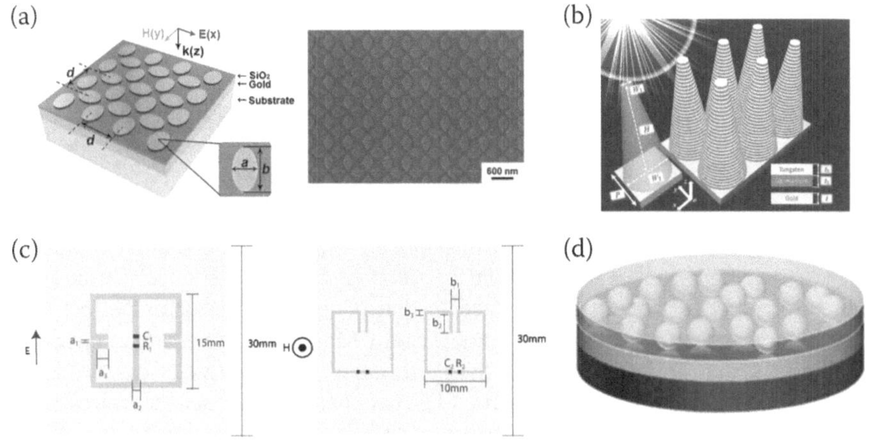

FIGURE 5.1 Conventional designs for broad MMPAs. (a) Planar arrangement. (Reprinted with permission from Zhang, B., Y. Zhao, Q. Hao, B. Kiraly, I.-C. Khoo, S. Chen, and T.J. Huang. 2011. "Polarization-independent dual-band infrared perfect absorber based on a metal-dielectric-metal elliptical nanodisk array." *Opt. Exp.* 19 (16):15221–15228. © The Optical Society). (b) Vertical arrangement. (Reprinted with permission from Lin, Y., Y. Cui, F. Ding, K.H. Fung, T. Ji, D. Li, and Y. Hao. 2017. "Tungsten based anisotropic metamaterial as an ultra-broadband absorber." *Opt. Mater. Exp.* 7 (2):606–617. © The Optical Society). (c) Welded with lumped elements. (Reprinted from Gu, S., J.P. Barrett, T.H. Hand, B.-I. Popa, and S.A. Cummer. 2010. "A broadband low-reflection metamaterial absorber." *J. Appl. Phys.* 108 (6):064913 with the permission of AIP Publishing). (d) Nanocomposites. (Reprinted under a Creative Commons license from Keshavarz Hedayati, M., M. Abdelaziz, C. Etrich, S. Homaeigohar, C. Rockstuhl, and M. Elbahri. 2016. "Broadband anti-reflective coating based on plasmonic nanocomposite." *Materials* 9 (8):636).

The absorption mechanism can be associated mainly with the dielectric loss of the second layer, or the Ohmic loss, caused by the top metallic periodic arrays. In the case of Ohmic loss, size and patterns on the top metallic layer can affect the bandwidth as a result of multiple resonances and cause broadening. At a lower frequency, metals have lower loss acting on a perfect conductor, i.e., the loss of MTMs is related mainly to the lossy dielectric layer [57]. On the other hand, at a higher frequency, absorption by MTMs is contributed predominantly by the metallic absorption than in the dielectric layer. For this reason, for MMPAs in the visible and the near-infrared regime, a relatively thin dielectric layer is sufficient, and a number of structural designs adopting top metallic periodic arrays for MTMs have been developed along this direction. The overall phenomena of perfect absorption can be described by various theories based on effective medium approximation [58,59], equivalent circuits [60] and interference [61].

A homogeneous absorbing medium with complex permittivity and permeability in contact with the free space (air) is the simplest model of a perfect absorber. When an EM wave is incident in the normal direction from the free space to the absorbing medium, the reflectance is given by

$$R = \left| \frac{Z(\omega) - Z_0(\omega)}{Z(\omega) + Z_0(\omega)} \right|^2,$$

where $Z(\omega) = \sqrt{\mu(\omega)\mu_0/\varepsilon(\omega)\varepsilon_0}$ is the impedance of the perfect absorber and $Z_0(\omega) = \sqrt{\mu_0/\varepsilon_0}$ is the free-space impedance. When we comprise the perfect absorber as an MTM to obtain zero reflectance, because a continuous metal film in the final MTM layer makes near-zero transmission, the impedance matching condition, $Z(\omega) = Z_0(\omega)$ and $\mu(\omega) = \varepsilon(\omega)$, should be satisfied [62]. In other words, the impedance of an MTM should be matched to that of the free space for perfect absorption. Therefore, the design and fabrication of MTMs for impedance matching are important factors for determining the performance of MMPAs.

Although the top metallic arrays for MTMs consist of distinct structures that are spatially inhomogeneous, MTMs may be regarded as homogeneous in a so-called effective medium approximation. The representative effective medium theories are Bruggeman's and Maxwell-Garnett's models. Bruggeman's model takes the condition as

$$\sum_{i=1}^{n} f_i \frac{\varepsilon_i - \varepsilon_{eff}}{\varepsilon_i + 2\varepsilon_{eff}} = 0.$$

Here, f_i and ε_i are, respectively, the volume fraction and permittivity of inclusion materials i, and ε_{eff} is the permittivity of the effective medium [63]. The total sum of volume fraction is unity, $\Sigma_i f_i = 1$. If we take the Maxwell-Garnet effective medium theory, the composite medium is regarded as a collection of small particles distributed in a host medium. Permittivity of the effective medium satisfies the condition as

$$\frac{\varepsilon_{eff} - \varepsilon_h}{\varepsilon_{eff} + 2\varepsilon_h} = f_i \frac{\varepsilon_i - \varepsilon_h}{\varepsilon_i + 2\varepsilon_h},$$

where ε_h indicates the permittivity of the host [64]. Effective media were also used to describe the excitation of SPP and LSP [65,66], and they may be obtained by fitting [67].

Another theory for understanding absorption mechanisms by MTMs is to use equivalent circuits. Although commercial software, which solves Maxwell's equations numerically, can calculate optical properties of MTMs, such as reflection and transmission spectra, it would be difficult to figure out a specific dependence of properties on structural parameters, such as the periodicity, length and width of MTMs. One of the analytical frames is the equivalent-circuit theory, which replaces MTMs with equivalent inductors and capacitors based on structural parameters of the metallic patterns. Fig. 5.2 shows a diagram of the equivalent inductor-capacitor circuit (LC model), where inductors exhibit mutual inductance between parallel layers (L_m) and the contribution of drifting electrons (L_e). In addition, equivalent capacitors are connected in series and composed of a capacitor (C_m) in two parallel

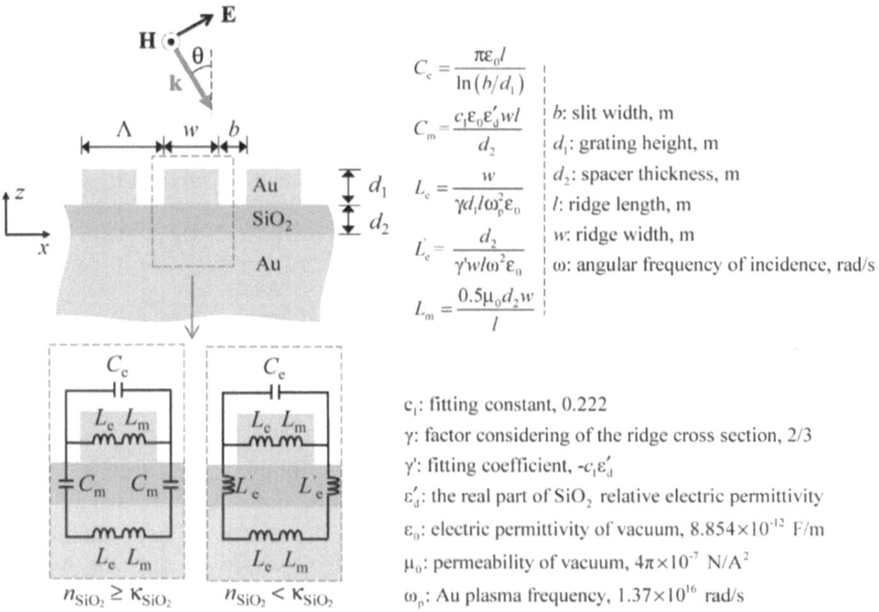

FIGURE 5.2 The schematic diagram of MMPA baseline and its equivalent inductor-capacitor circuits. (Reprinted with permission from Chen, Y.-B., and F.-C. Chiu. 2013. "Trapping mid-infrared rays in a lossy film with the Berreman mode, epsilon near zero mode, and magnetic polaritons." *Opt. Exp.* 21 (18):20771–20785. © The Optical Society).

metallic layers sandwiching the dielectric layers and a capacitor (C_e) to explain the gap-capacitance between neighboring metallic patterns [68,69]. Consequently, the total impedance of the circuit can be obtained as

$$Z_{tot} = \frac{i\omega(L_m + L_e)}{1 - \omega^2 C_e(L_m + L_e)} + \frac{2}{i\omega C_m} + i\omega(L_m + L_e).$$

The resonance frequency of the LC model for MTMs can be obtained from the condition that the total impedance Z_{tot}= 0. Zhou et al. used the equivalent circuit theory for the unit cell of the metal wire-pair structure [60]. The magnetic resonance frequency is expressed as

$$f_m = \frac{1}{2\pi\sqrt{LC}} = \frac{1}{\pi}\frac{1}{l\sqrt{\varepsilon_0\varepsilon_r\mu_0}} = \frac{c_0}{\pi l\sqrt{\varepsilon_r}},$$

where L is inductance, C is capacitance, l is the length of the short wires, ε_0 is the permittivity in a vacuum, ε_r is the relative permittivity of the area between the wires, μ_0 is the permeability in a vacuum, and c_0 is the speed of light in a vacuum. The length of the metal wire-pair is inversely proportional to the resonance frequency, which helps optimize structural parameters of MTMs. Matsuno et al. also applied an equivalent circuit model to MTMs, which results in the enhancement of understanding for underlying resonance mechanisms [70].

Interference between the upper metallic patterns and the bottom metal thin film was also suggested as another theoretical possibility for impedance matching in MMPAs [61]. Strong absorption is interpreted as arising from destructive interference between direct reflection and multiple reflections that cause light trapping in MMPAs.

5.2.2 Bandwidth

MMPAs can be classified by the resonance bandwidth of absorption, as described in Section 5.1. First, narrow-band MMPAs have been widely introduced in the visible and infrared regimes because of intrinsically narrow bandwidth of SPPs and LSPs caused by metallic MTMs. In particular, ultra-narrow-band MMPAs have been developed for applications of high spectral resolution infrared spectroscopy.

Kang et al. presented metal-dielectric-metal nanostructured absorbers with cross-shaped gold arrays [71]. The absorbers showed full width at half maximum (FWHM) of 180 nm with near-perfect absorption ($\eta > 99.7\%$) at 5.83 μm wavelength. This result demonstrated ultrahigh spectral resolution, which can be applied to enhance infrared sensing microsystems. He et al. also proposed MTMs for electromagnetically induced absorption to achieve an ultra-narrow absorption peak [72]. Fig. 5.3a shows the proposed structure – a metal bar stacked above a metal film with a long slot. The top bar, which plays the role of a broad-linewidth dipole antenna, and the bottom slot, which serves as a non-radiative magnetic quadrupole antenna, were oriented perpendicular to each other for optimal near-field coupling. Fig. 5.3b shows the simulated reflectance and absorption spectra with respect to the S parameter, which is the lateral displacement of the top bar with regard to the symmetry axis of the slot. When S is 70 nm, the absorption peak has a linewidth of 7.5 nm and exceeds 97% of absorption at 420 THz.

MMPA which consists of periodic arrays of infinite-length rectangular dielectric bars inside a semi-infinite metal was known to accomplish near-perfect absorption of 99.99% at the resonant wavelength of 1061.1 nm with a high-quality factor of 487 (obtained as the ratio of the peak frequency to the FWHM) [73]. Arrays of triangular gold nanoribbons on a thin gold layer were also shown to produce 95% absorption in the visible region with the FWHM of 1.82 nm [74]. Recently, the dielectric-dielectric-metal structure showed an ultra-narrow absorption with FWHM of 0.028 nm with a quality factor larger than 50,000 in the near-infrared regime [75]. These ultra-narrow-band MMPAs can be useful for refractive index sensing applications.

In contrary to the narrow single-band absorption, various applications, such as selective optical filters, multiplexing detector arrays and surface-enhanced infrared absorption sensing require dual and multiple bands. For example, multiple absorption bands may be used for detection and accurate identification of molecular species, such as cells or bacteria, with convenience. H-shaped nanoresonators with stacks of flexible polyimide and gold film substrate have demonstrated dual-band MMPAs with measured absorption greater than 90% over a range of ±50° incidence

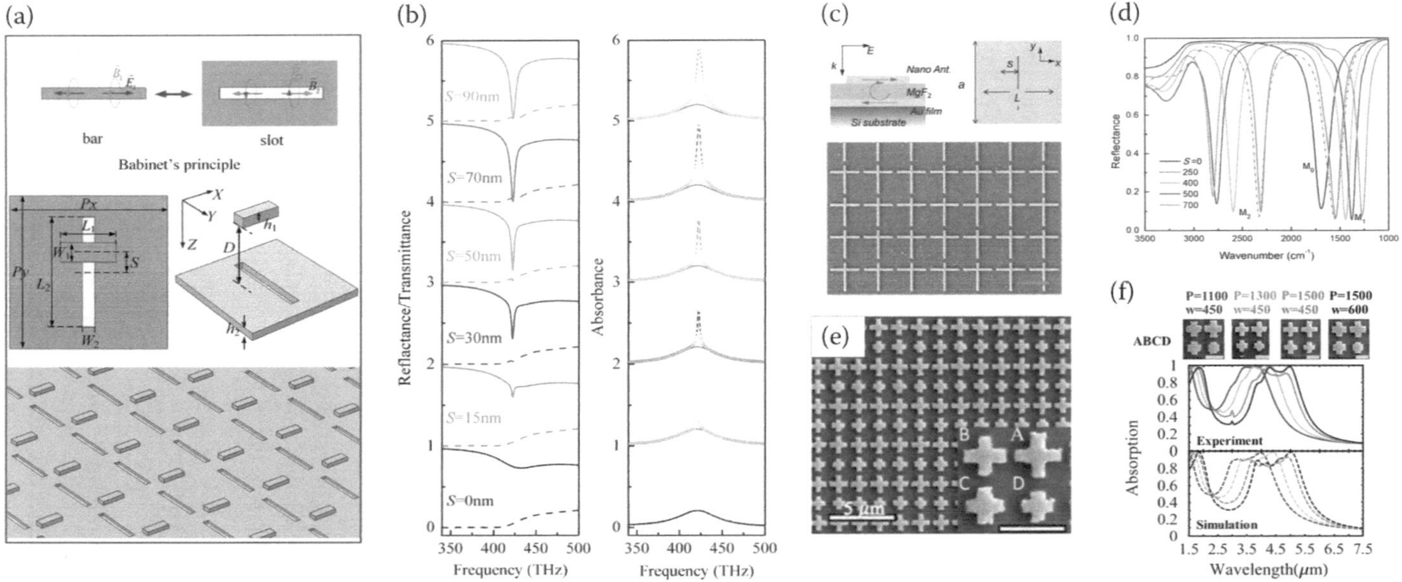

FIGURE 5.3 (a) Description of the electric and magnetic quadrupoles in a resonator and schematic of ultra-narrow-band stacked MMPA. (b) Simulated reflectance (solid line) and transmittance (dashed line) spectra and absorption spectra of the ultra-narrow-band MMPA for different values of displacement *S*. (Reprinted with permission from He, J., P. Ding, J. Wang, C. Fan, and E. Liang. 2015. "Ultra-narrow band perfect absorbers based on plasmonic analog of electromagnetically induced absorption." *Opt. Exp.* 23 (5):6083–6091. © The Optical Society). (c) General schematic of a dual-band MMPA and its SEM image. The scale bar is 2 μm in length. (d) Measured reflectance spectra of the cross-shaped MMPA at different values of *S*. (Reprinted with permission from Chen, K., R. Adato, and H. Altug. 2012. "Dual-band perfect absorber for multispectral plasmon-enhanced infrared spectroscopy." *ACS Nano* 6 (9):7998–8006. Copyright 2012 American Chemical Society). (e) SEM image of tungsten cross resonators. (f) Measured and simulated absorption spectra for cross resonator arrays. The scale bar of the SEM image is 1 μm. (Reprinted with permission from Li, Z., L. Stan, D.A. Czaplewski, X. Yang, and J. Gao. 2018. "Wavelength-selective mid-infrared metamaterial absorbers with multiple tungsten cross resonators." *Opt. Exp.* 26 (5):5616–5631. © The Optical Society).

angles in the mid-IR waveband between 3.3 μm and 3.9 μm [76]. An asymmetric cross-shaped structure shown in Fig. 5.3c, which consists of gold nanoparticles with a magnesium fluoride (MgF_2) film as a spacer layer, generates near-field coupling between two antiparallel currents at the antenna and the metal film surface. The structure was shown to enhance electric and magnetic dipole resonance, providing tunable dual-band MMPA [77]. Fig. 5.3d shows the measured reflectance spectra of the cross-shaped MMPA, which confirm tunable dual-band absorption with relatively high-quality factors throughout the mid-infrared range.

Because structural parameters of MMPAs are highly correlated to the resonance wavelength of single narrow resonators, a mixture of dual or multiple resonators with different designs of unit cells leads to multiple-band or broadband MMPA at visible and infrared regimes. The overall designs for the mixture can be based on cut wire [78], multi-sized disk arrays [79], hole array grating [80], periodic cubes arrays [81], multiple cross resonators [82], vertically aligned hyperbolic nanotubes [83], stacked double-ring resonators [84], 2D trapezoid cavities [85], nanostructured patterns [86] and tandem grating [87]. Fig. 5.3e shows the combination of four patterned tungsten cross resonators with different arm lengths, which causes a widely tunable absorption band ranging from 3.5 μm to 5.5 μm. A broadband near-perfect absorption over 94% was observed experimentally for the combination of four cross resonators with a period of 1,500 nm and a width of 450 nm, as shown in Fig. 5.3f. The results imply that the bandwidth of absorption bands may be tuned by adjusting the structural parameters of MMPAs.

5.3 PERFORMANCE OF VARIOUS MTMS FOR NEAR-PERFECT ABSORPTION

MTMs have been used to achieve near-perfect absorption characteristics, often in cooperation with electric and magnetic resonance coupling induced by the designed structure. Various structures have been suggested for near-perfect absorption, including periodic metal arrays, stacked structures and resonators. The combination of nanomaterials may also produce near-perfect absorption, while metal was recently substituted for dielectric materials in MTMs.

The performance of MMPAs can be estimated with an absorption ratio at the resonance wavelength and bandwidth that accompanies near-perfect absorption. These properties depend on structural parameters and material composition of MTMs. In this section, we explore the effect of these variations on the near-perfect absorption performance by investigating light-field resonance coupling.

5.3.1 Metallic Periodic Arrays

Most MMPAs are composed of metal-dielectric-metal thin film triple layers where the top metal layer is patterned with various periodic designs. Periodic metal arrays on the top and bottom metal film are coupled and give rise to magnetic resonance depending on the permittivity of the dielectric layer [13]. Fig. 5.4a shows the simple metal-dielectric-metal thin film structure of MMPAs, which consists of a periodic array of gold squares and gold thin film with a phase-change

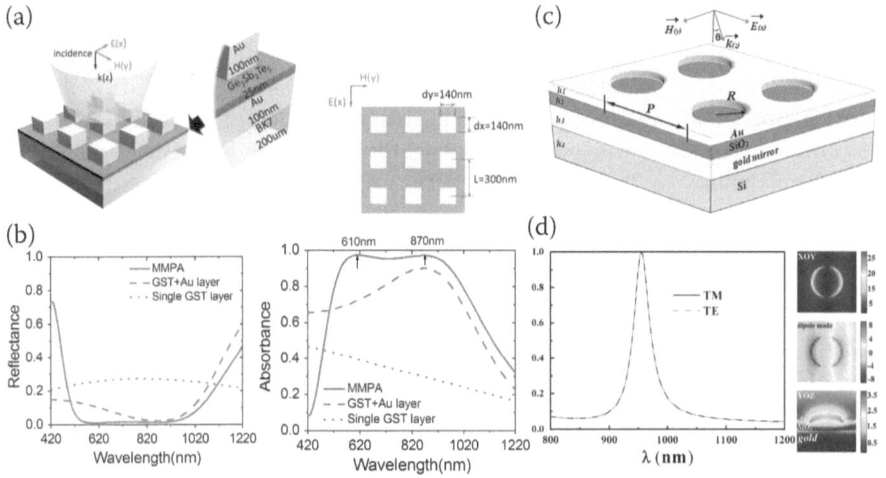

FIGURE 5.4 (a) Schematic and top view of periodic gold squares MMPA. (b) Simulated reflectance and absorption spectra for the MMPA, $Ge_2Sb_2Te_5$ dielectric layer with ground gold layer, and single $Ge_2Sb_2Te_5$ dielectric layer at normal incidence. (Reprinted under a Creative Commons license from Cao, T., C.-W. Wei, R.E. Simpson, L. Zhang, and M.J. Cryan. 2014. "Broadband polarization-independent perfect absorber using a phase-change metamaterial at visible frequencies." *Sci. Repts.* 4:3955). (c) Schematic of MMPA composed of nanocircle hole arrays. (d) Absorption spectra of the MMPA at the TE and TM modes. Electric field distributions and the dipole plasmon mode of metal-dielectric-metal structure. (Reprinted under a Creative Commons license from Wu, P., Z. Chen, H. Jile, C. Zhang, D. Xu, and L. Lv. 2020. "An infrared perfect absorber based on metal-dielectric-metal multi-layer films with nanocircle holes arrays." *Results in Phys.* 16:102952).

material ($Ge_2Sb_2Te_5$) layer [88]. Because a dielectric layer with a large imaginary part of the refractive index can significantly enhance the absorption by increased coupling to cavity resonance, the Ge-Sb-Te system was, therefore, chosen, which has a high extinction coefficient over the visible wavelength range [89]. The interaction between square arrays of gold on the top and the thin layer of gold at the bottom brings about localization of EM fields within $Ge_2Sb_2Te_5$. The size of gold patterns and the thickness of the metal-dielectric-metal layer influence dipolar resonances critically, which affects the impedance matching and can induce strong absorption, while the square arrays induce polarization-independent excitation of resonance [90]. Fig. 5.4b shows the simulated spectrum of reflectance and absorption by the proposed MMPAs at normal incidence. The broadband near-perfect absorption was obtained around 670 nm, with the absorption of 96.8% at λ_1=610 nm and 96.2% at λ_2 = 870 nm.

Recently, metal-shell/dielectric-core nanorod arrays were proposed as near-perfect absorbers in the near-infrared regime [91]. Ag-shell/dielectric-core nanorods with an Ag thin film layer produce SP resonance (SPR) and cavity plasmon resonance, which, in turn, give rise to effective coupled-mode, thereby providing near-perfect absorption. Strong LSP enhancement in the gap and cavity regions of nanorod results in near-perfect absorption with an extremely narrow bandwidth.

The proposed structure develops dual-band absorption spectrum as well as tunability of near-perfect absorption with various materials for dielectric core and diverse geometrical parameters of Ag-shell/dielectric-core nanorod, e.g., lattice constant, shell thickness, outer radius and height. When the proposed Ag-shell/dielectric-core nanorod was operated as an SPR sensor, sensitivity and figure of merit were obtained as 757.58 nm/RIU (RIU is the refractive index unit) and 50.51 (RIU^{-1}), respectively.

Periodic metal arrays include more than nanopost or nanorod arrays. Nanocircle hole arrays can be easily applied to the implementation of fundamental metal-dielectric-metal multi-layer perfect absorbers [92]. Fig. 5.4c shows a schematic diagram of a MMPA, which consists of a gold-SiO_2-gold multilayer with a nanocircle hole in each unit cell on an Si substrate. Radius of nanocircle holes is optimized to be R = 130 nm with the lattice constant set to be P = 500 nm. With an optimized structure, 99.8% near-perfect absorption was obtained at the resonance wavelength of 955 nm, as shown in Fig. 5.4d. Fig. 5.4d also shows the calculated electric field distribution of the proposed structure in the lateral (xy) and axial (yz) plane. With field enhancement within interface between gold and air at about 25 times, the resonant absorption was produced in connection with dipole plasmon modes and trapping of electric fields in SiO_2 and nanocircle hole as a result of LSP excitation. The absorption characteristic was determined to be independent of light polarization and incidence because of structural symmetry.

5.3.2 Stacked Structure

Broadband near-perfect absorbers in infrared and visible regimes have been investigated with stacked structure. Mao et al. have proposed a broadband near-perfect absorber, which is composed of alternating metallic grating and dielectric layers on the metallic substrate in the visible region [93]. This structure uses three symmetric Ag gratings and SiO_2 spacers with different widths. Average absorption was obtained as 95.2% in the range from 350 nm to 650 nm and reaches 99.9% at the wavelength of 370–450 nm. Near-perfect absorption is attributed to three mechanisms, i.e., excitation of SPPs at the Ag-air interface, Fabry-Perot resonances in the slits of grating and magnetic polaritons in the SiO_2 layer. The proposed structure can be applicable to solar energy harvesting.

A novel ultra-broadband near-perfect absorber has also been investigated based on stacked structure in the range of visible to middle infrared wavelength [94]. Fig. 5.5a shows the proposed structure, which consists of a top Ti circular disk, one single SiO dielectric layer, two Fe-SiO layers, and a bottom Fe mirror layer. The MMPA exhibits average absorption higher than 97.2% in the range of 400–6,000 nm (from visible to mid-infrared wavelength), as shown in Fig. 5.5b. The bandwidth of absorption over 90% attains almost 5,500 nm, and high absorption is maintained regardless of polarization for light incidence smaller than $\theta = 55°$. The result provides wider bandwidth than previously reported broadband MMPAs in the range of visible to mid-infrared wavelength [95–99]. Strong LSP induced around the two edges of the top Ti disk and two Fe layers exhibits broadband absorption at different resonant frequencies. EM oscillation within a short wavelength range of

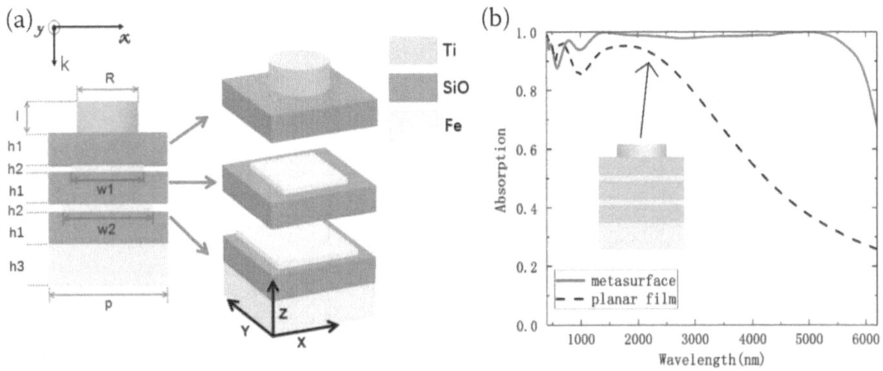

FIGURE 5.5 (a) Schematic of the stacked MMPA. (b) Absorption spectrum of the MMPA (red solid line) compared to that of the planar structure (black dashed line). (Reprinted under a Creative Commons license from Liu, J., W. Chen, J.-C. Zheng, Y.-S. Chen, and C.-F. Yang. 2020. "Wide-angle polarization-independent ultra-broadband absorber from visible to infrared." *Nanomater.* 10 (1):27).

about 400–1,500 nm is mainly caused by the top Ti disk. On the other hand, the lower Fe layer is responsible for enhanced EM oscillations at longer wavelength. The MMPAs may find potentials in photodetector, thermal emitters, infrared cloaking and solar energy harvesting.

5.3.3 Resonators

We now explore resonators as top metallic layers instead of simple metallic periodic arrays, such as gold square posts, nanorods and nanocircle hole arrays, as mentioned in Section 5.3.1. Landy et al. used electric ring resonators for the top periodic structure to construct MMPA at the microwave regime [14]. An electric ring resonator, which creates electric and magnetic resonance combined with cut wires, becomes highly absorptive over a narrow frequency range. The general principle can be extended to the range of visible and infrared wavebands.

We start by describing a simple cross resonator designed for MMPAs. Ma et al. theoretically and experimentally investigated MMPA composed of a gold cross resonator, which operates in the mid-infrared regime [100]. First, the absorption characteristics of periodic single crosses were calculated with an arm length and width of 1,000 nm and 200 nm, respectively, when deposited on 190 nm SiO_2 and the bottom gold ground plane layer. Strong coupling to the localized electric field in the dielectric spacer at the butt of the cross arm and magnetic response by antiparallel currents in the cross and the ground plane result in dielectric loss [28]. The maximum absorption reaches 99.7% at 4.5 μm due to the resonance, while FWHM is a mere 0.61 μm, corresponding to a quality factor set to 7.43. Furthermore, broadband absorption was also investigated using multiplexed cross resonators with four different arm lengths of 880 nm, 800 nm, 720 nm and 640 nm, as shown in Fig. 5.6a. Since near-perfect absorption was obtained at 4.04 μm, 3.73 μm, 3.43 μm and 3.14 μm for each cross resonator, a merged

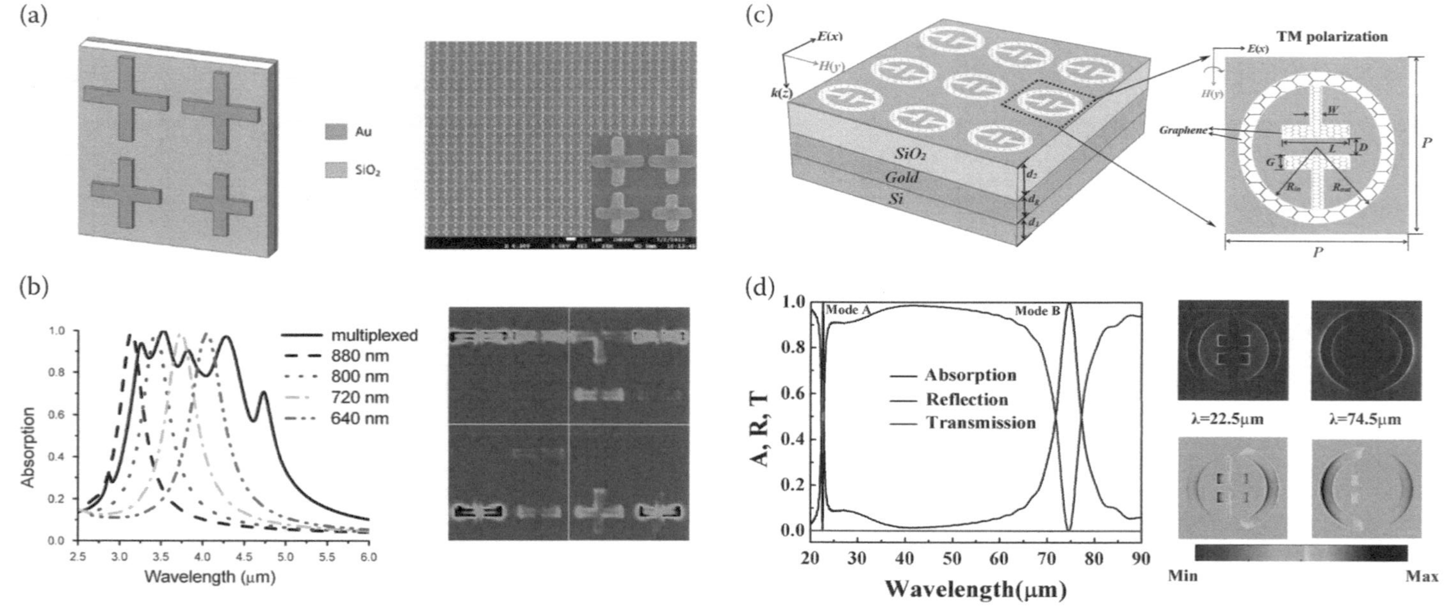

FIGURE 5.6 (a) Unit cell schematic diagram of multiplexed cross resonator and SEM image of the proposed MMPA. (b) Simulated absorption spectra of the proposed MMPA and four traditional non-multiplexed absorbers and power loss density at the four absorption peaks. (Reprinted with permission from Ma, W., Y. Wen, and X. Yu. 2013. "Broadband metamaterial absorber at mid-infrared using multiplexed cross resonators." *Opt. Exp.* 21 (25):30724–30730. © The Optical Society). (c) Schematic of graphene MMPA and the incident light polarization configuration. (d) Absorption, reflectance and transmission spectra of graphene MMPA and its electric field and surface charge density distributions at $\lambda_1 = 22.5$ μm and $\lambda_2 = 74.5$ μm. (Reprinted under a Creative Commons license from Yi, Z., C. Liang, X. Chen, Z. Zhou, Y. Tang, X. Ye, Y. Yi, J. Wang, and P. Wu. 2019. "Dual-band plasmonic perfect absorber based on graphene metamaterials for refractive index sensing application." *Micromachines* 10 (7):443).

structure with multiplexed cross resonators was shown to generate a much wider absorption band covering mid-infrared wavebands from 3 μm to 5 μm, with absorption higher than 50%. Fig. 5.6b shows the calculated absorption spectra of single and multiplexed cross resonators. The spectra present combined four peaks and power loss density at four absorption peaks at various wavelengths. In the case of multiplexed cross resonators, the absorption spectrum exhibits a small red shift of peaks compared with that of individual crosses [101]. The red shift is caused by the coupling effect between neighboring metallic crosses, which does not affect overall high-absorption characteristics.

In the visible range, broadband MMPA with a meander-ring-resonator was investigated [102]. The structure consists of a periodic tungsten meander-ring-resonator on dielectric SiO_2 layer and continuous tungsten film. The absorption covers the whole visible range at about 90% from 370–854 THz. The strong absorption is attributed to the excitation of LSP, propagating SP and guide-mode resonances [81,103,104]. Concentric multi-split circular ring resonators were also proposed as a broadband MMPA in the visible range [105]. The MMPA is composed of an array of gold concentric multi-split circular rings and plane mirror that are separated by a SiO_2 dielectric spacer layer. Numerical simulation shows that the absorption reaches 97.2% over the wavelength range from 580 nm to 800 nm regardless of the light polarization. Strong resonant coupling and overlapping of resonant frequencies in multi-rings accompanied near-perfect absorption. A gold mixed-slot resonator was also demonstrated for polarization-independent wide-angle broadband MMPAs in the visible range [106]. The strong broadband absorption stems from the excitation of SPP due to the dielectric layer arrangement and thickness, and also the LSP, which is mainly affected by the shape and size of lattice geometry. Propagation of SPP is also related to the interference of multiple waves scattered by slot arrays.

Recently, Yi et al. have demonstrated a dual-band MMPA in the infrared regime, which consists of four rectangular graphene resonators with a ring, as shown in Fig. 5.6c [107]. The proposed MMPA is based on a periodic graphene resonator and was experimentally implemented by depositing SiO_2 and gold mirror layers on a Si substrate [108]. Fig. 5.6d shows two resonance absorption modes observed at λ_1 = 2.5 μm and λ_2 =74.5 μm with 99.4% and 99.9% absorption, respectively. Localized electric field intensity appears mainly at the outer ring, inner edges and corners of the graphene structure. The distribution represents dipole and quadrupole resonance modes excited in the graphene resonator. The graphene-based MMPA can be used for tuning the absorption spectrum with different Fermi level (E_F) and relaxation time (τ) without changing geometric structures. The peak absorption wavelength is blue-shifted as the Fermi level E_F increases and the peak width narrows down with an increase of relaxation time τ.

5.3.4 Nanocomposites

The periodic resonators or stacked structures often need complicated fabrication processes such as electron-beam lithography or focused ion beam milling. Therefore, MMPAs based on these structures suffer from high cost and lack of flexibility. However, if nanocomposites that consist of nanoparticles of one material or multiple materials in a mixture are used in a stacked metal, MMPAs can present almost 100% absorption with relatively simple fabrication steps at low cost, and be compatible to current standard fabrication methods. Metallic nanoparticles excite LSPs in the course of Mie scattering, thereby producing high absorption. Resonance bandwidth can be controlled by size, shape, density and distribution of nanoparticles [109–113].

Basic structures of nearly percolated film of nanocomposites (Au/SiO_2) deposited on dielectric (SiO_2) and thick metallic layer (Au) on a Si or glass substrate have been widely investigated as MMPAs within the whole visible spectrum [51,114,115]. In addition to Au/SiO_2 nanocomposites, those composed of Copper-PTFE (polytetrafluoroethylene) were used to implement a broadband MMPA and showed an average absorption of 97.5% in the whole visible range [116]. Because nanocomposites can be prepared as an extremely thin film, MMPAs using nanocomposites have a potential to be fabricated on a flexible substrate and provide an outstanding candidate for anti-reflection coating.

Recently, a SiO_2 dielectric layer, which is limited when converting absorbed light energy into electron-hole pair generation, was alternated with Ge-Sb-Te phase-change materials, a family of chalcogenide semiconductors, in the photonics applications, such as multilevel storage, displays and integrated nanophotonic systems [117]. Chalcogenide semiconductor thin-film MMPAs have also been investigated in the mid-infrared and visible ranges [88,118,119]. Cao et al. introduced MMPA showing polarization-independent absorption at ~92% that operates in the spectral region from 400 nm to 1,000 nm using gold nanoparticles clustered on a 35 nm thick $Ge_2Sb_2Te_5$ (GST225) chalcogenide film [120]. Fig. 5.7a,b present a schematic diagram and a SEM image of MMPA, which consists of gold nanoparticles coupled to a GST225 thin film with a thin Si_3N_4 protective layer. Coupling between plasmon modes in gold nanoparticles and the lossy GST225 dielectric film results in high broadband absorption in the visible to near-infrared spectrum, as shown in Fig. 5.7c,d.

5.3.5 Dielectrics

Most MMPAs are based on metal-dielectric-metal nanostructures, as discussed in the previous section. However, semiconductors have recently been highlighted as favorable alternatives to metals in plasmonics in the infrared region. Among many semiconductor materials, silicon has a great chance to integrate MMPAs to photonic devices, such as sub-wavelength interconnects, modulators and emission sources [121,122].

Gorgulu et al. demonstrated ultra-broadband mid-infrared MMPA purely based on silicon [123]. The proposed MMPA is composed of periodic silicon grating that is arranged on silicon dioxide and silicon substrate, as shown in Fig. 5.8a. Fig. 5.8b shows simulated reflection, transmission and absorption spectra of the MMPA at normal incidence with a period of 7 μm and 8 μm. Two different samples produced broadband absorption in the wavelength range of 5–18.8 μm and 5.4–20 μm with a period of 7 μm and 8 μm, respectively. The absorption was higher than 90% on

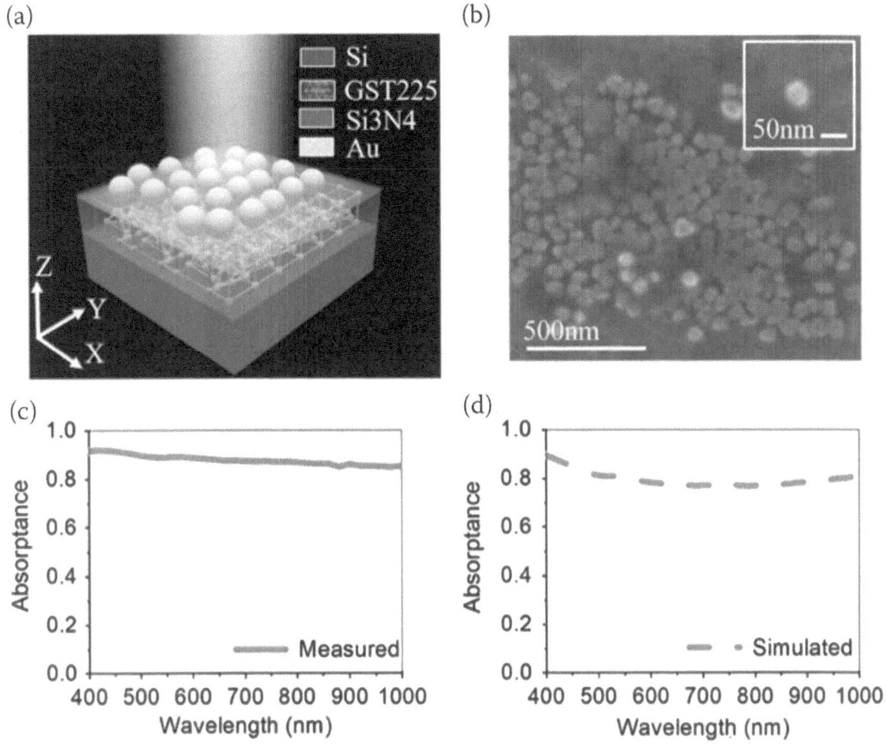

FIGURE 5.7 (a) Schematic of the nanocomposite MMPA composed of gold nanoparticles, a thin Si_3N_4 protective layer, and the amorphous GST225 film. (b) SEM image of fabricated device. (c) Measured and (d) simulated absorption spectra of the nanocomposite MMPA at normal incidence. (Reprinted with permission from Cao, T., K. Liu, L. Lu, H.-C. Chui, and R.E. Simpson. 2019. "Large-area broadband near-perfect absorption from a thin chalcogenide film coupled to gold nanoparticles." *ACS Appl. Mater. & Interf.* 11 (5):5176–5182., Copyright 2019 American Chemical Society).

average. Ultra-broadband absorption is associated with superposition of grating-induced silica-side propagating SP and gap-plasmon modes due to the cavity between silicon layers. Weng et al. also proposed an all-dielectric MMPA with two coupled sub-wavelength silicon nanodisk resonators embedded in a low dielectric constant silicon dioxide material [124]. The peak absorption of this MMPA was higher than 99%, with a bandwidth of 60 nm in the infrared regime. The silicon-based MMPAs have potentials as infrared sensors, imagers and enhanced spectroscopy elements with simple fabrication and optoelectronic integration compatible with the CMOS technology.

An all-dielectric Ge-based MMPA was also proposed as an ultra-thin semiconductor film because Ge has a high refractive index and, therefore, shows considerable absorption in the near-infrared region [125]. An all dielectric Ge MMPA composed of square arrays of Ge disks on a CaF_2 substrate was demonstrated to be independent of light polarization for an angle of light incidence above 28° with

(a)

(b)

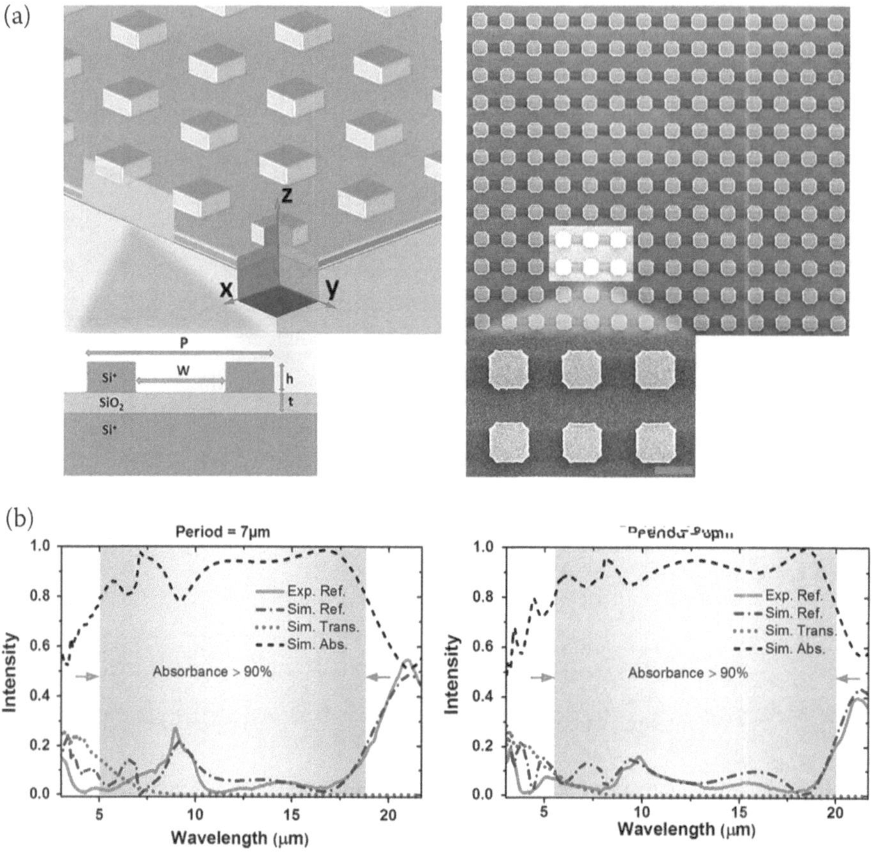

FIGURE 5.8 (a) Schematic of the all-dielectric MMPA and SEM image of one of the fabricated samples with a period P = 8 μm. The scale bar is 4 μm. (b) Measured and simulated reflectance patterns, transmission and absorption spectra of the all-dielectric MMPA with periods of P = 7 μm and P = 8 μm. (Reprinted under a Creative Commons license from Gorgulu, K., A. Gok, M. Yilmaz, K. Topalli, N. Bıyıklı, and A.K. Okyay. 2016. "All-silicon ultra-broadband infrared light absorbers." *Sci. Repts.* 6:38589).

absorption over 80% in the near-infrared region (800–1,600 nm) [126]. Near-perfect absorption is attributed to the destructive interference between the electric and magnetic dipoles excited inside each metasurface element and also to the interference between the scattered and incident fields.

5.4 APPLICATIONS OF NEAR-PERFECT ABSORPTION IN THE VISIBLE AND INFRARED RANGE

Until now, we have explored various MTM structures that exhibit near-perfect absorption characteristics. The MTM structures can find useful applications in various areas. Here, we describe some of the major applications that have used near-perfect absorbers, for example, sensing techniques where MTMs can be used to detect concentration changes of various materials. We also address MTMs in solar cells, light-emitting diodes and other applications.

5.4.1 Sensing

In this sub-section, we focus on sensing techniques using a near-perfect absorber based on sub-wavelength periodic nanostructures that may be difficult to fabricate. For this reason, many researchers have studied near-perfect absorbers in the gigahertz to terahertz range. In 2010, Liu et al. proposed a metal-insulator-metal structure including gold disk arrays, an MgF_2 spacer and a gold film mirror as a perfect absorber that is polarization independent and operates for a wide range of incidence angles at wavelength 1.6 μm, as shown in Fig. 5.9a,b [127].

In terms of figure of merit, the structure using gold disk arrays was suggested to outperform that of gold nanorods by almost 4 times [128]. They showed the potential of a metal-insulator-metal structure as a glucose sensor. After this pioneering work, various metal-insulator-metal structures, cross-shaped aperture [129], four-tined rod resonators [130], U-shaped antenna with nano-bar [131] and nanorod combinations [132,133] have been reported as refractive-index sensors in the frequency ranging from gigahertz to terahertz. As an example, the use of nanorod combinations for refractive-index sensing is shown in Fig. 5.9c,d. In 2011, although MTM structures for the visible wavelength are, in general, difficult to fabricate, Tittl et al. reported a palladium-based plasmonic perfect absorber for hydrogen sensing, as shown in Fig. 5.9e–h [134]. They used palladium grating to detect hydrogen and obtained enhanced sensing performance with metal-insulator-metal structures that consisted of a 65 nm MgF_2 spacer and a 200 nm gold film. Absorption of 99.5% at wavelength 650 nm was reported, with a detection limit of 0.5% H_2. Furthermore, many studies have used a near-perfect absorber for surface-enhanced Raman spectroscopy and surface-enhanced infrared spectroscopy to detect single molecules and chemical reactions [77,135–137].

5.4.2 Solar Cells

Near-perfect absorption can make a great contribution to improving the efficiency of solar cells. For this reason, broadband MTM absorbers were studied for solar-cell applications in the past decade [138–142].

In 2012, Wang et al. used amorphous silicon and perforated silver film to implement highly absorbed solar cells for the entire visible range [143]. Fig. 5.10a shows a honeycomb (top) and checkerboard array (middle), which act as a percolation threshold structure changing from islands to holes. The reflectance with respect to the wavelength and width of the perforated silver film is presented in Fig. 5.10b. The optimized checkerboard super-absorber structure shows low reflectance (see Fig. 5.10c) and high absorption (see Fig. 5.10d) in the visible range. In 2016, Azad et al. reported complex gold resonator arrays to demonstrate a broadband, wide-angle and polarization-independent absorber, which achieves absorption higher than 90% in the wavelength range from visible to near-infrared waveband; yet, it shows low extinction at a long wavelength [144]. Fig. 5.10e shows the

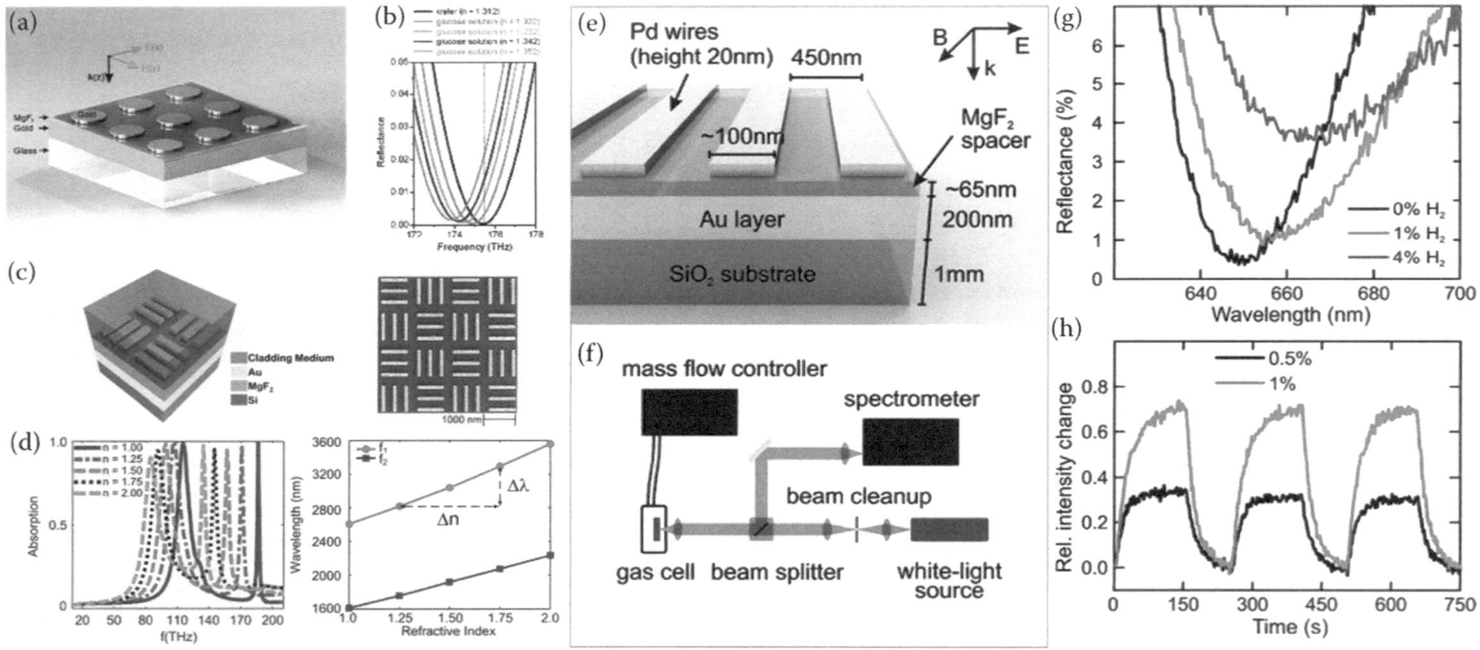

FIGURE 5.9 Near-perfect absorber for sensing applications. (a) Schematic of the near-perfect absorber structure with gold disk arrays. (b) Simulation results of an absorber sensor. (Reprinted with permission from Liu, N., M. Mesch, T. Weiss, M. Hentschel, and H. Giessen. 2010. "Infrared perfect absorber and its application as plasmonic sensor." *Nano Lett.* 10 (7):2342–2348. Copyright 2010 American Chemical Society). (c) The schematic and SEM of antisymmetric nanorod combination. (d) The absorption spectrum and resonance wavelength according to the refractive index of cladding. (Reprinted from Aslan, E., S. Kaya, E. Aslan, S. Korkmaz, O.G. Saracoglu, and M. Turkmen. 2017. "Polarization insensitive plasmonic perfect absorber with coupled antisymmetric nanorod array." *Sensors and Actuators B: Chemical*, 243:617–625, Copyright 2017 with permission from Elsevier). (e) Palladium grating with MgF_2 spacer and gold mirror. (f) Schematic of hydrogen sensing experiment. (g,h) The experimental results of hydrogen. (Reprinted with permission from Tittl, A., P. Mai, R. Taubert, D. Dregely, N. Liu, and H. Giessen. 2011. "Palladium-based plasmonic perfect absorber in the visible wavelength range and its application to hydrogen sensing." *Nano Lett.* 11 (10):4366–4369. Copyright 2011 American Chemical Society).

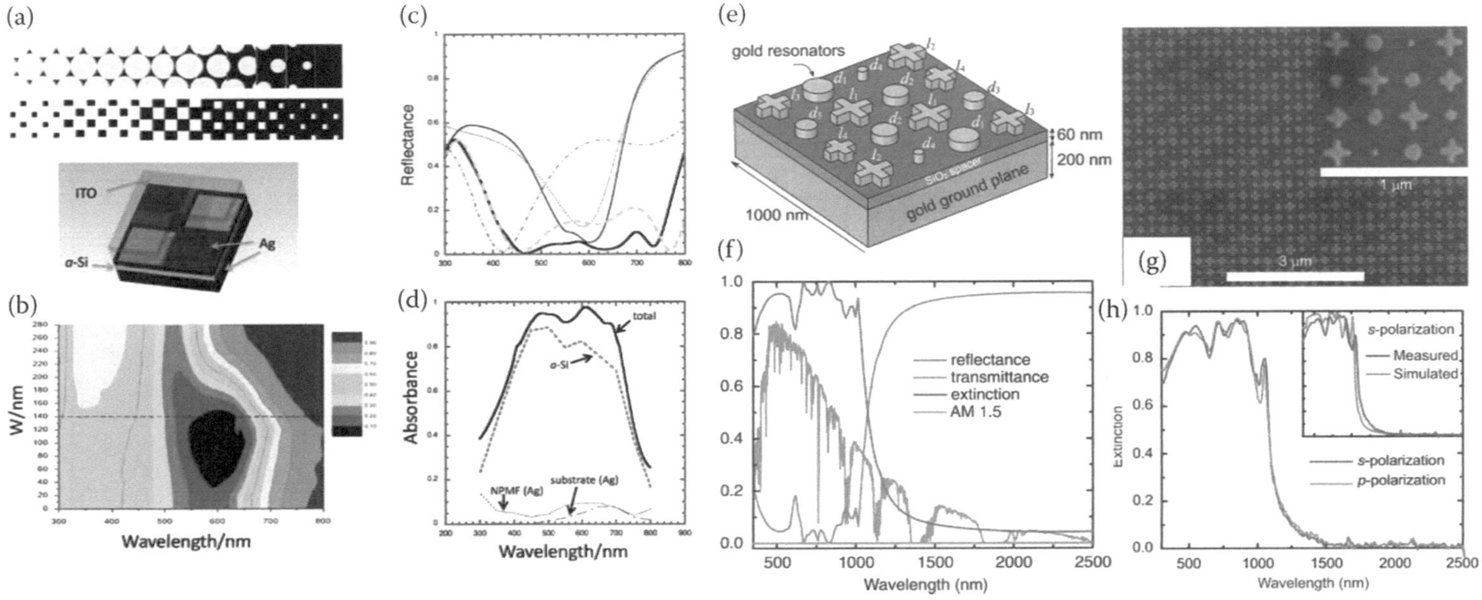

FIGURE 5.10 Near-perfect absorber for solar cell applications. (a) Honeycomb arrays evolving from islands to holes (top), checkerboard series (middle), and the schematic of optimized unit cell (bottom). (b) The color-encoded map of reflectance vs. width of nano perforated silver film and wavelength. (c) Reflection and (d) absorption spectra of the suggested structure. (Reprinted with permission from Wang, Y., T. Sun, T. Paudel, Y. Zhang, Z. Ren, and K. Kempa. 2012. "Metamaterial-plasmonic absorber structure for high efficiency amorphous silicon solar cells." *Nano Lett.* 12 (1):440–445. Copyright 2012 American Chemical Society). (e) The schematic of the broadband metamaterial absorber with gold resonator. (f) Simulated spectrum, (g) SEM image, and (h) measured extinction spectrum of the suggested gold resonator based absorber. (Reprinted under a Creative Commons license from Azad, A.K., W.J.M. Kort-Kamp, M. Sykora, N.R. Weisse-Bernstein, T.S. Luk, A.J. Taylor, D.A.R. Dalvit, and H.-T. Chen. 2016. "Metasurface broadband solar absorber." *Sci. Repts.* 6 (1):20347).

schematic of the suggested structure, while Fig. 5.10g is a SEM image of the fabricated absorber. The simulation results of the spectrum produced by the suggested complex gold resonator arrays are shown in Fig. 5.10f. Experimentally measured results that show absorption higher than 90% in the wavelength 450 nm < λ < 920 nm are in excellent agreement with the simulation, as shown in Fig. 5.10h.

5.4.3 Light-emitting Diodes

A light-emitting diode is another application for which MTMs can be useful. Plasmon resonance, induced by MTMs, allows the mechanism of light out-coupling to be controlled [145]. Al nanoparticle arrays embedded in the color conversion layer in solid state lighting devices were reported to have enhancement of light generation and out-coupling [146].

Recently, Al nanodisc arrays have been used to tune the emission color of organic light-emitting diodes (OLEDs) and increase the current efficiency by balancing the ohmic damping of LSPR and collective lattice resonances [147]. Fig. 5.11a shows the schematic of OLED with Al nanodisc arrays on ITO. A SEM image of Al nanodisc arrays is also presented in Fig. 5.11b. Al nanodisc arrays are overcoated on a 50 nm thick CuSCN layer, as shown in Fig. 5.11c. Fig. 5.11d shows the measured and simulated extinction spectra, which determine the spectral position of collective lattice resonance where the electroluminescence is selectively enhanced, as shown in Fig. 5.11e. Using this MTM structure, the current efficiency of blue-emitting phosphorescent OLED was obtained up to 35%. Ding et al. applied a plasmonic cavity with a subwavelength hole array, which shows the enhancement of light radiation and broadband absorption, to OLEDs achieving 1.57-fold enhancement of quantum efficiency and 2.5-fold higher ambient light absorption, compared to the conventional OLEDs [148].

5.4.4 Other Applications

MMPAs can be useful in applications other than sensing, solar cells and light-emitting diodes, for example, for implementation of spatial light modulators [149] and photodetectors [150] and in stealth technology [35,151] and imaging [152].

Chen and co-workers proposed a prototype spatial light modulator based on a pixelated hybrid graphene metasurface that includes gold antenna arrays, amorphous silicon spacers, Al_2O_3 gates and gold thin films, as shown in Fig. 5.12a. The optical response of these structures was modulated by switching graphene conductivity, as shown in Fig. 5.12a,b. They used the device for single pixel imaging. The setup and reconstructed images are shown in Fig. 5.12c,d. In plasmonic devices, hot electrons, induced by the non-radiative decay of SP, have been used in photocatalysis [153–156], photovoltaics [157–159] and photodetectors [160–162]. In 2014, Li and Valentine reported MMPA-based photodetectors with hot electrons [150]. The photodetectors improve optical absorption using 15 nm plasmonic film with an n-type silicon 2D grating substrate, as shown in Fig. 5.12e,f. It was reported that the efficiency of the hot electron transfer process was enhanced with high photo-responsivity. A broadband, omnidirectional and polarization-independent response was also demonstrated, as shown in Fig. 5.12g–i.

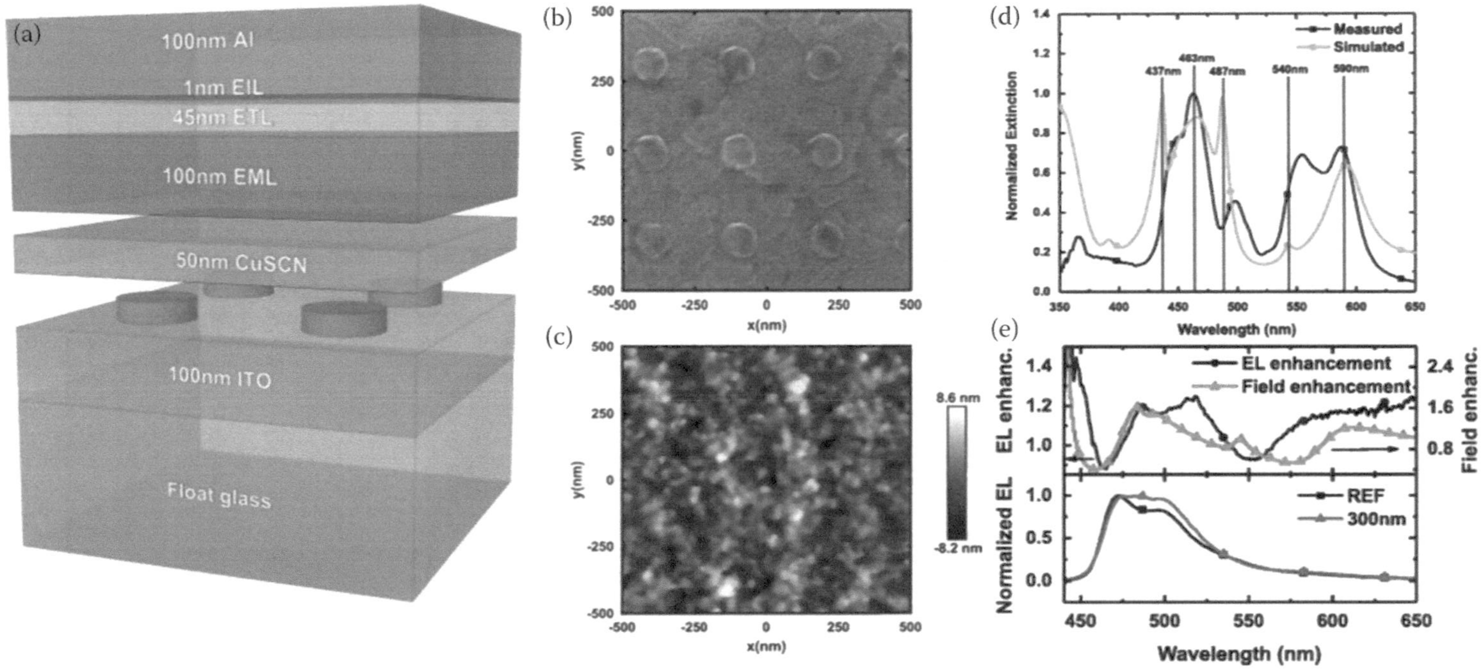

FIGURE 5.11 (a) Sketch view of the OLED with Al nanodics arrays. (b) SEM image of Al nanodics arrays. (c) Al nanodisc arrays overcoated on CuSCN layer. (d) Measured and simulated extinction spectra of the OLED. (e) Emission spectra (bottom) and measured enhancement and calculated field enhancement for recombination zone (top). (Reprinted from Auer-Berger, M., V. Tretnak, F.-P. Wenzl, J.R. Krenn, and E.J.W. List-Kratochvil. 2017. "Aluminum-nanodisc-induced collective lattice resonances: controlling the light extraction in organic light emitting diodes." *Appl. Phys. Lett.* 111 (17):173301., with the permission of AIP Publishing).

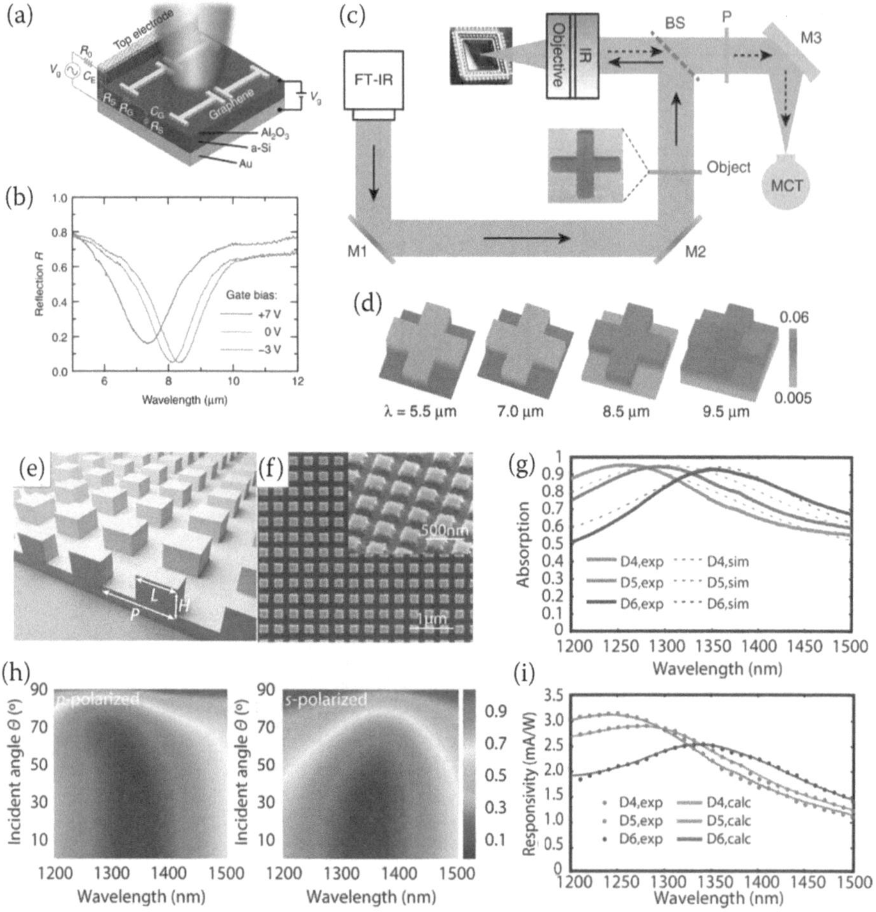

FIGURE 5.12 Various applications based on a near-perfect metamaterial absorber. (a) Schematic of the hybrid graphene perfect absorber. (b) Experimentally measured reflection spectra according to different gate voltages. (c) The single-pixel imaging system using the suggested spatial light modulator. (d) The imaging results of a cross-shaped object. (Reprinted under a Creative Commons license from Zeng, B., Z. Huang, A. Singh, Y. Yao, A.K. Azad, A.D. Mohite, A.J. Taylor, D.R. Smith, and H.-T. Chen. 2018. "Hybrid graphene metasurfaces for high-speed mid-infrared light modulation and single-pixel imaging." *Light: Sci. & Appl.* 7 (1):51.) (e) Schematic and (f) SEM image of the polarization-independent MMPA photodetectors. (g) Experimentally and simulated absorption spectrum of the photodetectors. (h) Numerically calculated optical absorption spectrum with *p*- and *s*-polarizations. (i) Experimentally and calculated photo-responsivity spectrum of the photodetectors. (Reprinted with permission from Li, W., and J. Valentine. 2014. "Metamaterial perfect absorber based hot electron photodetection." *Nano Lett.* 14 (6):3510–3514. Copyright 2014 American Chemical Society).

5.5 CONCLUDING REMARKS

In this chapter, we presented absorption characteristics of MMPAs and showed many applications in which MMPAs can be useful using diverse structures in the visible and infrared range. Metal-dielectric-metal multilayers have been widely investigated for MMPAs, and periodic top metallic structure plays an important role, which brings about the excitation of SP. Many studies of MMPA were conducted in the past with varied structural parameters such as periodicity, width, length and size of MTMs so that absorption wavelengths can be tuned for customization. Multiplexed structures have led to multiple-band and broadband near-perfect absorbers for various applications such as selective optical filters, multiplexing detector arrays, surface-enhanced infrared absorption sensing and solar cells. Moreover, all-dielectric MMPA shows near-perfect absorption characteristics in the infrared region substituting metallic structure, which allows simple fabrication and optoelectronic integration compatible with CMOS technology. The capability of near-perfect absorption by MTMs has been widely applied to diverse optoelectronic devices: for example, selective acquisition of a single molecule signal for sensors and spectroscopy, solar energy harvest for solar cells, enhancement of light emission in light-emitting diodes, optical imaging and photodetection. In the future, the properties and performance of MTMs for near-perfect absorption are expected to be topics of more intense investigation in wider research areas, combined with the development of nanotechnology, and to find more applications for photonic devices.

REFERENCES

1. D. R. Smith, W. J. Padilla, D. C. Vier, S. C. Nemat-Nasser, and S. Schultz, "Composite medium with simultaneously negative permeability and permittivity," *Phys. Rev. Lett.*, vol. 84, no. 18, p. 4184, 2000.
2. D. R. Smith, J. B. Pendry, and M. C. K. Wiltshire, "Metamaterials and negative refractive index," *Science*, vol. 305, no. 5685, p. 788, 2004.
3. D. Schurig, J. J. Mock, B. J. Justice, S. A. Cummer, J. B. Pendry, A. F. Starr, and D. R. Smith, "Metamaterial electromagnetic cloak at microwave frequencies," *Science*, vol. 314, no. 5801, pp. 977–980, 2006.
4. W. Cai, U. K. Chettiar, A. V. Kildishev, and V. M. Shalaev, "Optical cloaking with metamaterials," *Nature Photon.*, vol. 1, no. 4, pp. 224–227, 2007.
5. H. Chen, B.-I. Wu, B. Zhang, and J. A. Kong, "Electromagnetic wave interactions with a metamaterial cloak," *Phys. Rev. Lett.*, vol. 99, no. 6, p. 063903, 2007.
6. A. Alù and N. Engheta, "Achieving transparency with plasmonic and metamaterial coatings," *Phys. Rev. E*, vol. 72, no. 1, p. 016623, 2005.
7. N. Seddon and T. Bearpark, "Observation of the inverse Doppler effect," *Science*, vol. 302, no. 5650, pp. 1537–1540, 2003.
8. J. B. Pendry, "Negative refraction makes a perfect lens," *Phys. Rev. Lett.*, vol. 85, no. 18, p. 3966, 2000.
9. O. M. Ramahi, T. S. Almoneef, M. AlShareef, and M. S. Boybay, "Metamaterial particles for electromagnetic energy harvesting," *Appl. Phys. Lett.*, vol. 101, no. 17, p. 173903, 2012.
10. Z. Chen, B. Guo, Y. Yang, and C. Cheng, "Metamaterials-based enhanced energy harvesting: A review," *Physica B: Cond. Matt.*, vol. 438, pp. 1–8, 2014.

11. A. M. Hawkes, A. R. Katko, and S. A. Cummer, "A microwave metamaterial with integrated power harvesting functionality," *Appl. Phys. Lett.*, vol. 103, no. 16, p. 163901, 2013.
12. G. Moon, J. Choi, C. Lee, Y. Oh, K. H. Kim, and D. Kim, "Machine learning-based design of meta-plasmonic biosensors with negative index metamaterials," *Biosensors and Bioelectron.*, vol. 164C, p. 112335, 2020.
13. J. Hao, J. Wang, X. Liu, W. J. Padilla, L. Zhou, and M. Qiu, "High performance optical absorber based on a plasmonic metamaterial," *Appl. Phys. Lett.*, vol. 96, no. 25, p. 251104, 2010.
14. N. I. Landy, S. Sajuyigbe, J. J. Mock, D. R. Smith, and W. J. Padilla, "Perfect metamaterial absorber," *Phys. Rev. Lett.*, vol. 100, no. 20, p. 207402, 2008.
15. Y. Rádi, V. S. Asadchy, and S. A. Tretyakov, "Total absorption of electromagnetic waves in ultimately thin layers," *IEEE Trans. Antennas Propagat.*, vol. 61, no. 9, pp. 4606–4614, 2013.
16. C. M. Watts, X. Liu, and W. J. Padilla, "Metamaterial electromagnetic wave absorbers," *Adv. Mater.*, vol. 24, no. 23, pp. OP98–OP120, 2012.
17. R. Alaee, C. Menzel, C. Rockstuhl, and F. Lederer, "Perfect absorbers on curved surfaces and their potential applications," *Opt. Exp.*, vol. 20, no. 16, pp. 18370–18376, 2012.
18. J. R. Piper and S. Fan, "Total absorption in a graphene monolayer in the optical regime by critical coupling with a photonic crystal guided resonance," *ACS Photon.*, vol. 1, no. 4, pp. 347–353, 2014.
19. M. Pu, C. Hu, M. Wang, C. Huang, Z. Zhao, C. Wang, Q. Feng, and X. Luo, "Design principles for infrared wide-angle perfect absorber based on plasmonic structure," *Opt. Exp.*, vol. 19, no. 18, pp. 17413–17420, 2011.
20. M. Albooyeh, D. Morits, and S. A. Tretyakov, "Effective electric and magnetic properties of metasurfaces in transition from crystalline to amorphous state," *Phys. Rev. B.*, vol. 85, no. 20, p. 205110, 2012.
21. L. Huang, D. R. Chowdhury, S. Ramani, M. T. Reiten, S.-N. Luo, A. J. Taylor, and H.-T. Chen, "Experimental demonstration of terahertz metamaterial absorbers with a broad and flat high absorption band," *Opt. Lett.*, vol. 37, no. 2, pp. 154–156, 2012.
22. Y. Avitzour, Y. A. Urzhumov, and G. Shvets, "Wide-angle infrared absorber based on a negative-index plasmonic metamaterial," *Phys. Rev. B*, vol. 79, no. 4, p. 045131, 2009.
23. N. Liu, H. Guo, L. Fu, S. Kaiser, H. Schweizer, and H. Giessen, "Plasmon hybridization in stacked cut-wire metamaterials," *Adv. Mater.*, vol. 19, no. 21, pp. 3628–3632, 2007.
24. C. Hu, Z. Zhao, X. Chen, and X. Luo, "Realizing near-perfect absorption at visible frequencies," *Opt. Exp.*, vol. 17, no. 13, pp. 11039–11044, 2009.
25. K. Aydin, V. E. Ferry, R. M. Briggs, and H. A. Atwater, "Broadband polarization-independent resonant light absorption using ultrathin plasmonic super absorbers," *Nature Commun.*, vol. 2, no. 1, pp. 1–7, 2011.
26. B. Reinhard, K. M. Schmitt, V. Wollrab, J. Neu, R. Beigang, and M. Rahm, "Metamaterial near-field sensor for deep-subwavelength thickness measurements and sensitive refractometry in the terahertz frequency range," *Appl. Phys. Lett.*, vol. 100, no. 22, p. 221101, 2012.
27. Z. Yong, S. Zhang, C. Gong, and S. He, "Narrow band perfect absorber for maximum localized magnetic and electric field enhancement and sensing applications," *Sci. Repts.*, vol. 6, p. 24063, 2016.
28. X. Liu, T. Starr, A. F. Starr, and W. J. Padilla, "Infrared spatial and frequency selective metamaterial with near-unity absorbance," *Phys. Rev. Lett.*, vol. 104, no. 20, p. 207403, 2010.

29. D. Shrekenhamer, W. Xu, S. Venkatesh, D. Schurig, S. Sonkusale, and W. J. Padilla, "Experimental realization of a metamaterial detector focal plane array," *Phys. Rev. Lett.*, vol. 109, no. 17, p. 177401, 2012.
30. K.-T. Lin, H.-L. Chen, Y.-S. Lai, and C.-C. Yu, "Silicon-based broadband antenna for high responsivity and polarization-insensitive photodetection at telecommunication wavelengths," *Nature Commun.*, vol. 5, no. 1, pp. 1–10, 2014.
31. M. Gil, J. Bonache, and F. Martin, "Metamaterial filters: A review," *Metamaterials*, vol. 2, no. 4, pp. 186–197, 2008.
32. Y. Oh, W. Lee, and D. Kim, "Colocalization of gold nanoparticle-conjugated DNA hybridization for enhanced surface plasmon detection using nanograting antennas," *Opt. Lett.*, vol. 36, no. 8, pp. 1353–1355, 2011.
33. H. Yu, K. Kim, K. Ma, W. Lee, J.-W. Choi, C.-O. Yun, and D. Kim, "Enhanced detection of virus particles by nanoisland-based localized surface plasmon resonance," *Biosensors and Bioelectron.*, vol. 41, pp. 249–255, 2013.
34. Y. Kim, K. Chung, W. Lee, D. H. Kim, and D. Kim, "Nanogap-based dielectric-specific colocalization for highly sensitive surface plasmon resonance detection of biotin-streptavidin interactions," *Appl. Phys. Lett.*, vol. 101, no. 23, p. 233701, 2012.
35. J. Kim, K. Han, and J. W. Hahn, "Selective dual-band metamaterial perfect absorber for infrared stealth technology," *Sci. Repts.*, vol. 7, no. 1, p. 6740, 2017.
36. K. Kim, W. Lee, K. Chung, H. Lee, T. Son, Y. Oh, Y.-F. Xiao, D. H. Kim, and D. Kim, "Molecular overlap with optical near-fields based on plasmonic nanolithography for ultrasensitive label-free detection by light-matter colocalization," *Biosensors and Bioelectron.*, vol. 96, pp. 89–98, 2017.
37. K. Ma, D. J. Kim, K. Kim, S. Moon, and D. Kim, "Target-localized nanograting-based surface plasmon resonance detection toward label-free molecular biosensing," *IEEE J. Selected Topics in Quantum Electron.*, vol. 16, no. 4, pp. 1004–1014, 2010.
38. T. Son, D. Lee, C. Lee, G. Moon, G. E. Ha, H. Lee, H. Kwak, E. Cheong, and D. Kim, "Superlocalized three-dimensional live imaging of mitochondrial dynamics in neurons using plasmonic nanohole arrays," *ACS Nano*, vol. 13, no. 3, pp. 3063–3074, 2019.
39. K. Kim, J. Yajima, Y. Oh, W. Lee, S. Oowada, T. Nishizaka, and D. Kim, "Nanoscale localization sampling based on nanoantenna arrays for super-resolution imaging of fluorescent monomers on sliding microtubules," *Small*, vol. 8, no. 6, pp. 892–900, 2012.
40. W. Lee, Y. Kinosita, Y. Oh, N. Mikami, H. Yang, M. Miyata, T. Nishizaka, and D. Kim, "Three-dimensional superlocalization imaging of gliding Mycoplasma mobile by extraordinary light transmission through arrayed nanoholes," *ACS Nano*, vol. 9, no. 11, pp. 10896–10908, 2015.
41. P. Yu, L. V. Besteiro, J. Wu, Y. Huang, Y. Wang, A. O. Govorov, and Z. Wang, "Metamaterial perfect absorber with unabated size-independent absorption," *Opt. Exp.*, vol. 26, no. 16, pp. 20471–20480, 2018.
42. Y. P. Lee, P. V. Tuong, H. Y. Zheng, J. Y. Rhee, and W. H. Jang, "An application of metamaterials: Perfect absorbers," *J. of the Kor. Physical Soc.*, vol. 60, no. 8, pp. 1203–1206, 2012.
43. P. Pitchappa, C. P. Ho, P. Kropelnicki, N. Singh, D.-L. Kwong, and C. Lee, "Dual band complementary metamaterial absorber in near infrared region," *J. Appl. Phys.*, vol. 115, no. 19, p. 193109, 2014.
44. Y. J. Yoo, Y. J. Kim, P. Van Tuong, J. Y. Rhee, K. W. Kim, W. H. Jang, Y. H. Kim, H. Cheong, and Y. Lee, "Polarization-independent dual-band perfect absorber utilizing multiple magnetic resonances," *Opt. Exp.*, vol. 21, no. 26, pp. 32484–32490, 2013.

45. S. Bhattacharyya, S. Ghosh, and K. Vaibhav Srivastava, "Triple band polarization-independent metamaterial absorber with bandwidth enhancement at X-band," *J. Appl. Phys.*, vol. 114, no. 9, p. 094514, 2013.
46. G. Dayal and S. A. Ramakrishna, "Design of multi-band metamaterial perfect absorbers with stacked metal–dielectric disks," *J. Opt.*, vol. 15, no. 5, p. 055106, 2013.
47. L. Huang, and H. Chen, "Multi-band and polarization insensitive metamaterial absorber," *Prog. In Electromagn. Res.*, vol. 113, pp. 103–110, 2011.
48. B. Zhang, Y. Zhao, Q. Hao, B. Kiraly, I.-C. Khoo, S. Chen, and T. J. Huang, "Polarization-independent dual-band infrared perfect absorber based on a metal-dielectric-metal
elliptical nanodisk array," *Opt. Exp.*, vol. 19, no. 16, pp. 15221–15228, 2011.
49. Y. Lin, Y. Cui, F. Ding, K. H. Fung, T. Ji, D. Li, and Y. Hao, "Tungsten based anisotropic metamaterial as an ultra-broadband absorber," *Opt. Mater. Exp.*, vol. 7, no. 2, pp. 606–617, 2017.
50. S. Gu, J. P. Barrett, T. H. Hand, B.-I. Popa, and S. A. Cummer, "A broadband low-reflection metamaterial absorber," *J. Appl. Phys.*, vol. 108, no. 6, p. 064913, 2010.
51. M. Keshavarz Hedayati, M. Abdelaziz, C. Etrich, S. Homaeigohar, C. Rockstuhl, and M. Elbahri, "Broadband anti-reflective coating based on plasmonic nano-composite," *Materials*, vol. 9, no. 8, p. 636, 2016.
52. H. T. Miyazaki, K. Ikeda, T. Kasaya, K. Yamamoto, Y. Inoue, K. Fujimura, T. Kanakugi, M. Okada, K. Hatade, and S. Kitagawa, "Thermal emission of two-color polarized infrared waves from integrated plasmon cavities," *Appl. Phys. Lett.*, vol. 92, no. 14, p. 141114, 2008.
53. S. Ogawa, J. Komoda, K. Masuda, and M. Kimata, "Wavelength selective wideband uncooled infrared sensor using a two-dimensional plasmonic absorber," *Opt. Eng.*, vol. 52, no. 12, p. 127104, 2013.
54. S. Ogawa, K. Okada, N. Fukushima, and M. Kimata, "Wavelength selective uncooled infrared sensor by plasmonics," *Appl. Phys. Lett.*, vol. 100, no. 2, p. 021111, 2012.
55. S. Ogawa, Y. Takagawa, and M. Kimata, "Broadband polarization-selective uncooled infrared sensors using tapered plasmonic micrograting absorbers," *Sensors and Actuators A: Physical*, vol. 269, pp. 563–568, 2018.
56. Y. Takagawa, S. Ogawa, and M. Kimata, "Detection wavelength control of uncooled infrared sensors using two-dimensional lattice plasmonic absorbers," *Sensors*, vol. 15, no. 6, pp. 13660–13669, 2015.
57. J. W. Lee, M. A. Seo, J. Y. Sohn, Y. H. Ahn, D. S. Kim, S. C. Jeoung, Ch. Lienau, and Q.-H. Park, "Invisible plasmonic meta-materials through impedance matching to vacuum," *Opt. Exp.*, vol. 13, no. 26, pp. 10681–10687, 2005.
58. B. A. Slovick, Z. G. Yu, and S. Krishnamurthy, "Generalized effective-medium theory for metamaterials," *Phys. Rev. B.*, vol. 89, no. 15, p. 155118, 2014.
59. Y. Wu, J. Li, Z.-Q. Zhang, and C. T. Chan, "Effective medium theory for magnetodielectric composites: beyond the long-wavelength limit," *Phys. Rev. B.*, vol. 74, no. 8, p. 085111, 2006.
60. J. Zhou, L. Zhang, G. Tuttle, T. Koschny, and C. M. Soukoulis, "Negative index materials using simple short wire pairs," *Phys. Rev. B.*, vol. 73, no. 4, p. 041101, 2006.
61. H.-T. Chen, "Interference theory of metamaterial perfect absorbers," *Opt. Exp.*, vol. 20, no. 7, pp. 7165–7172, 2012.
62. T. Badloe, J. Mun, and J. Rho, "Metasurfaces-based absorption and reflection control: Perfect absorbers and reflectors," *J. Nanomater.*, vol. 2017, p. 2361042, 2017.

63. D. A. G. Bruggeman, "Berechnung verschiedener physikalischer Konstanten von heterogenen Substanzen. I. Dielektrizitätskonstanten und Leitfähigkeiten der Mischkörper aus isotropen Substanzen," *Annalen der Physik*, vol. 416, no. 7, pp. 636–664, 1935.
64. Garnett J. C. Maxwell, "XII. Colours in metal glasses and in metallic films," *Philos. Trans. Royal Soc. London. Series A, Containing Papers of a Mathematical or Physical Character*, vol. 203, no. 359–371, pp. 385–420, 1904.
65. K. Kang and D. Kim, "Effective optical properties of nanoparticle-mediated surface plasmon resonance sensors," *Opt. Exp.*, vol. 27, no. 3, pp. 3091–3100, 2019.
66. D. Kim and S. J. Yoon, "Effective medium-based analysis of nanowire-mediated localized surface plasmon resonance," *Appl. Opt.*, vol. 46, no. 6, pp. 872–880, 2007.
67. S. Moon and D. Kim, "Fitting-based determination of an effective medium of a metallic periodic structure and application to photonic crystals," *J. Opt. Soc. of Am. A*, vol. 23, no. 1, pp. 199–207, 2006.
68. Y.-B. Chen and F.-C. Chiu, "Trapping mid-infrared rays in a lossy film with the Berreman mode, epsilon near zero mode, and magnetic polaritons," *Opt. Exp.*, vol. 21, no. 18, pp. 20771–20785, 2013.
69. B. J. Lee, L. P. Wang, and Z. M. Zhang, "Coherent thermal emission by excitation of magnetic polaritons between periodic strips and a metallic film," *Opt. Exp.*, vol. 16, no. 15, pp. 11328–11336, 2008.
70. Y. Matsuno and A. Sakurai, "Perfect infrared absorber and emitter based on a large-area metasurface," *Opt. Mater. Exp.*, vol. 7, no. 2, pp. 618–626, 2017.
71. S. Kang, Z. Qian, V. Rajaram, S. D. Calisgan, A. Alù, and M. Rinaldi, "Ultra-narrowband metamaterial absorbers for high spectral resolution infrared spectroscopy," *Adv. Opt. Mater.*, vol. 7, no. 2, p. 1801236, 2019.
72. J. He, P. Ding, J. Wang, C. Fan, and E. Liang, "Ultra-narrow band perfect absorbers based on plasmonic analog of electromagnetically induced absorption," *Opt. Exp.*, vol. 23, no. 5, pp. 6083–6091, 2015.
73. X. Zhu, J. Fu, F. Ding, Y. Jin, and A. Wu, "Angle-insensitive narrowband optical absorption based on high-Q localized resonance," *Sci. Repts.*, vol. 8, no. 1, pp. 1–6, 2018.
74. D. Wu, R. Li, Y. Liu, Z. G. Yu, L. Yu, L. Chen, C. Liu, R. Ma, and H. Ye, "Ultra-narrow band perfect absorber and its application as plasmonic sensor in the visible region," *Nanoscale Res. Lett.*, vol. 12, no. 1, pp. 1–11, 2017.
75. Y.-L. Liao and Y. Zhao, "Ultra-narrowband dielectric metamaterial absorber for sensing based on cavity-coupled phase resonance," *Results in Phys.*, vol. 17, p. 103072, 2020.
76. Z. H. Jiang, S. Yun, F. Toor, D. H. Werner, and T. S. Mayer, "Conformal dual-band near-perfectly absorbing mid-infrared metamaterial coating," *ACS Nano*, vol. 5, no. 6, pp. 4641–4647, 2011.
77. K. Chen, R. Adato, and H. Altug, "Dual-band perfect absorber for multispectral plasmon-enhanced infrared spectroscopy," *ACS Nano*, vol. 6, no. 9, pp. 7998–8006, 2012.
78. S. Chen, H. Cheng, H. Yang, J. Li, X. Duan, C. Gu, and J. Tian, "Polarization insensitive and omnidirectional broadband near perfect planar metamaterial absorber in the near infrared regime," *Appl. Phys. Lett.*, vol. 99, no. 25, p. 253104, 2011.
79. C.-W. Cheng, M. N. Abbas, C.-W. Chiu, K.-T. Lai, M.-H. Shih, and Y.-C. Chang, "Wide-angle polarization independent infrared broadband absorbers based on metallic multi-sized disk arrays," *Opt. Exp.*, vol. 20, no. 9, pp. 10376–10381, 2012.

80. K. V. Sreekanth, M. ElKabbash, Y. Alapan, A. R. Rashed, U. A. Gurkan, and G. Strangi, "A multiband perfect absorber based on hyperbolic metamaterials," *Sci. Repts.*, vol. 6, p. 26272, 2016.
81. L. Lei, S. Li, H. Huang, K. Tao, and P. Xu, "Ultra-broadband absorber from visible to near-infrared using plasmonic metamaterial," *Opt. Exp.*, vol. 26, no. 5, pp. 5686–5693, 2018.
82. Z. Li, L. Stan, D. A. Czaplewski, X. Yang, and J. Gao, "Wavelength-selective mid-infrared metamaterial absorbers with multiple tungsten cross resonators," *Opt. Exp.*, vol. 26, no. 5, pp. 5616–5631, 2018.
83. C. T. Riley, J. S. T. Smalley, J. R. J. Brodie, Y. Fainman, D. J. Sirbuly, and Z. Liu, "Near-perfect broadband absorption from hyperbolic metamaterial nanoparticles," *Proc. National Acad. Sci.*, vol. 114, no. 6, pp. 1264–1268, 2017.
84. H. Deng, L. Stan, D. A. Czaplewski, J. Gao, and X. Yang, "Broadband infrared absorbers with stacked double chromium ring resonators," *Opt. Exp.*, vol. 25, no. 23, pp. 28295–28304, 2017.
85. D. Ji, H. Song, X. Zeng, H. Hu, K. Liu, N. Zhang, and Q. Gan, "Broadband absorption engineering of hyperbolic metafilm patterns," *Sci. Repts.*, vol. 4, p. 4498, 2014.
86. J. A. Bossard, L. Lin, S. Yun, L. Liu, D. H. Werner, and T. S. Mayer, "Near-ideal optical metamaterial absorbers with super-octave bandwidth," *ACS Nano*, vol. 8, no. 2, pp. 1517–1524, 2014.
87. S. Han, J. H. Shin, P. H. Jung, H. Lee, and B. J. Lee, "Broadband solar thermal absorber based on optical metamaterials for high-temperature applications," *Adv. Opt. Mater.*, vol. 4, no. 8, pp. 1265–1273, 2016.
88. T. Cao, C.-w. Wei, R. E. Simpson, L. Zhang, and M. J. Cryan, "Broadband polarization-independent perfect absorber using a phase-change metamaterial at visible frequencies," *Sci. Repts.*, vol. 4, p. 3955, 2014.
89. K., Shportko, S. Kremers, M. Woda, D. Lencer, J. Robertson, and M. Wuttig, "Resonant bonding in crystalline phase-change materials," *Nature Mater.*, vol. 7, no. 8, pp. 653–658, 2008.
90. G. Dayal and S. A. Ramakrishna, "Design of highly absorbing metamaterials for infrared frequencies," *Opt. Exp.*, vol. 20, no. 16, pp. 17503–17508, 2012.
91. Y.-F. C. Chau, C.-T. Chou Chao, C. M. Lim, H. J. Huang, and H.-P. Chiang, "Depolying tunable metal-shell/dielectric core nanorod arrays as the virtually perfect absorber in the near-infrared regime," *ACS Omega*, vol. 3, no. 7, pp. 7508–7516, 2018.
92. P. Wu, Z. Chen, H. Jile, C. Zhang, D. Xu, and L. Lv, "An infrared perfect absorber based on metal-dielectric-metal multi-layer films with nanocircle holes arrays," *Results in Phys.*, vol. 16, p. 102952, 2020.
93. Q. Mao, C. Feng, Y. Yang, and Y. Tan, "Design of broadband metamaterial near-perfect absorbers in visible region based on stacked metal-dielectric gratings," *Mater. Res. Exp.*, vol. 5, no. 6, p. 065801, 2018.
94. J. Liu, W. Chen, J.-C. Zheng, Y.-S. Chen, and C.-F. Yang, "Wide-angle polarization-independent ultra-broadband absorber from visible to infrared," *Nanomaterials*, vol. 10, no. 1, p. 27, 2020.
95. W. Xue, X. Chen, Y. Peng, and R. Yang, "Grating-type mid-infrared light absorber based on silicon carbide material," *Opt. Exp.*, vol. 24, no. 20, pp. 22596–22605, 2016.
96. Y. K. Zhong, S. M. Fu, W. Huang, D. Rung, J. Y.-W. Huang, P. Parashar, and A. Lin, "Polarization-selective ultra-broadband super absorber," *Opt. Exp.*, vol. 25, no. 4, pp. A124–A133, 2017.

97. J. Cong, Z. Zhou, B. Yun, L. Lv, H. Yao, Y. Fu, and N. Ren, "Broadband visible-light absorber via hybridization of propagating surface plasmon," *Opt. Lett.*, vol. 41, no. 9, pp. 1965–1968, 2016.
98. Y. Li, Z. Liu, H. Zhang, P. Tang, B.-I. Wu, and G. Liu, "Ultra-broadband perfect absorber utilizing refractory materials in metal-insulator composite multilayer stacks," *Opt. Exp.*, vol. 27, no. 8, pp. 11809–11818, 2019.
99. H. Lin, B. C. P. Sturmberg, K.-T. Lin, Y. Yang, X. Zheng, T. K. Chong, C. M. de Sterke, and B. Jia, "A 90-nm-thick graphene metamaterial for strong and extremely broadband absorption of unpolarized light," *Nature Photon.*, vol. 13, no. 4, pp. 270–276, 2019.
100. W. Ma, Y. Wen, and X. Yu, "Broadband metamaterial absorber at mid-infrared using multiplexed cross resonators," *Opt. Exp.*, vol. 21, no. 25, pp. 30724–30730, 2013.
101. X. Shen, T. J. Cui, J. Zhao, H. F. Ma, W. X. Jiang, and H. Li, "Polarization-independent wide-angle triple-band metamaterial absorber," *Opt. Exp.*, vol. 19, no. 10, pp. 9401–9407, 2011.
102. C. Cao and Y. Cheng, "A broadband plasmonic light absorber based on a tungsten meander-ring-resonator in visible region," *Appl. Phys. A*, vol. 125, no. 1, p. 15, 2019.
103. D. Wu, C. Liu, Y. Liu, L. Yu, Z. Yu, L. Chen, R. Ma, and H. Ye, "Numerical study of an ultra-broadband near-perfect solar absorber in the visible and near-infrared region," *Opt. Lett.*, vol. 42, no. 3, pp. 450–453, 2017.
104. X. Chen, Y. Chen, M. Yan, and M. Qiu, "Nanosecond photothermal effects in plasmonic nanostructures," *ACS Nano*, vol. 6, no. 3, pp. 2550–2557, 2012.
105. B. Tang, Y. Zhu, X. Zhou, L. Huang, and X. Lang, "Wide-angle polarization-independent broadband absorbers based on concentric multisplit ring arrays," *IEEE Photon. J.*, vol. 9, no. 6, pp. 1–7, 2017.
106. X. Duan, S. Chen, W. Liu, H. Cheng, Z. Li, and J. Tian, "Polarization-insensitive and wide-angle broadband nearly perfect absorber by tunable planar metamaterials in the visible regime," *J. Opt.*, vol. 16, no. 12, p. 125107, 2014.
107. Z. Yi, C. Liang, X. Chen, Z. Zhou, Y. Tang, X. Ye, Y. Yi, J. Wang, and P. Wu, "Dual-band plasmonic perfect absorber based on graphene metamaterials for refractive index sensing application," *Micromachines*, vol. 10, no. 7, p. 443, 2019.
108. B. G. Ghamsari, A. Olivieri, F. Variola, and P. Berini, "Enhanced Raman scattering in graphene by plasmonic resonant Stokes emission," *Nanophoton.*, vol. 3, no. 6, pp. 363–371, 2014.
109. S. Moon, D. J. Kim, K. Kim, D. Kim, H. Lee, K. Lee, and S. Haam, "Surface-enhanced plasmon resonance detection of nanoparticle-conjugated DNA hybridization," *Appl. Opt.*, vol. 49, no. 3, pp. 484–491, 2010.
110. S. Moon, Y. Kim, Y. Oh, H. Lee, H. C. Kim, K. Lee, and D. Kim, "Grating-based surface plasmon resonance detection of core-shell nanoparticle mediated DNA hybridization," *Biosens. Bioelectron.*, vol. 32, no. 1, pp. 141–147, 2012.
111. W. A. Murray and W. L. Barnes, "Plasmonic materials," *Adv. Mater.*, vol. 19, no. 22, pp. 3771–3782, 2007.
112. A. Biswas, H. Eilers, F. Hidden Jr., O. C. Aktas, and C. V. S. Kiran, "Large broadband visible to infrared plasmonic absorption from Ag nanoparticles with a fractal structure embedded in a Teflon AFR matrix," *Appl. Phys. Lett.*, vol. 88, no. 1, p. 013103, 2006.
113. C. F. Bohren and D. R. Huffman, *Absorption and Scattering of Light by Small Particles* New York, USA: John Wiley & Sons Inc., 1983.
114. M. K., Hedayati, M. Javaherirahim, B. Mozooni, R. Abdelaziz, A. Tavassolizadeh, V. S. K. Chakravadhanula, V. Zaporojtchenko, T. Strunkus, F. Faupel, and

M. Elbahri, "Design of a perfect black absorber at visible frequencies using plasmonic metamaterials," *Adv. Mater.*, vol. 23, no. 45, pp. 5410–5414, 2011.
115. Y.-S. Lin and W. Chen, "Perfect meta-absorber by using pod-like nanostructures with ultra-broadband, omnidirectional, and polarization-independent characteristics," *Sci. Repts.*, vol. 8, no. 1, pp. 1–9, 2018.
116. M. K. Hedayati, F. Faupel, and M. Elbahri, "Tunable broadband plasmonic perfect absorber at visible frequency," *Appl. Phys. A*, vol. 109, no. 4, pp. 769–773, 2012.
117. M. Wuttig, H. Bhaskaran, and T. Taubner, "Phase-change materials for non-volatile photonic applications," *Nature Photon.*, vol. 11, no. 8, pp. 465–476, 2017.
118. T. Cao, L. Zhang, R. E. Simpson, and M. J. Cryan, "Mid-infrared tunable polarization-independent perfect absorber using a phase-change metamaterial," *J. Opt. Soc. of Am. B*, vol. 30, no. 6, pp. 1580–1585, 2013.
119. W. Dong, Y. Qiu, J. Yang, R. E. Simpson, and T. Cao, "Wideband absorbers in the visible with ultrathin plasmonic-phase change material nanogratings," *J. Physical Chem. C*, vol. 120, no. 23, pp. 12713–12722, 2016.
120. T. Cao, K. Liu, L. Lu, H.-C. Chui, and R. E. Simpson, "Large-area broadband near-perfect absorption from a thin chalcogenide film coupled to gold nanoparticles," *ACS Appl. Mater. & Interf.*, vol. 11, no. 5, pp. 5176–5182, 2019.
121. R. J. Walters, R. V. A. van Loon, I. Brunets, J. Schmitz, and A. Polman, "A silicon-based electrical source of surface plasmon polaritons," *Nature Mater.*, vol. 9, no. 1, pp. 21–25, 2010.
122. J. A. Dionne, L. A. Sweatlock, M. T. Sheldon, A. P. Alivisatos, and H. A. Atwater, "Silicon-based plasmonics for on-chip photonics," *IEEE J. Selected Topics in Quantum Electron.*, vol. 16, no. 1, pp. 295–306, 2010.
123. K. Gorgulu, A. Gok, M. Yilmaz, K. Topalli, N. Bıyıklı, and A. K. Okyay, "All-silicon ultra-broadband infrared light absorbers," *Sci. Repts.*, vol. 6, p. 38589, 2016.
124. Z. Weng and Y. Guo, "Broadband perfect optical absorption by coupled semiconductor resonator-based all-dielectric metasurface," *Materials*, vol. 12, no. 8, p. 1221, 2019.
125. E. D. Palik, *Handbook of Optical Constants of Solids*, Vol. 3. New York, NY, USA: Academic Press, 1998.
126. J. Tian, H. Luo, Q. Li, X. Pei, K. Du, and M. Qiu, "Near-infrared super-absorbing all-dielectric metasurface based on single-layer germanium nanostructures," *Laser & Photon. Rev.*, vol. 12, no. 9, p. 1800076, 2018.
127. N. Liu, M. Mesch, T. Weiss, M. Hentschel, and H. Giessen, "Infrared Perfect Absorber and Its Application As Plasmonic Sensor," *Nano Lett.*, vol. 10, no. 7, pp. 2342–2348, 2010.
128. J. Becker, A. Trügler, A. Jakab, U. Hohenester, and C. Sönnichsen, "The optimal aspect ratio of gold nanorods for plasmonic bio-sensing," *Plasmonics*, vol. 5, no. 2, pp. 161–167, 2010.
129. F. Cheng, X. Yang, and J. Gao, "Enhancing intensity and refractive index sensing capability with infrared plasmonic perfect absorbers," *Opt. Lett.*, vol. 39, no. 11, pp. 3185–3188, 2014.
130. Y. Cheng, X. S. Mao, C. Wu, L. Wu, and R. Gong, "Infrared non-planar plasmonic perfect absorber for enhanced sensitive refractive index sensing," *Opt. Mater.*, vol. 53, pp. 195–200, 2016.
131. H. Durmaz, Y. Li, and A. E. Cetin, "A multiple-band perfect absorber for SEIRA applications," *Sensors and Actuators B: Chemical*, vol. 275, pp. 174–179, 2018.
132. C. Wu, A. B. Khanikaev, R. Adato, N. Arju, A. A. Yanik, H. Altug, and G. Shvets, "Fano-resonant asymmetric metamaterials for ultrasensitive spectroscopy and identification of molecular monolayers," *Nature Mater.*, vol. 11, no. 1, pp. 69–75, 2012.

133. E. Aslan, S. Kaya, E. Aslan, S. Korkmaz, O. G. Saracoglu, and M. Turkmen, "Polarization insensitive plasmonic perfect absorber with coupled antisymmetric nanorod array," *Sensors and Actuators B: Chemical*, vol. 243, pp. 617–625, 2017.
134. A. Tittl, P. Mai, R. Taubert, D. Dregely, N. Liu, and H. Giessen, "Palladium-based plasmonic perfect absorber in the visible wavelength range and its application to hydrogen sensing," *Nano Lett.*, vol. 11, no. 10, pp. 4366–4369, 2011.
135. K. Chen, T. D. Dao, S. Ishii, M. Aono, and T. Nagao, "Infrared aluminum metamaterial perfect absorbers for plasmon-enhanced infrared spectroscopy," *Adv. Funct. Mater.*, vol. 25, no. 42, pp. 6637–6643, 2015.
136. E. Aslan, E. Aslan, M. Turkmen, and O. G. Saracoglu, "Experimental and numerical characterization of a mid-infrared plasmonic perfect absorber for dual-band enhanced vibrational spectroscopy," *Opt. Mater.*, vol. 73, pp. 213–222, 2017.
137. Y. Li, L. Su, C. Shou, C. Yu, J. Deng, and Y. Fang, "Surface-enhanced molecular spectroscopy (SEMS) based on perfect-absorber metamaterials in the mid-infrared," *Sci. Repts.*, vol. 3, no. 1, p. 2865, 2013.
138. M. Bağmancı, M. Karaaslan, E. Ünal, O. Akgol, F. Karadağ, and C. Sabah, "Broad-band polarization-independent metamaterial absorber for solar energy harvesting applications," *Physica E: Low-Dimen. Syst. and Nanostr.*, vol. 90, pp. 1–6, 2017.
139. M. A. Shameli, P. Salami, and L. Yousefi, "Light trapping in thin film solar cells using a polarization independent phase gradient metasurface," *J. Opt.*, vol. 20, no. 12, p. 125004, 2018.
140. R. Vismara, N. O. Länk, R. Verre, M. Käll, O. Isabella, and M. Zeman, "Solar harvesting based on perfect absorbing all-dielectric nanoresonators on a mirror," *Opt. Exp.*, vol. 27, no. 16, pp. A967–A980, 2019.
141. Z. Liu, G. Liu, Z. Huang, X. Liu, and G. Fu, "Ultra-broadband perfect solar absorber by an ultra-thin refractory titanium nitride meta-surface," *Solar Energy Mater. Solar Cells*, vol. 179, pp. 346–352, 2018.
142. H. Lu, X. Guo, J. Zhang, X. Zhang, S. Li, and C. Yang, "Asymmetric metasurface structures for light absorption enhancement in thin film silicon solar cell," *J. Opt.*, vol. 21, no. 4, p. 045901, 2019.
143. Y. Wang, T. Sun, T. Paudel, Y. Zhang, Z. Ren, and K. Kempa, "Metamaterial-plasmonic absorber structure for high efficiency amorphous silicon solar cells," *Nano Lett.*, vol. 12, no. 1, pp. 440–445, 2012.
144. A. K. Azad, W. J. M. Kort-Kamp, M. Sykora, N. R. Weisse-Bernstein, T. S. Luk, A. J. Taylor, D. A. R. Dalvit, and H.-T. Chen, "Metasurface broadband solar absorber," *Sci. Repts.*, vol. 6, no. 1, p. 20347, 2016.
145. S. Murai, M. A. Verschuuren, G. Lozano, G. Pirruccio, S. R. K. Rodriguez, and J. G. Rivas, "Hybrid plasmonic-photonic modes in diffractive arrays of nanoparticles coupled to light-emitting optical waveguides," *Opt. Exp.*, vol. 21, no. 4, pp. 4250–4262, 2013.
146. G. Lozano, D. J. Louwers, S. R. K. Rodríguez, S. Murai, O. T. A. Jansen, M. A. Verschuuren, and J. Gómez Rivas, "Plasmonics for solid-state lighting: Enhanced excitation and directional emission of highly efficient light sources," *Light: Sci. & Appl.*, vol. 2, no. 5, p. e66, 2013.
147. M. Auer-Berger, V. Tretnak, F.-P. Wenzl, J. R. Krenn, and E. J. W. List-Kratochvil, "Aluminum-nanodisc-induced collective lattice resonances: Controlling the light extraction in organic light emitting diodes," *Appl. Phys. Lett.*, vol. 111, no. 17, p. 173301, 2017.
148. W. Ding, Y. Wang, H. Chen, and S. Y. Chou, "Plasmonic nanocavity organic light-emitting diode with significantly enhanced light extraction, contrast, viewing angle, brightness, and low-glare," *Adv. Funct. Mater.*, vol. 24, no. 40, pp. 6329–6339, 2014.

149. B. Zeng, Z. Huang, A. Singh, Y. Yao, A. K. Azad, A. D. Mohite, A. J. Taylor, D. R. Smith, and H.-T. Chen, "Hybrid graphene metasurfaces for high-speed mid-infrared light modulation and single-pixel imaging," *Light: Sci. & Appl.*, vol. 7, no. 1, p. 51, 2018.
150. W. Li and J. Valentine, "Metamaterial perfect absorber based hot electron photodetection," *Nano Lett.*, vol. 14, no. 6, pp. 3510–3514, 2014.
151. X.-l. Chen, C.-h. Tian, Z.-x. Che, and T.-p. Chen, "Selective metamaterial perfect absorber for infrared and 1.54 μm laser compatible stealth technology," *Optik*, vol. 172, pp. 840–846, 2018.
152. A. Tittl, A.-K. U. Michel, M. Schäferling, X. Yin, B. Gholipour, L. Cui, M. Wuttig, T. Taubner, F. Neubrech, and H. Giessen, "A switchable mid-infrared plasmonic perfect absorber with multispectral thermal imaging capability," *Adv. Mater.*, vol. 27, no. 31, pp. 4597–4603, 2015.
153. S. Mubeen, J. Lee, N. Singh, S. Krämer, G. D. Stucky, and M. Moskovits, "An autonomous photosynthetic device in which all charge carriers derive from surface plasmons," *Nature Nanotechnol.*, vol. 8, no. 4, pp. 247–251, 2013.
154. S. Mukherjee, F. Libisch, N. Large, O. Neumann, L. V. Brown, J. Cheng, J. B. Lassiter, E. A. Carter, P. Nordlander, and N. J. Halas, "Hot electrons do the Impossible: plasmon-induced dissociation of H_2 on Au," *Nano Lett.*, vol. 13, no. 1, pp. 240–247, 2013.
155. S. Mukherjee, L. Zhou, A. M. Goodman, N. Large, C. Ayala-Orozco, Y. Zhang, P. Nordlander, and N. J. Halas, "Hot-electron-induced dissociation of H_2 on gold nanoparticles supported on SiO_2," *J. Am. Chem. Soc.*, vol. 136, no. 1, pp. 64–67, 2014.
156. J. Li, S. K. Cushing, P. Zheng, F. Meng, D. Chu, and N. Wu, "Plasmon-induced photonic and energy-transfer enhancement of solar water splitting by a hematite nanorod array," *Nature Commun.*, vol. 4, no. 1, p. 2651, 2013.
157. E. W. McFarland and J. Tang, "A photovoltaic device structure based on internal electron emission," *Nature*, vol. 421, no. 6923, pp. 616–618, 2003.
158. C. Clavero, "Plasmon-induced hot-electron generation at nanoparticle/metal-oxide interfaces for photovoltaic and photocatalytic devices," *Nature Photon.*, vol. 8, no. 2, pp. 95–103, 2014.
159. H. A. Atwater and A. Polman, "Plasmonics for improved photovoltaic devices," *Nature Mater.*, vol. 9, no. 3, pp. 205–213, 2010.
160. Z. Fang, Z. Liu, Y. Wang, P. M. Ajayan, P. Nordlander, and N. J. Halas, "Graphene-antenna sandwich photodetector," *Nano Lett.*, vol. 12, no. 7, pp. 3808–3813, 2012.
161. I. Goykhman, B. Desiatov, J. Khurgin, J. Shappir, and U. Levy, "Locally oxidized silicon surface-plasmon schottky detector for telecom regime," *Nano Lett.*, vol. 11, no. 6, pp. 2219–2224, 2011.
162. M. W. Knight, H. Sobhani, P. Nordlander, and N. J. Halas, "Photodetection with active optical antennas," *Science*, vol. 332, no. 6030, p. 702, 2011.

6 Advances in Metamaterials in Conventional Low-frequency Perfect Absorbers: A Brief Review

Bui Xuan Khuyen[1,2], *Bui Son Tung*[1,2], *Nguyen Thanh Tung*[1], *Vu Dinh Lam*[2], *Liang Yao Chen*[3], *and YoungPak Lee*[3,4]

[1]Institute of Materials Science, Vietnam Academy of Science and Technology, Hanoi, Vietnam

[2]Graduate University of Science and Technology, Vietnam Academy of Science and Technology, Hanoi, Vietnam

[3]Department of Optical Science and Engineering, Fudan University, Shanghai, China

[4]Department of Physics, Quantum Photonic Science Research Center and RINS, Hanyang University, Seoul, Korea

6.1 INTRODUCTION

The innovation of the new generation of artificial materials has contributed to technological developments in materials science. The so-called metamaterials (MTMs), which owe their unnatural phenomena to light–matter interactions, were theoretically proposed by V. G. Veselago in 1968 [1]. So far, the mechanical MTMs [2], acoustic MTMs [3] and thermal MTMs [4] have enabled numerous practical applications in communication, energy harvesting, health care, recycling energy, self-driving motion, civil services and more [5–16].

A new branch, the so-called metamaterial perfect absorbers (MMPAs), which was first realized by Landy et al. in 2008 under the concept of sub-wavelength artificial materials, has grown because MMPAs can completely consume the energy of incident electromagnetic (EM) waves. In particular, the unit-cell size of MMPAs

DOI: 10.1201/9781003050162-6

was miniaturized to be far smaller than that of traditional absorbers (involving wedge absorber, ferrite, natural composite materials, etc.) [17–19]. Moreover, because of their flexibility of structure, MMPAs can be controlled from the microwave to the optical region of the EM spectrum.

Because of the revolution of 5G-communication technology, the development of telecommunication devices based on MMPAs has become a serious issue in the lower frequency bands. Low-frequency MMPAs (LFMAs) are not only used for camouflage, but also are applied to various fields, such as improved readability within radio-frequency identification (RFID) portals [20,21], power imaging [22], chip-less RFID tags [23] and sub-GHz wireless systems [24]. However, for real applications of LFMAs in the radio region, the scale-down of reported structures is a huge challenge due to the limitation of permitted dimensions for the unit cell. For instance, in Costa et al. [25], the LFMA has a periodic unit-cell dimension of ~$\lambda/2$ and a thickness of ~$\lambda/44$ at 686 MHz, where λ is the operation wavelength. In many situations, these sizes are still too large to be applied to the practical devices operating in lower bands [VHF-UHF TV, FM radio, navigation aids, cell phones and GPS (30 MHz – 3 GHz) gadgets]. Therefore, new technologies integrated with LFMAs are being investigated so that they can be used in the real-world products.

6.2 SCALING DOWN OF THE SIZE OF LFMAS BY OPTIMIZING STRUCTURES

Generally, the peak position of an LFMA is designed with the equivalent effective inductor of inductance L_{eff} and the equivalent effective capacitor of conductance C_{eff} by using formula $f = 1/2\pi\sqrt{L_{eff}C_{eff}}$ [26,27]. Thereafter, L_{eff} is defined by the unit-cell pattern, whereas C_{eff} is further optimized by the coupling between parallel metallic plates. However, broadband LFMAs are more complicated and harder to be achieved than single- or multi-band ones. The current miniaturization of LFMAs is mainly conducted under the condition of enhancing the values of C_{eff} and L_{eff} simultaneously. Following this idea, Cao et al. [28] proposed a model of thin LFMAs in the L-band (1–2 GHz), whose unit cell consists of a pair of electric ring resonators (ERRs) and the complementary ERR (CERR) with a high dielectric substrate sandwiched in between. As shown in Fig. 6.1, two key parameters are taken for the miniaturized LFMA: the short CERR and the high permittivity substrate (permittivity of 100 and low loss tangent of 0.004). Consequently, as presented in Fig. 6.2, the near perfect impedance matching is observed at 1.264 GHz, and the absorption coefficient is equal to near-unity, whereas the reflection and the transmission coefficients are close to zero. In this case, the ultrathin thickness is only around 0.4% of the wavelength with a quite narrow frequency band. Theoretically, this type of absorber can exhibit a good absorption (over 90%) in a wide incident angle range up to 49°.

Optimization of a small or thin pattern on the metasurface has emerged as an efficient tool that further reduces the unit-cell dimensions of LFMAs [29–33]. For instance, we demonstrated an ultrathin LFMA, which had a thickness of only 0.41% of the absorption wavelength, by increasing the effective capacitance through the

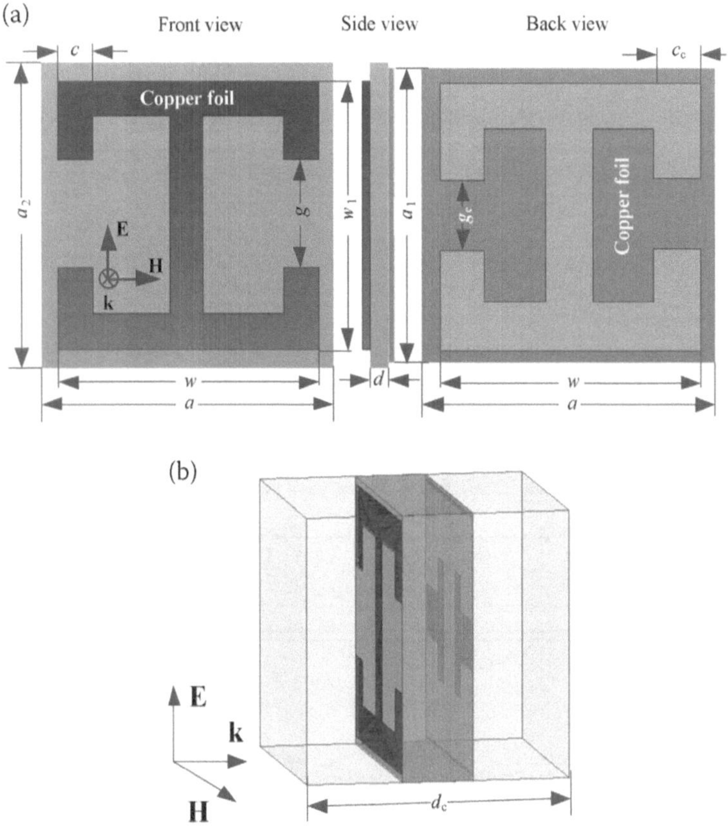

FIGURE 6.1 Schematic of proposed L-band LFMA for (a) ERR and CERR patterns and for (b) combined unit cell. The optimized geometrical values were $a = a_1 = 7.2$, $a_2 = 7.4$, $w = w_1 = 7$, $c = 1$, $c_c = 1.8$, $g = 2.5$, $g_c = 1.39$, $d = 1$, and $d_c = 7.4$ mm. The dielectric constant of substrate was chosen as $\varepsilon = 100 + 0.4i$ [28]. Reprinted with permission from Cao et al. [28].

large cross-section metasurface in the VHF band (30–300 MHz). The absorption was maintained at over 90% for the oblique incident angle (up to 45°) and for polarization-independent behavior [30]. In addition, we optimized a small-size LFMA operating at radio frequency using a spatial wire to enhance L_{eff} [34,35]. The absorption feature was created by the magnetic resonance appearing at a very low frequency, where the periodicity and thickness were miniaturized as 5.8% and 1.38% of the absorption wavelength at 377 MHz, respectively. We found that the length of the zig-zag wire, which induced a large L_{eff} due to the long-moving length of the surface current, plays an important role in obtaining the small-size absorbers [35].

6.3 FABRICATION OF LFMAS, ALLOWING MORE FUNCTIONALITIES

Most MMPAs have been realized generally on rigid substrates, such as FR-4 [17,36], silicon [37,38] and vanadium oxide [39]. Because these materials are

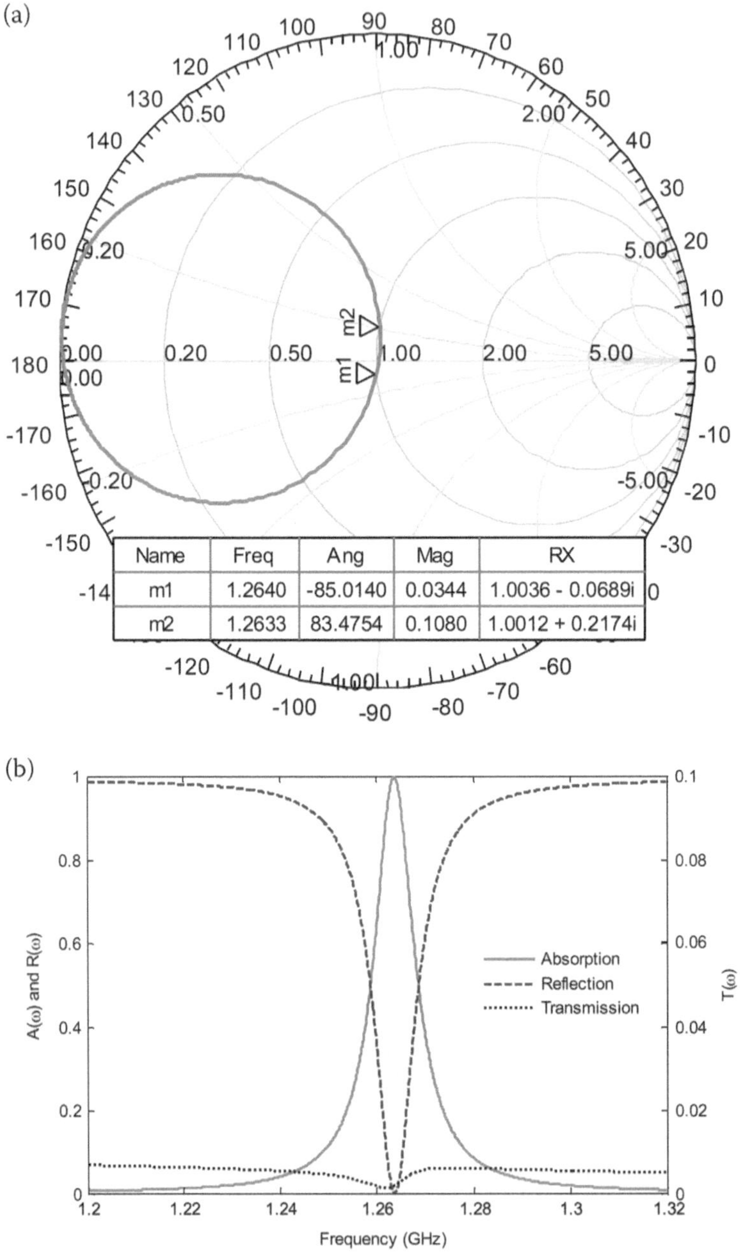

FIGURE 6.2 (a) Smith chart of the input impedance, and (b) absorption, reflection and transmission coefficients of the proposed LFMA [28]. Reprinted with permission from Cao et al. [28].

inflexible, recent hard MMPAs are limited to only planar surfaces. On the other hand, a few advanced MMPAs, based on flexible substrates (polyimide), have been proposed [40–42]. To fabricate the MA structure on a polyimide substrate or any other flexible material, researchers have used photo-lithography, where the traditional masks and laminated printed-circuit boards are preferred.

Inkjet printing technology (IPT) has been employed for the development of some LFMAs, since this led to quick, low-cost and easy fabrication for the low-frequency band and has been used for many relevant electronic components [43–60]. IPT has been featured in the literature by showing a series of successfully fabricated samples of antennas [46,49,50,54–56], RFID tags [43,44,59,59,60] and MTMs [47,51,56,57]. For IPT, both special material printers and modified conventional inkjet printers have been employed. By using very-low-cost inkjet printing setups, which comprise the standard printers loaded with metal-based inks, the fabrication errors need to be considered carefully, and these errors often put limitations on the design procedure [56]. The typical conductivity issues can be mitigated partially via post-processing (e.g., heat curing after printing), and very good flexibility is achieved by selecting the proper ink type (e.g., metal/liquid ratio) and substrate (e.g., type of photograph paper). Flexible, cheap and environmentally friendly components can be fabricated in such optimized fabrication setups, whereas the structures should still be carefully designed to overcome the inherent limitations in printing resolution [57].

One of the pioneering studies that applied the inkjet printing technique to LFMAs was suggested by Kim et al. [61]. In this work, they designed a flexible MMPA on a polymer film, and the conductive patterns were inkjet-printed using silver-nanoparticle inks. Compared with previous lithography-fabrication processes, the inkjet-printing one is very fast, simple and inexpensive because the pattern can be printed from a graphic file, even using a home printer. In addition, no chemical waste is generated, no post-processing is necessary, and no additional cost is needed. Therefore, IPT is very useful in realizing printing on flexible materials, such as polymer films and papers. To demonstrate the flexibility, the inkjet-printed MMPA was deployed on a cylindrical object in their study. At 9.21 GHz, the simulation and the measurement results for both the absorber models showed an absorption exceeding 99%, and the results were in good agreement at the resonant frequency. Afterward, another study was devoted to the fabrication of thin, polarization-independent and multi-band frequency-selective structures (typically in a low-frequency band from 2 GHz to 10 GHz) that were also suitable to be fabricated via very-low-cost inkjet printing [62]. These structures needed to be made of multiple (unconnected) elements that were easier to print and were based on resonating structures, as shown in Fig. 6.3.

Along with the innovation of optimizing LFMAs, origami and kirigami, Japanese art forms that date to at least the 17th century, could be regarded as potential candidates to motivate more unique patterns and shapes for LFMAs from paper folding (*ori-* means folding) and cutting (*kiri-* means cutting). Origami and kirigami inspire engineers to design active materials and smart structures that are bent, stretched and curved by overcoming traditional design constraints and by rendering products and systems with remarkable performance characteristics and

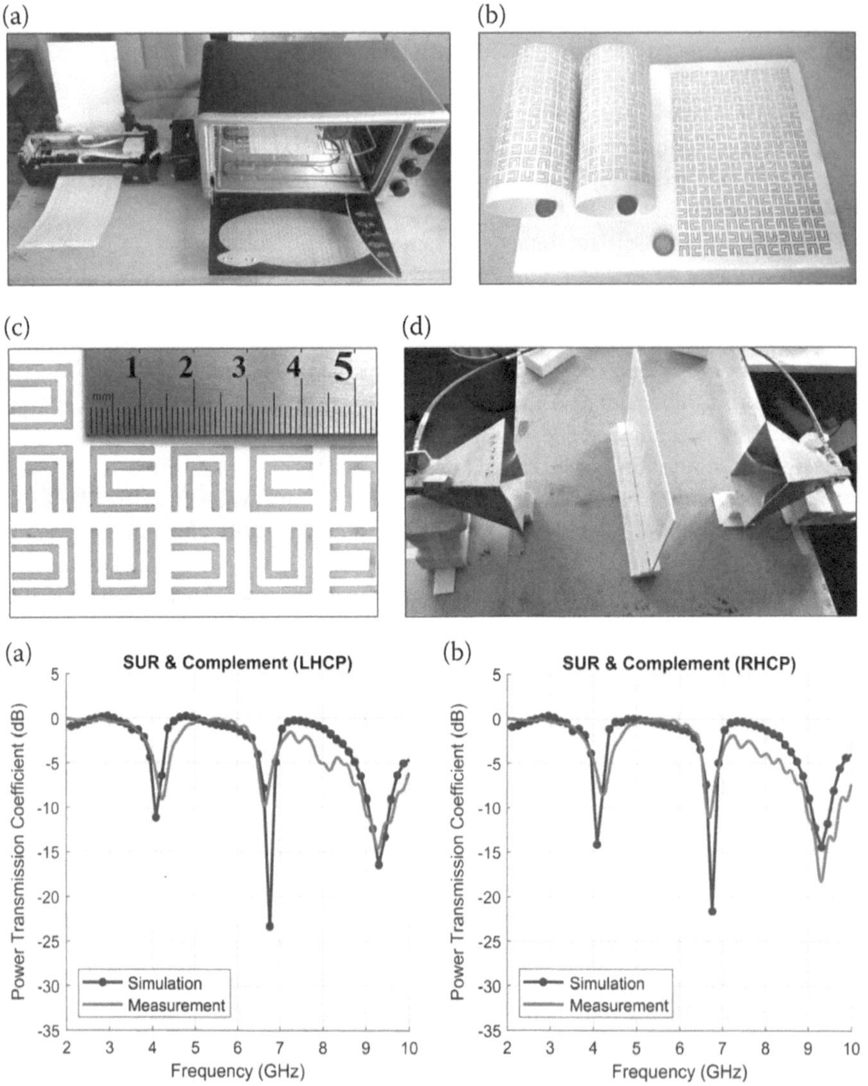

FIGURE 6.3 (Top panel) Fabrication and test of MA structures including the U-shaped resonators. (a) Inkjet printer and standard oven for heat curing. (b) Printed samples (layers). (c) A focused view on the printed U-shaped resonators. (d) Measurement setup, including two horn antennas. (Bottom panel) Simulated and measured power transmission coefficient for the structure. Both (a) left-handed and (b) right-handed circularly-polarized illuminations are considered [62]. Reprinted with permission from Eris et al. [62].

features. These techniques offer promising approaches to control mechanical, acoustic and EM properties through reconfigurable functionalities [63–67]. Due to shape change and simple tailoring of topology characteristics, origami- and kirigami-based MTMs have been proposed as a solution to create deployable continuous-state tunable structures, in which an origami pattern enables the change in the overall shape of the structure, thereby realizing the on-demand reconfigurability. For example, origami can be used to achieve tunable chirality and a frequency-selective surface, whereas kirigami brings interesting results of flexural wave control and reconfigurable toroidal circular dichroism, as shown in Fig. 6.4 [68].

Although researches have achieved early results, the current IPT still has some limitations that need to be assessed, such as the conductivity of ink after manufacturing and processing, the capability of different types of metallic nano-inks, and the dielectric loss and material behavior, based on high-order resonances and

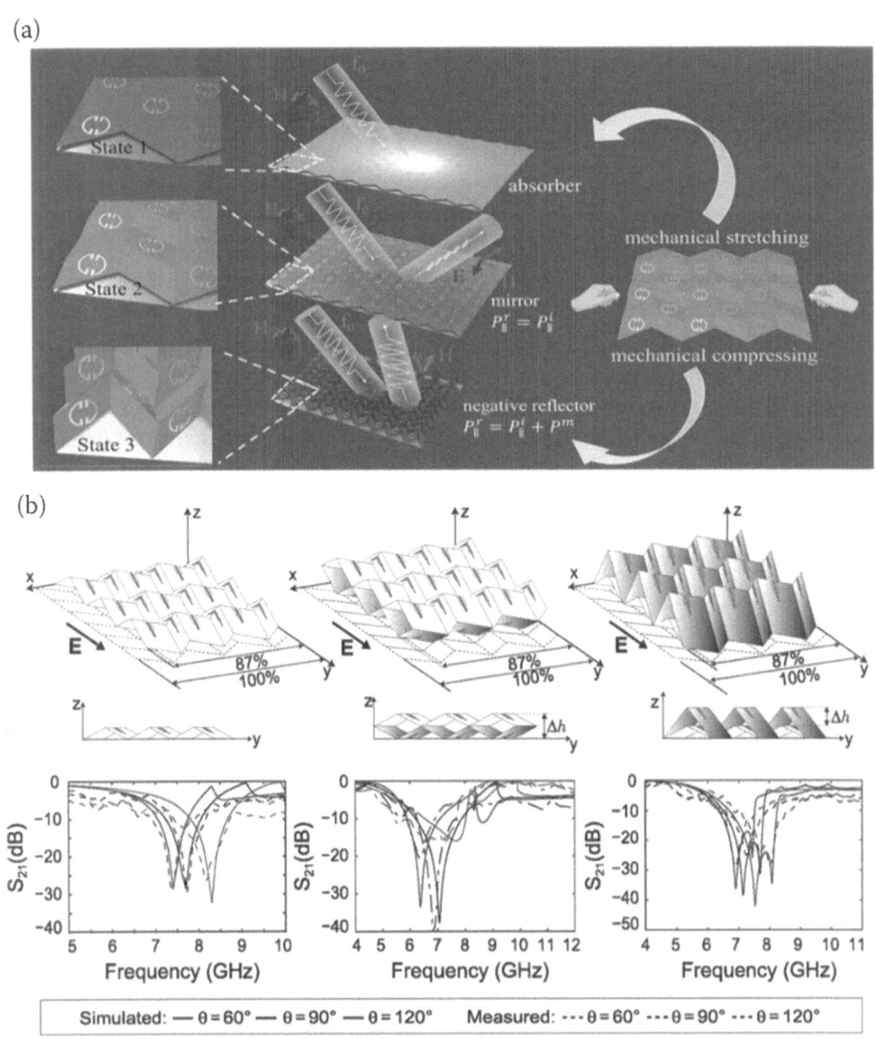

FIGURE 6.4 (a) Multi-functional and (b) active (tunable) MTMs by using origami tessellation [68]. Reprinted with permission from Li et al. [68].

integration with two-dimensional advanced materials (graphene, MoS_2, WS_2 and more). In addition, although the MTM structures based on origami and kirigami provide undeniable advantages, especially in the field of mechanical engineering, the application of origami and kirigami resonances is limited due to difficulty selecting the folding model or IPT with high practical capability, the mechanical support system, and tunability within the allowed time domain. Therefore, we predict that more effective models of origami and/or kirigami LFMAs will be explored soon, as well as improved manufacturing technology to bring MTM devices into practical operation in the low-frequency region.

6.4 MINIATURIZATION OF LFMAS BASED ON INTEGRATED PARASITIC ELEMENTS

During the past decades, LFMAs have been miniaturized quickly by several techniques in terms of integrated parasitic elements, including the hybridization method of multi-resonances [69–72,73], low-conductive materials [74–77], flexible substrates [61,78,79] and lumped electronic elements (resistor, capacitor, inductor, etc.) [80–89]. These efforts have been made toward improving future LFMAs, and broadband absorption and ultra-small unit cells are simultaneously expected at lower frequencies. Since the lumped LFMAs have more control over the internal models due to their unchanged EM properties, they are considered to be advantageous in creating new applications in civilian and military tasks. In this regard, Zhang et al. demonstrated a tunable LFMA by applying a bias voltage [89]. As shown in Fig. 6.5, they confirmed that the resonance state was excited and the absorption peak was tuned with a parasitic resistor (600 Ω) and capacitor (0.5–5 pF). Fig. 6.6 presents the measured results, which can be tuned from 0.68 GHz to 2.13 GHz (reflection less than –10 dB). In particular, the total thickness (6 mm) is only 3.3% of the center wavelength.

Meanwhile, we also proposed some efficient models of perfect absorption by MTM by means of simultaneous optimization of perfect impedance matching and fundamental magnetic resonances in the VHF and UHF bands (0.1–3.0 GHz). These works explain how we reduce effectively the size of single-/dual-/triple-/ broadband LFMAs by integrating parasitic capacitors through interconnects and low-conductive polymers [74,75,81,81,87].

As shown in Fig. 6.7, we commenced an efficient solution for an ultrathin, flexible and broadband LFMA operating from the VHF radio to Bluetooth/LTE band (200–4,000 MHz) [74]. The active role of low-conductivity material was integrated in our proposed absorber, instead of using only the usual passive role of continuous metallic layer in the traditional sandwiched structures. This new improvement for common MMPAs let the energy of incoming EM waves (from 0.2 GHz to 3.6 GHz with a fractional bandwidth of 179%) be consumed totally inside an extremely small unit cell (the thickness and periodicity were 1,362 and 72 times smaller than the longest operating wavelength, respectively). Additionally, the proposed hybrid LFMA could absorb the incoming EM wave in a broadband frequency, up to 40° of the incident angle and for the overall polarization angles of normal-incident wave.

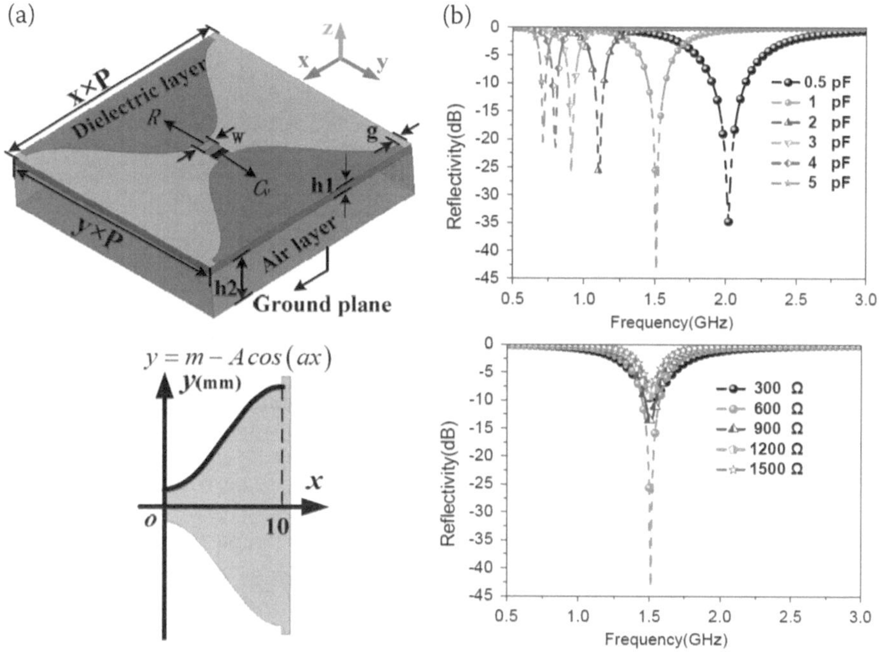

FIGURE 6.5 [Left panel] Schematic geometry parameters of the proposed tunable LFMA with (a) three-dimensional view of single unit cell, and (b) curve of metallic pattern. Optimized parameters are: $p = 30$, $h_1 = 1$, $h_2 = 5$, $w = 2$, $g = 1$, $m = 14$, $A = 12$, $x = 0.8$, $y = 1$, $a = \pi/10$ mm. [Right panel] Dependence of the simulated reflection on a parasitic element: (a) R and (b) C_v [89]. Reprinted with permission from Zhang et al. [89].

Among these suggestions, the next generation of ultrathin and angularly stable LFMAs, which have an extremely thin thickness (0.12% of the operating wavelength at 102 MHz, for example), has received the most attention [75,81]. All proposed LFMAs meet high requirements of a wide range of incident angles (from 30° to 55°) and polarization-independent performance, and they satisfy the practical applications.

6.5 EM BEHAVIOR IN CONVENTIONAL LFMAS

First of all, investigations of MTMs were mostly demonstrated from the frequency regions ranging from GHz to optical. The most pressing issue, then, was how to devise MTMs working in the low-frequency region, such as a few GHz or MHz, and attract more interests. Unfortunately, since the operational wavelength is increased in the low-frequency range, the unit-cell size of MTMs is also larger. To overcome this obstacle, researchers review the nature of resonance in MTMs, which is described by the equivalent LC-circuit model.

In some pioneer works, researchers exploited the dependence of resonance frequency on inductance by optimizing the geometry of MMPAs [34,35]. The key idea was that, if the MMPA structure could support a long motion path of induced

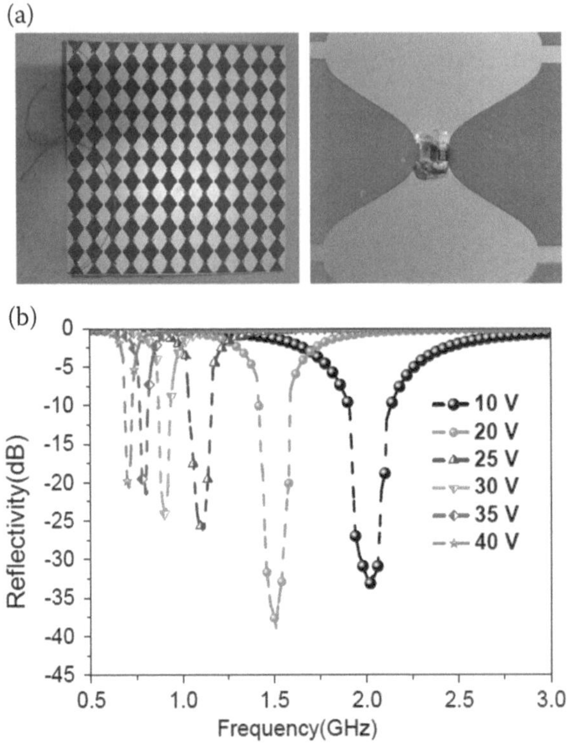

FIGURE 6.6 [Top panel] Photos of (a) whole fabricated sample, and (b) magnified picture of single unit cell. [Bottom panel] Dependence of measured reflection spectra on the applied bias voltages [89]. Reprinted with permission from Zhang et al. [89].

surface current, a large effective inductance would be created. Consequently, the resonance frequency, which was inversely proportional to inductance, was decreased to a lower frequency. Fig. 6.8 shows a typical LFMA structure, based on the aforementioned idea. The structure consisted of a teflon dielectric layer between a continuous copper layer and a copper layer patterned in a snake-like structure. As shown in Fig. 6.9, when the snake length m was increased, the induced surface current flowed on a longer motion path, leading to the decrease of resonance frequency. The absorption peak occurred at a low frequency of 1.99 GHz when the snake length was largest in the investigation, $m = 7.5$ mm.

According to the equivalent circuit model, the resonance frequency is also dependent partly on the dielectric constant of material. By exploiting this aspect, Cao et al. proposed a LFMA operating in the L-band based on high-permittivity substrates, as shown in Fig. 6.1 [28]. Fig. 6.9 presents the calculated effective parameters of the LFMA, including impedance, permittivity, permeability and refractive index in the investigated frequency region. Due to the substrate with high permittivity of 100, the matching condition of the absorber, which was responsible for the eliminated reflection, was achieved at 1.264 GHz with the real and imaginary parts of impedance of 1.023 and 0.002, respectively. In addition,

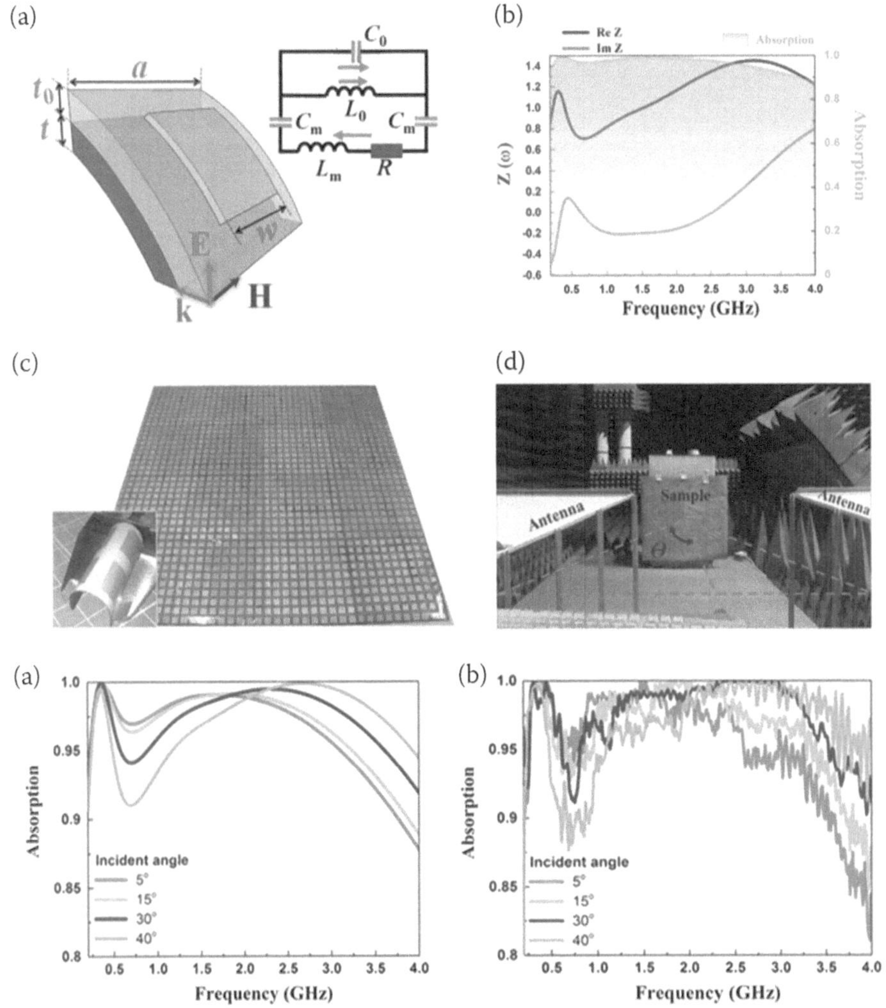

FIGURE 6.7 [Top panel] (a) A unit cell of flexible broadband LFMA integrated with a low-conductivity layer and its equivalent LC-circuit model (right-top inset with direction of flowing current, denoted by red arrows). (b) Extracted effective impedance and absorption spectrum of the proposed LFMA. (c) Photos of fabricated sample, whose inset shows zoomed flexible capability for 2 × 2 unit cells. (d) Setup for the reflection-coefficient measurement. [Bottom panel] (a) Simulated and (b) measured absorption spectra according to the incident angle of EM wave for the TE-polarization. [74]. Reprinted with permission from Khuyen et al. [74].

the imaginary part of the refractive index was also high at this frequency, which caused loss and low transmission. Consequently, the near-perfect absorption occurred at 1.264 GHz. It is noteworthy that the absorption frequency deviates slightly from the frequency of the highest imaginary part of the refractive index. However, 15.83 of the imaginary part of refractive index at 1.264 GHz still

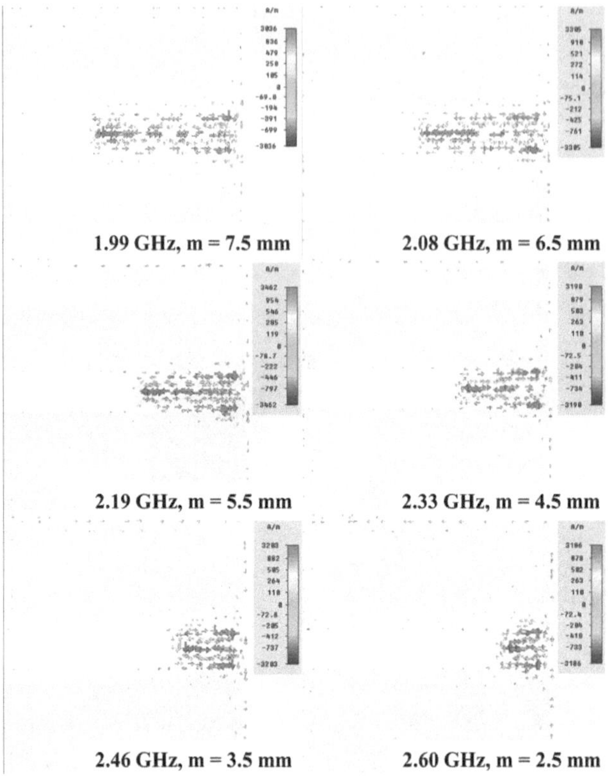

FIGURE 6.8 Simulated induced surface currents at absorption frequencies according to snake length [35]. Reprinted with permission from Yoo et al. [35].

satisfied to create extremely low transmission. In the LFMA, the calculated permeability and permittivity showed that there was a magnetic resonance at 1.264 GHz and two electric resonances at 1.08 GHz and 1.35 GHz. The results indicate that the magnetic resonance is the main mechanism of perfect absorption at 1.264 GHz, which is in agreement with most of the reported MMPAs.

Another interesting idea is to add an extra coupling between the MTM and the external EM wave in the available design space, which enhances the values of C_{eff} and L_{eff} simultaneously [89]. For example, for the proposed unit cell in Fig. 6.5, the resonance frequency f of the simplified circuit (as shown in Fig. 6.10) is designated as

$$f = \frac{\sqrt{1/(C'L' - C'^2R'^2}}{2\pi} \tag{6.1}$$

where

$$R' = \frac{R}{(C_d + C_V)^2R^2\omega^2 + 1} \tag{6.2}$$

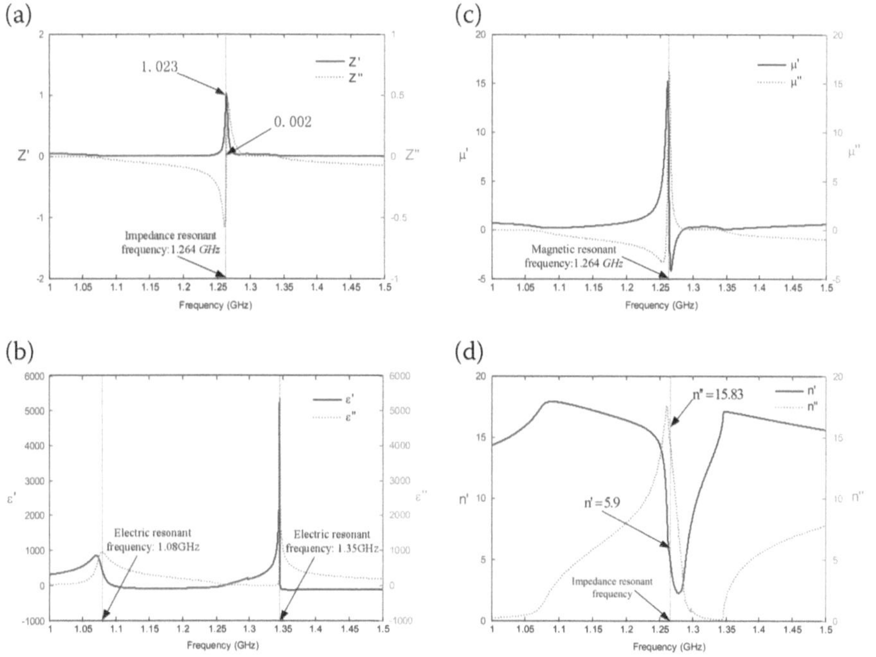

FIGURE 6.9 Calculated effective parameters of L-band LMPA: (a) impedance, (b) permittivity, (c) permeability and (d) refractive-index [28]. Reprinted with permission from Cao et al. [28].

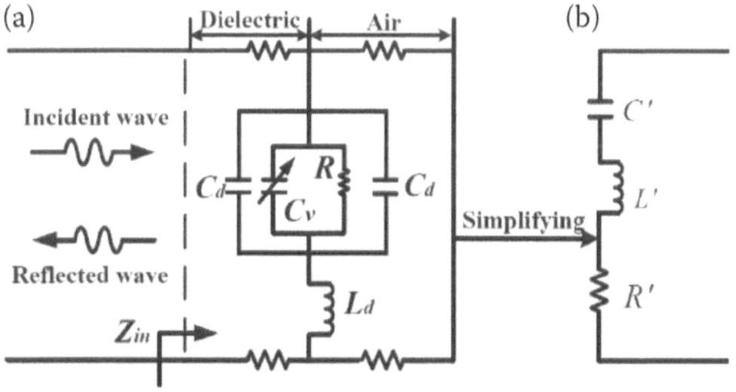

FIGURE 6.10 Equivalent circuit of typical LFMA for (a) detailed principle, and (b) simplified model [89]. Reprinted with permission from Zhang et al. [89].

and

$$C' = 1/\left[\frac{C_d + C_V}{(C_d + C_V)^2 + 1/R^2\omega^2} - L_d\omega^2\right]. \tag{6.3}$$

In this case, the resonance frequency f is dependent strongly on the intrinsic effective values (L_d and C_d) and also on the lumped elements (C_v and R).

Physically meaning, the LFMA (thickness of d) can be regarded as a typical inhomogeneous asymmetric structure. Therefore, its reflection coefficient $r(\omega)$ and transmission coefficient $t(\omega)$ are acceptably calculated by using the conventional method [28]:

$$t(\omega) = \frac{e^{-ikd\cos\theta_m}}{\cos(nkd\cos\theta_m) - \frac{i}{2}(Z(\omega,\theta_m) + Z(\omega,\theta_m)^{-1})\sin(nkd\cos\theta_m)} \tag{6.4}$$

$$r(\omega) = \frac{i}{2}[Z(\omega,\theta_m) - Z(\omega,\theta_m)^{-1}]\sin(nkd\cos\theta_m). \tag{6.5}$$

Since θ_m is the refraction angle of the EM wave inside the LFMA, k denotes the wave number of air, and the normalized impedance $Z(\omega,\theta_m)$ tends to unity, Eqs. (6.4) and (6.5) can be further simplified to

$$t(\omega) \approx e^{i(n-1)kd\cos\theta_m}, \tag{6.6}$$

$$r(\omega) \approx 0 \tag{6.7}$$

Here, the complex effective refractive index and the refraction and transmission coefficients are defined as $n(\omega) = n'(\omega) + n''(\omega)$, $|r(\omega)|^2 = R(\omega)$ and $|t(\omega)|^2 = T(\omega)$, respectively. In other words, since the effective impedance of LFMA matches perfectly with that of the surrounding environment, there is no reflection on the metasurface; then, the absorption $A(\omega) = T(\omega) - R(\omega) \approx e^{-\sigma}$, where $\sigma = 2n''kd\cos\theta_m$. Therefore, the physical mechanism of the so-called conventional LFMAs is described by both the perfect impedance matching and the extreme enhancement of attenuation factor σ around the low-resonant frequency.

6.6 CONCLUSIONS AND PERSPECTIVE

We briefly reviewed how to reduce the size of single-/multi-/broadband LFMAs by increasing the effective inductance and the effective capacitance of conventional meta-patterns. Simultaneously, by integrating the parasitic elements (capacitors, resistors, inductors and low-conductivity materials), the next generation of ultrathin and angularly stable LFMAs has been further realized with more advantages. The creation of small-size LMPAs is important and fulfills the requirements for the applications of MTMs in the low-frequency region, especially at the MHz regime. Particularly, these results have gained plenty of interests, caused by their high-potential applications in the radio band, such as radio-frequency shielding devices, single-/dual-frequency filters and single-/multi-mode switching devices. In addition, these LFMAs might also be promising candidates for military and civilian tasks or future health care applications since they can absorb harmful EM radiation from handheld wireless telephones, cellular wireless radio services, PCS (personal communications services), Wi-Fi, smart home devices, etc. However, research in areas of three-dimensional/flexible printed LFMAs and reconfigurable origami-/kirigami-meta patterns is still a huge challenge to be explored. Besides, for innovative development, the adoption of LFMAs in real-world applications should be extensively performed, such as AI (artificial intelligence), machine learning, 5G/6G and IoT (internet of things), to promote multiple-user MIMO (multiple-input and multiple-output) approaches at lower frequencies, and many more.

ACKNOWLEDGMENTS

This research was funded by Fudan University, China, by Vietnam National Foundation for Science and Technology Development (NAFOSTED) under the grant number 103.99-2018.332, and by Korea Evaluation Institute of Industrial Technology (ProjectNo. 20016179).

REFERENCES

1. V. G. Veselago, "The electrodynamics of substances with simultaneously negative values of ε and μ," *Sov. Phys. Usp.*, vol. 10, p. 509, 1968.
2. K. Bertoldi, V. Vitelli, J. Christensen, and M. V. Hecke, "Flexible mechanical metamaterials," *Nat. Rev. Mater.*, vol. 2, p. 17066, 2017.
3. Z. J. Wong, Y. Wang, K. O'Brien, J. Rho, X. Yin, S. Zhang, N. Fang, T. J. Yen, and X. Zhang, "Optical and acoustic metamaterials: Superlens, negative refractive index and invisibility cloak," *J. Opt.*, vol. 19, p. 084007, 2017.
4. S. R. Sklan and B. Li, "Thermal metamaterials: Functions and prospects," *Natl. Sci. Rev.*, vol. 5, p. 138, 2018.
5. Z. Chen, B. Guo, Y. Yang, and C. Cheng, "Metamaterials-based enhanced energy harvesting: A review," *Physica B.*, vol. 438, pp. 1, 2014.
6. O. Altintas, M. Aksoy, O. Akgol, E. Unal, M. Karaaslan, and C. Sabah, "Fluid, strain and rotation sensing applications by using metamaterial based sensor," *J. Electrochem. Soc.*, vol. 164, p. B567, 2017.
7. D. Schurig, J. J. Mock, B. J. Justice, S. A. Cummer, J. B. Pendry, A. F. Starr, and D. R. Smith, "Metamaterial electromagnetic cloak at microwave frequencies," *Science*, vol. 314, p. 977, 2006.
8. J. B. Pendry, "Negative refraction makes a perfect lens," *Phys. Rev. Lett.*, vol. 85, p. 3966, 2000.
9. S. L. Zhang, Y. C. Lai, X. He, R. Liu, Y. Zi, and Z. L. Wang, "Auxetic foam-based contact-mode triboelectric nanogenerator with highly sensitive self-powered strain sensing capabilities to monitor human body movement," *Adv. Func. Mater.*, vol. 27, p. 1606695, 2017.
10. A. M. Hawkes, A. R. Katko, and S. A. Cummer, "A microwave metamaterial with integrated power harvesting functionality," *Appl. Phys. Lett.*, vol. 103, p. 163901, 2013.
11. S. Qi, M. Oudich, Y. Li, and B. Assouar, "Acoustic energy harvesting based on a planar acoustic metamaterial," *Appl. Phys. Lett.*, vol. 108, p. 263501, 2016.
12. T. J. Cui, M. Q. Qi, X. Wan, J. Zhao, and Q. Cheng, "Coding metamaterials, digital metamaterials and programmable metamaterials," *Light Sci. Appl.*, vol. 3, p. e218, 2014.

13. K. Chen, L. Cui, Y. Feng, J. Zhao, T. Jiang, and B. Zhu, "Coding metasurface for broadband microwave scattering reduction with optical transparency," *Opt. Express.*, vol. 25, p. 5571, 2017.
14. H. Yang, X. Cao, F. Yang, J. Gao, S. Xu, M. Li, X. Chen, Y. Zhao, Y. Zheng, and S. Li, "A programmable metasurface with dynamic polarization, scattering and focusing control," *Sci. Repts.*, vol. 6, p. 35692, 2016.
15. X. G. Zhang, W. X. Tang, W. X. Jiang, G. D. Bai, J. Tang, L. Bai, C. W. Qiu, and T. J. Cui, "Light-controllable digital coding metasurfaces," *Adv. Sci.*, vol. 5, p. 1801028, 2018.
16. M. Ozturk M, U. K. Sevim, O. Akgol, M. Karaaslan, and E. Unal, "An electromagnetic non-destructive approach to determine dispersion and orientation of fiber reinforced concretes," *Measurement*, vol. 138, p. 356, 2019.
17. M. Johansson, C. Holloway, and E. Kuester, "Effective electromagnetic properties of honeycomb composites, and hollow-pyramidal and alternating-wedge absorbers," *IEEE Trans. Antennas Propag.*, vol. 53, pp. 728–736, 2005.
18. T. Deng, Y. Yu, Z. Shen, and Z. Chen, "Design of 3-D multilayer ferrite-loaded frequency-selective rasorbers with wide absorption bands," *IEEE Trans. Microwave Theory Tech.*, vol. 67, pp. 108–117, 2019.
19. Z. Wang, Z. Zhang, X. Quan, and P. Cheng, "A perfect absorber design using a natural hyperbolic material for harvesting solar energy," *Sol. Energy.*, vol. 159, pp. 329–336, 2018.
20. W. Zuo, Y. Yang, X. He, D. Zhan, and Q. Zhang, "A miniaturized metamaterial absorber for ultrahigh-frequency RFID system," *IEEE Antennas Wireless Propag. Lett.*, vol. 10, pp. 329–332, 2017.
21. G. W. Zhang, J. Gao, X. Y. Cao, S. J. Li, and H. H. Yang, "Wideband miniaturized metamaterial absorber covering L-frequency range," *Radioengineering.*, vol. 28, pp. 154–160, 2019.
22. S. Yagitani, K. Katsuda, M. Nojima, Y. Yoshimura, and H. Sugiura, "Imaging radio-frequency power distributions by an EBG absorber," *IEICE Trans. Commun.*, vol. E94-B, p. 2306, 2011.
23. F. Costa, S. Genovesi, and A. Monorchio, "A chipless RFID based on multiresonant high-impedance surfaces," *IEEE Trans. Microwave Theory Tech.*, vol. 61, p. 146, 2013.
24. F. Costa, S. Genovesi, A. Monorchio, and G. Manara, "Low-cost metamaterial absorbers for sub-GHz wireless systems," *IEEE Antennas Wireless Propag. Lett.*, vol. 13, p. 27, 2014.
25. F. Costa, S. Genovesi, A. Monorchio, and G. Manara, "Low-cost metamaterial absorbers for sub-ghz wireless systems," *IEEE Antennas Wirel. Propag. Lett.*, vol. 13, p. 27, 2014.
26. J. Zhou, E. N. Economou, T. Koschny, and C. M. Soukoulis, "Unifying approach to left-handed material design," *Opt. Lett.*, vol. 31, p. 3620, 2006.
27. N. Zhang, P. Zhou, S. Wang, X. Weng, J. Xie, and L. Deng, "Broadband absorption in mid-infrared metamaterial absorbers with multiple dielectric layers," *Opt. Commun.*, vol. 338, p. 388, 2015.
28. Z. X. Cao, F. G. Yuan, and L. H. Li, "A super-compact metamaterial absorber cell in L-band," *J. Appl. Phys.*, vol. 115, p. 184904, 2014.
29. B. X. Khuyen, B. S. Tung, N. V. Dung, Y. J. Yoo, Y. J. Kim, K. W. Kim, V. D. Lam, J. G. Yang, and Y. P. Lee, "Size-efficient metamaterial absorber at low frequencies: Design, fabrication, and characterization," *J. Appl. Phys.*, vol. 117, p. 243105, 2015.
30. B. X. Khuyen, B. S. Tung, Y. J. Yoo, Y. J. Kim, V. D. Lam, J. G. Yang, and Y. P. Lee, "Ultrathin metamaterial-based perfect absorbers for VHF and THz bands," *Curr. Appl. Phys.*, vol. 16, p. 1009, 2016.

31. F. Erkmen, T. S. Almoneef, and O. M. Ramahi, "Scalable electromagnetic energy harvesting using frequency-selective surfaces," *IEEE Trans. Microwave Theory Tech.*, vol. 66, pp. 2433–2441, 2018.
32. H.-X. Xu, G.-M. Wang, M.-Q. Qi, J.-G. Liang, J.-Q. Gong, and Z.-M. Xu, "Triple-band polarization-insensitive wide-angle ultra-miniature metamaterial transmission line absorber," *Phys. Rev. B.*, vol. 86, p. 205104, 2012.
33. S. Costanzo and F. Venneri, "Polarization-insensitive fractal metamaterial surface for energy harvesting in IoT applications," *Electronics*, vol. 9, p. 959, 2020.
34. B. S. Tung, B. X. Khuyen, N. V. Dung, Y. J. Yoo, K. W. Kim, V. D. Lam, and Y. P. Lee, "Small-size metamaterial perfect absorber operating at low frequency," *Adv. Nat. Sci.: Nanosci. Nanotechnol.*, vol. 5, p. 045008, 2014.
35. Y. J. Yoo, H. Y. Zheng, Y. J. Kim, J. Y. Rhee, J. H. Kang, K. W. Kim, H. Cheong, Y. H. Kim, and Y. P. Lee, "Flexible and elastic metamaterial absorber for low frequency, based on small-size unit cell," *Appl. Phys. Lett.*, vol. 105, p. 041902, 2014.
36. B. Zhu, Z. Wang, C. Huang, Y. Feng, J. Zhao, and T. Jiang, "Polarization insensitive metamaterial absorber with wide incident angle," *Prog. Electromagnetics Res.*, vol. 101, pp. 231–239, 2010.
37. H. Tao, A. C. Strikwerda, K. Fan, W. J. Padilla, X. Zhang, and R. D. Averitt, "MEMS based structurally tunable metamaterials at terahertz frequencies," *J. Infrared Millim. THz Waves.*, vol. 32, no. 5, pp. 580–595, 2011.
38. N. Landy, C. Bingham, T. Tyler, N. Jokerst, D. Smith, and W. Padilla, "Design, theory, and measurement of a polarization-insensitive absorber for terahertz imaging," *Phys. Rev. B.*, vol. 79, no. 12, p. 125104, 2009.
39. Q. Wen, H. Zhang, Q. Yang, Z. Chen, Y. Long, Y. Jing, Y. Lin, and P. Zhang, "A tunable hybrid metamaterial absorber based on vanadium oxide films," *J. Phys. D.*, vol. 45, no. 23, p. 235106, 2012.
40. K. Iwaszczuk, A. C. Strikwerda, K. Fan, X. Zhang, R. D. Averitt, and P. U. Jepsen, "Flexible metamaterial absorbers for stealth applications at terahertz frequencies," *Opt. Exp.*, vol. 20, no. 1, pp. 635–643, 2012.
41. P. K. Singh, K. A. Korolev, M. N. Afsar, and S. Sonkusale, "Single and dual band 77/95/110 GHz metamaterial absorbers on flexible polyimide substrate," *Appl. Phys. Lett.* vol. 99, no. 26, p. 264101, 2011.
42. H. Tao, C. Bingham, D. Pilon, K. Fan, A. Strikwerda, D. Shrekenhamer, W. Padilla, X. Zhang, and R. Averitt, "A dual band terahertz metamaterial absorber," *J. Phys. D.*, vol. 43, no. 22, p. 225102, 2010.
43. P. V. Nikitin, S. Lam, and K. V. S. Rao, "Low cost silver ink RFID tag antennas," *Proc. IEEE Antennas and Propagation Soc. Int. Symp.*, vol. 2, pp. 353–356, 2005.
44. L. Yang, A. Rida, R. Vyas, and M. M. Tentzeris, "RFID tag and RF structures on a paper substrate using inkjet-printing technology," *IEEE Trans. Microw. Theory Tech.*, vol. 55, pp. 2894–2901, 2007.
45. J. C. Batchelor, E. A. Parker, J. A. Miller, V. Sanchez-Romaguera, and S. G. Yeates, "Inkjet printing of frequency selective surfaces," *Electron. Lett.*, vol. 45, pp. 7–8, 2009.
46. A. Rida, L. Yang, R. Vyas, and M. M. Tentzeris, "Conductive inkjet-printed antennas on flexible low-cost paper-based substrates for RFID and WSN applications," *IEEE Antennas Propag. Mag.*, vol. 51, pp. 13–23, 2009.
47. M. Walther, A. Ortner, H. Meier, U. Loffelmann, P. J. Smith, and J. G. Korvink, "Terahertz metamaterials fabricated by inkjet printing," *Appl. Phys. Lett.*, vol. 95, p. 251107, 2009.
48. J. R. Cooper, S. Kim, and M. M. Tentzeris, "A novel polarization-independent, free-space, microwave beam splitter utilizing an inkjet-printed, 2-D array frequency selective surface," *IEEE Antennas Wireless Propag. Lett.*, vol. 11, pp. 686–688, 2012.

49. A. R. Maza, B. Cook, G. Jabbour, and A. Shamim, "Paper-based inkjet-printed ultra-wideband fractal antennas," *IET Microwaves, Antennas & Propagation*, vol. 6, pp. 1366–1373, 2012.
50. H. Subbaraman, D. T. Pham, X. Xu, M. Y. Chen, A. Hosseini, X. Lu, and R. T. Chen, "Inkjet-printed two-dimensional phased-array antenna on a flexible substrate," *IEEE Antennas Wireless Propag. Lett.*, vol. 12, pp. 170–173, 2013.
51. M. Yoo, H. K. Kim, S. Kim, and M. M. Tentzeris, "Silver nanoparticle-based inkjet-printed metamaterial absorber on flexible paper," *IEEE Antennas Wireless Propag. Lett.*, vol. 14, pp. 1718–1721, 2015.
52. F. B. Ashraf, T. Alam and M. T. Islam, "A printed xi-shaped left-handed metamaterial on low-cost flexible photo paper," *Materials.*, vol. 10, p. 752, 2017.
53. S. N. Zabri, R. Cahill, G. Conway, and A. Schuchinsky, "Inkjet printing of resistively loaded FSS for microwave absorbers," *Electron. Lett.*, vol. 51, pp. 999–1001, 2015.
54. T. Ciftci, B. Karaosmanoglu, and Ö. Ergül, "Low-cost inkjet antennas for RFID applications,"*IOP Conf. Ser.: Mater. Sci. Eng.*, vol. 120, pp. 1, 2016.
55. S. Guler, B. Karaosmanoglu, and Ö. Ergül, "Design and fabrication of low-cost inkjet-printed metamaterials," *Prog. Electromagn. Res. Lett.*, vol. 64, pp. 51–55, 2016.
56. F. Mutlu, C. Onol, B.Karaosmanoglu, and Ö. Ergül, "Inkjet-printed cage-dipole antennas for radio-frequency applications," *IET Microwaves, Antennas & Propagation*, vol. 11, pp. 2016–2020, 2017.
57. H. Ibili and Ö. Ergül, "Very low-cost inkjet-printed metamaterials: progress and challenges," *Proc. IEEE Int. Microwave Workshop Series Adv. Mater. Processes*, pp. 1–3, 2017.
58. H. Ibili, B. Karaosmanoglu, and Ö. Ergül, "Demonstration of negative refractive index with low-cost inkjet-printed microwave metamaterials," *Microw. Opt. Technol. Lett.*, vol. 60, pp. 187–191, 2017.
59. E. Çetin, M. B. Sahin, and Ö. Ergül, "Low-cost inkjet-printed multiband frequency-selective structures consisting of u-shaped resonators," *Proc. IEEE Intl. Symp. Antennas Propagation Soc.*, pp. 783–786, 2018.
60. M. A. Demir, F. Mutlu, and Ö. Ergül, "Design of highly distinguishable letters for inkjet-printed chipless RFID tags," *Proc. IEEE-APS Topical Conf. Antennas Propagation Wireless Commun.*, pp. 783–786, 2018.
61. H. K. Kim, K. Ling, K. Kim, and S. Lim, "Flexible inkjet-printed metamaterial absorber for coating a cylindrical object," *Opt. Express.*, vol. 23, p. 5898, 2015.
62. O. Eris, H. Ibili, and Ö. Ergül, "Low-cost inkjet-printed multiband frequency-selective structures consisting of u-shaped resonators," *Prog. Electromagn. Res. C.*, vol. 98, pp. 31–44, 2020.
63. E. T. Filipov, T. Tachi, and G. H. Paulino, "Origami tubes assembled into stiff, yet reconfigurable structures and metamaterials," *Proc. Natl. Acad. Sci. USA.* vol. 112, no. 40, p. 12321, 2015.
64. S. Li, H. Fang, S. Sadeghi, P. Bhovad, and K.-W. Wang, "Architected origami materials: how folding creates sophisticated mechanical properties," *Adv. Mater.*, vol. 31, no. 5, p. 1805282, 2019.
65. X. Yu, H. Fang, F. Cui, L. Cheng, and Z. Lu, "Origami-inspired foldable sound barrier designs," *J. Sound Vib.*, vol. 442, pp. 514–526, 2019.
66. Z. Wang, L. Jing, K. Yao, Y. Yang, B. Zheng, C. M. Soukoulis, H. Chen, and Y. Liu, "Origami-based reconfigurable metamaterials for tunable chirality," *Adv. Mater.*, vol. 29, no. 27, p. 1700412, 2017.
67. S. A. Nauroze, L. S. Novelino, M. M. Tentzeris, and G. H. Paulino, "Continuous-range tunable multilayer frequency-selective surfaces using origami and inkjet printing," *Proc. Natl. Acad. Sci. USA*, vol. 115, no. 52, p. 13210, 2018.

68. M. Li, L. Shen, L. Jing, S. Xu, B. Zheng, X. Lin, Y. Yan, Z. Wang, and H. Chen, "Origami metawall: mechanically controlled absorption and deflection of light," *Adv. Sci.*, vol. 6, p. 1901434, 2019.
69. D. Q. Vu, D. H. Le, H. T. Dinh, T. G. Trinh, L. Yue, D. T. Le, and D. L. Vu, "Broadening the absorption bandwidth of metamaterial absorber by coupling three dipole resonances," *Physica B: Condensed Matter.*, vol. 532, pp. 90–94, 2018.
70. Y. Cheng, B. He, J. Zhao, and R. Gong, "Ultra-thin low-frequency broadband microwave absorber based on magnetic medium and metamaterial," *J. Electron. Mater.*, vol. 46, pp. 1293–1299, 2017.
71. H. Jeong and S. Lim, "Broadband frequency-reconfigurable metamaterial absorber using switchable ground plane," *Sci. Repts.*, vol. 8, pp. 9226, 2018.
72. C. Y. Wang, J. G. Liang, T. Cai, H. P. Li, W. Y. Ji, Q. Zhang, and C. W. Zhang, "High-performance and ultra-broadband metamaterial absorber based on mixed absorption mechanisms," *IEEE Access.*, vol. 7, pp. 57259–57266, 2019.
73. D. H. Tiep, B. X. Khuyen, B. S. Tung, Y. J. Kim, J. S. Hwang, V. D. Lam, and Y. P. Lee, "Enhanced-bandwidth perfect absorption based on a hybrid metamaterial," *Opt. Mat. Exp.*, vol. 8, p. 2751, 2018.
74. B. X. Khuyen, B. S. Tung, Y. J. Kim, K. W. Kim, J. Y. Rhee, V. D. Lam, L. Y. Chen, X. Guo, and Y. P. Lee, "Broadband and ultrathin metamaterial absorber fabricated on a flexible substrate in the long-term evolution band," *J. Electron. Mater.*, vol. 48, pp. 7937–7943, 2019.
75. Y. J. Kim, J. S. Hwang, B. X. Khuyen, B. S. Tung, K. W. Kim, J. Y. Rhee, L. Y. Chen, and Y. P. Lee, "Flexible ultrathin metamaterial absorber for wide frequency band, based on conductive fibers," *Sci. Tech. Adv. Mater.*, vol. 19, pp. 711–717, 2018.
76. W. Li, L. Lin, C. Li, Y. Wang, and J. Zhang, "Radar absorbing combinatorial metamaterial based on silicon carbide/carbon foam material embedded with split square ring metal," *Results Phys.*, vol. 12, pp. 278–286, 2019.
77. W. Zuo, Y. Yang, X. He, C. Mao, and T. Liu, "An ultrawideband miniaturized metamaterial absorber in the ultrahigh-frequency range," *IEEE Antennas Wireless Propag. Lett.*, vol. 16, pp. 928–931, 2017.
78. D. Simon, T. Ware, R. Marcotte, B. R. Lund, D. W. Smith, M. D. Prima, R. L. Rennaker, and W. Voit, "A comparison of polymer substrates for photolithographic processing of flexible bioelectronics," *Biomed Microdevices.*, vol. 15, pp. 925–939, 2013.
79. D. Zha, J. Dong, Z. Cao, Y. Zhang, F. He, R. Li, Y. He, L. Miao, S. Bie, and J. Jiang, "A multimode, broadband and all-inkjet-printed absorber using characteristic mode analysis," *Opt. Express.*, vol. 28, pp. 8609–8618, 2020.
80. Y. Shen, Z. Pei, Y. Pang, J. Wang, A. Zhang, and S. Qu, "An extremely wideband and lightweight metamaterial absorber," *J. Appl. Phys.*, vol. 117, pp. 224503, 2015.
81. B. X. Khuyen, B. S. Tung, Y. J. Kim, J. S. Hwang, K. W. Kim, J. Y. Rhee, V. D. Lam, and Y. P. Lee, "Ultra-subwavelength thickness for dual/triple-band metamaterial absorber at very low frequency," *Sci. Repts.*, vol. 8, p. 11632, 2018.
82. Y. J. Kim, J. S. Hwang, Y. J. Yoo, B. X. Khuyen, J. Y. Rhee, X. Chen, and Y. P. Lee, "Ultrathin microwave metamaterial absorber utilizing embedded resistors," *J. Phys. D: Appl. Phys.*, vol. 50, p. 405110, 2017.
83. H. Jeong, T. T. Nguyen, and S. Lim, "Subwavelength metamaterial unit cell for low-frequency electromagnetic absorber applications," *Sci. Repts.*, vol. 8, p. 16774, 2018.
84. B. X. Khuyen, B. S. Tung, T. T. Nguyen, N. T. Hien, Y. J. Kim, L. Y. Chen, Y. P. Lee, P. T. Linh, and V. D. Lam, "Realization for dual-band high-order perfect absorption, based on metamaterial," *J. Phys. D: Appl. Phys.*, vol. 53, p. 105502, 2020.

85. W. Yuan and Y. Cheng, "Low-frequency and broadband metamaterial absorber based on lumped elements: design, characterization and experiment," *Appl. Phys. A.*, vol. 117, pp. 1915–1921, 2014.
86. H. F. Zhang, J. Yang, H. Zhang, and J. X. Liu, "Design of an ultra-broadband absorber based on plasma metamaterial and lumped resistors," *Opt. Mater. Exp.*, vol. 8, pp. 2103–2113, 2018.
87. B. X. Khuyen, B. S. Tung, Y. J. Yoo, Y. J. Kim, K. W. Kim, L.-Y. Chen, V. D. Lam, and Y. P. Lee, "Miniaturization for ultrathin metamaterial perfect absorber in the VHF band, " *Sci. Repts.*, vol. 7, p. 45151, 2017.
88. S. J. Li, P. X. Wu, H. X. Xu, Y. L. Zhou, Y. Y. Cao, J. F. Han, C. Zhang, H. H. Yang, and Z. Zhang, "Ultra-wideband and polarization-insensitive perfect absorber using multilayer metamaterials, lumped resistors, and strong coupling effects," *Nano Res. Lett.*, vol. 13, p. 386, 2018.
89. G. W. Zhang, J. Gao, X. Cao, H. H. Yang, and L. Jidi, "Improving radar absorbing capability of polystyrene nanocomposites: preparation and investigation of microwave absorbing properties," *Radioengineering*, vol. 28, p. 579, 2019.

7 Photonic Metamaterials

Tatjana Gric[1,2,3]
[1]Department of Electronic Systems, Vilnius Gediminas Technical University, Vilnius, Lithuania
[2]Aston Institute of Photonic Technologies, Aston University, Birmingham, UK
[3]Semiconductor Physics Institute, Center for Physical Sciences and Technology, Vilnius, Lithuania

7.1 INTRODUCTION

Recently, the field of plasmonics has been the subject of extensive research, giving rise to a perfect match of electronics and photonics at the nanoscale. Plasmons are collective oscillations of charge carriers in metal and semiconductors. The former can either be localized as the eigen-mode of nanocomposite or propagate over extended surfaces. Doing so, the phenomenon of sub-wavelength light confinement, thereby resulting in giant electric field enhancement into specific nano-sized regions (called hotspots), is enabled [1]. These attractive features open wide avenues for a range of applications, such as optical sensing [2,3], quantum electrodynamics [4,5], non-linear optics [6,7], photovoltaic technologies [8], and medical diagnosis and treatment [9,10]. A deep physical insight has been gained to engineer plasmonic devices with tunable plasmon frequencies and associated electric field spatial distribution. This can be reached by designing different metallic micro- and nanostructures [11–14] and by exploiting widely studied phenomena, such as Fano interactions [15,16] with other excitations, plasmon hybridization [17,18] and electromagnetically induced transparency [19,20].

The system is treated as non-local if its behavior at a given point is affected by its state at another spatially divided area. Quantum states of light and matter are inherently non-local. The former reflects the fundamental wave–particle duality [21]. Quantum entanglement is treated as one of the most amazing illustrations of non-localities in nature [22,23]. Inherently weak photon–photon interactions enable their efficient demonstration in optics with photons [24,25].

Metamaterials, composed of sub-wavelength artificial unit cells, open up wide avenues in optical applications, such as perfect lenses [26] and invisibility cloaks [27,28]. Unusual hyperbolic dispersion curves [29–39] supporting radiative modes with unbounded wave numbers give rise to tremendous interest in hyperbolic metamaterials (HMMs). Because of the extreme anisotropy in the uniaxial permittivity tensor, HMMs give rise to unique properties. The hyperbolicity occurs when one of the diagonal tensor components is opposite in sign from the other two

DOI: 10.1201/9781003050162-7

components. This unique material dispersion is attributed to the novel optical properties of HMMs. This includes a negative refractive index in a waveguiding geometry [40], nominally unbounded density of photonic states [41], and the existence of an ENZ (epsilon-near-zero) wavelength range for one of the permittivity components [42].

It is worthwhile to note that the scale on which metamaterials can be patterned is entering the nanometer regime in light of novel nanofabrication techniques. Therefore, the non-local response of metal is important [43–53]. The localized surface plasmon polariton (SPP) resonance peak can be significantly blue-shifted due to the non-local response. Also, the non-local response can cause the modification of field enhancement of a nanoscale plasmonic structure [46–50,54].

Graphene has attracted significant attention because of its remarkable features, thereby giving rise to many interesting phenomena in the field of metamaterials, such as a negative refractive index, cloaking, perfect focusing, phase discontinuities and polarization manipulation. In recent years, the fabrication and characterization of graphene-based metamaterials (designed for THz and optical frequencies) have shown rapid progress, thereby promising applications in many fields, such as security, biology and chemistry. This chapter studies active graphene metamaterials, graphene nanocomposites with tunable and switchable properties, and novel functionalities.

The electromagnetic (plasmonic) properties of single- and multilayer graphene systems (including analytic expression for dynamic conductivity and dispersion equations) have been the subject of intense study [55–63]. Single- and double-layer structures have been proposed as "promising lasing media" for THz radiation. Terahertz emission from an optically pumped graphene was reported, for instance, in Karasawa et al. [64].

One may consider a composite of graphene sheets, divided by sub-wavelength dielectric inserts, as a composite material with uniaxial electric features. Thus, hyperbolic graphene-based metamaterials stand for artificially designed materials. The main goal of the former design is related to attaining strong anisotropic effects. The permittivities related to different polarization directions exhibit opposite signs [65–69] in the frame of this phenomenon. For instance, it has been suggested that these structures possess unique properties, such as imaging [65]. Moreover, further dramatic manifestations are shown by optical topological transitions recently found in electromagnetic metamaterials. It is worth noting that these optical properties are not subject to the restrictions of the Fermi exclusion principle [66]. A periodic metal-dielectric nanostructure metamaterial is an example of one of the probable realizations of hyperbolic medium. In this relation, the hyperbolic nature of isofrequency contours takes place because of the excitation of near-field plasmon Bloch waves [68]. Moreover, HMMs have been considered from theoretical perspectives as electromagnetic absorbers for scattered fields. The described phenomenon was experimentally shown by placing scatterers on top of HMMs. Consequently, enhanced absorption was demonstrated; however, this design was neither perfect nor narrow-band absorption [69]. The isofrequency wave vector dispersion in a uniaxially anisotropic medium can be elliptic for some polarization. Moreover, a hyperbolic relation leading to a number of fascinating physical features can evolve [70,71]. The transition from the elliptical to hyperbolic regime takes place in the case of metamaterials. A simple way for realizing such transitions is offered by graphene-based metamaterials. Investigations of hyperbolic media and HMMs have provided a fertile ground for many applications recently because of their rather simple geometry and a number of fascinating features, such as high density of states, all-angle negative refraction and hyperlens imaging beyond the diffraction limit [71,72]. For instance, graphene metamaterials and metasurfaces have been proposed to make amplifiers and lasing devices [73–75]. Our approach, on the contrary, allows for narrowing the quasibound leaky part of dispersion diagrams by using the enhanced system.

This chapter consists of four main parts. First, we analyze the propagation of SPPs – coupled oscillation of surface charges and electromagnetic field [1] – at the boundary of two-layered and multilayered nanostructured metamaterials. Inspired by the recent exploration of SPPs in nanostructured metamaterials [76], we deal with several remarkable matters, namely (i) we calculate the dispersion properties and propagation lengths of SPPs at the boundary of nanostructured metamaterials (employing realistic material parameters; the former allows comparison of potentials of these materials in plasmonics); (ii) we determine crucial features influencing the propagation length and absorption enhancement and prove that the latter can be enhanced by two orders of magnitude in some cases (the former can give rise to practical applications); and (iii) we investigate the outcomes of stacking of graphene layers and dielectrics into one-dimensional (1D) superlattice aiming to modify the dispersion relation of SPPs [77]. Due to the dependence of the optical response of materials on its bulk dielectric function $\varepsilon(\omega)$, it is imperative to take its possible variation into account. We approximate $\varepsilon(\omega)$ by exploiting the Drude model, aiming to properly address questions i–iii. In the second part, we consider metamaterials from the perspective of non-local effective medium approximation. Moreover, we analyze two different instances, i.e., the nanowired and nanostructured metamaterials cases. We study a non-local environment employing a plasmonic nanowire metamaterial system. The former was employed to allow for a topological transition between elliptic and hyperbolic dispersions.

The work is systematized in the following way. We describe the effective medium formalism in the case of local and non-local approaches of a composite metamaterial. Then, we present an analytical approach providing sufficient explanation of electromagnetism in nanowire metamaterials, considering the non-local optical response taking its origin from the homogenization technique. The formalism might be used to characterize optics of coaxial-cable-like media [78–80] and other uniaxial composites. Based on the presented theoretical approach, one may reconcile the local and non-local effective medium approaches frequently employed to characterize the optics of nanowire media in different limits [81]. Moreover, we relate the origin of optical non-locality to the collective (averaged over many nanorods) plasmonic excitation of nanowire media due to the presented formalism. Doing so, we provide the methodology to fulfill the additional boundary conditions in nanowire composites. We illustrate the established approach using the instance of plasmonic nanowire metamaterials. These are designed employing an assembly of systemized plasmonic nanowires [82–85] embedded in a dielectric host. To get a full picture, we vary the unit cell geometry of the composite. Moreover, we have considered metamaterials with different unit cell types, viz. the rectangular and hexagon. We make an assumption that the system is operated in the effective medium regime (its unit cell $S \ll \lambda_0$ with λ_0 being the free-space wavelength). In our previous works, the anisotropic multilayer metamaterials were designed in order to realize propagation of SPPs mainly based on the effective medium theory. However, the effective permittivity tensor, including the optical non-locality, is not well studied in the analysis. Therefore, in the present work, new kinds of metal-dielectric multilayer metamaterials are designed to have the extremely anisotropic effective permittivity tensor for the demonstration of the propagation of SPPs, based on a theoretical analysis of the effective permittivity tensor, including the optical non-locality. We discuss the effects of non-local optical response of HMMs. We conclude by presenting an application of nanostructured metamaterial. Here, we systematically investigate the dispersion relations supported by the HMM model based on a graphene-dielectric composite heterostructure [86–90]. The goal is to model and create a new generation of active plasmonic graphene-based metamaterials with the optical features electrically engineered by two-dimensional (2D) heterostructures [87].

Plasmonic crystals, a class of particularly interesting metamaterials, consist of stacked graphene layers arranged periodically with sub-wavelength distance and embedded in a dielectric host [88]. The HMM based on a semi-infinite system of graphene-dielectric multilayers was considered in Iorsh et al. [89]. The study was conducted at a temperature of 4 K. Therefore, the graphene losses were neglected, and purely real permittivity and wave vectors were considered. In this study, graphene was treated as a 2D honeycomb system with the thickness of a monolayer of carbon atoms. Thus, it was considered to be a type of gapless semiconductor [91,92]. The surface waves at the boundary of graphene-based HMM and isotropic media were discussed in Xiang et al. [93]. Hence, we consider the interface separating the anisotropic media and dielectric.

We also discuss recent approaches to incorporate active photo-excited graphene layers into the HMM and explore opportunities based on this configuration. We consider the presented problem from different perspectives and start our analysis from the simplified model. We suggest the enhancement of SPP properties by considering the improved structure with air spacers.

7.2 METAMATERIAL HETEROSTRUCTURES

The starting point for our consideration is a multilayer metamaterial made of stacked graphene sheets separated by dielectric layers, as depicted schematically in Fig. 7.1a, along with the simplified two-layered composites (Fig. 7.1b). It is worth noting that the ribbon-array structure possesses an intrinsic optical anisotropy. For light polarization parallel to ribbons, the single-particle optical response of Dirac electrons takes place, while in the case of perpendicular polarization, plasmons can be excited.

The effective-medium approach is applied, aiming to describe the optical response of such a system. The former proposition is justified if the wavelength of radiation considered is much larger than the thickness of any layer. It is based on averaging the structure parameters. Hence, we consider in this chapter the effective

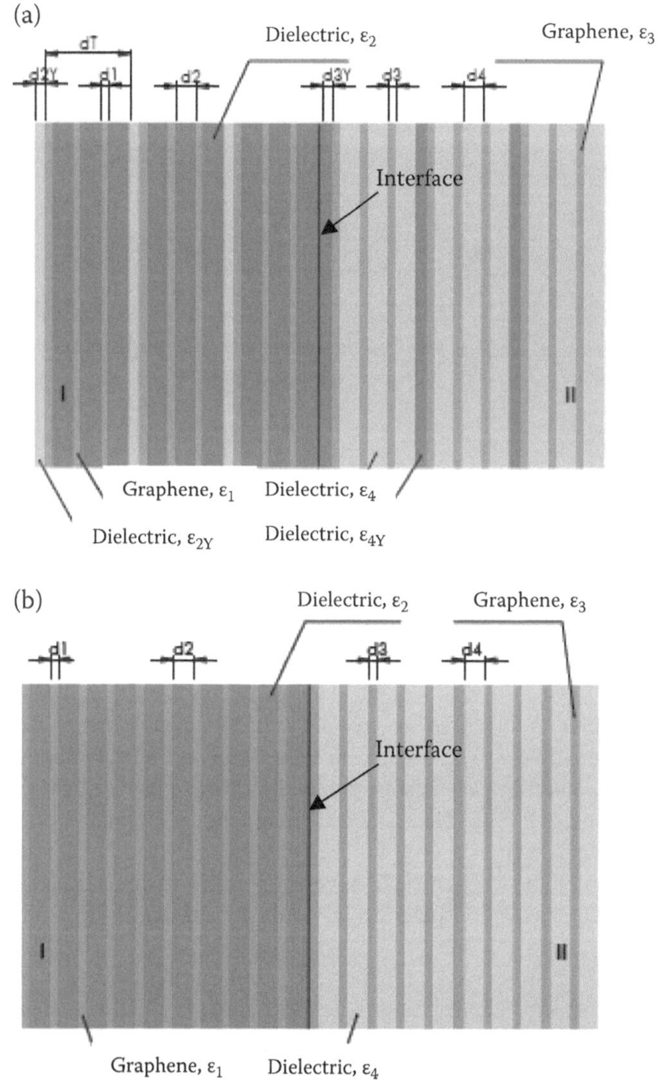

FIGURE 7.1 An interface separating two different infinite heterostructures designed by alternating graphene and dielectric layers; (a) multilayered structures, and (b) two-layered structures. Herein, the indexes 1 and 2 correspond to the graphene and dielectric layers, respectively.

homogeneous media for two semi-infinite (multilayered) periodic structures. The effective permittivities are as follows [23–25]:

$$\varepsilon_{\parallel}^{I} = \frac{d_{2Y}\varepsilon_{2Y} + N(\varepsilon_1 d_1 + \varepsilon_2 d_2)}{d_{2Y} + N(d_1 + d_2)} \tag{7.1}$$

$$\varepsilon_{\perp}^{I} = \frac{\varepsilon_1 \varepsilon_2 \varepsilon_{2Y} (d_{2Y} + Nd_1 + Nd_2)}{d_{2Y} \varepsilon_1 \varepsilon_2 + Nd_1 \varepsilon_2 \varepsilon_{2Y} + Nd_2 \varepsilon_1 \varepsilon_{2Y}} \tag{7.2}$$

Here, $N = \frac{d_T}{d_1 + d_2}$

If $\varepsilon_{2Y} = \varepsilon_2$, $d_{2Y} = d_2$

$$\varepsilon_{\parallel}^{I} = \varepsilon_2 + \frac{Nd_1 (\varepsilon_1 - \varepsilon_2)}{Nd_1 + d_2 (N + 1)} \tag{7.3}$$

$$\varepsilon_{\perp}^{I} = \frac{\varepsilon_1 \varepsilon_2 (d_2 + Nd_1 + Nd_2)}{d_2 \varepsilon_1 + Nd_1 \varepsilon_2 + Nd_2 \varepsilon_1} \tag{7.4}$$

If $N = 1$

$$\varepsilon_{\parallel}^{I} = \frac{d_1 \varepsilon_1 + 2d_2 \varepsilon_2}{d_1 + 2d_2} \tag{7.5}$$

$$\varepsilon_{\perp}^{I} = \frac{\varepsilon_1 \varepsilon_2 (d_1 + 2d_2)}{d_1 \varepsilon_2 + 2d_2 \varepsilon_1} \tag{7.6}$$

where I and II refer to the first and second (left and right) metamaterials under consideration, respectively. The dispersion relation for the surface modes localized at the boundary separating two anisotropic media [26] is proposed by matching the tangential components of the electrical and magnetic fields at the boundary. We assume that the permittivities $\varepsilon_{1,3}(\omega)$ are frequency dependent as the corresponding layers are represented by graphene. We have temporarily assumed that the thickness of the graphene layer is d_1, associating an equivalent permittivity with this single layer.

Eqs. (7.5) and (7.6) serve as the perfect proof of the effective medium approximation [21] applied for the multilayer structure (Fig. 7.1a) under consideration. As a matter of fact, we obtain a predictable result: the permittivity components in the case of $\varepsilon_{2Y} = \varepsilon_2$, $d_{2Y} = d_2$ and $N = 1$ coincide with the relations for the case of the two-layered metamaterial structure (Fig. 7.1b).

We will treat graphene as isotropic with the surface conductivity written as [27,28] $\sigma = ie^2\mu/\pi h^2(\omega + i/\tau)$, where ω, h, e, μ and τ are angular frequency, Planck constant, charge of an electron, chemical potential (i.e., Fermi energy) and phenomenological scattering rate, respectively. The Fermi energy μ can be obtained from the carrier density n_{2D} in a graphene sheet; $\mu = hv_F\sqrt{\pi n_{2D}}$, v_F is the Fermi velocity of electrons. The carrier density n_{2D} can be electrically controlled by an applied gate voltage, thus leading to a voltage-controlled Fermi energy μ [29]. Here, we made an assumption that the electronic band structure of a graphene sheet is not influenced by the neighboring layers. Doing so, the effective permittivity ε_g of graphene can be calculated as [30] $\varepsilon_g = 1 + i\sigma/\varepsilon_0\omega d_g$, where d_g is the thickness of the graphene sheet, and ε_0 is the permittivity in the vacuum. Fig. 7.2 displays the

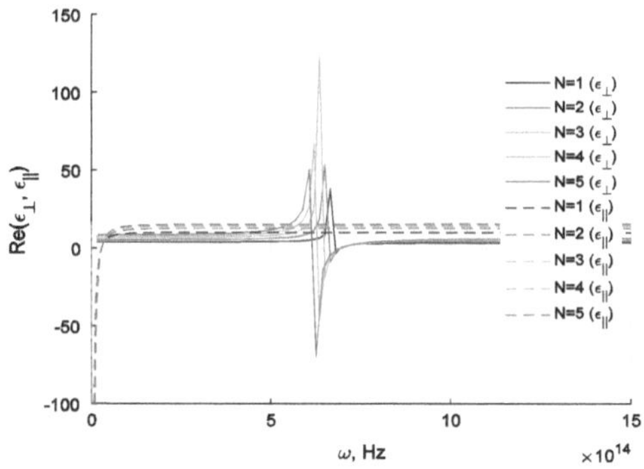

FIGURE 7.2 The influence of N on the real part of $\varepsilon_\perp$ (solid lines) and $\varepsilon_\parallel$ (dashed lines).

contribution of the term N (geometry) in the permittivity functions. The calculations were performed if $d_1 = 0.35$ nm, $d_{2Y} = d_2 = 10$ nm, $\varepsilon_2 = 18.8$ and $\varepsilon_{2Y} = 2.25$.

One can dramatically control the frequency range of surface wave existence by modifying the permittivities and thicknesses of layers [32] employed in the nanocomposites. To get a deep physical insight into the problem, evaluate the tangential components of electric and magnetic fields at the interface, and obtain a single surface mode with the propagation constant, seeking to get the unique dispersion relation for surface modes confined at the interface between two metamaterials [26].

$$\beta = k\sqrt{\frac{(\varepsilon_\parallel^{II} - \varepsilon_\parallel^{I})\varepsilon_\perp^{II}\varepsilon_\perp^{I}}{\varepsilon_\perp^{II}\varepsilon_\parallel^{II} - \varepsilon_\perp^{I}\varepsilon_\parallel^{I}}}, \tag{7.7}$$

where k is the absolute value of the wave vector in a vacuum and β is the component of the wave vector parallel to the interface.

It is important that Eq. (7.5) is valid only under the condition of surface confinement, which can be presented in the following form:

$$\begin{cases} k_{z,I}^2 = (k^2 - \beta^2/\varepsilon_\parallel^{I})\varepsilon_\perp^{I} < 0 \\ k_{z,II}^2 = (k^2 - \beta^2/\varepsilon_\parallel^{II})\varepsilon_\perp^{II} < 0 \end{cases}$$

A clear insight into the wave propagation depends on the concept of propagation length referred to as the distance that the wave travels along the interface until the energy has been reduced to e^{-1} of its original value. As the energy is proportional to the square of field ($energy \sim |H|^2 \sim e^{-2\mathrm{Im}(\beta)x}$), the propagation length is [31,32]

$$L_p = \frac{1}{2\mathrm{Im}(\beta)} \tag{7.8}$$

7.2.1 Dispersion Properties

In this section, we compute the dispersion relation and evaluate the propagation lengths and absorption for SPPs at the boundary of both, the two-layered nano-composite and multilayered metamaterial, following the approach of de Broglie [21]. A two-layered heterostructure supports several SPP modes (see Poddubny et al. [33]), and we will concentrate on the higher-frequency optical mode.

In Fig. 7.3, the dispersion curves for the case $\varepsilon_1 = \varepsilon_3$, $\varepsilon_2 = \varepsilon_4$ and $d_1 \neq d_2 \neq d_3 \neq d_4$ are reported. Fig. 7.3a presents the dispersion curves for different Fermi levels, whereas Fig. 7.3b,c presents the propagation lengths along with the absorption enhancement for the same parameters. The circles and dashed lines correspond to the two different types of nanocomposites using different values of Fermi energy.

Despite the presence of visible changes in the dispersion properties, the most dramatic feature to take into consideration is the drastic enhancement of the propagation length followed by the decrease of Fermi level. Thus, the increase of frequency range of the surface waves' existence is particularly notable, if we raise

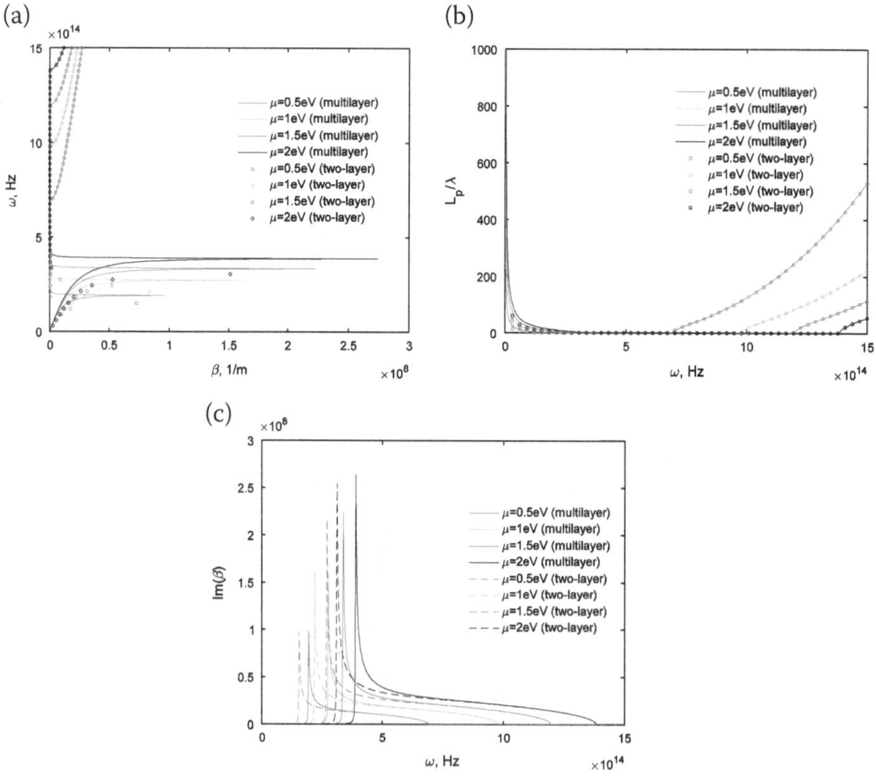

FIGURE 7.3 (a) Dispersion relations of SPPs at the boundary of nanocomposites for different Fermi levels E_F, if $\varepsilon_1 = \varepsilon_3$, $\varepsilon_2 = \varepsilon_4$ and $d_1 \neq d_2 \neq d_3 \neq d_4$; (b) propagation lengths and (c) absorption.

the Fermi energy. Doing so, we can conclude that the propagation length of SPP in the case of nanocomposites is strongly influenced by the concentration of both the surface and bulk carriers. Because the Fermi level can be engineered by gating, the sensitivity provides a fertile ground for tuning the propagation length by almost two orders of magnitude without making a dramatic impact on the SPP dispersion relation.

It is worthwhile to note that we have studied the simplest case of the discussed phenomenon. We chose the considered case because it can be solved analytically [21] and provides the physical insight, though the effect will hold for other cases analogously. This can be explored further by considering other cases. Therefore, we next perform calculations of the dispersion curves and propagation lengths for the case $\varepsilon_1 = \varepsilon_3$, $\varepsilon_2 \neq \varepsilon_4$ and $d_1 \neq d_2 \neq d_3 \neq d_4$; Fig. 7.4 shows the obtained results.

Compared to the two-layered structure, the multilayer nanocomposite demonstrates 1) greater dependence of dispersion curves on the layer thickness, and 2) dramatically increased frequency ranges for the surface wave existence.

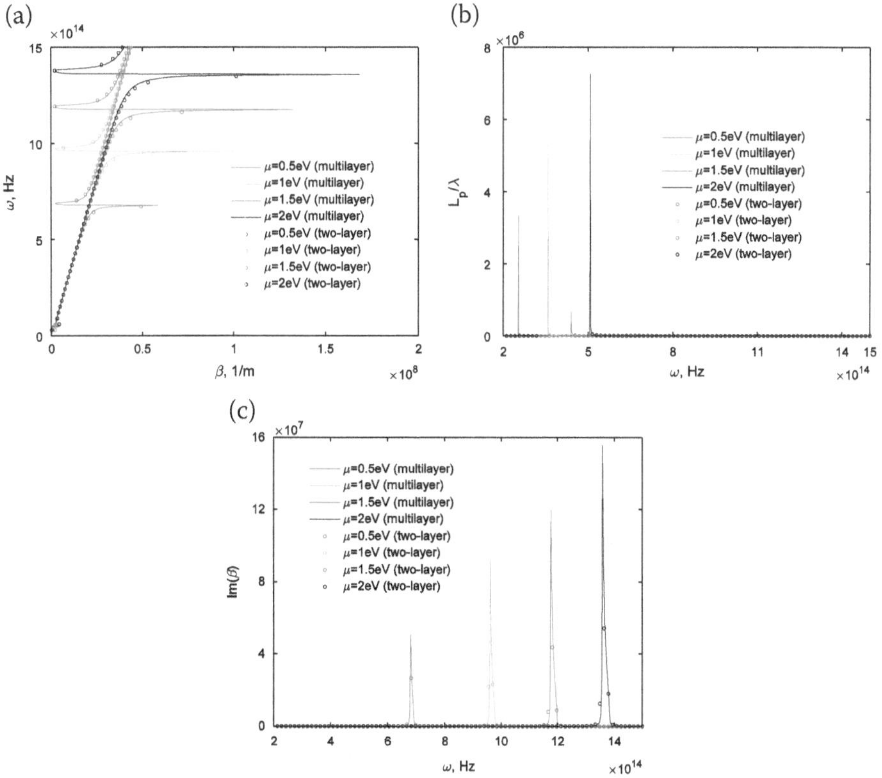

FIGURE 7.4 (a) Dispersion relations of SPPs at the boundary of nanocomposites for different Fermi levels E_F, if $\varepsilon_1 = \varepsilon_3$, $\varepsilon_2 \neq \varepsilon_4$ and $d_1 \neq d_2 \neq d_3 \neq d_4$; (b) propagation lengths and (c) absorption.

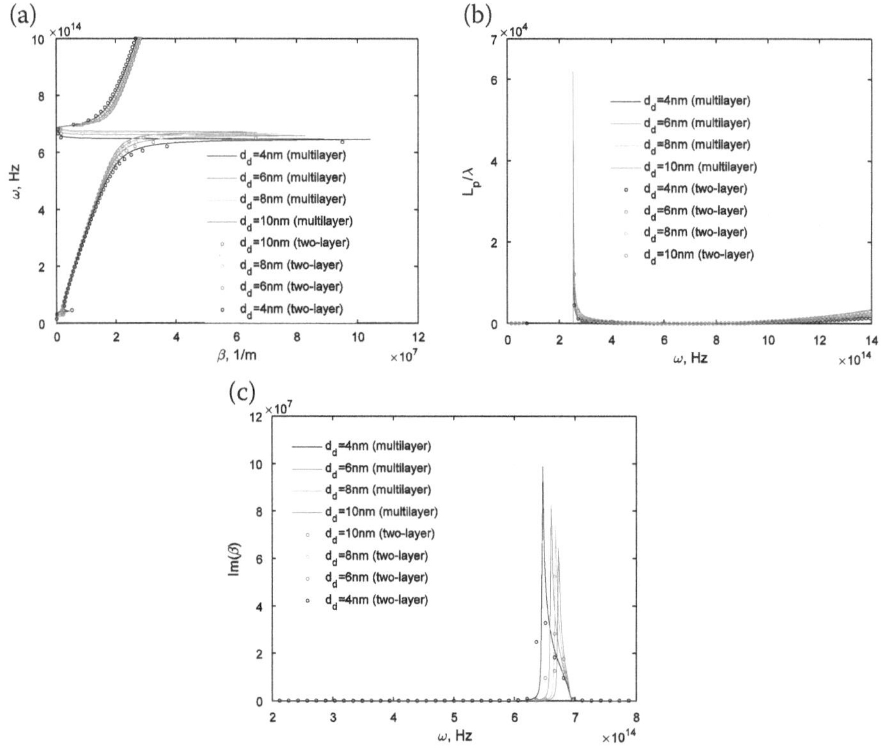

FIGURE 7.5 (a) Dispersion relations of SPPs at the boundary of nanocomposites for different thicknesses of dielectric layer d_d, if $\varepsilon_1 = \varepsilon_3$, $\varepsilon_2 \neq \varepsilon_4$ and $d_1 = d_3$, $d_2 = d_4$; (b) propagation lengths and (c) absorption.

In Figs. 7.5 and 7.6, graphical representations of dispersion relation are presented for both the structures under consideration. Calculations are performed for the cases $\varepsilon_1 = \varepsilon_3$, $\varepsilon_2 \neq \varepsilon_4$ and $d_1 = d_3$, $d_2 = d_4$. In addition to the good agreement between the dispersion equation, these figures show the presence of Ferrell-Berreman modes in the low frequency range.

This investigation of SPPs in multilayered graphene waveguides [34] demonstrates that effective optical conductance linearly depends on the number of layers in the stack in the limit of long wavelengths. Consequently, it seems reasonable to apply heterostructures made of alternating layers of thin films of graphene and dielectric, aiming to engineer the dispersion curve of a fundamental mode of SPP by increasing the number of layers. The growth of such superlattice employing Bi_2Se_3 films has been reported in Tumkur et al. [35]. Since the SPP frequency depends on conductance, one can shift it up for a given wave vector β, increasing the number of unit cells in the superlattice. The illustration of the contrary effect is presented in Figs. 7.5 and 7.6 for the case $\varepsilon_1 = \varepsilon_3$, $\varepsilon_2 \neq \varepsilon_4$ and $d_1 = d_3$, $d_2 = d_4$ allowing for the increase of frequency range of surface wave existence with the thinner graphene layers.

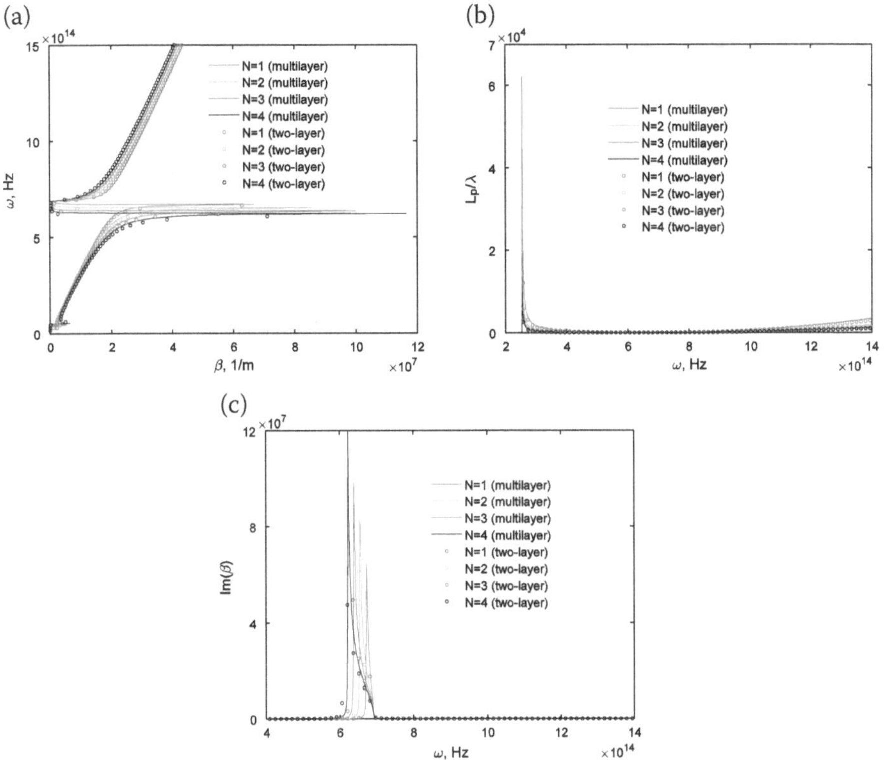

FIGURE 7.6 (a) Dispersion relations of SPPs at the boundary of nanocomposites for different numbers of graphene layers N, if $\varepsilon_1 = \varepsilon_3$, $\varepsilon_2 \neq \varepsilon_4$ and $d_1 = d_3$, $d_2 = d_4$; (b) propagation lengths and (c) absorption.

7.3 NON-LOCAL RESPONSE OF METAMATERIALS

7.3.1 NANOWIRES

The optical response of nanowire composites resembles that of uniaxial media with the optical axis parallel to the direction (z) of nanowires. Consequently, the dielectric permittivity tensor characterizing the features of modes propagating in the nanowire composite is diagonal with the components $\varepsilon_x = \varepsilon_y = \varepsilon_{x,y}$ and ε_z.

One may apply the effective permittivity aiming to describe the optical features of metamaterial. The former approach can be applied to evaluate the absorption spectra of the composite. The described formalism is valid if all the characteristic sizes of composite, such as the nanorod radius and period, are much smaller than the wavelength of light. The optical features of a nanorod composite are similar to the properties of a homogeneous uniaxial anisotropic medium with the optical axis parallel to the nanorods (Fig. 7.7) because of the symmetry. Macroscopically, a diagonal permittivity tensor $\hat{\varepsilon}$ with $\varepsilon_{xx} = \varepsilon_{yy} \equiv \varepsilon_{\perp} \neq \varepsilon_{zz}$ is used to describe these optical properties. Aiming to conduct the analysis, a metamaterial sheet was treated as homogeneous, characterized by either the local [14] or non-local [15] effective medium approaches.

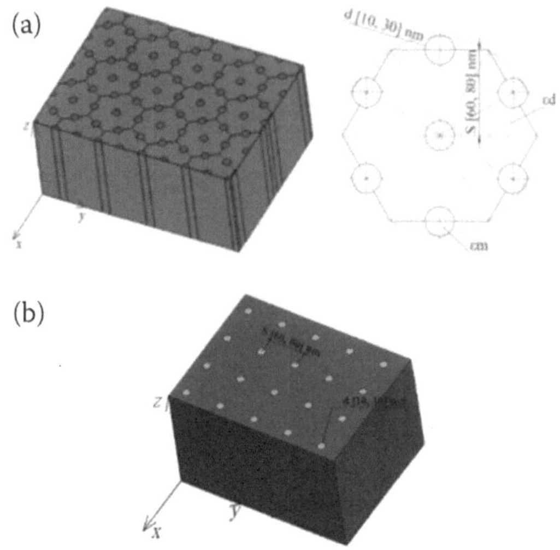

FIGURE 7.7 Schematic illustration of the hexagonal (a) and rectangular (b) unit cell of a nanowire composite.

7.3.1.1 Local EMT

Following the conventional, local effective medium characterization, the permittivity of metamaterial is presented by the Maxwell Garnett approach [16–19]:

$$\varepsilon_{\perp}^{mg} = \varepsilon_d \frac{(1+p)\varepsilon_{Ag} + (1-p)\varepsilon_d}{(1+p)\varepsilon_d + (1-p)\varepsilon_{Ag}} \tag{7.9}$$

$$\varepsilon_{zz}^{mg} = p\varepsilon_{Ag} + (1-p)\varepsilon_d \tag{7.10}$$

Here, ε_d and ε_{Ag} are the permittivities of the dielectric and silver [20], respectively, and p is the fill coefficient of the metal inside the unit cell. It is worth noting that $p = \frac{\pi d^2}{2\sqrt{3}S^2}$ in the case of the hexagon unit cell, and it becomes $p = \frac{\pi d^2}{4S^2}$ in the case of the rectangular unit cell. The spectral properties of the components of effective permittivity tensor of metamaterial are depicted in Fig. 7.8.

7.3.1.2 Non-local EMT

A more complicated, non-local effective medium response should be considered if the optical loss in Ag nanorods is negligible. In this case, a local effective medium approach of metamaterial becomes inappropriate. It is worth mentioning that the components of permittivity tensor perpendicular to the optical axis are still calculated by the Maxwell Garnett approach ($\varepsilon_{xx} = \varepsilon_{yy} = \varepsilon_{\perp}^{MG}$) in the case of non-local EMT. On the other hand, the component of permittivity tensor along the optical axis becomes dramatically influenced by the wave vector as follows:

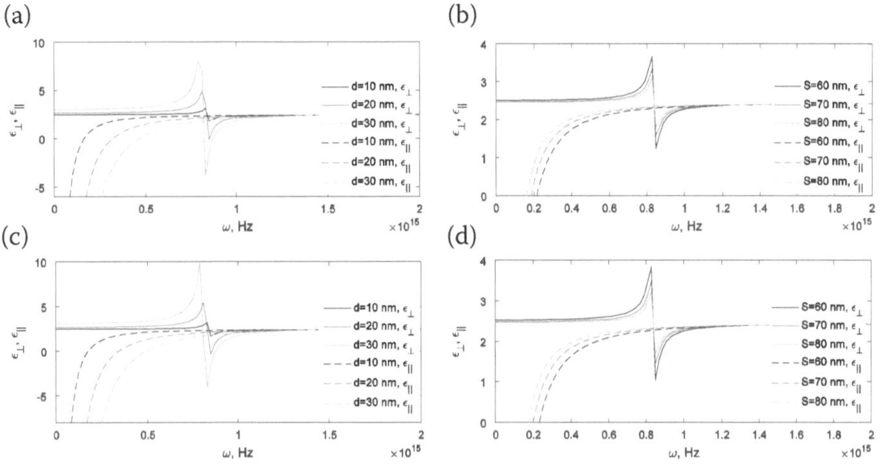

FIGURE 7.8 Local effective medium parameters of metamaterials (Ag nanowires in dielectric ($\varepsilon_d = 2.4$)) calculated based on the geometry of nanorod metamaterial: rectangular unit cell (a, b); hexagon unit cell (c, d). $S = 80$ nm in (a, c); $d = 10$ nm in (b, d).

$$\varepsilon_{zz}(k_z) = \xi\left(k_z^2\frac{c^2}{\omega^2} - \left(n_z^l\right)^2\right) \tag{7.11}$$

$$\xi = p\frac{\varepsilon_{Ag} + \varepsilon_h}{\varepsilon_h - (n_\infty^l)^2} \tag{7.12}$$

where n_z^l is the effective refractive coefficient of the cylindrical surface plasmons that propagate in a nanowire metamaterial with the nanorod permittivity ε_{Ag}, and n_∞^l is the limit of n_z^l for perfectly conducting nanorods.

The extended considerations presented above can be applied for the case of propagation of waves at an angle to the optical axis. Here, the case $k_y = 0$, $k_x \neq 0$ is considered. It is comparatively straightforward to obtain a set of two uncoupled dispersion relations. In the case of nanorod metamaterials, the first of these, $k_x^2 + k_z^2 = \varepsilon_{x,y}^{mg}\omega^2/c^2$, defines the propagation of transverse-electric (TE) polarized waves. The second, $\varepsilon_z(k_z)\left(k_z^2 - \varepsilon_{x,y}^{mg}\frac{\omega^2}{c^2}\right) = -\varepsilon_{x,y}^{mg}k_x^2$, defines the propagation of transverse-magnetic (TM) waves. The latter relation can be rewritten as [21]

$$(k_z^2 - k_z^{l2})\left(k_z^2 - \varepsilon_{x,y}^{mg}\frac{\omega^2}{c^2}\right) = -\frac{\varepsilon_{x,y}^{mg}}{\xi}\frac{\omega^2}{c^2}k_x^2 \tag{7.13}$$

The former proves the fact that the nanorod metamaterials support two TM-polarized waves propagating with different coefficients. The similar mode pattern takes place in the case of other non-local materials [22].

It is clearly demonstrated by Eq. (7.13) that the off-angle ($k_x \neq 0$) propagation of the two TM-polarized waves in anisotropic wire media can be mapped to the microscopic model of the optical features of a nanowire metamaterial. Following the presented formalism, the two TM modes are characterized by the components of the effective permittivities taking their origin from the (i) transverse (electron oscillations perpendicular to the nanowire axes) and (ii) longitudinal (electron oscillation and the wave vector parallel to the nanowire axes) parts of the cylindrical plasmons of wires. The off-angle wave vector stands for the coupling constant. This nonlocality is considered only in the effective medium model because of the homogenization technique. It is worth noting that, in the microscopic model of the nanorod metamaterial, all the quantities are local.

We used three different numerical techniques, providing the full-wave solutions of Maxwell equations aiming to analyze the problem under consideration. First, we employed the finite-difference time-domain method incorporating the approximate dispersion model, given by Eq. (7.11). Second, we used a more accurate description of non-local EMT (Eq. (7.13)), aiming to solve Maxwell's equations in the frequency domain with the non-local transfer matrix formalism. Finally, we employed the finite-element method (FEM) to assess the validities of effective medium description of this complex media. Due to the utilization of FEM calculations, one may consider the full three-dimensional (3D) structure of the composite without assuming any homogenization. From the implementation perspective, the FEM results use significantly more memory and time to make calculations than the local or non-local EMTs. Understanding propagation of light through the interface between the local and non-local layers requires one additional boundary condition (ABC). As shown in de Broglie [21], the continuity of E_z and D_x can be used to introduce such ABCs from the first principles. Doing so, we extend the technique first introduced in de Broglie [21]. Thus, the fields of the two waves propagating in the wire media are represented as

$$E_z(y,\ z) = E_z^{mg} + (\gamma^{mg} + \gamma^l)E_z^l|_{(x=0)} \tag{7.14}$$

$$E_x(y,\ z) = \gamma^{mg}E_x^{mg} + \gamma^l E_z^l|_{(x=0)} \tag{7.15}$$

where $\gamma^{mg} = -\frac{\varepsilon_\perp^{mg} k_z}{\varepsilon_x k_x}\frac{\varepsilon_x - \varepsilon^l}{\varepsilon_x^{mg} - \varepsilon^l}$, $\gamma^l = -\frac{\varepsilon_\perp^{mg} k_x}{\varepsilon_z k_z}\frac{\varepsilon_z - \varepsilon_z^e}{\varepsilon^l - \varepsilon_z^e}$.

We increase diameter d and the distance between nanowires S, aiming to figure out the origin of the SPP modes. The outputs of these investigations are depicted in Fig. 7.9. When S is increased, the energy band of SPP modes gets closer and eventually converges to the light line because the coupling between nanorods becomes weaker. The modes in Fig. 7.9a and 7.9c are still distinguishable, while the modes in Fig. 7.9b and 7.9d almost overlap with each other.

7.3.2 Nanostructures

A similar multilayer HMM geometry, as in the experiments by Tumkur et al. [10] is considered in Fig. 7.10. Here the unit cell is considered as a sub-wavelength

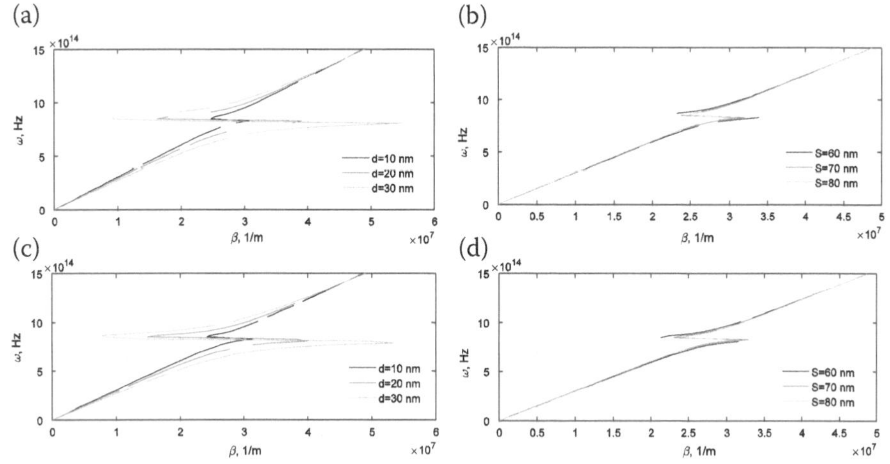

FIGURE 7.9 The influence of pore diameter d (a, c) and distance S (b, d) on the mode patterns for different metamaterial geometries: rectangular unit cell (a, b); hexagon unit cell (c, d).

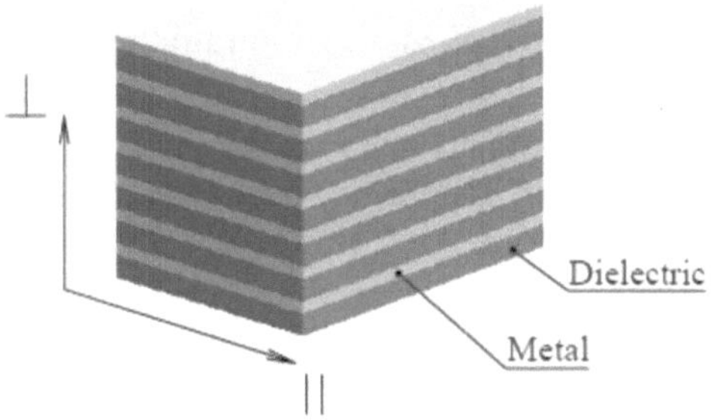

FIGURE 7.10 Schematic of multilayered metamaterial (MM).

metal-dielectric nanostructure. The former is relatively simple and cheap to fabricate [10] and allows analytical investigation. Aiming to describe a former metamaterial from the perspective of an effective medium theory, a local-response approximation (LRA) is usually employed, i.e., the spatial dispersion is neglected. This gives the effective dispersion relation as

$$\frac{k_z^2}{\varepsilon_{zz}^{loc}} + \frac{k_\parallel^2}{\varepsilon_\parallel^{loc}} = \frac{\omega^2}{c^2}, \tag{7.16}$$

where $\varepsilon_{zz}^{loc} = a(a_d/\varepsilon_d + a_m/\varepsilon_m)^{-1}$, $a\varepsilon_\parallel^{loc} = a_d\varepsilon_d + a_m\varepsilon_m$, and $k_\parallel = (k_x^2 + k_y^2)^{1/2}$.

Despite the fact that we discuss metal-dielectric bilayer structures, the theory under consideration may also be extended to structures where metal is replaced by other materials with a Drude response [30].

Here, we go beyond the LRA. Thus, the optical properties of HMMs are studied by exploiting the linearized hydrodynamic Drude model (HDM) within the Thomas-Fermi approximation [18–20]. In the HDM, the metal supports both the usual divergence-free ("transverse") and rotation-free ("longitudinal") waves. Above the plasma frequency, both types of waves can propagate. The dispersion $k_T(\omega)$ of the transverse waves is given by $\varepsilon_m^T(\omega)\omega^2 = k^2c^2$ while $k_L(\omega)$ of the longitudinal waves follows from $\varepsilon_m^L(k, \ \omega) = 0$; in terms of the dielectric functions

$$\varepsilon_m^T(\omega) = 1 - \frac{\omega_p^2}{\omega + i\omega\gamma} \tag{7.17}$$

$$\varepsilon_m^L(k, \ \omega) = 1 - \frac{\omega_p^2}{\omega + i\omega\gamma - \beta^2 k^2} \tag{7.18}$$

Here, γ is the Drude damping, and ω_p is the plasma frequency. The non-local parameter β is equal to $\sqrt{3/5}\,\upsilon_F$, with υ_F representing the Fermi velocity. While ε_m^T is the familiar Drude dielectric function, ε_m^L depends on υ_F and describes the non-local response.

7.3.2.1 Dispersion and Effective Material Parameters

In this section, we employ a transfer-matrix method for both the transverse and longitudinal waves, combined with the aim to calculate the exact dispersion equation for the infinitely extended HMM. The method under consideration is quite similar to the one developed by Moch´an et al. [31]. However, we corrected the ABC that was employed for simplicity in Jacob et al. [31]. An ABC is required to complement the usual Maxwell boundary conditions, and all boundary conditions together make the solution to the coupled Maxwell and hydrodynamic equations unique.

We find the exact analytical dispersion relation considering the infinite HMM with arbitrary unit cell size a and metal and dielectric filling fractions. The effective material parameters of the HMM are derived by a mean-field theory and can be applied to many geometries. The effective material parameters are obtained in the limit of vanishing unit cell size, as follows:

$$\varepsilon_{zz}^{nloc} = \varepsilon_{zz}^{d}, \quad \varepsilon_{\|}^{nloc} = \varepsilon_{\|}^{d}\,\frac{k_L^2\varepsilon_{\|}^{loc}/\varepsilon_{\|}^{d} - k_{\|}^2\varepsilon_m^T}{k_L^2 - k_{\|}^2\varepsilon_m^T} \tag{7.19}$$

where ε_{zz}^d and $\varepsilon_{\|}^d$ represent the effective parameters of metamaterial when the metal layer is replaced by a free-space layer, with $\varepsilon_{zz}^d = a(a_d/\varepsilon_d + a_m)^{-1}$, and $a\varepsilon_{\|}^d = a_d\varepsilon_d + a_m$. It is worth noting that $k_L^2 = (\omega^2 + i\gamma\omega - \omega_p^2)\beta^2$. Both the non-local effective material parameters in Eq. (7.19) differ from the corresponding parameters for the local response.

Dealing with the arbitrary unit cell size a and metal and dielectric filling fractions, we concluded that the exact dispersion relation for the infinite HMM is

$$\beta = \frac{\sqrt{\varepsilon_{zz}^{loc}}\sqrt{-c^2k_z^2 + \varepsilon_{\|}^{nloc}\omega^2}}{c\sqrt{\varepsilon_{\|}^{nloc}}} \tag{7.20}$$

7.3.2.2 Surface Plasmon Polariton Supported by the Nanostructured Metamaterial

Jacob et al. [32] demonstrated SPPs supported by a single metal layer. To understand this better, we will investigate SPPs supported by nanostructured metamaterials dealing with different properties of single metal layers in various theories. The bulk modes of hyperbolic metamaterial result from the coupling of SPPs of neighboring metal layers. We will investigate the SPPs supported by the nanostructured metamaterial analytically in the quasistatic limit.

In Fig. 7.11, we analyze the effect of retardation on the SPP dispersion of the nanostructured metamaterial for non-local response. All the modes depicted in Fig. 7.11 have finite-frequency solutions with β approaching infinity, which leads to the characteristic hyperbolic curve of the HMM that extends to infinitely large wave vectors.

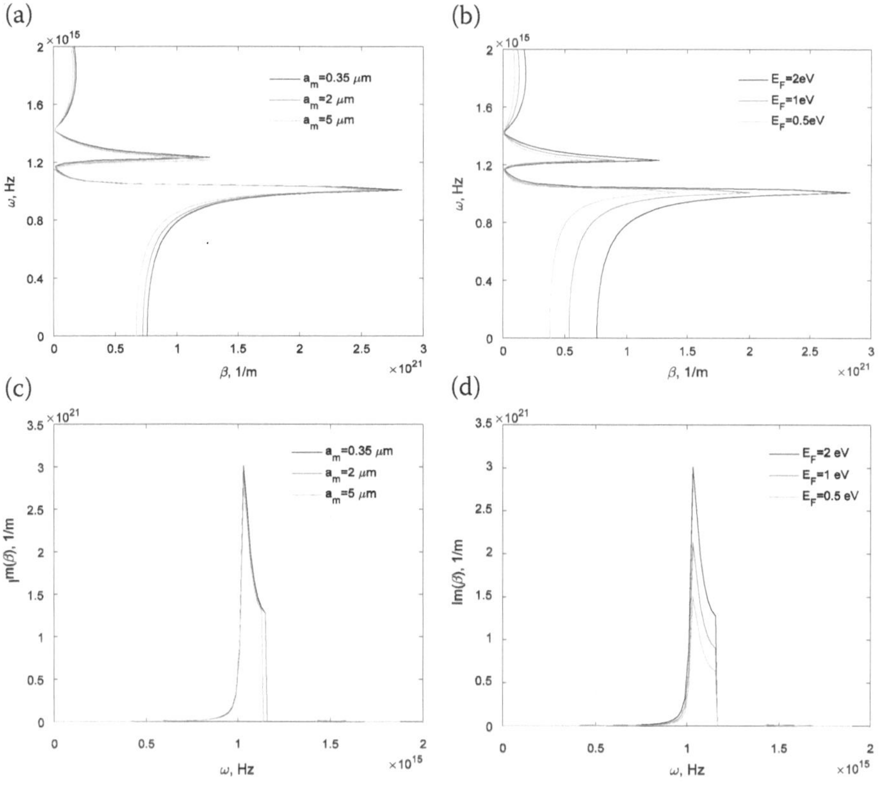

FIGURE 7.11 Real (a), (b) and imaginary (c), (d) dispersion curves of the SPP mode supported by the nanostructured metamaterial in the non-local response approximation. (a), (c) E_F = 2 eV, (b), (d) a_m = 0.35 μm.

Dealing with the TM-polarization first, the quasilocal case shown in Fig. 7.11 exhibits a dispersion relation similar to the one obtained by Ruppin in Poddubny [33] but for a lossy medium. The ability to obtain the curve is either by a very low non-locality or the classical derivation going from the purely local wave equation, and the interface condition is denoted as the quasilocality phenomenon. Instead of two divergent branches in the lossless case, it is now one connected curve that exhibits back-bending in the place of divergence. Different from the lossless case, following this scenario not all solutions outside the light cone fulfill the SPP conditions. Scaling up to a variable Fermi energy case (Fig. 7.11b), the dispersion relation stays similar to the case (a), although a slight broadening can be observed for the dispersion curves. Fig. 7.11d shows that the variable E_F has led to the almost vanishing peak, but it did not alter the peak frequency much.

The frequency ranges of the surface wave can be tuned by modifying the Fermi energy, which is consistent with the dependence of $\varepsilon_\perp$ on the Fermi energy, as shown in Noginov et al. [34]. It should be mentioned that the decrease in Fermi energy μ will move the dispersion curves to lower frequencies; in contrast, the increase moves the curves to higher frequencies. As seen in Fig. 7.11, the smallest asymptotic frequency is achieved by employing the smallest Fermi energy. The former tunability property suggests that the surface wave can be engineered by the Fermi energy of graphene sheets.

It is of particular interest to examine the effect of controlling the fill factors of dielectric and graphene sheets, as shown in Fig. 7.11b. First, we discuss the influence of the thickness a_m of dielectric on the dispersion curve (see Fig. 7.11b). We find that the upper limit moves to the higher frequencies as a_m is increased. This is consistent with the effect of a_m on the frequency range of $\varepsilon_\perp$.

7.4 TUNABLE THZ STRUCTURE BASED ON GRAPHENE HMMS

Fig. 7.12 schematically demonstrates HMM heterostructure made of stacked graphene layers divided by dielectric spacers.

The electron-hole pairs split the Fermi level into two quasi-Fermi levels $E_{Fn} = \varepsilon_F$ and $E_{Fp} = -\varepsilon_F$, respectively, in the case of a graphene monolayer being optically pumped. Accordingly, one can model the dynamic conductivity of graphene by means of non-equilibrium Green's function, presented as follows [30]:

$$\sigma = j\frac{q^2}{\pi\hbar^2}\frac{1}{\omega - j\tau^{-1}}\left[\int_0^\infty \varepsilon\left(\frac{\partial f_1(\varepsilon)}{\partial\varepsilon} - \frac{\partial f_2(-\varepsilon)}{\partial\varepsilon}\right)d\varepsilon\right] - j\frac{q^2}{\pi\hbar^2}(\omega - j\tau^{-1})\int_0^\infty \frac{f_2(-\varepsilon) - f_1(\varepsilon)}{(\omega - j\tau^{-1})^2 - 4\varepsilon^2/\hbar^2}d\varepsilon \tag{7.21}$$

where $f_1(\varepsilon) = [1 + e^{(\varepsilon - E_{Fn})/k_B T}]^{-1}$, $f_2(\varepsilon) = [1 + e^{(\varepsilon - E_{Fp})/k_B T}]^{-1}$, q is the electron charge, ε is energy, k_B is Boltzmann's constant, T is temperature, ω is the angular frequency, and τ is the relaxation time of charge carriers. It is assumed that $\tau = 2\cdot 10^{-12}$ s. The contribution of intra-band transition is represented by the first term of

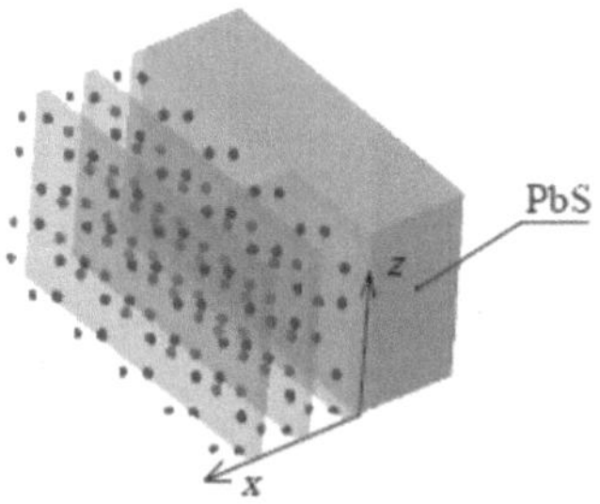

FIGURE 7.12 Schematic of an interface separating infinite-layered nanostructure metamaterial designed by alternating graphene and dielectric sheets and dielectric medium; waves propagate in the z-direction.

Eq. (7.21). On the other hand, the impact of inter-band transition is described by the second part of the expression. In the case of absence of optical excitation, one can model a graphene layer by the local surface conductivity formula [31,32]:

$$\sigma = \frac{-jq^2 k_B T}{\pi\hbar^2(\omega - j\tau^{-1})}\left(\frac{E_F}{k_B T} + 2\ln\left(e^{-E_F/(k_B T)} + 1\right)\right) - \frac{jq^2(\omega - j\tau^{-1})}{\pi\hbar^2}\int_0^\infty \frac{f_D(-\varepsilon) - f_D(\varepsilon)}{(\omega - j\tau^{-1})^2 - 4\varepsilon^2/\hbar^2}d\varepsilon \qquad (7.22)$$

where $f_D(\varepsilon) = [e^{(\varepsilon - E_F)/(k_B T)} + 1]^{-1}$ is the Fermi-Dirac distribution, and E_F is the Fermi energy of the doping level in the passive graphene case.

The computed frequency dependence of the real part of conductivity is shown in Fig. 7.13 for a graphene monolayer with $T = 77$ K. Increasing the Fermi level leads to the increase of the absolute value of the real and imaginary parts of conductivity.

Graphene shows significant loss due to the possible scattering owing to the intra-band free-carrier absorption. In general, graphene naturally is treated as a lossy material at the THz spectral region [33]. Fig. 7.13 demonstrates the existence of the spectral range such that the imaginary part of conductivity

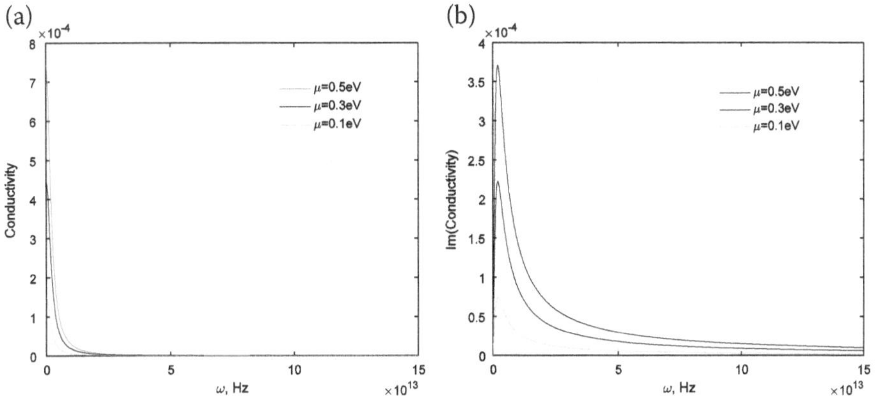

FIGURE 7.13 Dependence of the real (a) and imaginary (b) components of graphene conductivity on the frequency.

exceeds the real part. Because of this frequency region lying in the mid-infrared (IR) range, graphene-based HMM demonstrates an enhanced performance in the mid-IR regime than the THz range. In fact, the performance of graphene for realizing HMM can be improved by doping, as shown in Fig. 7.13. A large E_F can turn off the inter-band absorption by the Pauli blocking and increase the Im(σ) required for achieving negative $\varepsilon_{eff,\perp}$. Additionally, doping can suppress the intra-band scattering by screening charged impurities [33,34].

Here, we study a multilayer HMM made of clustered graphene monolayers and dielectric spacers. The proposed design is schematically shown in the inset of Fig. 7.14, aiming to enhance the light–matter interaction. One can model this structure as an anisotropic uniaxial medium with the effective permittivity ε_{eff} calculated as $\varepsilon_{eff} = \varepsilon_t(\hat{x}\hat{x} + \hat{y}\hat{y}) + \varepsilon_z\hat{z}\hat{z}$ [35] by means of the effective medium approximation (EMA) theory. The transverse relative permittivity ε_t is $\varepsilon_t = \varepsilon_t' - j\varepsilon_t'' = \varepsilon_d - j\frac{\sigma}{\omega\varepsilon_0 d}$ [36], with ε_d and d being the permittivity and thickness of each dielectric spacer, and σ is the conductivity of graphene expressed by Eqs. (7.21) and (7.22). The longitudinal permittivity ε_z is equal to ε_d. The mentioned formalism is widely applied to study the surface waves at the graphene-based metamaterial interface [37–39]. This is valid because of the graphene having insignificant thickness compared to the dielectric spacers. In the case of graphene monolayers, the hyperbolic dispersion is attained at the frequency range with the real component of the transverse effective permittivity ε_t' being negative, while ε_z is always positive. It is worth mentioning that only the TM-polarized waves can demonstrate hyperbolic dispersion, as shown in Podolskiy and Narimanov [40].

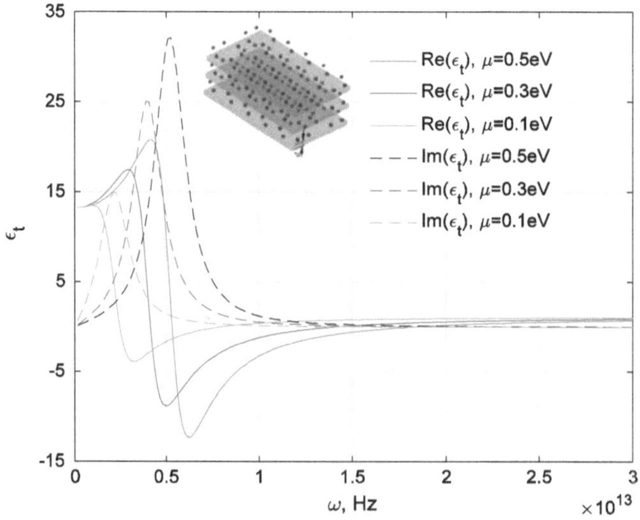

FIGURE 7.14 Real and imaginary components of the effective transverse relative permittivity of the graphene (in the inset) HMM multilayer structure. Dielectric spacers with thicknesses d and permittivities ε_d separate the graphene layers. $\varepsilon_d = 2.2$ and $d = 0.1$ μm.

The real and imaginary components of the effective transverse relative permittivity, respectively, are depicted in Fig. 7.14 for the THz HMM made of graphene (presented in the inset of Fig. 7.14) with $\varepsilon_d = 2.2$ and $d = 0.1$ μm. It is assumed that every graphene monolayer is uniformly photo-doped to have the same quasi-Fermi level. This takes place because of the graphene being transparent to the near-IR and visible pump waves, without the loss of generality. Fig. 7.14 demonstrates that the photo-pumped graphene-based HMM has a negative imaginary part of the effective transverse permittivity ε_t'' (gain) over a broad frequency range, where $\mathrm{Re}(\varepsilon_t')$ and $\mathrm{Re}(\varepsilon_z)$ have opposite signs (i.e., hyperbolic dispersion).

The non-local permittivity becomes negative for higher frequencies. This leads to a hyperbolic shape of the isofrequency contours because the component ε_z is always positive. However, the real part ε_t' changes its sign to positive in the lower frequency range. Thus, an effective hyperbolic medium is changed into the elliptic one. This is direct evidence of strong non-local properties of the graphene-based metamaterials.

We study the dispersion diagrams obtained after the homogenization of HMM. Consequently, in Fig. 7.15, the dispersion diagrams for the structure (of Fig. 7.12) are presented. One can engineer the frequency ranges of the surface wave by modifying the Fermi energy. This is consistent with the dependence of $\varepsilon_\perp$ on the Fermi energy.

One can conclude from Figs. 7.15 and 7.16 that the bound SPPs approach a maximum, finite wave vector at the surface plasmon frequency ω_{sp} of the system. A lower bound is set by this limitation on both the wavelength $\lambda_{sp} = 2\pi/\mathrm{Re}[\beta]$ of surface plasmon and the amount of mode confinement normal to the interface. The described phenomenon takes place because of the specific nature of SPP fields in the dielectric, that is, $e^{-|k_x||x|}$. Moreover, the quasibound, leaky part of the dispersion relation between ω_{sp} and ω_p is allowed, with ω_p being the plasmon frequency. It is possible to narrow the quasibound leaky part by using the enhanced

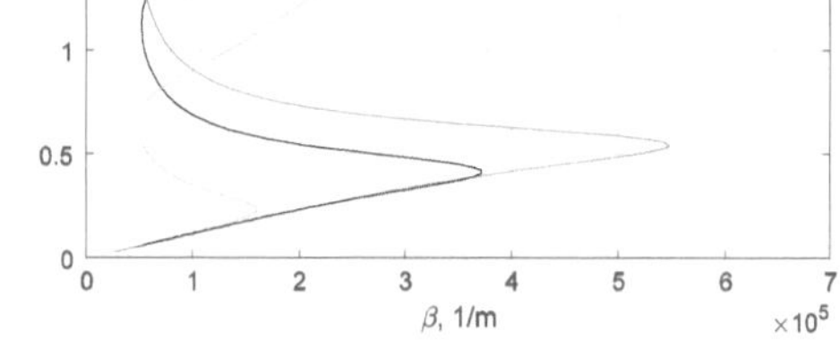

FIGURE 7.15 The impact of Fermi energy on the dispersion of surface waves. $\varepsilon_d = 2.2$ and $d = 0.1$ μm.

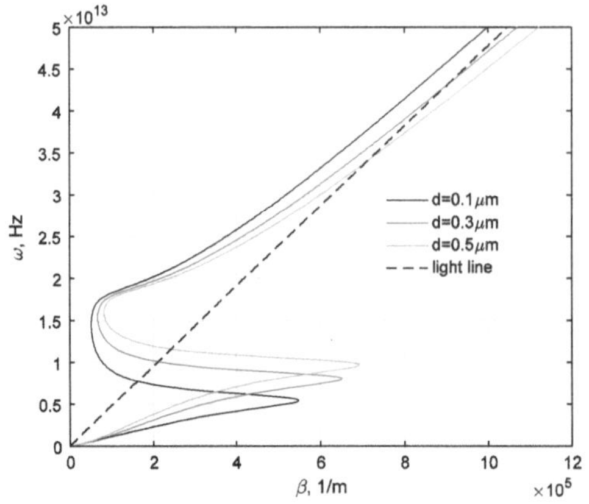

FIGURE 7.16 The impact of thickness of dielectric layers upon the dispersion of surface waves. $\varepsilon_d = 2.2$ and $\mu = 0.1$ μm.

system. Doing so, we aim to investigate an infinite periodic multilayer composite, as shown in Fig. 7.17. Its unit cell is composed of a graphene sheet and a dielectric layer of atomic t and relative permittivity ε_t. The demonstration of narrowed quasibound leaky part is clearly observed in Fig. 7.18. To describe the optical response of such a system, we apply the effective medium approach, which is justified if the wavelength of radiation considered is much larger than the thickness of any layer. It is based on averaging the structure parameters [36], i.e.,

$$\varepsilon_{\|}' = \frac{\varepsilon_t t + \varepsilon_{tg} t_g}{t + t_g} \tag{7.23}$$

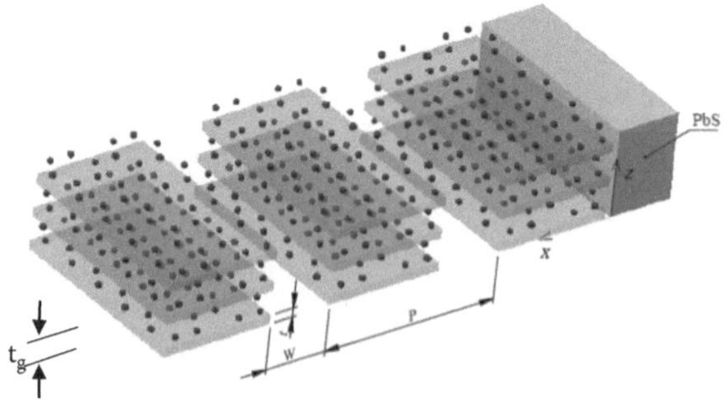

FIGURE 7.17 The infinite HMM structure made of patterned graphene sheets (described by ε_{tg}) stacks between dielectric spacers. P is the period, W is the width of air slots, and t is the thickness of each dielectric spacer layer. Wave propagation takes place along the z-direction.

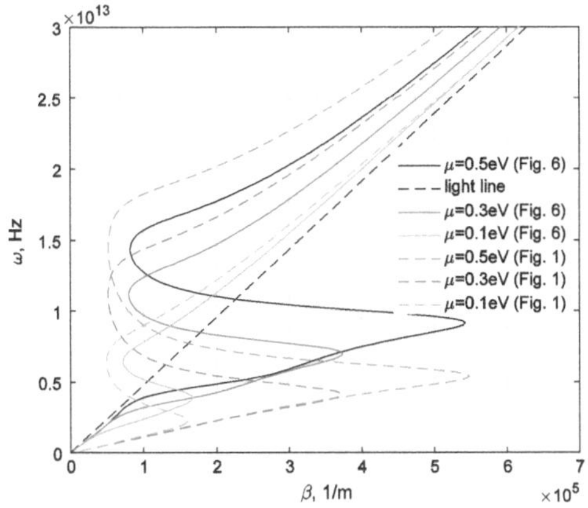

FIGURE 7.18 Dependence of the dispersion property on the Fermi energy with the structure dimensions as $P = 1.9$ μm and $W = 0.5$ μm.

$$\varepsilon_{\perp}' = \frac{\varepsilon_t \varepsilon_{tg} (t + t_g)}{\varepsilon_t t_g + \varepsilon_{tg} t} \tag{7.24}$$

Aiming to deal with the enhanced system (Fig. 7.17), the nanostructured unit cell is treated as the metamaterial layer. By following it with the air gap, the effective medium approximation is applied again, i.e.,

$$\varepsilon_{\parallel} = \frac{\varepsilon_{air} W + \varepsilon_{\perp}' (P - W)}{P} \tag{7.25}$$

$$\varepsilon_{\perp}' = \frac{\varepsilon_{air} \varepsilon_{\parallel}' P}{\varepsilon_{air} (P - W) + \varepsilon_{\parallel}' W} \tag{7.26}$$

The dispersion-related diagrams of the proposed structure are plotted in Figs. 7.18–7.20. One can tune the frequency of the near-zero group velocity using different mechanisms, including (i) the quasi-Fermi level of graphene, (ii) the geometry of HMM and (iii) the dielectric material that constitutes the HMM.

By presenting a modal analysis, we now show that the physical origin of bulk absorption in metamaterials is due to the excitation of leaky bulk polaritons, called as the Ferrel-Berreman (FB) modes [41–43]. As a matter of fact, the volume charge oscillations at the ENZ of metal (bulk or volume plasmons) are formed by the bulk metal. The important property to note now is that these excitations are in the form of a completely longitudinal wave and, hence, cannot be excited with the free-space light (which is a transverse wave). The top and bottom interface couple in the case of metal films of thicknesses less than the metal skin depth, permitting for collective charge oscillations across the film. In such a case, the bulk plasmon is no longer purely longitudinal and can interact with the free-space light at frequencies near the

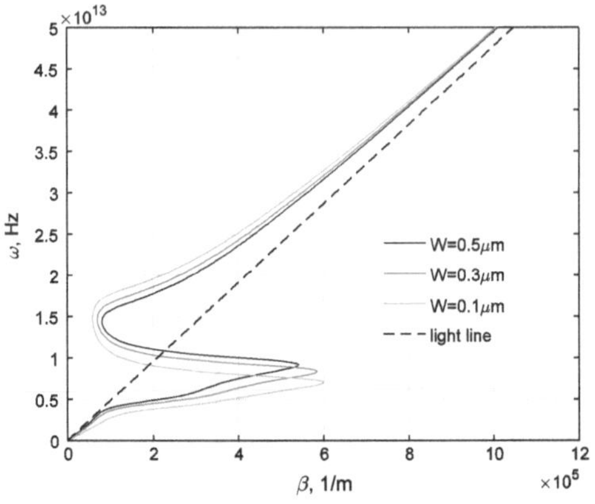

FIGURE 7.19 Dependence of dispersion property on the thickness of air slots with the structure dimensions as P = 1.9 μm and μ = 0.5 eV.

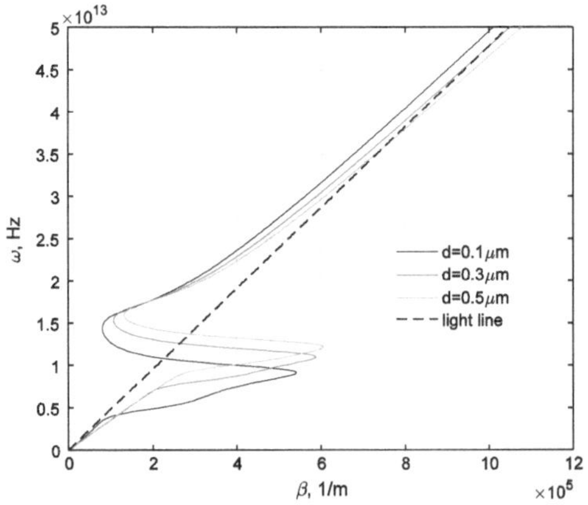

FIGURE 7.20 Dependence of the dispersion properties on the thickness of the dielectric layer with the structure dimensions as P = 1.9 μm, W = 0.5 μm, μ = 0.5 eV.

metal ENZ [44]. Ferrell addressed this approach for metallic foils in Raza et al. [45] and Berreman for polar dielectric films in David and García de Abajo [46]. Radiative excitations, which we call the FB modes, are supported by multilayer metamaterials with the employed graphene layers. It is worth mentioning that these FB modes are different from SPPs supported by metal foils. In the case of SPP modes, the energy propagates along the surfaces of metal, whereas in the FB modes, the volume charge oscillations are set up across the foil, and the energy propagates within the bulk of the metal. The bulk polaritons under consideration have transverse wave vectors similar to the free-space light and exist to the left of the light line. Thus, in Figs. 7.18–7.20, it is interesting to observe the FB modes.

The dispersion relations of the SPP modes in nanostructured metamaterials have been studied before [37]. Here, the distance d_W has been increased to find the origins of the SPP modes. The outputs of these computations are demonstrated in Fig. 7.19. As the wave vector becomes smaller, the dispersion curves of all the modes meet together for all the cases (Figs. 7.18–7.20). Moreover, the dispersion lines appearing on the left-hand side of the light lines in Figs. 7.15, 7.16, 7.18–7.20 are the radiative modes. This fact indicates the radiation of terahertz in free space. This is very important for the generation of THz waves. It should be mentioned that, it is possible to control the frequency range of radiative modes by choosing the appropriate structure.

7.5 CONCLUSIONS

In the first part of this chapter, we reported on the dispersion properties of heterostructures. We showed different examples of plasmon structures fabricated from graphene layers. In particular, the results of this work can be summarized as follows:

- Dispersion relation and propagation lengths of SPPs at the boundary of nanocomposites were investigated using realistic material parameters. The multilayered heterostructure is identified as the material of choice if the increase of frequency range of surface waves existence is desired.
- The key parameters influencing the propagation length were found to be (i) the Fermi level E_F, and (ii) layer thicknesses. Lowering E_F by gating is a feasible way to control the propagation lengths of SPPs. Additional enhancement of propagation lengths can be implemented when dealing with the case of $\varepsilon_1 = \varepsilon_3$, $\varepsilon_2 \neq \varepsilon_4$ and $d_1 \neq d_2 \neq d_3 \neq d_4$.
- Because the structures under consideration possess similar properties for the cases $\varepsilon_1 = \varepsilon_3$, $\varepsilon_2 \neq \varepsilon_4$ and $d_1 \neq d_2 \neq d_3 \neq d_4$; $\varepsilon_1 = \varepsilon_3$, $\varepsilon_2 \neq \varepsilon_4$ and $d_1 = d_3$, $d_2 = d_4$, it is possible to obtain the desired functionality using the thinner structures.

In the second part of chapter, we discussed a formalism to treat the optics of non-local nanowire metamaterials across the whole optical spectrum. The described theory can be applied to characterize the optics of other wire-like composites together with the coated-wire structures and coax-cable-based systems [78–80]. We investigated the SPP modes in silver nanowire systems by means of non-local EMT. The calculated dispersion relations of the SPP modes were considered for different types of metamaterial unit cells, i.e., either rectangular or hexagonal. We presented the dispersion relations for varying geometries of metamaterial unit cells.

In the third part, extremely anisotropic metal-dielectric multilayer metamaterials were designed based on the transfer-matrix method, including optical nonlocality. We discussed equations for an interface between a non-local, homogeneous and isotropic metamaterial and dielectric, and we derived the dispersion expression for TM polarization. Using the results, we obtained the dispersion relation for SPPs and discussed the effect of gradually increasing Fermi energy E_F on it. We observed that the tuning parameter has no significant effect in the case of TM-polarized light.

In the fourth part of this chapter, we studied a lasing graphene HMM system, considering patterned photo-pumped graphene-dielectric nanocomposites. The graphene nanocomposites were treated from different perspectives, i.e., starting with the simplified model and ending up with the system of higher complexity, aiming to increase the frequency range of the SPP existence. Our approach allowed for narrowing the quasibound leaky part of dispersion diagrams by using the enhanced system. We investigated the system from both the analytical and numerical perspectives. We presented that super properties can occur in the THz spectral region because of the studied patterned graphene-based HMM device. Furthermore, the geometry and quasi-Fermi levels of the investigated graphene HMM can make a dramatic impact on the properties of the system under consideration. Additionally, we attained the engineerable THz properties in a broad frequency region by changing the dimensions and geometry of the considered HMM structure. The modeled system has an atomic thickness. This makes it an ideal candidate to be used as a compact THz source in the envisioned THz on-chip integrated circuits.

REFERENCES

1. S. A. Maier, *Plasmonics: Fundamentals and Applications*, New York: Springer, 2007.
2. O. Limaj, et al., "Field distribution and quality factor of surface plasmon resonances of metal meshes for midinfrared sensing," *Plasmonics*, vol. 10, pp. 45–50, 2015.
3. A. Toma, et al., "Squeezing terahertz light into nanovolumes: Nanoantenna enhanced terahertz spectroscopy (NETS) of semiconductor quantum dots," *Nano Lett.*, vol. 15, pp. 386–391, 2015.
4. D. E. Chang, et al., "Quantum optics with surface plasmons," *Phys. Rev. Lett.*, vol. 97, p. 053002, 2006.
5. A. V. Akimov, et al., "Generation of single optical plasmons in metallic nanowires coupled to quantum dots," *Nature*, vol. 450, pp. 402–406, 2007.
6. M. Danckwerts, et al., "Optical frequency mixing at coupled gold nanoparticles," *Phys. Rev. Lett.*, vol. 98, p. 026104, 2007.
7. A. R. Davoyan, et al., "Nonlinear plasmonic slot waveguide," *Opt. Express*, vol. 16, pp. 21209–21214, 2008.
8. H. A. Atwater, et al., "Plasmonics for improved photovoltaic devices," *Nat. Mater.*, vol. 9, pp. 205–213, 2010.
9. D. P. O'Neal, et al., "Photo-thermal tumor ablation in mice using near infrared-absorbing nanoparticles," *Cancer Lett.*, vol. 209, pp. 171–176, 2004.
10. P. K. Jain, et al., "Au nanoparticles target cancer," *Nanotoday*, vol. 2, pp. 18–29, 2007.
11. P. Nordlander, et al., "Plasmon hybridizaton in nanoparticle dimers," *Nano Lett.*, vol. 4, pp. 899–903, 2004.
12. L. M. Liz-Marzán, "Tailoring surface plasmon through the morphology and assembly of metal nanoparticles," *Langmuir*, vol. 22, pp. 32–41, 2006.
13. P. S. Kumar, et al., "High-yield synthesis and optical response of gold nanostars," *Nanotechnology*, vol. 19, p. 015606, 2008.
14. M. Grzelczak, et al., "Shape control in gold nanoparticle synthesis," *Chem. Soc. Rev.*, vol. 37, pp. 1783–1791, 2008.
15. S. Lupi, et al., "Fano effect in the a-b plane of Nd1.96Ce0.04CuO4+y: Evidence of phonon interaction with a polaronic background," *Phys. Rev. B.*, vol. 57, pp. 1248–1252, 1998.

16. S. Lupi, et al., "Infrared active aibrational-modes strongly coupled to carriers in high-Tc superconductors," *Europhys. Lett.*, vol. 31, pp. 473–478, 1995.
17. E. Prodan, et al., "A hybridization model for the plasmon response of complex nanostructures," *Science*, vol. 302, pp. 419–422, 2003.
18. Z. Fang, et al., "Gated tunability and hybridization of localized plasmons in nanostructured graphene," *ACS Nano*, vol. 7, p. 2388, 2013.
19. O. Limaj, et al., "Superconductivity-induced transparency in terahertz metamaterials," *ACS Photonics*, vol. 1, pp. 570–575, 2014.
20. S. Zhang, et al., "Plasmon-induced transparency in metamaterials," *Phys. Rev. Lett.*, vol. 101, p. 047401, 2008.
21. L. de Broglie, "A tentative theory of light quanta," *Philos. Mag. Lett.*, vol. 86, pp. 411–423, 2006.
22. A. Einstein, B. Podolsky, and N. Rosen, "Can quantum-mechanical description of physical reality be considered complete?" *Phys. Rev.*, vol. 47, pp. 777 –780, 1935.
23. J. S. Bell, "On the problem of hidden variables in quantum mechanics," *Rev. Mod. Phys.*, vol. 38, pp. 447–452, 1966.
24. S. J. Freedman and J. F. Clauser, "Experimental test of local hidden-variable theories," *Phys. Rev. Lett.*, vol. 28, pp. 938 –941, 1972.
25. A. Aspect, P. Grangier, and G. Roger, "Experimental realization of Einstein-Podolsky-Rosenbohm gedankenexperiment: A new violation of Bell's inequalities," *Phys. Rev. Lett.*, vol. 49, pp. 91–94, 1982.
26. J. B. Pendry, "Negative refraction makes a perfect lens," *Phys. Rev. Lett.*, vol. 85, p. 3966, 2000.
27. J. B. Pendry, D. R. Smith, and D. Schurig, "Controlling electromagnetic fields," *Science*, vol. 312, p. 1780, 2006.
28. U. Leonhardt, "Optical conformal mapping," *Science*, vol. 312, p. 1777, 2006.
29. D. R. Smith and D. Schurig, "Electromagnetic wave propagation in media with indefinite permittivity and permeability tensors," *Phys. Rev. Lett.*, vol. 90, p. 077405, 2003.
30. I. I. Smolyaninov, "Vacuum in a strong magnetic field as a hyperbolic metamaterial," *Phys. Rev. Lett.*, vol. 107, p. 253903, 2011.
31. Z. Jacob, I. Smolyaninov, and E. Narimanov, "Broadband Purcell effect: Radiative decay engineering with metamaterials," *Appl. Phys. Lett.*, vol. 100, p. 181105, 2012.
32. Z. Jacob, J. Y. Kim, G. V. Naik, A. Boltasseva, E. E. Narimanov, and V. M. Shalaev, "Engineering photonic density of states using metamaterials," *Appl. Phys. B.*, vol. 100, p. 215, 2010.
33. A. N. Poddubny, P. A. Belov, G. V. Naik, and Y. S. Kivshar, "Spontaneous radiation of a finite-size dipole emitter in hyperbolic media," *Phys. Rev. A.*, vol. 84, p. 023807, 2011.
34. M. A. Noginov, H. Li, Y. A. Barnakov, D. Dryden, G. Nataraj, G. Zhu, C. E. Bonner, M. Mayy, Z. Jacob, and E. E. Narimanov, "Controlling spontaneous emission with metamaterials," *Opt. Lett.*, vol. 35, p. 1863, 2010.
35. T. Tumkur, G. Zhu, P. Black, Y. A. Barnakov, C. E. Bonner, and M. A. Noginov, "Control of spontaneous emission in a volume of functionalized hyperbolic metamaterial," *Appl. Phys. Lett.*, vol. 99, p. 151115, 2011.
36. O. Kidwai, S. V. Zhukovsky, and J. E. Sipe, "Dipole radiation near hyperbolic metamaterials: Applicability of effective-medium approximation," *Opt. Lett.*, vol. 36, p. 2530, 2011.
37. M. A. Noginov, Y. A. Barnakov, G. Zhu, T. Tumkur, H. Li, and E. E. Narimanov, "Bulk photonic metamaterial with hyperbolic dispersion," *Appl. Phys. Lett.*, vol. 94, p. 151105.

38. J. Yao, Z. W. Liu, Y. M. Liu, Y. Wang, C. Sun, G. Bartal, A. M. Stacy, and X. Zhang, "Optical negative refraction in bulk metamaterials of nanowires," *Science*, vol. 321, p. 930, 2008.
39. M. Yan and N. A. Mortensen, "Hollow-core infrared fiber incorporating metal-wire metamaterial," *Opt. Express*, vol. 17, p. 14851, 2009.
40. V. A. Podolskiy and E. E. Narimanov, "Strongly anisotropic waveguide as a non-magnetic left-handed system," *Phys. Rev. B*, vol. 71, p. 201101, 2005.
41. Z. Jacob, J. Y. Kim, G. V. Naik, A. Boltasseva, E. E. Narimanov, and V. M. Shalaev, "Engineering photonic density of states using metamaterials," *Appl. Phys. B*, vol. 100, p. 215, 2010.
42. P. Ginzburg, F. J. Rodriguez Fortuno, G. A. Wurtz, W. Dickson, A. Murphy, F. Morgan, R. J. Pollard, I. Iorsh, A. Atrashchenko, P. A. Belov, Y. S. Kivshar, A. Nevet, G. Ankonina, M. Orenstein, and A. V. Zayats, "Manipulating polarization of light with ultrathin epsilon-near-zero metamaterials," *Opt. Express*, vol. 21, p. 14907, 2013.
43. F. Bloch, "Bremsvermögen von Atomen mit mehreren Elektronen," *Z. Phys. A*, vol. 81, p. 363, 1933.
44. A. D. Boardman, *Electromagnetic Surface Modes*, Chichester, UK: John Wiley and Sons, 1982.
45. S. Raza, G. Toscano, A.-P. Jauho, M. Wubs, and N. A. Mortensen, "Unusual resonances in nanoplasmonic structures due to nonlocal response," *Phys. Rev. B*, vol. 84, p. 121412(R), 2011.
46. C. David and F. J. Garc´ıa de Abajo, "Spatial nonlocality in the optical response of metal nanoparticles," *J. Phys. Chem. C*, vol. 115, p. 19470, 2011.
47. F. J. Garc´ıa de Abajo, "Nonlocal effects in the plasmons of strongly interacting nanoparticles, dimers, and waveguides," *J. Phys. Chem. C*, vol. 112, p. 17983, 2008.
48. G. Toscano, S. Raza, A.-P. Jauho, M. Wubs, and N. A. Mortensen, "Modified field enhancement and extinction by plasmonic nanowire dimers due to nonlocal response," *Opt. Express*, vol. 13, p. 4176, 2012.
49. G. Toscano, S. Raza, S. Xiao, M. Wubs, A.-P. Jauho, S. I. Bozhevolnyi, and N. A. Mortensen, "Surface-enhanced Raman spectroscopy: nonlocal limitations," *Opt. Lett.*, vol. 37, p. 2538, 2012.
50. A. I. Fern´andez-Dom´ınguez, A. Wiener, F. J. Garc´ıa-Vidal, S. A. Maier, and J. B. Pendry, "Transformation-optics description of nonlocal effects in plasmonic nanostructures," *Phys. Rev. Lett.*, vol. 108, p. 106802, 2012.
51. I. C. Cirac`, E. Poutrina, M. Scalora, and D. R. Smith, "Origin of second-harmonic generation enhancement in optical split-ring resonators," *Phys. Rev. B*, vol. 85, p. 201403(R), 2012.
52. A. Wiener, A. I. Fern´andez-Dom´ınguez, A. P. Horsfield, J. B. Pendry, and S. A. Maier, "Nonlocal effects in the nanofocusing performance of plasmonic tips," *Nano Lett.*, vol. 12, p. 3308, 2012.
53. I. C.Cirac`, R. T. Hill, J. J. Mock, Y. Urzhumov, A. I. Fern´andez-Dom´ınguez, S. A. Maier, J. B. Pendry, A. Chilkoti, and D. R. Smith, "Probing the ultimate limits of plasmonic enhancement," *Science*, vol. 337, p. 1072, 2012.
54. J. A. Scholl, A. L. Koh, and J. A. Dionne, "Quantum plasmon resonances of individual metallic nanoparticles," *Nature (London)*, vol. 483, p. 421, 2012.
55. V. Ryzhii, "Terahertz plasma waves in gated graphene heterostructures," *Jpn. J. Appl. Phys.*, vol. 45, p. L923, 2006.
56. V. Ryzhii, A. Satou, and T. Otsuji, "Plasma waves in two-dimensional electron-hole system in gated graphene heterostructures," *J. Appl. Phys.*, vol. 101, p. 024509, 2007.

57. D. Svintsov, V. Vyurkov, V. Ryzhii, and T. Otsuji, "Voltage-controlled surface plasmon-polaritons in double graphene layer structures," *J. Appl. Phys.*, vol. 113, p. 053701, 2013.
58. V. Ryzhii, A. A. Dubinov, V. Y. Aleshkin, M. Ryzhii, and T. Otsuji, "Injection terahertz laser using the resonant inter-layer radiative transitions in double-graphene-layer structure," *Appl. Phys. Lett.*, vol. 103, p. 163507, 2013.
59. K. Batrakov and S. Maksimenko, "Graphene layered systems as a terahertz source with tuned frequency," *Phys. Rev. B*, vol. 95, p. 205408, 2017.
60. K. Batrakov and V. Saroka, in Nanomaterials Imaging Techniques, Surface Studies, and Applications, vol. 146, edited by O. Fesenko, L. Yatsenko and M. Brodin, New York: Springer Science+Business Media, pp. 291–307, 2013.
61. K. G. Batrakov, V. A. Saroka, S. A. Maksimenko, and C. Thomsen, "Plasmon polariton deceleration in graphene structure," *J. Nanophotonics*, vol. 6, p. 061719, 2012.
62. A. Hill, S. A. Mikhailov, and K. Ziegler, "Dielectric function and plasmons in graphene," *EPL Europhysics Lett.*, vol. 87, p. 27005, 2009.
63. S. A. Mikhailov, "Quantum theory of third-harmonic generation in graphene," *Phys. Rev. B*, vol. 90, p. 241301, 2014.
64. H. Karasawa, T. Komori, T. Watanabe, A. Satou, H. Fukidome, M. Suemitsu, V. Ryzhii, and T. Otsuji, "Observation of amplified stimulated Terahertz emission from optically pumped heteroepitaxial graphene-on-silicon materials," *J. Infrared, Millimeter, Terahertz Waves*, vol. 32, p. 655, 2010.
65. A. J. Hoffman, L. Alekseyev, S. S. Howard, K. J. Franz, D. Wasserman, V. A. Podolskiy, E. E. Narimanov, D. L. Sivco, and C. Gmachl, "Negative refraction in semiconductor metamaterials," *Nat. Mater.*, vol. 6, no. 12, pp. 946–950, 2007.
66. H. N. S. Krishnamoorthy, Z. Jacob, E. Narimanov, I. Kretzschmar, and V. M. Menon, "Topological transitions in metamaterials," *Science*, vol. 336, no. 6078, pp. 205–209, 2012.
67. C. Argyropoulos, N. M. Estakhri, F. Monticone, and A. Alù, "Negative refraction, gain and nonlinear effects in hyperbolic metamaterials," *Opt. Express*, vol. 21, no. 12, pp. 15037–15047, 2013.
68. S. Campione, T. S. Luk, S. Liu, and M. B. Sinclair, "Optical properties of transiently-excited semiconductor hyperbolic metamaterials," *Opt. Express*, vol. 5, no. 11, pp. 2385–2394, 2015.
69. C. Guclu, S. Campione, and F. Capolino, "Hyperbolic metamaterial as super absorber for scattered fields generated at its surface," *Phys. Rev. B*, vol. 86, no. 20, p. 205130, 2012.
70. L. B. Felsen and N. Marcuvitz, *Radiation and Scattering of Waves*, Hoboken, New Jersey, USA: Prentice Hall, 1973.
71. D. R. Smith and D. Schurig, "Electromagnetic wave propagation in media with indefinite permittivity and permeability tensors," *Phys. Rev. Lett.*, vol. 90, no. 7, p. 077405, 2003.
72. A. Poddubny, I. Iorsh, P. Belov, and Y. Kivshar, "Hyperbolic metamaterials," *Nat. Photonics*, vol. 7, no. 12, pp. 948–957, 2013.
73. T. Guo et al., "Tunable terahertz amplification based on photoexcited active graphene hyperbolic metamaterials," *Optical Materials Express*, vol. 8, no. 12, pp. 3941–3952, 2018.
74. P. Y. Chen and J. Jung, "P T Symmetry and Singularity-Enhanced Sensing Based on Photoexcited Graphene Metasurfaces," *Physical Review Applied*, vol. 5, no. 6, p. 064018, 2016.
75. M. Sakhdari et al., "PT-symmetric metasurfaces: Wave manipulation and sensing using singular points," *New Journal of Physics*, vol. 19, no. 6, p. 065002, 2018.

76. T. Gric and O. Hess, "Tunable surface waves at the interface separating different graphene-dielectric composite hyperbolic metamaterials," *Opt. Express*, vol. 25, no. 10, pp. 11466–11476, 2017.
77. B. A. Bacha, T. Khan, N. Khan, S. A. Ullah, M. S. A. Jabar, and A. U. Rahman, "Hybrid modes propagation of SPPs at the interface of cesium and grapheme," *Eur. Phys. J. Plus*, vol. 133, p. 509, 2018.
78. S. P. Burgos, R. de Waele, A. Polman, and H. A. Atwater, "A single-layer wide-angle negative-index metamaterial at visible frequencies," *Nat. Mater.*, vol. 9, pp. 407–412, 2010.
79. S. M. Prokes, J. Glembocki Orest, J. E. Livenere, T. U. Tumkur, J. K. Kitur, G. Zhu, B. Wells, V. A. Podolskiy, and M. A. Noginov, "Hyperbolic and plasmonic properties of Silicon/Ag aligned nanowire arrays," *Opt. Express*, vol. 21, pp. 14962–14974, 2013.
80. A. Murphy et al., "Fabrication and optical properties of large-scale arrays of gold nanocavities based on rod-in-a-tube coaxials," *Appl. Phys. Lett.*, vol. 102, p. 103103, 2013.
81. R. J. Pollard, A. Murphy, W. R. Hendren, P. R. Evans, R. Atkinson, G. A. Wurtz, A. V. Zayats, and A. Podolskiy Viktor, "Optical nonlocalities and additional waves in epsilon-near-zero metamaterials," *Phys. Rev. Lett.*, vol. 102, p. 127405, 2009.
82. L. M. Custodio, C. T. Sousa, J. Ventura, J. M. Teixeira, P. V. S. Marques, and J. P. Araujo, "Birefringence swap at the transition to hyperbolic dispersion in metamaterials," *Phys. Rev. B*, vol. 85, p. 165408, 2012.
83. S. Biswas, J. Duan, D. Nepal, R. Pachter, and R. Vaia, "Plasmonic resonances in self-assembled reduced symmetry gold nanorod structures," *Nano Lett.*, vol. 13, pp. 2220–2225, 2013.
84. A. Olivieri, A. Akbari, P. Berin, "Surface plasmon waveguide Schottky detectors operating near breakdown," *Phys. Status Solidi RRL*, vol. 4, pp. 283–285, 2010.
85. B. Fan, F. Liu, Y. Li, Y. Huang, Y. Miura, and D. Ohnishi, "Refractive index sensor based on hybrid coupler with short-range surface plasmon polariton and dielectric waveguide," *Appl. Phys. Lett.*, vol. 100, p. 111108, 2012.
86. Y.-C. Chang, C.-H. Liu, C.-H. Liu, S. Zhang, S. R. Marder, E. E. Narimanov, Z. Zhong, and T. B. Norris, "Realization of mid-infrared graphene hyperbolic metamaterials," *Nat. Commun.*, vol. 7, p. 10568, 2016.
87. M. A. K. Othman, C. Guclu, and F. Capolino, "Graphene-based tunable hyperbolic metamaterials and enhanced near-field absorption," *Opt. Express*, vol. 21, no. 6, pp. 7614–7632, 2013.
88. M. A. K. Othman, C. Guclu, and F. Capolino, "Graphene-dielectric composite metamaterials: evolution from elliptic to hyperbolic wavevector dispersion and the transverse epsilon-near-zero condition," *J. Nanophotonics*, vol. 7, no. 1, p. 073089, 2013.
89. I. V. Iorsh, I. S. Mukhin, I. V. Shadrivov, P. A. Belov, and Y. S. Kivshar, "Hyperbolic metamaterials based on multilayer graphene structures," *Phys. Rev. B*, vol. 88, no. 3, p. 039904, 2013.
90. K. V. Sreekanth, A. De Luca, and G. Strangi, "Negative refraction in graphene-based hyperbolic metamaterials," *Appl. Phys. Lett.*, vol. 103, no. 2, p. 023107, 2013.
91. K. S. Novoselov, A. K. Geim, S. V. Morozov, D. Jiang, Y. Zhang, S. V. Dubonos, I. V. Grigorieva, and A. A. Firsov, "Electric field effect in atomically thin carbon films," *Science*, vol. 306, no. 5696, pp. 666–669, 2004.
92. F. H. Koppens, D. E. Chang, and F. J. García de Abajo, "Graphene plasmonics: A platform for strong light–matter interactions," *Nano Lett.*, vol. 11, no. 8, pp. 3370–3377, 2011.
93. Y. Xiang, J. Guo, X. Dai, S. Wen, and D. Tang, "Engineered surface Bloch waves in graphene-based hyperbolic metamaterials," *Opt. Express*, vol. 22, no. 3, pp. 3054–3062, 2014.

8 Active Hyperbolic Metamaterials and Their Applications: From Visible to Terahertz Frequencies

Kandammathe Valiyaveedu Sreekanth and Ranjan Singh
Division of Physics and Applied Physics, School of Physical and Mathematical Sciences, Nanyang Technological University, Singapore

8.1 INTRODUCTION

Hyperbolic metamaterials (HMMs) are highly anisotropic, uniaxial, sub-wavelength artificial nanostructures characterized by indefinite or hyperbolic dispersion relation, where the principal components of electric permittivity or magnetic permeability tensors have different signs [1]. Since these anisotropic materials exhibit hyperbolic isofrequency (k_x, k_y, k_z) contour, they are called HMMs. The hyperbolic isofrequency surface with the existence of high-k modes is unique to HMMs and is not observed in other types of anisotropic materials. The high-k modes of HMMs are the propagating optical modes across the structure with infinitely large momentum in the effective medium limit.

HMMs are usually composed of metallic and dielectric elements with nanoscale dimensions, whose interplay endows the overall metamaterial with remarkable anisotropic properties, and they can easily be fabricated using conventional thin film deposition techniques [2]. Among different geometries showing hyperbolic dispersion [3], two systems are particularly interesting. They are metallic nanowires immersed in a dielectric matrix and stacked metal/dielectric multilayers [1–3]. However, metal/dielectric multilayers are the most studied HMMs due to their straightforward fabrication, and they have been used as the technological core in many applications. To date, various metal-dielectric combinations have been proposed to realize hyperbolic dispersion over a wide spectral range from ultraviolet to terahertz (THz). In comparison to conventional metamaterials, HMMs show lower optical losses due to single resonance.

DOI: 10.1201/9781003050162-8

Therefore, these effective bulk metastructures with extraordinary optical properties are exciting for a range of new applications, including sensing, single-photon sources, nano-imaging, negative refraction and focusing [4–15]. The thermal flux engineering and the occurrence of super-Planckian thermal energy transfer have been demonstrated using these mediums [16,17]. Moreover, HMMs revealed themselves to be a stimulating platform for the investigation of sophisticated phenomena belonging to fields that are traditionally very far from them, configuring themselves to be a theoretical and experimental toolbox for exploring new frontiers in cosmology and astrophysics [18].

In recent years, tunable features of nanophotonic devices have become a popular research area because they enable reconfigurability and miniaturization. In particular, the components of future active photonic devices could be tunable and reconfigurable, since it is important for existing applications and for future applications in artificial intelligence and smart processing networks, such as optical neural networks, to meet new integrated requirements. In this direction, the field of HMM appears to be following a similar trend [19]. A number of ways to tune the optical properties of HMMs has been demonstrated by incorporating different functional materials in the HMM geometry, which are liquid crystals, graphene, topological insulators, hexagonal boron nitride, phase transition materials and chalcogenide phase change materials (PCMs).

In this chapter, we discuss the recent progress in active HMMs including different tuning mechanisms and materials. We start the chapter by discussing the basic physics of HMMs and related physical effects. Then, we discuss the central theme of the chapter as active HMMs. We provide a critical review of this field, realizing active HMMs in different spectral regions from visible to THz frequencies using different functional materials. We further present the applications of active HMMs in biosensing, supercollimation of light, THz modulators and reconfigurable photonic devices. Finally, we conclude the chapter with an outlook of HMMs.

8.2 PHYSICS OF MULTILAYERED HYPERBOLIC METAMATERIALS

The overall optical response of an HMM can be qualitatively predicted in the framework of the so-called *effective medium theory* (EMT) because the individual components of HMMs are deeply sub-wavelength [20]. It is worth noticing that the simple approximation provided by the EMT does not allow consideration of the effect of strong non-local response of these structures, so many corrections have been introduced in recent times [21]. However, the EMT is still able to capture some specific remarkable features of wave propagation inside HMMs. Since HMM is a one-dimensional sub-wavelength crystal by stacking multiple metal/dielectric bilayers, a birefringent optical response is artificially induced, as a result an extraordinary optical axis arises in the direction perpendicular to the surface plane, and the optical response of HMM in the plane parallel to its surface results dramatically different from the one detected in its bulk. If the number of bilayers is large enough, and their thickness is deeply sub-wavelength, it is possible to homogenize the optical response of HMM. According to EMT, the uniaxial overall optical response of HMM can be homogenized; hence, the components parallel and perpendicular to the HMM surface plane can be calculated by applying simple boundary conditions [20,22].

Consider a case in which the electric field is polarized along the HMM surface plane, which leads to the calculation of $\varepsilon_{||}$. Here, the electric field must be continuous at the interfaces between the metal and dielectric. That is,

$$E_m = E_d = E_{eff} \tag{8.1}$$

with the subscripts *m, d* and *eff* standing for metal, dielectric and "effective", respectively. The effective electric flux can be expressed as the sum of flux densities in the metal and dielectric, i.e.,

$$D_{eff,||} = \tilde{\varepsilon}_{||} E_{eff} = f_m\, D_m + f_d\, D_d \tag{8.2}$$

where f is the fill-fraction of individual layers; $f_j = (t_j/(t_{tot}))$, with t_j being the thickness of the jth material ant t_{tot} the total thickness of the unit cell.

We obtain the expression for $\tilde{\varepsilon}_{||}$ by placing Eq. (8.1) in Eq. (8.2),

$$\tilde{\varepsilon}_{||} = \frac{\tilde{\varepsilon}_d t_d + \tilde{\varepsilon}_m t_m}{t_d + t_m} \tag{8.3}$$

To calculate $\tilde{\varepsilon}_{\perp}$ we consider that the electric flux D_j is continuous at the interfaces when the electromagnetic field is polarized perpendicularly to the HMM surface plane,

$$D_{eff,\perp} = \tilde{\varepsilon}_{\perp} E_{eff} = D_m + D_d \tag{8.4}$$

The effective electric field can be expressed as the fill-fraction averaged electric field in every single component, that is

$$E_{eff} = f_m\, E_m + f_d\, E_d \tag{8.5}$$

And, the expression for $\tilde{\varepsilon}_{\perp}$ can be obtained by placing Eq. (8.5) in Eq. (8.4),

$$\tilde{\varepsilon}_{\perp} = \frac{\tilde{\varepsilon}_d \tilde{\varepsilon}_m (t_d + t_m)}{\tilde{\varepsilon}_m t_d + \tilde{\varepsilon}_d t_m} \tag{8.6}$$

The two components of the *effective local permittivities* of a multilayered HMM are shown in Eqs. (8.3) and (8.6).

Now, the dispersion relation describing the wavelength dependence of wave vector can be calculated by starting from Maxwell equations in the harmonic regime:

$$\bar{k} \times \bar{E} = \omega \mu_0 \bar{H} \tag{8.7}$$

$$\bar{k} \times \bar{H} = -\omega \overleftrightarrow{\varepsilon} \bar{E} \tag{8.8}$$

In Eqs. (8.7) and (8.8), $\bar{k}$ is the wave vector, ω is the frequency, μ_0 is the vacuum permittivity and $\overleftrightarrow{\varepsilon}$ is the complex dielectric permittivity tensor. When the dielectric permittivity of material changes in all the three spatial directions, $\overleftrightarrow{\varepsilon}$ assumes the form of a diagonal matrix with components ε_{xx}, ε_{yy} and ε_{zz}. Multiplying both the terms on the left- and right-sides of Eq. (8.7) by $\bar{k}\times$, we get

$$\bar{k} \times \bar{k} \times \bar{E} = \omega \mu_0 \bar{k} \times \bar{H} \tag{8.9}$$

where $\omega = k_0 c$, with c as the speed of light in vacuum, given as $c=(\varepsilon_0 \mu_0)^{-1/2}$. Substituting Eq. (8.8) in Eq. (8.9) leads to

$$\bar{k} \times \bar{k} \times \bar{E} - k_0^2 \overleftrightarrow{\varepsilon} \bar{E} = 0 \tag{8.10}$$

In general, a uniaxial crystal with the extraordinary optical axis along the z-direction is characterized by a dielectric permittivity tensor, given as

$$\overleftrightarrow{\varepsilon} = \begin{bmatrix} \varepsilon_{xx} & 0 & 0 \\ 0 & \varepsilon_{xx} & 0 \\ 0 & 0 & \varepsilon_{zz} \end{bmatrix} \tag{8.11}$$

Since Eq. (8.10) constitutes the eigenvalue equation for the electric field, it can be manipulated to obtain the general dispersion relation for a uniaxial crystal as

$$\frac{k_x^2 + k_y^2}{\varepsilon_{zz}} + \frac{k_z^2}{\varepsilon_{xx}} = \frac{\omega^2}{c^2} \tag{8.12}$$

Since $\varepsilon_x = \varepsilon_y = \varepsilon_{\|}$ and $\varepsilon_z = \varepsilon_{\perp}$ for HMMs, Eq. (8.12) becomes

$$\frac{k_x^2 + k_y^2}{\varepsilon_{\perp}} + \frac{k_z^2}{\varepsilon_{\|}} = \frac{\omega^2}{c^2}; \tag{8.13}$$

Equation (8.13) represents the dispersion relation of an HMM modeled via the effective parameters $\varepsilon_{\|}$ and $\varepsilon_{\perp}$. Four cases can be distinguished based on the signs of $\varepsilon_{\|}$ and $\varepsilon_{\perp}$; they are

1. The HMM behaves as an *effective anisotropic dielectric* when$\varepsilon_{\perp} > 0$ and $\varepsilon_{\|} > 0$.
2. The HMM manifests the so-called *type I* anisotropy, behaving as a metal in the bulk and as a dielectric in the plane, when $\varepsilon_{\perp} < 0$ and $\varepsilon_{\|} > 0$.

3. The HMM manifests the so-called *type II* anisotropy, behaving as a dielectric in the bulk and as a metal in the plane, when $\varepsilon_\perp > 0$ and $\varepsilon_\parallel < 0$.
4. The HMM behaves as an *effective anisotropic metal* when $\varepsilon_\perp < 0$ and $\varepsilon_\parallel < 0$.

The *equifrequency contours* of Eq. (8.13) can be plotted by fixing the frequency and varying the wave vectors. First, we consider a case of a conventional anisotropic dielectric, where the resulting *equifrequency contour* is an ellipsoid in the wave vector space. In this case, the permitted wave vectors result in a confined ellipsoid whose larger and smaller semi-axes are determined by the extraordinary and ordinary refractive index values for a precise frequency. This case is illustrated in point 1, where the HMM behaves as an *effective dielectric*. Equation (8.13) assumes the shape of an open hyperboloid for the *type I* anisotropy, and of a closed hyperboloid for the *type II* anisotropy, when the cases mentioned in point 2 and 3 are satisfied. As mentioned in point 4, the case of anisotropic metal has no solution in the real wave vector plane, but it leads to a dispersion as that calculated in the case of effective dielectric, in the complex wave vector space.

Fig. 8.1a illustrates the schematic of a typical multilayered HMM with the above-mentioned four cases. In Fig. 8.1b, we show the bi-dimensional conical section in the K_x-K_z plane of the equifrequency contours for the hyperbolic anisotropies. Here, the *type I* regime is characterized by an open hyperboloid, whereas for the *type II* regime, the hyperboloid is closed. In both the cases, there are ideally no upper limits in the values of K_x and K_z, which indicates that extremely high wave vectors are allowed in these materials. These modes with such huge wave vectors are usually called *high-k* modes [2]. Engineering of high-*k* modes has been extensively used to modify the spontaneous emission enhancement of fluorophores placed close to the HMMs and, hence, to achieve the broadband Purcell effect [6]. Moreover, they have been used to achieve extreme sensitivity in biosensing [5].

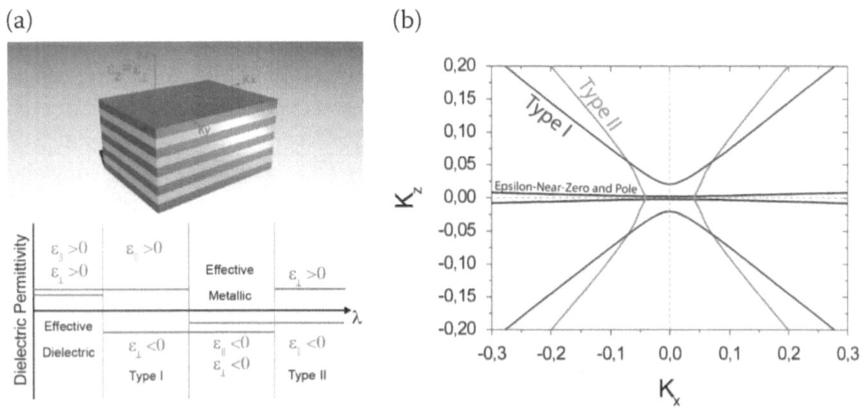

FIGURE 8.1 (a) Schematic of a multilayered hyperbolic metamaterial with references to all three homogenized dielectric permittivities and a scheme of the four possible combinations with relative anisotropy type. (b) The bidimensional conical section in the *x*–*z* plane representing the equifrequency contour for the *type II* (red) and *type I* (blue) hyperbolic dispersion. Reproduced with permission from Sreekanth et al. [22].

8.3 ACTIVE HYPERBOLIC METAMATERIALS

Conventional HMMs are made of alternating multilayers of metallic and dielectric materials, which can only provide a fixed hyperbolic dispersion relation over a wide spectral band from UV to infrared. For these HMMs, the hyperbolic dispersion is fixed to a specific spectral band and cannot be changed because their property is predetermined by the layer materials and filling fraction during fabrication. However, tunable hyperbolic dispersion over a wide spectral range, without changing the components and/or filling fraction, is required for future reconfigurable photonic device applications. One way to realize active HMM is by incorporating functional materials into HMMs. In the following sections, we discuss the tuning of hyperbolic spectral band in different spectral bands by using various functional materials.

8.3.1 Visible to Near-infrared Spectral Band

Realizing tunable hyperbolic dispersion in the visible and near infrared (NIR) wavelengths is particularly important for various applications. It is possible by using active materials, such as liquid crystals (LC), phase transition materials (VO_2) and chalcogenide PCMs. In the NIR wavelengths, a nanosphere dispersed liquid crystal (NDLC)-based HMM has been proposed for tunable hyperbolic dispersion [23]. By applying an external electric field, the orientation of a nematic liquid crystal (NLC) director can be changed, and, therefore, the effective permittivity and the dispersion relation of HMM can be altered. For this active HMM, the reflected and refracted waves can be tuned from normal to negative reflection/refraction when a TM-polarized wave is incident at the boundary between NDLC and the vacuum. In addition, a liquid crystal-based tunable HMM composed of NLC and silver nanowires was proposed for visible wavelengths, where the hyperbolic dispersion was tuned between 454 nm and 475 nm [24]. However, the large spectral tunability is not possible with LC-based HMMs due to a smaller refractive index contrast between the extraordinary and ordinary refractive indices ($\Delta n = \Delta n_e - \Delta n_o$). Phase transition materials, such as VO_2, undergo a metal-insulator transition at 68 °C. This volatile phase transition property is attractive for tuning the response of HMMs in the NIR frequencies. By stacking VO_2/TiO_2 multilayers and using the metal-insulator phase transition in VO_2, the dispersion was tuned from elliptic to hyperbolic [25].

8.3.1.1 *Chalcogenide Phase Change Material-based Active HMMs*

The interesting optical properties of chalcogenide-based PCMs have opened the door for photonic data storage again because the optical constants of these materials can be rapidly activated thermally, optically or electrically [26–28]. For visible and NIR wavelengths, PCMs offer one way to tune the properties of optical structures. In comparison to LC- and VO_2-tuned photonic devices, PCMs maintain their structural state, and the energy is only required during the switching process. This versatile and rapid modulation of such materials makes them suitable for a wide range of reconfigurable photonic device applications. The ideal PCMs, such as

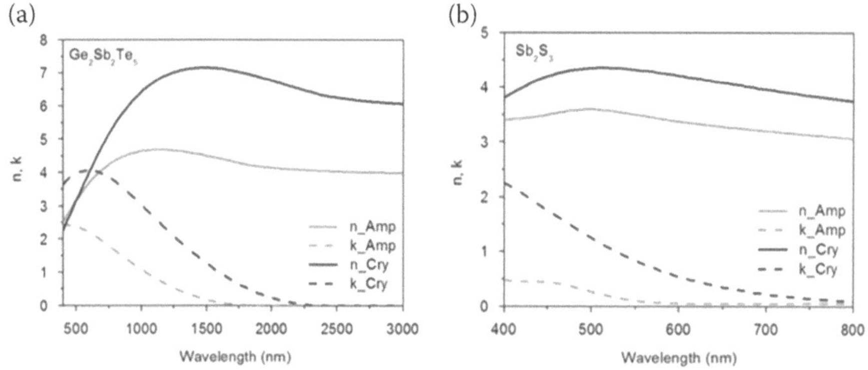

FIGURE 8.2 Experimentally determined optical constants of PCMs in amorphous and crystalline phases (a) GST and (b) Sb_2S_3. Reproduced with permission from Sreekanth et al. [22].

$Ge_2Sb_2Te_5$ (GST), exhibit a substantial optical constant change when switched between the states from amorphous to crystalline. More importantly, GST can be reversibly switched on a sub-nanosecond time scale billions of times. The real and imaginary refractive indices of GST in amorphous and crystalline phases are shown in Fig. 8.2a. As can be seen, GST shows a large refractive index (n) contrast in the infrared wavelengths by switching the phase from amorphous to crystalline, and low-loss (k) at >1,500 nm wavelength. However, GST is a lossy dielectric material with low refractive index contrast in the visible regime. Therefore, GST can be used as a potential tunable material for realizing reconfigurable photonic devices in the infrared spectral band.

An active HMM made of GST/Ag/Ge multilayers has been proposed to tune the hyperbolic to elliptical dispersion in the NIR wavelengths (Fig. 8.3a) [29]. As shown in Fig. 8.3b, this HMM shows type I hyperbolic dispersion for a wide wavelength range from 400 nm to 1,200 nm when the GST is in amorphous phase, and the hyperbolic dispersion band in the NIR wavelength can be changed to elliptical dispersion by switching the GST phase to crystalline. For TM-polarization, the sign reversal of the Poynting vector component parallel to the interface, resulting in the negative refraction of the wave, can be changed into the positive refraction by switching the phase of the GST layer from amorphous to crystalline (Fig. 8.3c). However, many of the GST tunable structures are not viable due to inter-diffusion of noble metals (Au and Ag) with phase-changing chalcogenides, when noble metals are in direct contact with chalcogenides [30].

Since GST is a lossy PCM in the visible wavelengths, and noble metal-based HMMs are severely limited due to strong energy dissipation in noble metals at visible wavelengths, we proposed a low-loss and tunable Sb_2S_3-TiN HMM for visible wavelengths [31]. In this HMM, Sb_2S_3 is a low-loss PCM with a bandgap of 2 eV, which renders it transmissive in the visible spectrum. As shown in Fig. 8.2b, the dielectric function of Sb_2S_3 exhibits a substantial change in the visible frequencies when the structure of material is switched from amorphous to

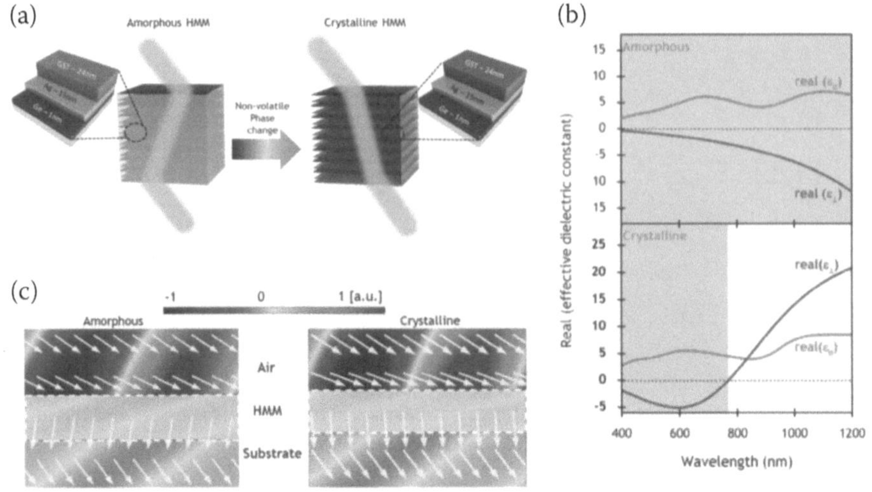

FIGURE 8.3 (a) Schematic of PCM-based active HMM when GST is in amorphous and crystalline phases. (b) Real parts of effective permittivities of HMM in amorphous and crystalline phase. (c) Simulated electric field amplitude and the direction of power flow when the TM-polarized plane wave incident at an angle incidence of 60° on the amorphous and crystalline HMM. Reproduced with permission from [Krishnamoorthy et al., Adv. Opt. Mater. 6: 1800332 (2018)] John Wiley and Sons.

crystalline phase. To avoid typical diffusion problems in noble metals, TiN has been selected as a plasmonic component in the designed HMM. The fabricated active Sb_2S_3-TiN HMM consists of 10 alternating thin layers of Sb_2S_3 and TiN on a glass substrate. In Fig. 8.4a, we show the cross-sectional scanning electron microscope (SEM) image of Sb_2S_3-TiN HMM and the sharp TiN-Sb_2S_3 interfaces clearly visible, which confirms that there is no diffusion of TiN into Sb_2S_3. The deposited HMM sample is in amorphous phase. To switch the phase of Sb_2S_3 from amorphous to crystalline, the sample was annealed at 300 °C for 30 min under vacuum environment (at low-pressure argon environment), since the crystallization temperature of Sb_2S_3 is > 285 °C.

Since the HMM consists of five bilayers of Sb_2S_3-TiN, and their thickness is deeply sub-wavelength, we used EMT to obtain the effective permittivity components. In Fig. 8.4b, we show the real parts of effective permittivities of amorphous and crystalline HMM. It is evident that the amorphous HMM shows type I hyperbolic dispersion at $\lambda \geq 580$ nm, and the hyperbolic spectral band is slightly blue-shifted to 564 nm wavelength with the crystallization of Sb_2S_3. The observed blue shift is due to the decrease in effective index of crystalline HMM compared to the amorphous HMM.

We further demonstrated the tunable Brewster angle in Sb_2S_3-TiN HMM [32]. We also performed angular reflectance measurements for *p*- and *s*-polarizations using a spectroscopic ellipsometer. Fig. 8.4c depicts the obtained results for the amorphous and crystalline HMMs. As can be seen, in the elliptical region ($\lambda < 600$ nm), HMM shows minimum reflection for both polarizations, especially at higher angles of

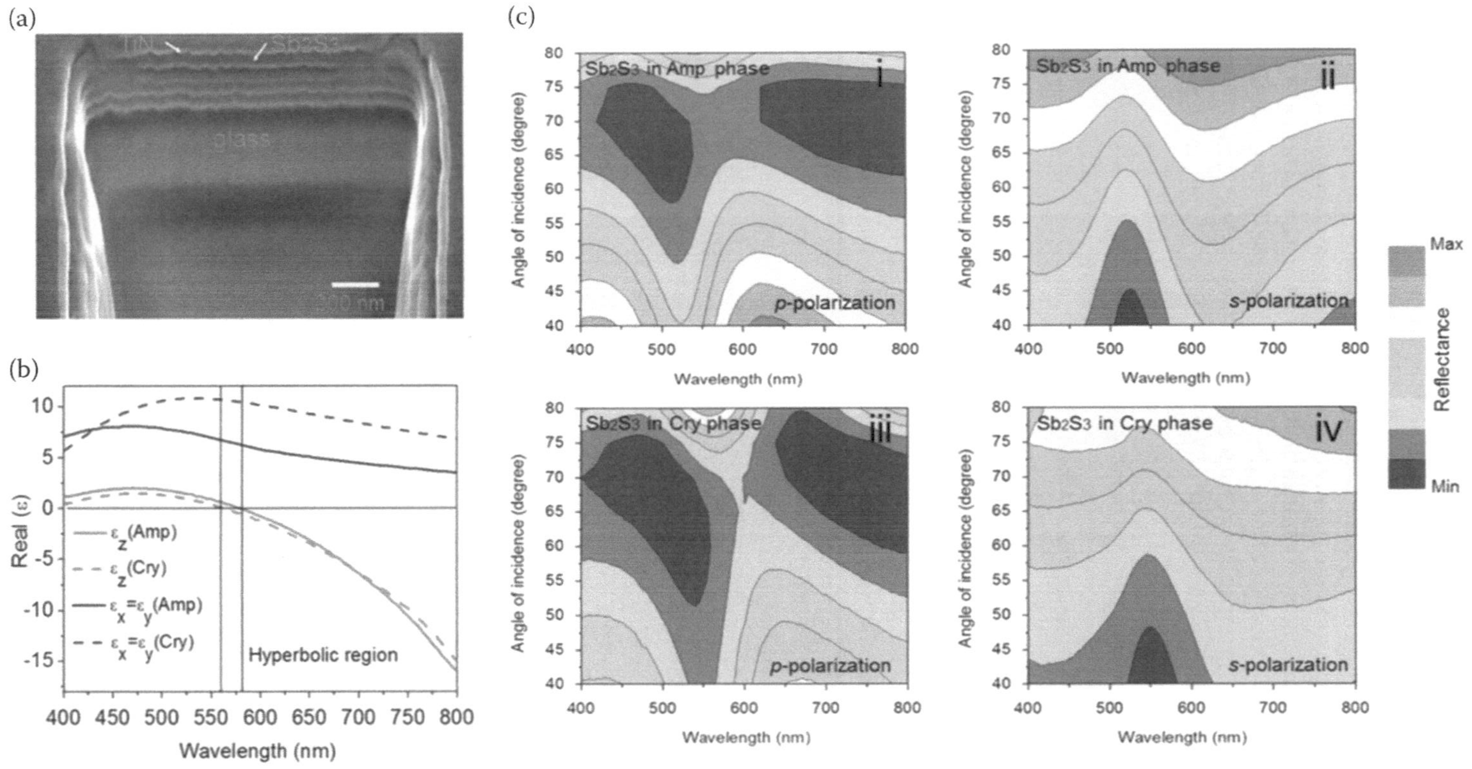

FIGURE 8.4 (a) A cross-sectional SEM image of the fabricated Sb_2S_3-TiN active HMM consisting of five pairs of Sb_2S_3 and TiN. (b) Real parts of uniaxial permittivity components of Sb_2S_3-TiN HMM when Sb_2S_3 is in the amorphous and crystalline phases, and (c) Measured angular reflectance spectrum of active HMM for *p*- and *s*-polarizations. Reproduced with permission from [Sreekanth et al., Adv. Opt. Mater. 7:1900081 (2019) & Sreekanth et al., Adv. Opt. Mater. 7:1900080 (2019)] John Wiley and Sons.

incidence for *p*-polarization (Fig. 8.4c(i)–(iii)) and lower angles of incidence for *s*-polarization (Fig. 8.4c(ii),(iv)). However, HMM shows minimum reflection only for *p*-polarization at higher angles of incidence in the hyperbolic region (λ > 600 nm). It evidences that the Sb_2S_3-TiN HMM supports Brewster modes due to hyperbolic dispersion. Interestingly, a red-shift in reflectance spectra and broadening of mode are observed after switching the Sb_2S_3 structural phase from amorphous to crystalline. It shows that the tuning of the Brewster angle is possible with phase change of Sb_2S_3 layers in the HMM. The obtained Brewster angles of HMM, when Sb_2S_3 is in the amorphous and crystalline states, are 70° at 725 nm and 69° at 750 nm, respectively.

8.3.2 Mid-infrared to THz Spectral Band

8.3.2.1 *Graphene-based Active HMMs*

Since graphene is a plasmonic material in the MIR to THz frequencies, it has been widely used to design active HMMs for this spectral band. Iorsh et al. [33] first proposed an active HMM based on multilayer graphene, where two-dimensional graphene sheets are separated with dielectric spacers. The dispersion relation of graphene-based HMM can be tuned from elliptic to hyperbolic by using an external gate voltage via controlling graphene-free carrier density. In particular, it concomitantly changes the in-plane HMM conductivity. This ability to control the electrical conductivity of graphene offers an easy mechanism to control the HMM dispersion and its resultant optical properties in the MIR to THz frequencies. Many graphene-based tunable HMMs have been theoretically and numerically demonstrated [33–35]. Recently, a graphene/Al_2O_3 multilayered HMM has been experimentally realized in the MIR wavelengths [36].

We proposed and numerically demonstrated a graphene-based HMM for tuning the negative group index spectral band [37]. As shown in Fig. 8.5a, for graphene/dielectric stack, we modeled each single-layer graphene (SLG) as a surface conducting sheet with conductivity $\sigma_g = ie^2\mu/(\pi\hbar^2(\omega + i/\tau))$ with ω as the angular frequency, e as the electron charge, τ as the electron relaxation time and μ as the chemical potential, which is a function of carrier density ($\mu \propto \sqrt{N_c}$). The complex dielectric constants of graphene are obtained from the expression, $\varepsilon_g = 1 + i(\sigma_g\eta_0/k_0t_g)$, where $\eta_0 \approx 377\ \Omega$ is the impedance of air, t_g is the effective graphene thickness and $k_0 = \omega/c$ is the vacuum wave vector with c being the speed of light. We considered SLG to have an effective thickness of 0.5 nm, and the electron relaxation time was assumed to be 1 ps. The dielectric constant and thickness of dielectric layer were considered to be 2.25 and 50 nm, respectively.

The EMT-derived real parts of effective permittivity component $\varepsilon_{//}$ with frequency are shown in Fig. 8.5a. The hyperbolic dispersion is obtained for certain frequencies in which the values of real parts of $\varepsilon_{//}$ are negative. However, in the elliptical region, the values of $\varepsilon_{//}$ result positive and approach the permittivity of the dielectric layer at higher frequencies. For a chemical potential, $\mu = 0.2$ eV, the transition from elliptical to hyperbolic dispersion occurred at a critical frequency of 24 THz. Fig. 8.5b shows the tuning of the hyperbolic spectral band with the increase in chemical potential of graphene. We observe a broadband tuning of critical

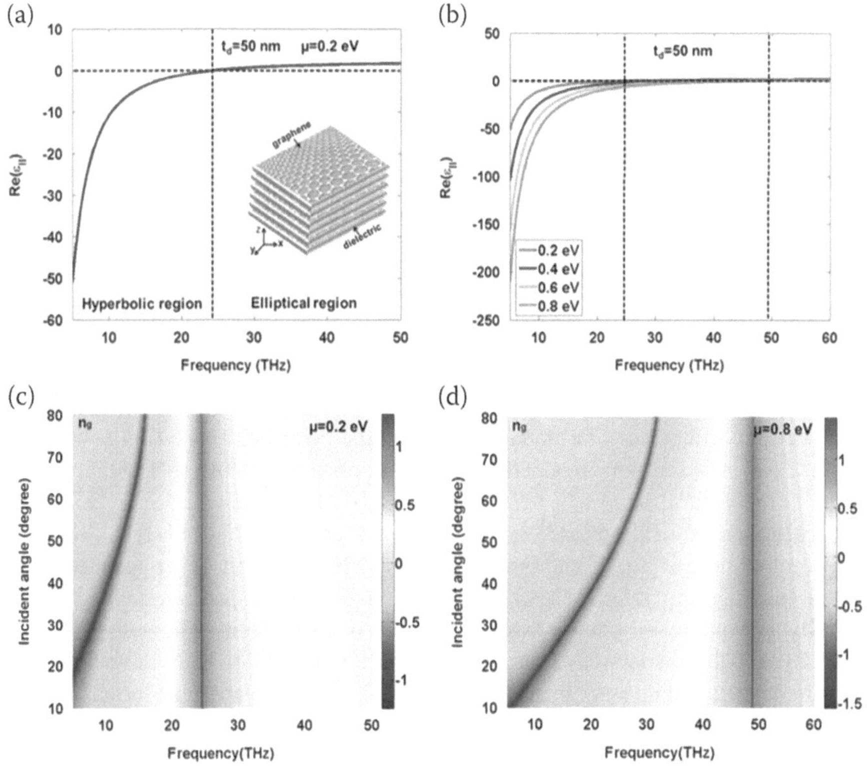

FIGURE 8.5 (a) Schematic of graphene-based HMM and the frequency-dependent variation of real parts of $\varepsilon_{//}$. (b) The frequency-dependent variation of real parts of $\varepsilon_{//}$ with chemical potential. 2D map (incident angle vs. frequency) of group-index of refraction for (c) $\mu = 0.2$ eV, and (d) $\mu = 0.8$ eV. Reproduced from [Sreekanth et al., Appl. Phys. Lett. 103, 023107 (2013)], with the permission of AIP Publishing.

frequency by slightly changing the chemical potential of graphene, as the obtained critical frequency for $\mu = 0.2$ eV and $\mu = 0.8$ eV is 24 THz and 49 THz, respectively.

To demonstrate the negative energy refraction in THz frequencies, we calculate the effective phase (n_p) and group (n_g) indexes of an anisotropic medium. In fact, the effective phase and group indexes are related to the angle of incidence (θ_{inc}). For low material absorption, these parameters can be calculated using

$$n_p = \sqrt{\varepsilon_{||} + \left(1 - \frac{\varepsilon_{||}}{\varepsilon_{\perp}}\right)\varepsilon_0 sin^2_{\theta_{inc}}} \tag{8.14}$$

$$n_g = \frac{\varepsilon_{\perp}}{\varepsilon_{||}}\sqrt{\varepsilon_{||} - \frac{\varepsilon_{||}}{\varepsilon_{\perp}}\left(1 - \frac{\varepsilon_{||}}{\varepsilon_{\perp}}\right)\varepsilon_0 sin^2_{\theta_{inc}}} \tag{8.15}$$

where ε_0 is the permittivity of air.

In Fig. 8.5c,d, we show the 2D map of group-index of refraction for chemical potentials $\mu = 0.2$ eV and $\mu = 0.8$ eV, respectively. As can be seen, the group-index results positive in the elliptical region and negative in the hyperbolic region. However, the calculated phase index is completely positive in both the hyperbolic and elliptical regions (not shown here). It indicates that the graphene-based HMM provides negative group and positive phase indexes of refraction in the THz frequencies at oblique incidence, which allows forward wavefront propagation and negative energy refraction. This is because the electric field vector $\boldsymbol{E}$ and electric displacement vector $\boldsymbol{D}$ are not usually parallel in HMM. As a result, the Poynting vector $\boldsymbol{S}$ (direction of energy flow) and wave vector $\boldsymbol{k}$ (along the wavefront normal) are not parallel. Therefore, HMM exhibits negative refraction with respect to $\boldsymbol{S}$ but positive refraction with respect to $\boldsymbol{k}$ when the boundary condition in which tangential component of k is conserved at the interface. Moreover, the directions of $\boldsymbol{k}$ and $\boldsymbol{S}$ depend on the effective phase (n_p) and group (n_g) indexes of refraction, respectively. It is important to note that the negative group index spectral band can be widened to MIR frequencies by increasing the chemical potential of graphene.

A combination of hexagonal boron nitride (hBN) and monolayer graphene can be used to realize tunable HMMs [38]. Hexagonal boron nitride is a dielectric that naturally forms vdW-bonded layered atomic planes. More importantly, hBN shows hyperbolic dispersion because the optical constants of hBN have different signs in the basal plane and normal plane since the out-of-plane vdW bonds do not conduct, but the in-plane covalent bonds are conductive. The dispersion tuning ability of graphene/hBN heterostructure is due to the hybridization of SPPs in graphene with the hyperbolic phonon polaritons in hBN. Also, note that the modes are dominated by hyperbolic plasmon-phonon polaritons, and the tunability is achieved by changing the gate voltage of the top graphene layer.

8.3.2.2 *Topological Insulator-based Active HMMs*

In recent years, topological insulators (TIs) have received much attention for THz plasmonics [39]; thus, they have been investigated for tunable HMM applications [40]. The common TIs are Bi_2Se_3 and Bi_2Te_3, which are characterized by unusual gapless edge or surface states, and a full insulating gap in the bulk. At THz frequencies, quintuple-layered Bi_2Se_3 was theoretically investigated to realize hyperbolic dispersion. In general, TI-based HMMs support highly directional and deeply sub-diffractional hyperbolic phonon polaritons [40]. More importantly, the dispersion relation of TI-based HMMs can be tuned by doping their surface states. Thus, high-speed dynamically tunable THz photonic devices could be realized using TI-based active HMMs.

We proposed a topological insulator-dielectric insulator-based superlattice structure and showed that the negative group index spectral band can be tuned from THz to MIR frequencies [41]. In Fig. 8.6a, we illustrate the proposed Bi_2Te_3-GeTe HMM superlattice structure, which comprises alternating layers of Bi_2Te_3 and GeTe. We used Bi_2Te_3 as the TI because the crystal structure of Bi_2Te_3 is similar to that of Bi_2Se_3, and GeTe is a PCM. For the effective permittivity calculations, the permittivities of Bi_2Te_3 were obtained from the Drude-Lorentz model

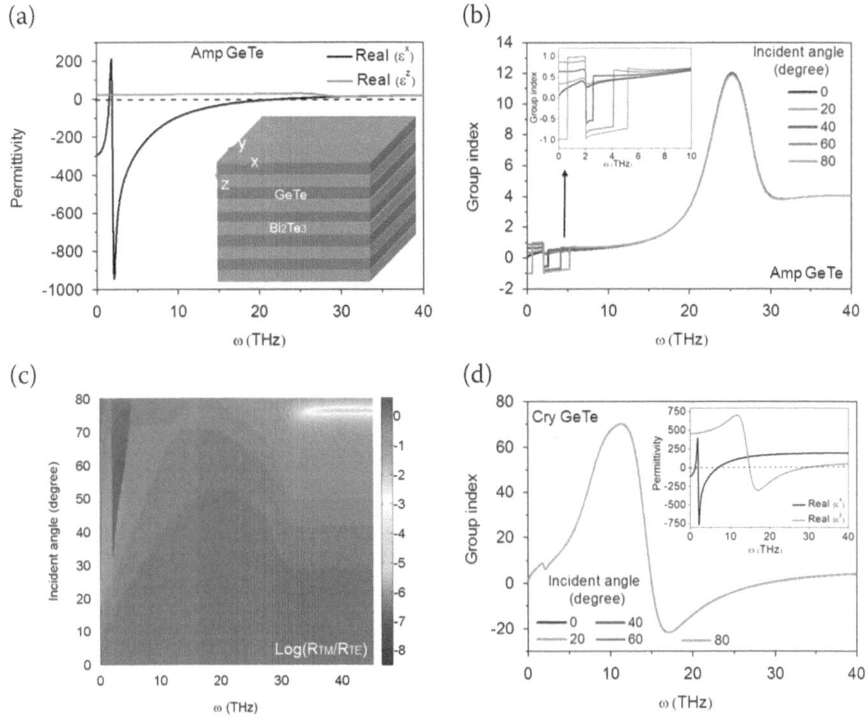

FIGURE 8.6 (a) Schematic of proposed Bi_2Te_3-GeTe HMM and real parts of the uniaxial permittivity components. (b) Group index spectrum for different angles of incidence when GeTe is in amorphous phase. (c) Calculated 2D map of reflection spectra as a function of incident angle and frequency. (d) Real parts of the uniaxial permittivity components when GeTe is in crystalline phase and group index spectrum for different angles of incidence. Reproduced with permission from Sreekanth et al., Opt. Commun. 440, 150 (2019).

$$\varepsilon(\omega) = \varepsilon_\infty - \frac{\omega_D^2}{\omega^2 + i\omega\gamma_D} + \sum_{j=1} \frac{\omega_{p,j}^2}{\omega_{0,j}^2 - \omega^2 - i\omega\gamma_j}$$

with $\varepsilon_\infty = 51$, $\omega_D = 207.06$ THz, $\gamma_D = 5.507$ THz, $\omega_{p,1} = 44.91$ THz, $\omega_{0,1} = 2.0095$ THz and $\gamma_1 = 0.2998$ THz. The dielectric permittivity of amorphous GeTe at THz frequencies is 15. In Fig. 8.6a, we plot the calculated real uniaxial permittivities of Bi_2Te_3-GeTe HMM for a TI fill fraction of 0.35, which demonstrates broadband *type II* hyperbolic dispersion from THz to MIR frequencies, where $\varepsilon^x(\omega) < 0$ and $\varepsilon^z(\omega) > 0$.

In Fig. 8.6b, we show the calculated group index of refraction with different angles of incidence for the amorphous phase of GeTe. One can see that the negative group index of refraction is obtained at incident angles greater than 35°, and the negative group index spectral band is widened with increase of incident angle. We observed a broad group index spectral band (2 THz to 5 THz) due to the broadband *type II* hyperbolic dispersion of the Bi_2Te_3-GeTe HMM. Fig. 8.6c shows the calculated fraction of incident light reflected at the air/Bi_2Te_3-GeTe HMM interface as a function of frequency and incident angle. Since Bi_2Te_3-GeTe HMM is an isotropic dielectric medium above 30 THz, a solid blue-yellow curve obtained above 30 THz represents the Brewster angle condition for TM-polarization. From 2 THz to 5 THz, a discontinuity in the reflection spectra (reflection peak) with incident angle (35° to 80°) is obtained in the negative group index spectral band.

For the crystalline phase of GeTe, the uniaxial permittivity data of HMM is shown in the inset of Fig. 8.6d, where we used the dielectric permittivity of crystalline GeTe at THz frequencies. It shows that both the *type I* and *type II* hyperbolic bands are present in Bi_2Te_3-GeTe HMM. As shown in Fig. 8.6d, the angle-independent positive and negative group index of refraction is obtained in the *type II* and *type I* hyperbolic bands, respectively. The change of negative to positive group index in the *type II* hyperbolic band is due to the fact that the uniaxial permittivity component, $\varepsilon^z(\omega)$ of crystalline Bi_2Te_3-GeTe HMM is considerably larger when the GeTe phase is switched from amorphous to crystalline. These results indicate that the Bi_2Te_3-GeTe superlattice structure is an alternative reconfigurable HMM for THz and MIR frequencies. An important advantage of these HMMs is that the optical properties can be tuned by switching the phase of GeTe from amorphous to crystalline as well as by doping the surface states of Bi_2Te_3.

8.3.2.3 *Superconductor-based Active HMMs*

High-Tc superconductors (HTS) can be used to realize hyperbolic dispersion in the infrared and THz frequencies since the superconductivity in most HTS materials could be achieved above the liquid nitrogen temperature. The superconductor-based HMM increases the superconducting critical temperature, which could be considered to be a metamaterial route to high temperature superconductivity [42]. Here, we discuss the realization of tunable HMMs using the superconductor-based multilayered geometry [43,44].

As shown in Fig. 8.7a, the proposed superconductor-based HMM consists of alternating thin layers of high-Tc superconductor, such as YBCO, and dielectric material, such as LAO. We selected this combination because YBCO is the widely used high-Tc superconductor with the transition temperature above liquid nitrogen temperature, and LAO is a suitable dielectric material for lattice-matched deposition of YBCO thin film. We used EMT to calculate the complex uniaxial permittivity components for the THz frequency band from 0.3 THz to 1.1 THz, where the total thickness of HMM was set to 1.6 μm, and the experimentally determined permittivities of YBCO and LAO were used. Note that the LAO permittivity is ~25, which is constant even with the temperature change, and the permittivity of YBCO significantly changes with frequency and temperature. In our calculations, we used the experimentally determined complex permittivities of 25 nm thick YBCO thin film with a fill-fraction of 50%.

The calculated real and imaginary parts of parallel and perpendicular permittivity components for different temperatures are shown in Fig. 8.7b,c, respectively. We consider a temperature range from 100 K to 20 K because the superconducting transition temperature of YBCO is 80–93 K. One can see that the real and

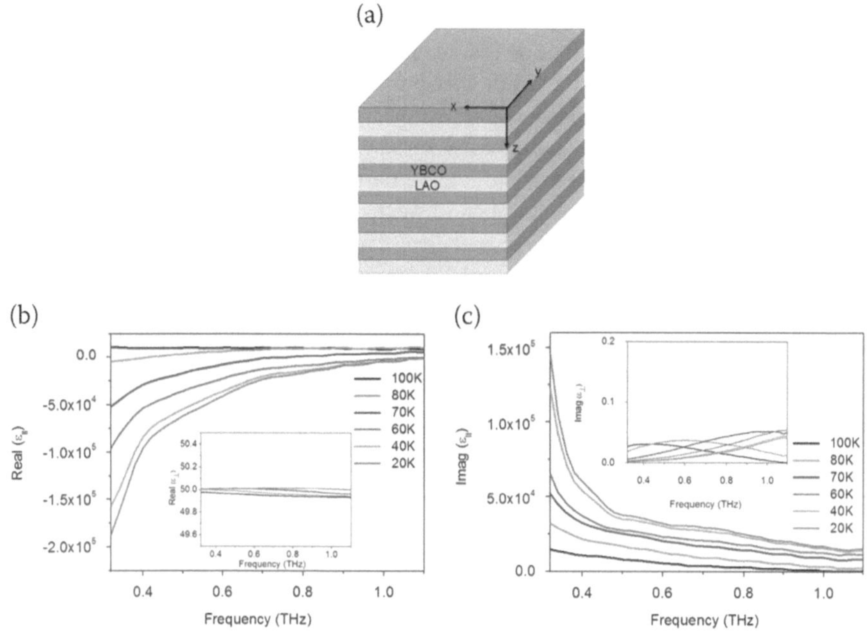

FIGURE 8.7 (a) Schematic of YBCO-LAO HMM. EMT-derived permittivity components for different temperatures. (b) Real parts of parallel and perpendicular permittivity components. (c) Imaginary parts of parallel and perpendicular permittivity components. Reproduced from [Sreekanth et al., J. Appl. Phys. (2020)], with the permission of AIP Publishing.

imaginary permittivities of both components are positive above the transition temperature (100 K), and real ($\varepsilon_{||}$) shows negative permittivity when the temperature is varied below the transition temperature of YBCO. It shows that the YBCO-LAO HMM is a dielectric material (elliptical dispersion) above the superconducting transition temperature of YBCO. Interestingly, the YBCO-LAO HMM shows *type II* hyperbolic dispersion in the superconducting phase of YBCO. More importantly, the *type II* hyperbolic dispersion bandwidth increases with decreasing temperature as the frequency-dependent conductivity of YBCO increases with decreasing temperature in the superconducting phase. These results evidence the transition of elliptical to hyperbolic dispersion with broadband tuning of hyperbolic dispersion bandwidth by changing the external stimulus as temperature.

8.3.3 Applications of Active HMMs

The tunable optical properties of HMMs can find various potential applications. Here, we show potential applications of active HMMs in different spectral bands, such as reconfigurable sensing, supercollimation of THz light and THz modulators.

8.3.3.1 *Reconfigurable Sensing*

We show the reconfigurable sensing of an active HMM with extreme sensitivity biosensing for small molecule detection using the Goos-Hänchen (G-H) shift interrogation scheme [31]. The G-H shift describes the lateral displacement of the reflected beam from the interface of two mediums when the angles of incidence are close to the coupling angle [45]. For this purpose, we used an active Sb_2S_3-TiN HMM. The BPP mode of HMM is excited using the prism coupling principle. A *p*-polarized light with wavelength 632.8 nm was used as the excitation source, because this wavelength belongs to the hyperbolic region of HMM, and the effective index of HMM is less than that of the used BK7 prism index (1.5), so that the momentum matching condition is satisfied (Fig. 8.4b). For both phases of Sb_2S_3, the excited BPP mode of Sb_2S_3-TiN HMM is shown in Fig. 8.8a. Due to the non-volatile phase change of Sb_2S_3 from amorphous to crystalline state, the effective index of HMM decreased. As a result, (i) the minimum reflected intensity at resonance angle is declined, (ii) the linewidth of reflection spectrum is decreased and (iii) the coupling angle is slightly shifted. We related the excited mode with the dispersion relation of HMM and confirmed that the excited mode is the fundamental BPP mode of Sb_2S_3-TiN HMM, which is a low-*k* mode [31]. The phase singularity at the coupling angle can be actively tuned by switching the phase of Sb_2S_3

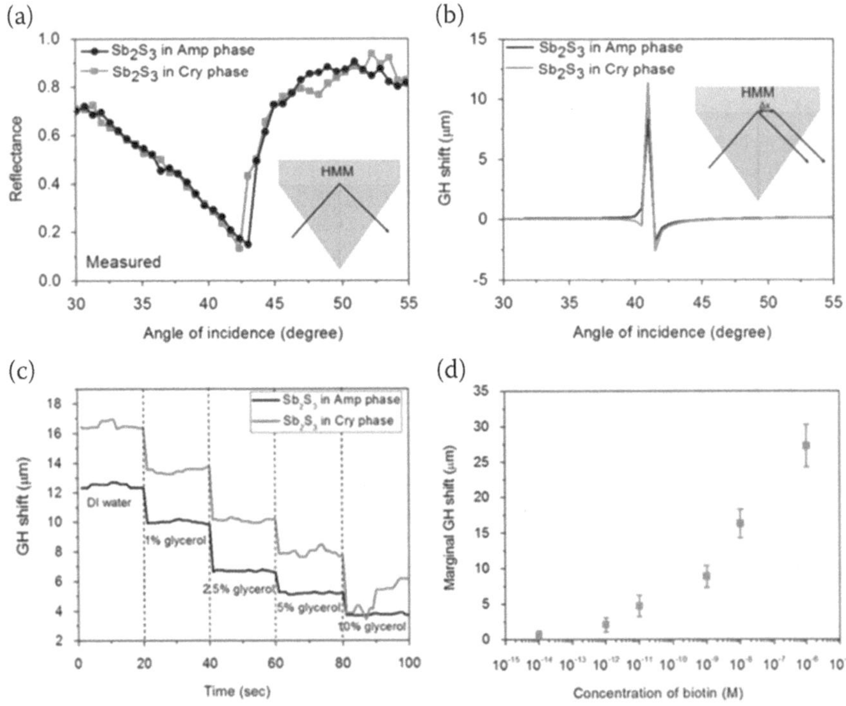

FIGURE 8.8 (a) Excited BPP mode of Sb_2S_3-TiN HMM for both phases of Sb_2S_3, (b) Calculated tunable G-H shift of HMM, (c) Real-time tunable refractive index sensing by injecting different weight percentage concentration of glycerol in distilled water, and (d) Measured marginal G-H shift for different concentrations of biotin in PBS (10 fM to 1 μM). Reproduced with permission from [Sreekanth et al., Adv. Opt. Mater. 7:1900081 (2019)] John Wiley and Sons.

from amorphous to crystalline since the phase difference between the TM- and TE-polarized light experiences a sharp singularity at the coupling angle.

We realized enhanced and tunable G-H shift at BPP mode excitation angle of Sb_2S_3-TiN HMM, because the phase derivative at the coupling angle determines the magnitude of the G-H shift. It is evident from Fig. 8.8b that the maximum G-H shift is possible at the coupling angle where there is a sharp change in phase difference. More importantly, tunable G-H shift is realized by switching the phase of Sb_2S_3 from amorphous to crystalline. To show the phase change induced tunable G-H shift experimentally, we performed refractive index sensing using a differential phase-sensitive setup [31]. We recorded the G-H shifts by injecting different weight ratios (1%–10% w/v) of aqueous solutions of glycerol into the sensor channel. In Fig. 8.8c, we show the real-time refractive index sensing data, where we report the measured G-H shift change with time due to the refractive index change of glycerol solutions. The clear step function in G-H shifts indicates that small refractive index changes can be recorded. Also, note that the tunable G-H shift is possible by switching the Sb_2S_3 structural phase from amorphous to crystalline. As can be seen, the maximum G-H shift is possible for crystalline HMM because the crystalline phase of Sb_2S_3 provides higher phase change at the coupling angle. The obtained maximum and minimum refractive index sensitivity for the crystalline and amorphous phases of HMM is 13.4×10^{-7} RIU/nm and 16.3×10^{-7} RIU/nm, respectively. By exploiting this tunable sensing feature, a reconfigurable sensor device can be developed for future intelligent sensing applications. We note that the obtained tunable range is relatively smaller; however, the G-H shift tunability can be further improved using longer wavelength sources and higher refractive index prisms.

Since the G-H shift interrogation scheme provides enhanced refractive index sensitivity, it can be used to detect small biomolecules at extremely low concentrations. In order to demonstrate this, we first sputtered an ultrathin film (10 nm) of gold on the top of the crystalline HMM sample. We followed the well-known streptavidin-biotin affinity model protocol for the capture of biotin by functionalizing streptavidin on the gold surface [5]. The refractive index change, caused by the capture of biotin on the sensor surface, was recorded by measuring the G-H shift. We injected different concentrations (10 fM to 1 μM) of biotin prepared in PBS into the sensor channel with a sample volume of 98 μl, and the corresponding G-H shift with increasing concentration was recorded. For biotin concentrations from 10 fM to 1 μM, the response of sensor with time was analyzed by calculating the marginal G-H shifts ($\delta GH = |GH_{pbs} - GH_{biotin}|$). Fig. 8.8d shows the recorded marginal G-H shifts after a reaction time of 40 min, where an increase in marginal G-H shifts with increasing biotin concentration is obtained. The detection limit of sensor device is less than 1 pM. In short, we envision that small molecules, such as exosomes, can be detected even from bodily fluids using the proposed HMM-based plasmonic platforms.

8.3.3.2 *Supercollimation of Light*

Realizing the supercollimation effect, the sub-wavelength focusing in THz frequencies is particularly important for developing high-resolution THz imaging devices for potential applications including security screening, biodetection and weather navigation. As mentioned above, the dispersion relation of TI-based HMMs can be tuned by doping their surface states. Here, we show that Bi_2Se_3 supports highly directional hyperbolic phonon polaritons, and it can be used to realize supercollimation of light in THz frequencies.

Bi_2Se_3 is a 3D TI composed of Se-Bi-Se-Bi-Se quintuple atomic planes that are separated by insulating Van der Waals bonds [41]. The layered structure of Bi_2Se_3 is depicted in the inset of Fig. 8.9b. Importantly, this layered structure exhibits strong anisotropy of their phonon modes. We note that the dominant (x-y)-axis and z-axis phonon frequencies of Bi_2Se_3 are 1.92 THz and 4.05 THz, respectively [40]. Bi_2Se_3 exhibits extremely anisotropic dielectric permittivity since these phonon mode frequencies are separated by a factor of two. As a result, the real part of the uniaxial permittivity components, $\varepsilon^x(\omega)$ and $\varepsilon^z(\omega)$, are indefinite for a certain range of THz frequency band, and Bi_2Se_3 exhibits hyperbolic dispersion in the THz frequencies for TM-polarization, $\frac{(k^x)^2+(k^y)^2}{\varepsilon^z(\omega)}+\frac{(k^z)^2}{\varepsilon^x(\omega)}=\frac{\omega^2}{c^2}$.

We used the following model and parameters to determine the uniaxial permittivity components of Bi_2Se_3 [40]:

$$\varepsilon^\alpha(\omega)=\varepsilon^\alpha_\infty+\sum_{j=1,2}\frac{\omega^{\alpha 2}_{p,j}}{\omega^{\alpha 2}_{to,j}-\omega^2-i\gamma^\alpha_j\omega},\quad \alpha=x,z \tag{8.16}$$

where $\varepsilon^x_\infty=29$, $\varepsilon^z_\infty=17.4$, $\omega^x_{to,1}=1.92$ THz, $\omega^x_{p,1}=21.1$ THz, $\omega^x_{to,2}=3.75$ THz, $\omega^x_{p,2}=1.65$ THz, $\omega^z_{to,1}=4.05$ THz, $\omega^z_{p,1}=8.5$ THz, $\omega^z_{to,2}=4.61$ THz, $\omega^z_{p,2}=4.67$ THz and $\gamma^\alpha_j=0.105$ THz. Fig. 8.9a,b, respectively, show the real and imaginary parts of uniaxial permittivity components ($\varepsilon^x(\omega)$ and $\varepsilon^z(\omega)$). As can be seen, the real permittivity components alter sign from positive to negative with frequency, while both the imaginary components are always positive. It is clear from the inset of Fig. 8.9a that the

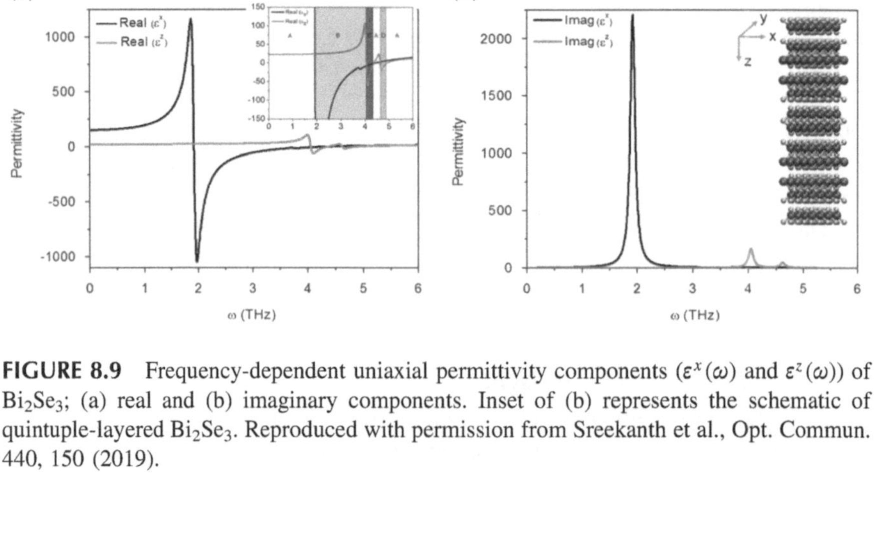

FIGURE 8.9 Frequency-dependent uniaxial permittivity components ($\varepsilon^x(\omega)$ and $\varepsilon^z(\omega)$) of Bi_2Se_3; (a) real and (b) imaginary components. Inset of (b) represents the schematic of quintuple-layered Bi_2Se_3. Reproduced with permission from Sreekanth et al., Opt. Commun. 440, 150 (2019).

region A represents dielectric band ($\varepsilon^x(\omega) > 0$ and $\varepsilon^z(\omega) > 0$), region B represents the *type II* hyperbolic band ($\varepsilon^x(\omega) < 0$ and $\varepsilon^z(\omega) > 0$), region C represents the Reststrahlen band ($\varepsilon^x(\omega) < 0$ and $\varepsilon^z(\omega) < 0$) and region D represents the *type I* hyperbolic band ($\varepsilon^x(\omega) > 0$ and $\varepsilon^z(\omega) < 0$). It shows that Bi_2Se_3 supports both the *type I* and *type II* hyperbolic dispersions at THz frequencies; however, it is not possible with graphene-based HMMs [37]. We note that the *type II* hyperbolic dispersion has a maximum bandwidth of 2.14 THz that belongs to the bandwidth difference between two phonon frequencies at 1.91 THz and 4.05 THz. In addition, the Reststrahlen band (4.05 THz to 4.4 THz) and *type I* hyperbolic band (4.6 THz to 4.9 THz) are narrow with a bandwidth of 0.35 THz and 0.3 THz, respectively.

Highly directional and deeply sub-diffractional hyperbolic phonon-polariton mode supported by Bi_2Se_3 can be used for supercollimation of light and superlensing. It is known that the propagation of wave inside HMM within the hyperbolic dispersion spectral band is allowed only at the resonance cone (RC) angle. The half of the resonance cone angle is given by $\alpha_{RC} = a\ \tan[-(\varepsilon^x(\omega)/\varepsilon^z(\omega))^{1/2}]$. The supercollimation of light inside HMM happens when $\alpha_{RC} = 0$, and it is possible when $\varepsilon^x(\omega) \approx 0$ or $\varepsilon^z(\omega) \approx \infty$. In Fig. 8.10a, we show the calculated half of the RC angle of Bi_2Se_3 HMM. The minimum angle close to zero is obtained at 4.05 THz, which is the inversion point of *type II* and Reststrahlen band.

To demonstrate the supercollimation of THz light at 4.05 THz, we performed numerical simulation using 2D finite difference time domain (FDTD) method. In

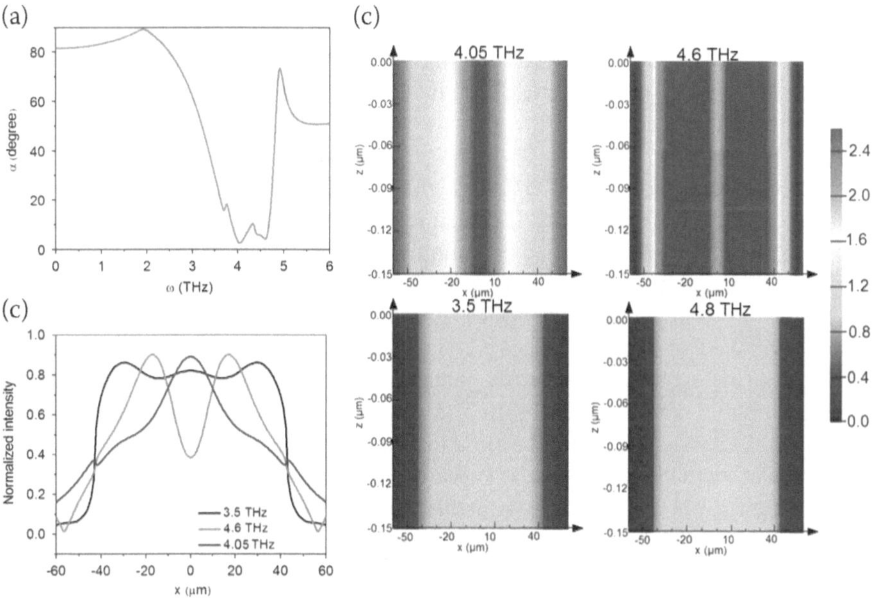

FIGURE 8.10 (a) Half of the resonance cone angle as a function of THz frequencies. (b) Propagation of plane wave through Bi_2Se_3 HMM at four THz frequencies. (c) Line profile of normalized intensity recorded at the bottom of quintuple layered Bi_2Se_3. Reproduced with permission from Sreekanth et al., Opt. Commun. 440, 150 (2019).

our simulation, light coming from an 85 μm slit placed on the top of a Bi_2Se_3 HMM for different frequencies was analyzed. Here, we considered the quintuple-layered Bi_2Se_3 as a single layer with thickness 150 nm and uniaxial permittivity components. The periodic and perfectly matched layer boundary condition was implemented along the *x*- and *z*-axes, respectively, and the incident polarization of light was set to TM. In Fig. 8.10b, we show the simulated propagation of plane wave at four different frequencies, corresponding to *type II* (3.5 THz), the inversion point of *type II* and Reststrahlen band (4.05 THz), the inversion point of Reststrahlen band and *type I* (4.6 THz) and *type I* (4.8 THz). In addition, the line profile of normalized intensity recorded at the bottom of Bi_2Se_3 is shown in Fig. 8.10c. It is evident that the supercollimation effect only occurred at 4.05 THz, because the RC angle is close to zero. The intensity spectrum broadens with the increase in RC angle. It is important to note that the supercollimation frequency can be changed by tuning the inversion point of *type II* and Reststrahlen bands, which is possible by doping the surface states of Bi_2Se_3 [40].

8.3.3.3 *THz Modulator*

In THz technology, THz modulators are one of the important components that can find applications in communications, spectroscopy and imaging. By integrating photonic structures with tunable conductive materials, such as semiconductors, superconductors, perovskites, 2D materials (graphene, MoS_2, etc.), phase materials and liquid crystals, various solid state and non-solid-state THz modulators with both broad and narrow bandwidth responses have been demonstrated [45–48]. We note that most of the narrow bandwidth THz modulators rely on narrow-band resonance effects, such as using metamaterials and plasmonic systems, which required intense lithography steps. However, broad bandwidth operation is important for many applications, and it can be achieved with non-resonant graphene-based solid-state devices [49].

The active control of THz modulators with high intensity and phase tunability is essential for various applications. It can be realized by developing THz modulators with high modulation depth, broad operation bandwidth and high modulation speed. Moreover, active control of polarization state of light in different spectral bands can find a plethora of applications in optics. It is important to note that two conditions must be satisfied to realize effective reflective type active polarization switches [50]; that is, (i) *s*-polarized reflected intensity should be unity with and without external stimulus, and (ii) the absolute modulation depth of *p*-polarized reflected intensity at the resonance should be close to 100%. The second condition essentially allows active switching. However, it is a challenging task to realize 100% modulation depth in THz frequencies. In this context, we propose and numerically demonstrate a THz modulator or an active polarization switch using a high Tc superconductor-based HMM. We show that above 98% intensity modulation and 100% phase tunability is possible by exciting the Brewster mode of HMM and varying the external stimulus as temperature.

As shown in Fig. 8.4, the *type I* HMMs support Brewster mode since the real ($\varepsilon_{\|}$) components are positive. However, the *type II* HMMs do not support Brewster mode since the real ($\varepsilon_{\|}$) components are negative. Since the YBCO-LAO HMM is a *type II* HMM (Fig. 8.7), it can be used for the intensity and phase modulation of THz light by selecting the incidence angle close to Brewster angle. The basic principle of the proposed THz modulator is based on the optical topological transition of YBCO-LAO HMM from elliptical to *type II* hyperbolic dispersion. The YBCO-LAO HMM supports Brewster mode in the dielectric phase of YBCO. This means that both the intensity and phase of THz light can be modulated at Brewster angle by varying the temperature above and below the superconductor transition temperature.

By using TMM calculations, we simulated the angular reflection spectra for *p*- and *s*-polarizations by considering the HMM as a multilayer of YBCO and LAO thin films, and using a single-layer effective medium approach. We also simulated the reflection spectra using COMSOL simulation software. We note that the results obtained using three approaches are almost similar. Therefore, we present the results obtained by TMM, where we consider the HMM as a multilayer of YBCO and LAO thin films. In Fig. 8.11, we show the 2D map of angular reflectance spectra for two temperatures, one above (100 K) and one below (20 K) the transition temperature. It is clear that the HMM supports Brewster angle (86°) at 0.98 THz in the dielectric phase of HMM (100 K), where the *p*-polarized reflected intensity is almost zero (Fig. 8.11a) and *s*-polarized reflected intensity is 100% (Fig. 8.11b).

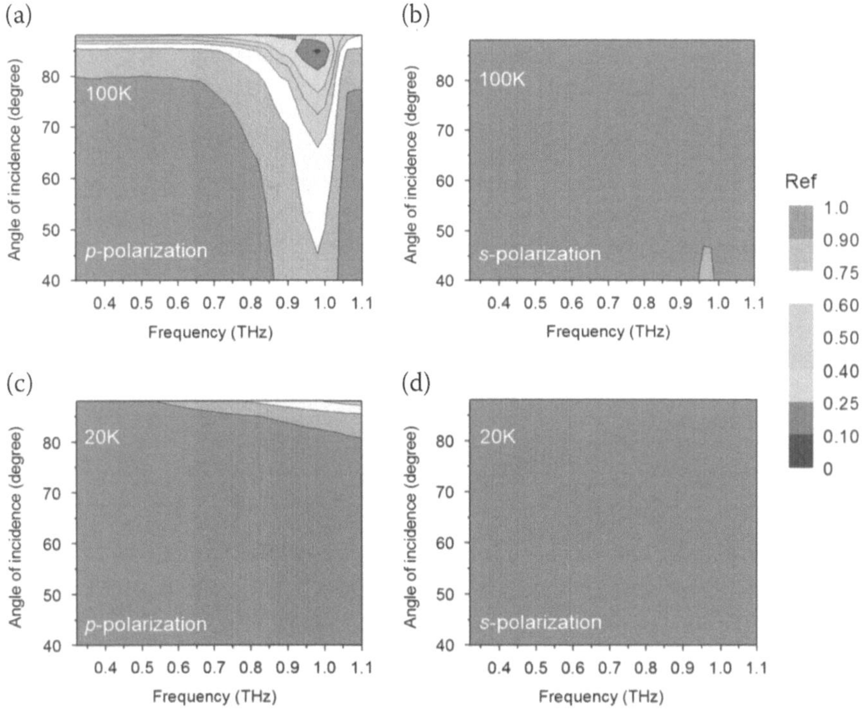

FIGURE 8.11 Simulated angular reflectance spectra for (a) *p*-polarization at 100 K, (b) *s*-polarization at 100 K, (c) *p*-polarization at 20 K and (d) *s*-polarization at 20 K. Reproduced from [Sreekanth et al., J. Appl. Phys. (2020)], with the permission of AIP Publishing.

However, when the temperature is reduced to 20 K, the p-polarized reflected intensity is increased to 77% (Fig. 8.11c) while keeping 100% s-polarized reflected intensity (Fig. 8.11d). For all temperatures, we found that the s-polarized reflected intensity keeps 100%. More importantly, the reflected intensity of p-polarized light can be tuned from zero to a maximum value at the Brewster angle by varying the temperature from 100 K to 20 K. The modulation depth (MD) of p-polarized reflected intensity is given by [49]

$$MD = \left(1 - \frac{|r|^2}{|r_{max}|^2}\right) \times 100\% \tag{8.17}$$

with r being the minimum reflectance at off-state and r_{max} being the maximum reflectance at on-state.

Fig. 8.12a illustrates the active tuning of p-polarized reflected intensity at the Brewster angle (86°) by varying the temperature. One can see that close to zero reflectance is obtained for 100 K at 0.98 THz, and the reflected intensity increased with decreasing temperature. In addition, the reflected intensity saturates at lower temperatures because of the saturation behavior of permittivity components of HMM. The calculated modulation depth of p-polarized reflected intensity at 0.98 THz is over 98%, where we considered off-state reflectance at 100 K and on-state reflection at 20 K. The proposed YBCO-LAO HMM fulfills the two conditions required to realize efficient reflective-type active polarization switches since YBCO-LAO HMM shows active control of polarization state of light by operating around the Brewster angle.

Since the reflection phase undergoes ~π phase shift at the Brewster angle [32], YBCO-LAO HMM can be used for the modulation of phase of light. Fig. 8.12b shows the simulated reflected phase difference between p- and s-polarized light for different temperatures at 86°. In the dielectric phase of HMM at 100 K, around 180° phase shift is obtained, whereas the phase shift is reduced to 0° when the temperature is decreased to 70 K, due to the *type II* conducting phase of HMM.

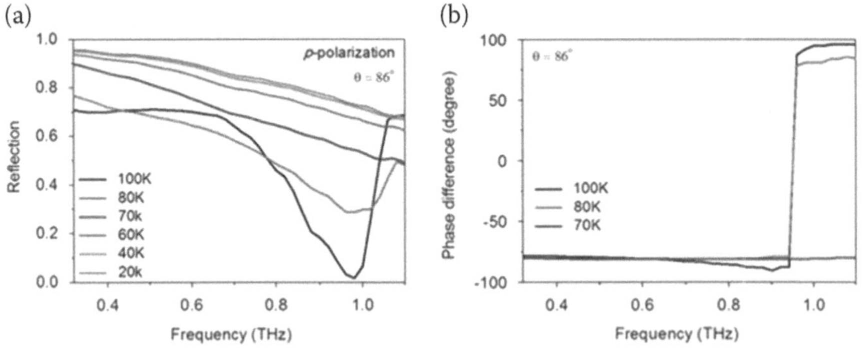

FIGURE 8.12 (a) Simulated p-polarized reflectance spectrum at 86° for different temperatures. (b) Simulated phase difference at 86° for different temperatures. Reproduced from [Sreekanth et al., J. Appl. Phys. (2020)], with the permission of AIP Publishing.

Interestingly, 100% tunability in reflection phase is possible when the temperature is slightly tuned below the transition temperature. Moreover, the calculated insertion loss ($log_{10}|r_{max}|^2$) of the YBCO-LAO HMM-based THz modulator is ~–2.3 dB. It is important to note that the realized phase tunability and insertion loss are much better compared to a graphene-controlled Brewster angle device-based THz modulator [49].

The observed 98% intensity modulation and 100% phase tunability are only due to the phase transition of YBCO-LAO HMM from the elliptical (dielectric) to the *type II* hyperbolic dispersion. To further emphasize this, we show in Fig. 8.13 the electromagnetic field distribution of the air-HMM-air system simulated using TMM. Fig. 8.13a,b shows the simulated magnetic field (H_z) distribution for 100 K and 20 K, respectively. The simulation is performed for the *p*-polarized light at 0.98 THz and 86°. It is visible that the H_z-field transmits through the HMM when the temperature is 100 K, which is due to vanishing reflection at Brewster angle. However, for 20 K, the H_z-component becomes almost zero beyond HMM due to maximum reflection. The simulated field distribution further indicates the disappearance of Brewster angle when the dispersion of HMM is changed from elliptical (dielectric) to *type II* hyperbolic. Even though the YBCO-LAO HMM shows maximum modulation depth at the Brewster angle, above 90% intensity modulation is still possible by selecting an operating angle around the Brewster angle. For the experimental realization of this concept, the multilayered HMM can be fabricated by physical vapor deposition of alternating thin films of YBCO and LAO since the lattice matched deposition of YBCO thin films on LAO is possible (Ceramic coating GmbH, Germany), and an optical cryostat equipped fiber laser-based THz time-domain spectroscopy setup can be used to perform the angular reflection measurements at cryogenic temperatures [51].

8.4 SUMMARY AND OUTLOOK

To design active HMMs for different spectral bands and their applications, it is important to consider suitable functional materials and approaches to achieve the tunability. Liquid crystals, phase transition materials (VO_2), and phase change chalcogenides-based active HMMs exhibit hyperbolic dispersion in the visible and NIR spectral bands. Even though high-quality plasmonic resonances are possible by combining LC with plasmonic materials and LC is transparent across a broad spectrum from the visible to NIR, it is difficult to form multilayered LC structures. Phase transition material-based active HMMs such as VO_2/TiO_2 can be tuned by changing their temperature. However, since VO_2 has a volatile phase switching mechanism, the energy needs to be supplied to maintain the optical state. Therefore, phase change chalcogenides-based active HMM is the only design where no energy is required to maintain the optical state due to their non-volatile switching mechanism. However, prototype PCM, such as $Ge_2Sb_2Te_5$, is not a good choice for visible wavelengths because it is highly lossy. The inter-diffusion of noble metals with chalcogenide is a critical issue. The combination of low-loss PCM, such as Sb_2S_3 with plasmonic TiN, can solve these issues at greater extent. Since $Ge_2Sb_2Te_5$ is a low-loss PCM with a large refractive index contrast between its

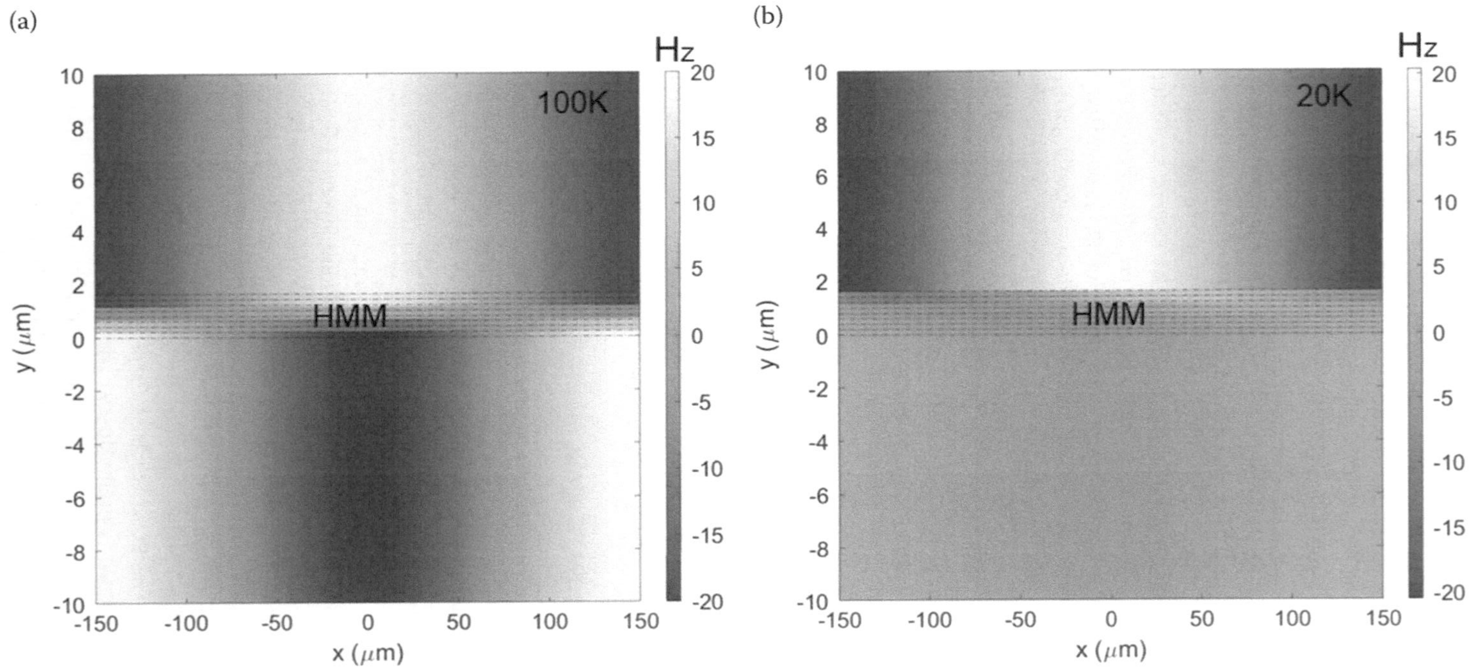

FIGURE 8.13 Simulated magnetic field (H_z) distribution for *p*-polarization at 0.98 THz and 86° for (a) 100 K and (b) 20 K. Reproduced from [Sreekanth et al., J. Appl. Phys. (2020)], with the permission of AIP Publishing.

structural states in the infrared regime, it can be combined with heavily doped semiconductors to form ideal active HMMs because heavily doped semiconductors have metallic properties at infrared wavelengths. However, it is important to consider the thermal design when using phase changing chalcogenides to design active HMMs, because PCM must be annealed above its melting temperature, and suddenly quenched at $\sim$109Ks^{-1} to freeze in the amorphous state to switch reversibly.

To date, tunable materials, such as graphene, hBN, and Tls, have been used to tune hyperbolic dispersion in the MIR and THz spectral bands, by changing the chemical potential of the surface state through an external gate voltage. While most metallic materials are not suitable to form HMMs in this spectral range, graphene is the best choice. However, the active tuning of individual layers of graphene in the stack is quite challenging. In this context, the hBN-based active HMMs can be made practically tunable by stacking it with graphene. It has a multilayered structure with monolayer graphene on the top, and exhibits a hyperbolic dispersion relation with tunable features. TIs are also potential tunable materials to achieve tunable hyperbolic dispersion across the similar wavelength band. A multilayer stack made of TI and PCM can also be used to tune the hyperbolic dispersion spectral band from MIR to THz. High-Tc superconductors are also promising functional materials to tune the hyperbolic dispersion in the THz frequencies thermally.

The dynamic control of spontaneous emission enhancement in different spectral bands is possible with the discussed active HMMs [52]. The tunable HMMs are particularly important for reconfigurable photonic structures. Realizing efficient tunable hyperbolic metamaterials from visible to THz spectral band will extend the research scope of hyperbolic dispersion phenomenon and its applications. The discussed active HMMs in different spectral bands can be used to realize cost-effective reconfigurable photonic devices for future applications, such as optical switches, modulators, single photon sources, beam splitters and sensors.

REFERENCES

1. A. Poddubny, I. Iorsh, P. Belov, and Y. Kivshar, "Hyperbolic metamaterials," *Nat. Photon.*, vol. 7, pp. 948–957, 2013.
2. C. L. Cortes, W. Newman, S. Molesky, and Z. Jacob, "Quantum nanophotonics using hyperbolic metamaterials," *J. Opt.*, vol. 14, p. 063001, 2013.
3. L. Lu, R. E. Simpson, and K. V. Sreekanth, "Active hyperbolic metamaterials: Progress, materials and design," *J. Opt.*, vol. 20, p. 103001, 2018.
4. A. J. Hoffman, L. Alekseyev, S. S. Howard, K. J. Franz, D. Wasserman, V. A. Podolskiy, E. E. Narimanov, D. L. Sivco, and C. Gmachi, "Negative refraction in semiconductor metamaterials," *Nat. Mater.*, vol. 6, pp. 946–950, 2007.
5. K. V. Sreekanth, Y. Alapan, M. Elkabbash, E. Ilker, M. Hinczewski, U. A. Gurkan, A. De Luca, and G. Strangi, "Extreme sensitivity biosensor platform based on hyperbolic metamaterials," *Nat. Mater.*, vol. 15, pp. 621–627, 2016.
6. H. N. S. Krishnamoorthy, Z. Jacob, E. Narimanov, I. Kretzschmar, and V. M. Menon, "Topological transitions in metamaterials," *Science*, vol. 336, pp. 205–209, 2012.
7. K. V. Sreekanth, Y. Alapan, M. Elkabbash, A. M. Wen, E. Ilker, M. Hinczewski, U. A. Gurkan, N. F. Steinmetz, and G. Strangi, "Enhancing the angular sensitivity of plasmonic sensors using hyperbolic metamaterials," *Adv. Opt. Mater.*, vol. 4, p. 1659, 2016.

8. D. Lu, J. J. Kan, E. E. Fullerton, and Z. Liu, "Enhancing spontaneous emission rates of molecules using nanopatterned multilayer hyperbolic metamaterials," *Nat. Nanotech.*, vol. 9, pp. 48–53, 2014.
9. K. V. Sreekanth, K. H. Krishna, A. De Luca, and G. Strangi, "Large spontaneous emission rate enhancement in grating coupled hyperbolic metamaterials," *Sci. Repts.*, vol. 4, p. 6340, 2014.
10. V. Caligiuri, R. Dhama, K. V. Sreekanth, G. Strangi, and A. De Luca, "Dielectric singularity in hyperbolic metamaterials: the inversion point of coexisting anisotropies," *Sci. Repts.*, vol. 6, p. 20002, 2016.
11. K. H. Krishna, K. V. Sreekanth, and G. Strangi, "Dye-embedded and nanopatterned hyperbolic metamaterials for spontaneous emission rate enhancement," *J Opt. Soc. Am. B*, vol. 33, pp. 1038–1043, 2016.
12. C. Guclu, S. Campione, and F. Capolino, "Hyperbolic metamaterial as super absorber for scattered fields generated at its surface," *Phys. Rev. B*, vol. 86, p. 205130, 2012.
13. T. Xu, A. Agrawal, M. Abashin, K. J. Chau, and H. J. Lezec, "All-angle negative refraction and active flat lensing of ultraviolet light," *Nature*, vol. 497, p. 470, 2013.
14. K. V. Sreekanth, Y. Alapan, A. R. Rashed, U. A. Gurkan, and G. Strangi, "A multiband perfect absorber based on hyperbolic metamaterials," *Sci. Repts.*, vol. 6, p. 26272, 2016.
15. H. Lee, Z. Liu, Y. Xiong, C. Sun, and X. Zhang, "Development of optical hyperlens for imaging below the diffraction limit," *Opt. Exp.*, vol. 15, p. 1588, 2007.
16. Y. Guo and Z. Jacob, "Thermal hyperbolic metamaterials," *Opt. Exp.*, vol. 21, p. 15014, 2013.
17. S. A. Biehs, M. Tschikin, and P. Ben-Abdallah, "Hyperbolic metamaterials as an analog of a blackbody in the near field," *Phys. Rev. Lett.*, vol. 109, p. 104301, 2012.
18. I. I. Smolyaninov, "Quantum electromagnetic "black Holes" in a strong magnetic field," *J. Phys. G Nucl. Part. Phys.*, vol. 40, p. 015005, 2013.
19. J. S. T. Smalley, F. Vallini, X. Zhang, and Y. Fainman, "Dynamically tunable and active hyperbolic metamaterials," *Adv. Opt. Photon.*, vol. 10, p. 354, 2018.
20. O. Kidwai, S. V. Zhukovsky, and J. E. Sipe, "Effective-medium approach to planar multilayer hyperbolic metamaterials: Strengths and limitations," *Phys. Rev. A*, vol. 85, p. 053842, 2012.
21. G. Castaldi, V. Galdi, A. Alù, and N. Engheta, "Nonlocal transformation optics," *Phys. Rev. Lett.*, vol. 108, p. 063902, 2012.
22. K. V. Sreekanth, M. Elkabbash, V. Caligiuri, R. Singh, A. De Luca, and G. Strangi, *New Directions in Thin Film Nanophotonics*. Singapore: Springer, 2019.
23. G. Pawlik, K. Tarnowski, W. Walasik, A. C. Mitus, and I. C. Khoo, "Liquid crystal hyperbolic metamaterial for wide-angle negative-positive refraction and reflection," *Opt. Lett.*, vol. 39, pp. 1744–1747, 2014.
24. Z. Cao, X. Xiang, C. Yang, Y. Zhang, Z. Peng, and L. Xuan, "Analysis of tunable characteristics of liquid-crystal-based hyperbolic metamaterials," *Liq. Cryst.*, vol. 43, pp. 1753–17539, 2016.
25. H. N. S. Krishnamoorthy, Y. Zhou, S. Ramanathan, E. Narimanov, and V. M. Menon, "Tunable hyperbolic metamaterials utilizing phase change heterostructures," *Appl. Phys. Lett.*, vol. 104, p. 121101, 2014.
26. M. Wuttig, H. Bhaskaran, and T. Taubner, "Phase-change materials for non-volatile photonic applications," *Nat. Photon.*, vol. 11, pp. 465–476, 2017.
27. K. V. Sreekanth, S. Han, and R. Singh, "$Ge_2Sb_2Te_5$-based tunable perfect absorber cavity with phase singularity at visible frequencies," *Adv. Mater.*, vol. 30, p. 706696, 2018.

28. W. Dong, H. Liu, J. K. Behera, L. Lu, R. J. H. Ng, K. V. Sreekanth, X. Zhou, J. K. W. Yang, and R. E. Simpson, "Wide bandgap phase change material tuned visible photonics," *Adv. Funct. Mater.*, vol. 29, p. 1806181, 2019.
29. H. N. S. Krishnamoorthy, B. Gholipour, N. I. Zheludev, and C. Soci, "A non-volatile chalcogenide switchable hyperbolic metamaterial," *Adv. Opt. Mater.*, vol. 6, p. 1800332, 2018.
30. L. Lu, W. Dong, J. K. Behera, L. Chew, and R. E. Simpson, "Inter-diffusion of plasmonic metals and phase change materials," *J. Mater. Sci.*, vol. 54, pp. 2814–2823, 2019.
31. K. V. Sreekanth, Q. Ouyang, S. Sreejith, S. Zeng, W. Lishu, E. Ilker, W. Dong, M. El Kabbash, Y. Ting, C. T. Lim, M. Hinczewski, G. Strangi, K. T. Yong, R. E. Simpson, and R. Singh, "Phase change material based low loss visible frequency hyperbolic metamaterials for ultra-sensitive label-free biosensing," *Adv. Opt. Mater.*, vol. 7, p. 1900081, 2019.
32. K. V. Sreekanth, P. Mahalakshmi, S. Han, M. S. Mani Rajan, P. K. Choudhury, and R. Singh, "Brewster mode-enhnaced sensing with hyperbolic metamaterials," *Adv. Opt. Mater.*, vol. 7, p. 1900680, 2019.
33. I. V. Iorsh, I. S. Mukhin, I. V. Shadrivov, P. A. Belov, and Y. S. Kivshar, "Hyperbolic metamaterials based on multilayer graphene structures," *Phys. Rev. B*, vol. 87, p. 075416, 2013.
34. M. A. K. Othman, C. Guclu, and F. Capolino, "Graphene based tunable hyperbolic metamaterials and enhanced nearfield absorption," *Opt. Exp.*, vol. 21, pp. 7614–7632, 2013.
35. K. V. Sreekanth, and Y. Ting, "Long range surface plasmons in a symmetric graphene system with anisotropic dielectrics," *J. Opt.*, vol. 15, p. 055002, 2013.
36. Y. C. Chang, C. H. Liu, C. H. Liu, S. Zhang, S. R. Marder, E. E. Narimanov, Z. Zhong, and T. B. Norris, "Realization of mid-infrared graphene hyperbolic metamaterials," *Nat. Commun.*, vol. 7, p. 10568, 2016.
37. K. V. Sreekanth, A. De Luca, and G. Strangi, "Negative refraction in graphene-based hyperbolic metamaterials," *Appl. Phys. Lett.*, vol. 103, p. 023107, 2013.
38. S. Dai, Q. Ma, M. K. Liu, Z. Fei, M. D. Goldflam, M. Wagner, K. Watanabe, T. Taniguchi, F. Thiemens, F. Keilmann, G. C. A. M. Janssen, S. E. Zhu, P. Jarillo-Herrero, M. M. Fogler, and D. N. Basov, "Graphene on hexagonal boron nitride as a tunable hyperbolic metamaterial," *Nat. Nanotech.*, vol. 10, pp. 682–686, 2015.
39. P. Di Pietro, M. Ortolani, O. Limaj, A. Di Gaspare, V. Giliberti, F. Giorgianni, M. Brahlek, N. Bansal, N. Koirala, S. Oh, P. Calvani, and S. Lupi, "Observation of Dirac plasmons in a topological insulator," *Nat. Nanotech.*, vol. 8, pp. 556–560, 2013.
40. J. S. Wu, D. N. Basov, and M. M. Fogler, "Topological insulators are tunable waveguides for hyperbolic polaritons," *Phys. Rev. B*, vol. 92, p. 205430, 2015.
41. K. V. Sreekanth and R. E. Simpson, "Super-collimation and negative refraction in hyperbolic Van der Waals superlattices," *Opt. Commun.*, vol. 440, pp. 150–154, 2019.
42. I. I. Smolyaninov and V. N. Smolyaninova, "Is there a metamaterial route to high temperature superconductivity?" *Adv. Cond. Matt. Phys.*, vol. 2014, p. 479635, 2014.
43. K. V. Sreekanth, P. Mahalakshmi, S. Han, M. S. Mani Rajan, D. Vigneswaran, R. Jha, and R. Singh, "A terahertz Brewster switch based on superconductor hyperbolic metamaterial," *J. Appl. Phys.*, 128,2020.
44. K. V. Sreekanth, Q. Ouyang, S. Han, K. T. Yong, R. Singh, "Giant enhancement in Goos-Hänchen shift at the singular phase of a nanophotonic cavity," *Appl. Phys. Lett.*, vol. 112, p. 161109, 2018.
45. M. Manjappa, Y. K. Srivastava, L. Cong, I. Al-Naib, and R. Singh, "Active photoswitching of sharp Fano resonances in THz metadevice," *Adv. Mater.*, vol. 29, p. 1603355, 2017.

46. Y. K. Srivastava, M. Manjappa, L. Cong, N. S. Krishnamoorthy, V. Savinov, P. Pitchappa, and R. Singh, "A superconducting dual-channel photonic switch," *Adv. Mater.*, vol. 30, p. 1801257, 2018.
47. P. Q. Liu, I. J. Luxmoore, S. A. Mikhailov, N. A. Savostianova, F. Valmorra, J. Faist, and G. R. Nash, "Highly tunable hybrid metamaterials employing split-ring resonators strongly coupled to graphene surface plasmons," *Nat. Commun.*, vol. 6, p. 8969, 2015.
48. P. Pitchappa, A. Kumar, S. Prakash, H. Jani, T. Venkatesan, and R. Singh, "Chalcogenide phase change material for active Terahertz photonics," *Adv. Mater.*, vol. 31, p. 1808157, 2019.
49. Z. Chen, X. Chen, L. Tao, K. Chen, M. Long, X. Liu, K. Yan, R. I. Stantchev, E. Pickwell-MacPherson, and J.-B. Xu, "Graphene controlled Brewster angle device for ultra-broadband terahertz modulation". *Nat. Commun.*, vol. 9, p. 4909, 2018.
50. Y. Yang, K. Kelley, E. Sachet, S. Campione, T. S. Luk, J.-P. Maria, M. B. Sinclair, and I. Brener, "Femtosecond optical polarization switching using a cadmium oxide-based perfect absorber," *Nat. Photon.*, vol. 11, p. 390, 2017.
51. L. Cong, P. Pitchappa, N. Wang, and R. Singh, "Electrically programmable terahertz diatomic metamolecules for chiral optical control," *Research*, vol. 2019, p. 7084251, 2019.
52. S. K. Chamoli, M. El Kabbash, J. Zhang, and C. Guo, "Dynamic control of spontaneous emission rate using tunable hyperbolic metamaterials," *Opt. Lett.*, vol. 45, p. 1671, 2020.

9 Graphene-Supported Nanoengineered Metamaterials – A Mini Review

P. K. Choudhury
Institute of Microengineering and Nanoelectronics, Universiti Kebangsaan Malaysia, UKM Bangi, Selangor, Malaysia

9.1 INTRODUCTION

Today, nanotechnology is in the research limelight. Much of the attention has been paid to the formation of novel engineered nanomaterials, followed by their characteristic features and possible applications. Formation of nanoengineered materials certainly introduces complications due to the need for miniaturized material structures with architectures and morphologies of interest. As the relevant materials and/or devices are nanoscale sized, their synthesis attracts interest for scientific investigations. The procedure to form nanoengineered mediums is very different from the ways adopted for designing bulk materials of specific shapes. The technique exploited for bulk materials begins with a larger structure from which smaller structures are derived. In contrast, nanostructured materials can begin at the atomic level and then build up to the nanoscale size.

Carbon is one of the most abundant elements found in nature, and it forms many compounds. In addition, it is one of the most common resource materials to form nanostructured composites. Carbon-based compounds have a variety of applications. In particular, carbon forms the basis of all known lives as well as all organic chemistry. Because of the flexibility of the carbon bond, carbon-based systems show an unlimited number of structures, with a variety of physical properties due to the dimensionality of these structures. Owing to the versatile bonding ability of carbon, it possesses the ability to react with many other elements, thereby allowing carbon-based compounds to find a wide range of applications in everyday life.

In the context of carbon-based (or carbon-derived) mediums, graphene – a two-dimensional (2D) allotrope of carbon – has become a revolutionary material because its charge carriers behave like massless Dirac-Fermions, the feature that results in many exotic properties of graphene [1,2], which can be utilized in a variety of electronic and optical applications. Graphene plays an important role as

DOI: 10.1201/9781003050162-9

it is the basis for the understanding of the electronic properties in other allotropes of carbon. It is made out of carbon atoms arranged in a hexagonal structure, thereby forming a one-atom-thick, tightly packed layer of carbon atoms, which is recognized to be the thinnest known lattice structure. As such, it constitutes a conceptually new class of 2D materials as these extend in only two dimensions, i.e., the length and width; the third dimension, the height, is assumed to be vanishing. Apart from its being an extremely lightweight material, graphene exhibits many amazing characteristics, such as having high electron mobility (100 times faster than silicon) and being an excellent conductor of heat (2 times better than diamond) – the hallmark that allows it to have great potential in the current thrust of nanotechnology research [3].

In the context of photonics technology, graphene behaves extraordinarily, which makes it a promising optical material [4]. For instance, the surface plasmons excited in graphene are confined much more strongly than those in noble metals, and they can be coupled with electrons, photons or phonons [5,6]. Furthermore, the tunability of plasmons in graphene-based structures may be achieved with better ease because the charge carrier can be easily controlled by electrical gating and doping with low loss [2]. The presence of graphene can even help the tuning of plasmonic waves in noble metals [4].

Metamaterials can be engineered by amalgamating noble metals and dielectric mediums in order to observe varieties of optical properties that can be useful for different photonic applications [7–15]. Interestingly, the exotic characteristics allow graphene-based structures to be useful in designing metamaterials to realize several optical components and/or devices [16–20]. Graphene has the property to enhance the absorption of light waves [21,22]. Within this context, graphene-based nanostructured slab waveguides can sustain slow light [23,24] and also can be used to suppress propagating modes [25]. Both of these phenomena have great importance in photonics technology. Investigators also reported the effect of plasmon-induced transparency [26–28] in graphene-based engineered metamaterials [29–31]. Apart from these, the use of graphene-supported structures comprising phase change mediums has also been quite attractive in applications like absorbers and sensors [32–34]. In most of these applications, graphene serves the purpose of tuning the spectral characteristics.

Pivoted to the aforementioned technological importance of graphene, this chapter touches upon research results of a few forms of graphene-supported metamaterials for various applications, such as filtering, wideband absorption and sensing.

9.2 FUNDAMENTAL PROPERTIES OF GRAPHENE

As discussed in Section 9.1, 2D graphene-based metasurfaces have attracted the R&D community because of possible control over the conductivity of graphene – the feature that offers tunability in the spectral bandwidth of metamaterials. Graphene-supported metamaterials are more conveniently used in the terahertz (THz) regime of the electromagnetic (EM) spectrum, in which the tunability of graphene can be achieved with better ease [35–37]. The conductivity can be achieved by adjusting the external parameters, such as chemical doping, mechanical stretching or electrical gating, which adjust the Fermi level of graphene [2]. The use of monolayer as well as multilayer graphene [38] has been widely reported in developing varieties of narrow-band and broadband absorbers [39–41]. Apart from these, many forms of metasurface-substrate combinations have been investigated to achieve on-demand spectral characteristics for various technological applications [42–45].

As to the complex conductivity of graphene, which can be written as $\sigma_g(\omega) = \sigma_{gr} + j\sigma_{gi}$, it depends on the radian frequency ω, scattering rate of charged particle (representing the loss mechanism), temperature and chemical potential. The imaginary part of graphene conductivity can attain negative and positive values in different frequency ranges, depending on the chemical potential [2]. The unconventional electrical conductivity of graphene remains the determining factor in the absorption of EM waves. The complex-valued surface conductivity $\sigma_g(\omega)$ of graphene is determined by the well-established and experimentally valid theoretical model [46,47]. $\sigma_g(\omega)$ consists of the algebraic sum of the two parts, namely the inter-band and intra-band conductivities [48,49], i.e.,

$$\sigma_g(\omega) = \sigma_{\text{inter}}(\omega) + \sigma_{\text{intra}}(\omega) \tag{9.1}$$

The conductivity of graphene is given by the Kubo formula [25]:

$$\begin{aligned}\sigma(\omega, \mu_c) = {} & \frac{2e^2 k_B T}{\pi\hbar^2}\left\{\frac{-j}{(\omega - j\Gamma)}\right\}\ln\left\{2\cosh\left(\frac{\mu_c}{2k_B T}\right)\right\} \\ & - \frac{je^2}{4\pi\hbar}\ln\left\{\frac{2\mu_c - \hbar(\omega - j\Gamma)}{2\mu_c + \hbar(\omega - j\Gamma)}\right\}\end{aligned} \tag{9.2}$$

where $\hbar$ (= $h/2\pi$) is the reduced Planck's constant, k_B is Boltzmann's constant, T is the absolute temperature, e is the electronic charge, and Γ is the rate of scattering, which is inversely proportional to the electron relaxation time, and μ_c is the chemical potential of graphene [2]. The two terms in Eq. (9.2) relate to the intra- and inter-band transitions of charge carriers. The inter-band transition of electrons mainly happens at frequencies above the Pauli Blocking frequency (i.e., $\omega > 2\mu_c/\hbar$) [50], in the vicinity of which the real part (of the inter-band term) increases and attains the universal value of conductivity $\sigma_0 = e^2/4\hbar$. It is noteworthy that the inter-band transitions dominate in the visible frequency range. On the other hand, the intra-band transitions significantly exceed the inter-band kind in the frequency range of far-infrared (IR) to the mid-IR. In other words, when the Fermi level is more than half of the photon energy, the intra-band conductivity dominates because the inter-band conductivity is insignificant (due to Pauli blocking) [32].

According to the Drude model, the surface conductivity $\sigma_g(\omega)$ of graphene is related to the Fermi energy level E_f through the relationship [2]

$$\sigma_g(\omega) = \frac{e^2 E_f}{\pi \hbar^2}\left(\frac{j}{\omega + j\tau^{-1}}\right) \tag{9.3}$$

with τ being the momentum relaxation time. Also, the relative permittivity $\varepsilon_g(\omega)$ of graphene is related to its surface conductivity $\sigma_g(\omega)$ as

$$\varepsilon_g(\omega) = 1 + j\frac{\sigma_g(\omega)}{t_g \varepsilon_0} \tag{9.4}$$

where t_g is the thickness of the 2D graphene layer and ε_0 is the free-space permittivity.

The chemical potential of graphene depends on the carrier density, which can be controlled by the external gate voltage, electric field, magnetic field and/or chemical doping [2]. The electronic modeling of graphene characterizes its conductivity along the surface plane (of graphene), which is a scalar quantity under the DC electrostatic field. In the absence of DC magnetostatic bias, the application of DC electrostatic bias (E_0) makes the conductivity to be [50]

$$\sigma_{xx} = \sigma_{yy} = \sigma_g[\mu(E_0)] \tag{9.5}$$

Eq. (9.5) shows that the conductivity σ_g of graphene is a function of its chemical potential μ_c, which, in turn, is a function of the electrostatic bias E_0. As such, the value of μ_c depends on the DC gate voltage V_g because of its relation to the field through $V_g = E_0 d$, d being the thickness of graphene medium.

The relationship of the 2D surface charge density n_s to the gate capacitance C of graphene is described as [51]

$$n_s = \frac{1}{e} C E_0 = \frac{2}{\pi \hbar^2 v_f^2} \int_0^\infty \varepsilon \left[f_d(\varepsilon_g - \mu_c) - f_d(\varepsilon_g + \mu_c) \right] d\varepsilon \tag{9.6}$$

In this equation, v_f is the Fermi velocity of electrons in graphene medium, which is usually taken to be 10^6 m/s; the meanings of other symbols are as used before.

As discussed above, metamaterials exploit monolayer or multilayer graphene mediums to realize varieties of absorber devices. A monolayer graphene has a thickness of ~0.49 nm. However, the structure of bilayer graphene can have two different forms: AA and AB. In the case of the AA type, the layers are exactly aligned [52], whereas in the AB type, half of the atoms lie directly over the center of a hexagon in the lower graphene layer, and the remaining half (of the atoms) lie over an atom [53]. Lin et al. [54] showed that the N-layer AA-stacked graphene sheets have a band structure consisting of N Dirac bands, which are shifted in energy. Consequently, the optical conductivity of an AA-stacked graphene in the THz region can be determined as

$$\sigma_{AA}(\omega) = \frac{2e^2 k_B T}{\pi \hbar^2 N}\left\{\frac{-j}{(\omega - j\gamma)}\right\} \sum_{m=1}^{N} \ln\left[2\cosh\left(\frac{\mu_c + 2\alpha_1 \cos m\pi/N + 1}{2k_B T}\right)\right] \tag{9.7}$$

with N as the number of graphene layers and α_1 as the interaction energy of misoriented (or AA-stacked) graphene layers. The value of α_1 is 217 meV, while the separation distance (of the AA-stacked layers) is 3.6 Å [54]. The authors demonstrated that the results derived using this formula remain in good agreement with the experimentally obtained values.

Since the use of graphene facilitates the tuning mechanism owing to varying electrical properties through altering gate voltage, which, in turn, changes the chemical potential (of graphene), it has now become indispensable in designing micro- and nanoengineered mediums that contribute to this class of metamaterials. In the following sections, some of these metamaterial configurations are touched upon, wherein graphene is amalgamated to the structure, in some way or the other, in order to achieve spectral characteristics suitable for certain photonic applications. In general, the discussions are pivoted to the use of graphene-based photonic structures in the areas of filters, absorbers and sensors.

9.3 WIDEBAND THZ FILTERING

Because it is easier to tune graphene in the THz frequency regime, it has been highly useful in designing THz devices. Several kinds of graphene-embedded absorbers have been reported [55–57]. As discussed before, mono- and/or multilayer kinds of graphene can be used with suitable biasing in metamaterials for a variety of applications. Pivoted to this, the transmission properties of certain forms of periodic multilayer structures have been studied. To be more explicit, the periodic configuration is composed of SiO_2 dielectric and bilayer graphene mediums. Such periodic structures have also been studied when a nanolayer of magnesium fluoride (MgF_2) medium was inserted as a defect medium. The SiO_2-based micro- and nanostructured mediums have been of immense use in THz devices owing to relatively small tangent loss at the THz frequencies [58,59]. Apart from this, MgF_2 is a low refractive index (RI) medium and exhibits broadband transmission in this frequency regime [60].

A variety of methods have been reported to analyze propagation through multilayered mediums, including the plane wave expansion [61], finite difference time domain (FDTD) [62], transfer-matrix [63,64] and Green function [4,65] methods. Among these analytical techniques, the transfer-matrix method (TMM) remains highly attractive owing to its simplicity, and it also can be implemented in both the isotropic and anisotropic mediums with defects [66,67]. To analyze the aforementioned problem, we will explore the TMM technique [68].

As to the schematic of the structure, a periodic stack of SiO_2 and bilayer graphene sheets may be taken, as shown in Fig. 9.1a. Fig. 9.1b exhibits the same configuration with the difference that a layer of MgF_2 is inserted at the midpoint of the structure. In Fig. 9.1a, the unit cell of the periodic structure is composed of a SiO_2 layer and bilayer graphene sheet, and, therefore, the overall structure comprises $N + 1$ bilayer graphene sheets sandwiching N number of SiO_2 layers.

Fig. 9.2 exhibits the frequency dependence of relative permittivity of the multilayer structure in Fig. 9.1a. In order to plot these figures, the SiO_2 layer has been taken as 6.9 μm thick, along with its relative permittivity ϵ_r to be 2.225. Also, the

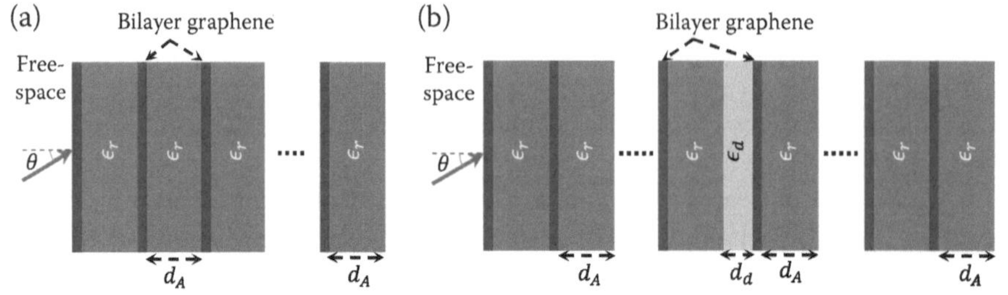

FIGURE 9.1 Schematic of the periodic structure comprising (a) bilayer graphene and SiO_2 mediums, and (b) the same as (a) with a MgF_2 defect layer.

thickness of bilayer graphene sheets is 0.9 nm. Fig. 9.2a,b depicts the frequency dependence of the real (ϵ') and imaginary (ϵ'') parts of transverse permittivity ϵ_t in the 0–50 THz frequency range upon exploiting the effective medium approximation, and using the parametric values as $\gamma = 1.32$ meV, $T = 300$ K, and the μ_c-values as 1.0 eV, 1.25 eV and 1.5 eV [68].

Notice in Fig. 9.2 that the real part of permittivity $\epsilon' > 0$ is above frequency 5.6 THz; below this frequency, $\epsilon' < 0$. The permittivity component ϵ' comes close to a saturation value of ~2.0 near 20 THz frequency and above. The figure shows that the effect of altering graphene chemical potential on ϵ' is more obvious for frequencies <10 THz. However, it remains less significant for ϵ'', which is positive valued. To be more specific, Fig. 9.2a exhibits ϵ' to be negative at ~5.5 THz, corresponding to $\mu_c = 1.0$ eV, whereas $\mu_c = 1.5$ eV increases this frequency value to ~8.0 THz. Beyond these frequency values, and upon reaching the Pauli blocking frequencies at $\omega = 2\mu_c/\hbar$, waves can propagate through the multilayer structure (of Fig. 9.2a) with low loss. The real ϵ' part of transverse permittivity approaches

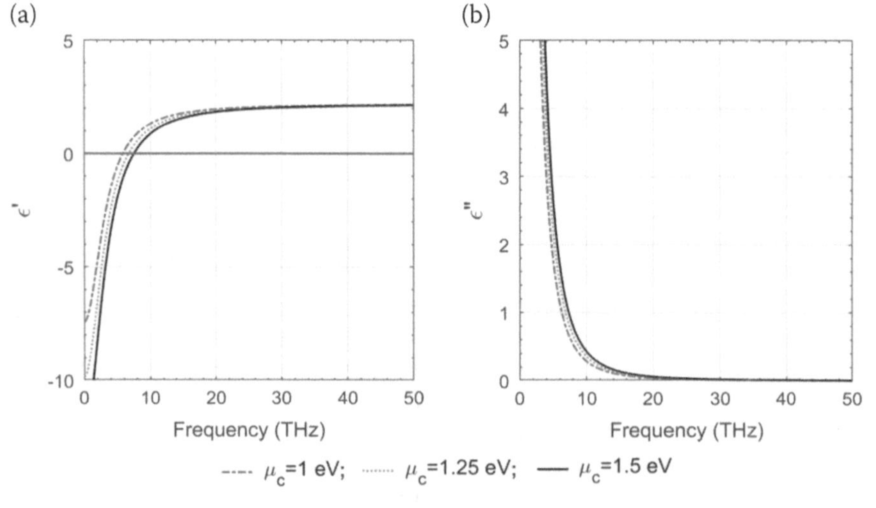

FIGURE 9.2 Effective permittivity vs. frequency plots; (a) real part, and (b) imaginary part.

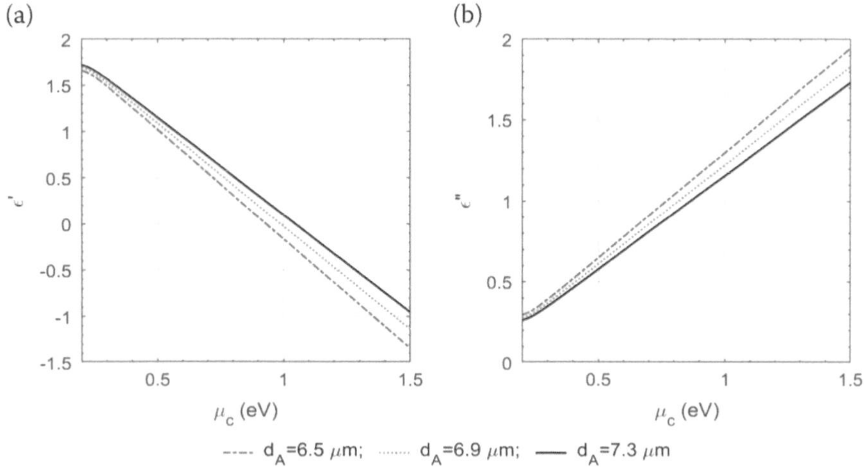

FIGURE 9.3 Effective permittivity vs. μ_c plots for different d_A-values; (a) real part, and (b) imaginary part.

the dielectric permittivity at high frequencies due to a lower contribution of graphene conductivity compared to ϵ_r – the relative permittivity of SiO_2. On the other hand, ϵ'' exhibits a sharp decrease in the low-frequency regime and becomes almost vanishing above 28 THz frequency.

In connection with Fig. 9.1, apart from the chemical potential of graphene, the effective permittivity value of multilayer configuration greatly depends on the thickness d of dielectric medium as well. With this viewpoint in mind, Fig. 9.3 depicts the dependence of ϵ' and ϵ'' components on graphene chemical potential corresponding to different values of d [68]. It can be seen that ϵ' and ϵ'' show gradual decrease and increase in their values with increasing μ_c.

Upon exploiting the TMM, the effect of bilayer graphene sheets on the spectral characteristics of the proposed multilayer configuration (of Fig. 9.1a) can be observed. Within this context, Fig. 9.4 illustrates the obtained transmission characteristics in the 0–30 THz frequency range considering 12 periods of G|A layers in the design with the chosen three values of μ_c, viz. 1.0 eV, 1.25 eV and 1.5 eV [68]. Furthermore, the dielectric layer thickness has been kept as 6.9 μm, and the incidence wave has been assumed to fall upon the surface vertically. Notice in Fig. 9.4 that μ_c = 1.0 eV yields the width of the first stop-band as ~4.5 THz, which increases to ~6 THz corresponding to μ_c = 1.5 eV. The second stop-band, μ_c = 1.0 eV, gives the width to be ~2 THz, which increases to ~3 THz corresponding to μ_c = 1.5 eV. Clearly, such a wide stop-band can be considered as the one of its kind reported among recent research results.

Upon introducing a MgF_2 defect layer of 5.8 μm thickness in the multilayer structure (as shown in Fig. 9.2b), the transmission characteristics corresponding to the values as 1.0 eV, 1.25 eV and 1.50 eV, as obtained upon exploiting the TMM, can be seen in Fig. 9.5 [68]. To plot these, the incidence waves have been considered to fall upon the structure surface vertically. Notice that the increase in μ_c

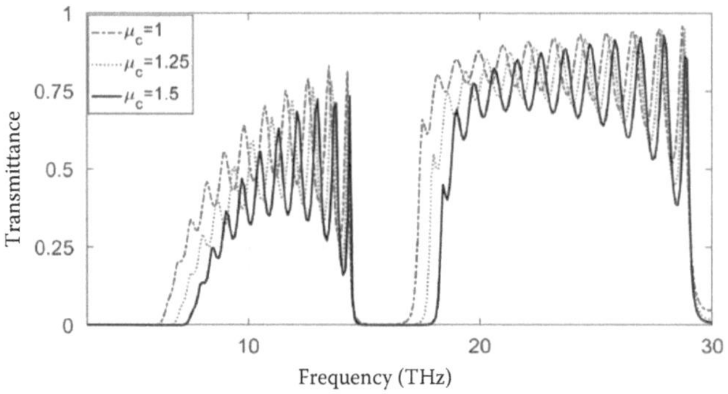

FIGURE 9.4 Transmittance vs. frequency plots for different μ_c-values considering $N = 12$.

results in larger width of the stop-band, and the position of defect mode moves toward higher frequencies. Furthermore, the peaks in transmission spectra are suppressed due to high reflections as these move near the center of stop-band. Clearly, μ_c = 1.5 eV yields the width of the stop-band ~3 THz.

The above results reveal that the optical response of periodically arranged dielectric and bilayer graphene sheets yield fairly wide stop-bands, the width and position of which can be governed by suitably controlling the graphene chemical potential. The obtained spectral characteristics show that the use of 12 periods of bilayer graphene-SiO_2 combination along with 1.0 eV graphene chemical potential yields a ~3 THz stop-band, which is fairly large. Apart from this, the incorporation of a MgF_2 defect layer in the middle of the periodic arrangement essentially yields alterations in the transmission properties, which is attributed to the presence of defect modes. In this case as well, the increase in chemical potential results in relatively wider stop-band. All these results indicate the possible prudent photonic applications of the structure in designing tunable wideband filters, absorbers and sensors.

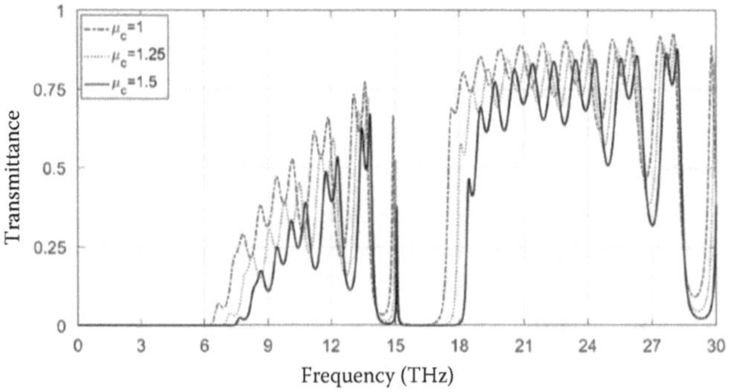

FIGURE 9.5 Transmittance vs. frequency plots for different μ_c-values considering $N = 6$.

9.4 OPTICAL FILTER

Surface plasmon polariton (SPP) waves propagating along the metal-dielectric interface can be effectively exploited to devise optical components [12,14,15,69–72]. This becomes possible owing to the surface plasmon resonance (SPR) condition being highly responsive to the light-matter interaction – the phenomenon precisely governed by the shape, size, type of medium and operating conditions. The magnetic response of metamaterials due to SPP has also been reported [73,74], with the emphasis on the effect of tailoring the SPP waves by altering the metasurface configuration.

As stated before, metamaterials have been vital in optical filtering applications [75,76]. The geometry of unit cell (in metamaterials) plays vital role in tailoring spectral properties. Since graphene is capable of manipulating EM waves through proper biasing, thereby directly influencing the optical response, varieties of graphene-based filters have been reported in the literature [71,77–79]. In line with this, a new kind of graphene-over-graphite nanorods type of metamaterial configuration has been reported, which can be utilized for wideband filtering in the optical regime [80]. In constructing the device, CYTOP glass substrate has been used, which is embedded with graphite nanorods in a periodic arrangement; and the top surface (of substrate) being coated with bilayer graphene. Fig. 9.6 shows the schematic, the spectral features (i.e., the transmission characteristics) of which have been studied (in the visible regime of EM spectra) in terms of the effects due to the chemical potential of graphene as well as the excitation incidence obliquity.

In the schematic, the top bilayer of graphene is of 0.9 nm thickness, which is placed over a planar CYTOP glass substrate. It can be seen in Fig. 9.6 that the substrate has nanoholes (of radius r), which are filled up with graphite. As such, the structure becomes like a graphite-nanorods-embedded substrate, the top surface of which has been coated with bilayer graphene. This has been simulated to extract the transmission characteristics in the visible and NIR regimes [80]. Since the resonant frequencies can be manipulated by appropriately changing the graphene chemical potential, transmission spectra of graphene-embedded design can be altered – the feature that enables one to obtain an electrically tunable filter without reconfiguring the structural parameters.

Fig. 9.7 illustrates the effective permittivity ϵ_{eff} plots of graphite nanorods embedded CYTOP glass substrate. Within the context, both the SiO_2 and graphite are

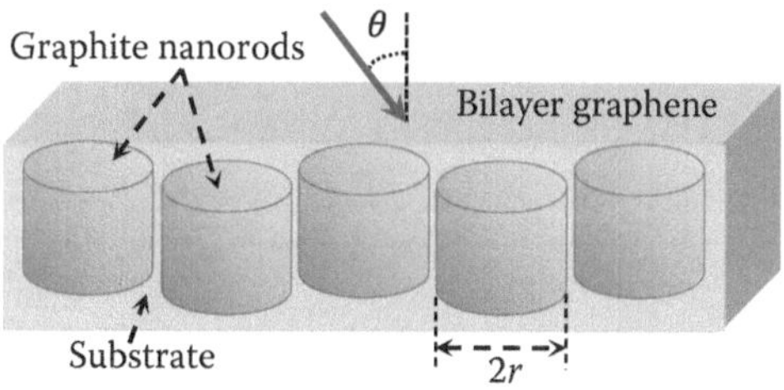

FIGURE 9.6 Graphene-over-graphite metamaterial structure.

non-dispersive in nature, and therefore, the relevant dispersion characteristics are expected to yield straight lines parallel to the λ-axis, as shown in this figure [80]. The x- and z-components of permittivity can be written as

$$\epsilon_x = \epsilon_g \rho + \epsilon_{CYTOP}(1 - \rho) \tag{9.8}$$

and

$$\frac{1}{\epsilon_z} = \frac{\rho}{\epsilon_g} + \frac{(1-\rho)}{\epsilon_{CYTOP}}, \tag{9.9}$$

respectively. Herein, ρ is the volume fraction element of graphite in the metasurface, which is defined as the ratio of total volume occupied by graphite nanorods in the dielectric substrate to the total volume of substrate. It can be seen in Fig. 9.7 that the increase in ρ results only in a larger value of effective permittivity, the value of which remains unchanged with increasing wavelength. This is due to the non-dispersive behavior of graphite and dielectric substrate.

The transmission characteristics of such a metamaterial have been studied considering the radius of the embedded graphite nanorods and their volume fraction element ρ to be of 25 nm and 0.013, respectively [80]. Apart from these, four different values of graphene chemical potential μ have been considered; those are 0.5 eV, 1.0 eV, 1.5 eV and 2.0 eV. It is worth mentioning that the RI of CYTOP glass substrate is 1.36.

Now, focusing on the transmission characteristics, Fig. 9.8 shows the wavelength dependence of transmittance in the span of 500–900 nm considering the normal incidence of waves [80]. This figure demonstrates achieving fairly wideband transmission, though it does not remain perfectly flat. It can also be seen here that the increase in chemical potential results in a little increased transmission (the inset in this figure, which is the magnified plot of a section,

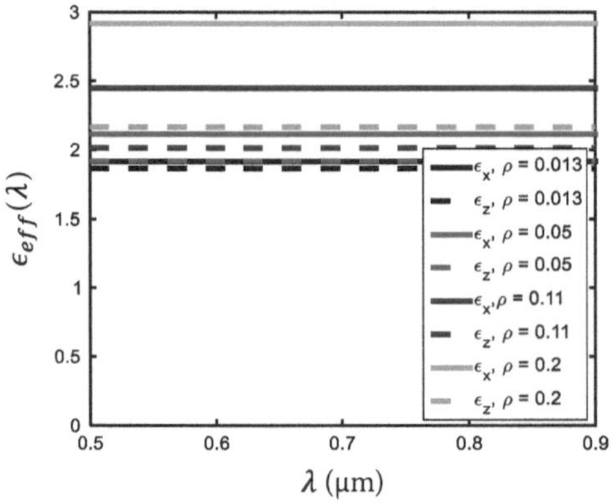

FIGURE 9.7 ϵ_{eff} vs. λ plots.

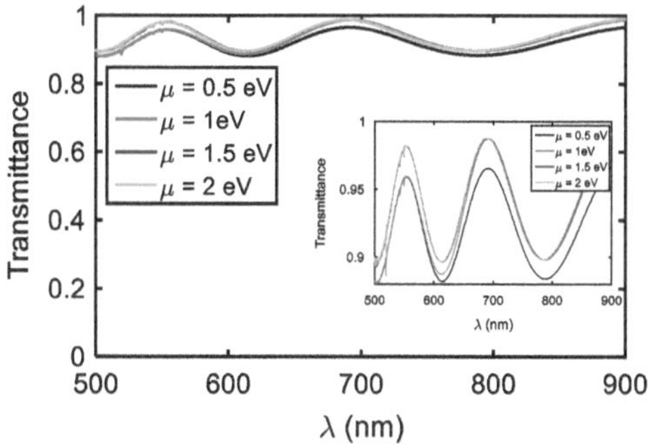

FIGURE 9.8 Transmittance vs. λ plots corresponding to 25 nm graphite nanorod radius and $\theta = 0°$.

makes it more obvious). Since the transmission spectra corresponding to 1.5 eV and 2.0 eV chemical potentials are almost similar, the use of 0.5 eV and 1.5 eV only should be fine, so far as the biasing of graphene is concerned; the following studies assume this.

In order to observe the impact of incidence obliquity, Fig. 9.9 has been plotted with the incidence angle varying in a range of 0°–60° in a step of 20°, considering the two values of μ as 0.5 eV (Fig. 9.9a) and 1.5 eV (Fig. 9.9b) [80]. It is obvious that the transmission generally remains high, and the change in chemical potential introduces minor alterations only in the transmission spectra. For high obliquity (e.g., $\theta = 60°$), transmission decreases slowly with increasing wavelength as such an incidence situation introduces more anisotropy to the waves impinging upon the top graphene layer. As Fig. 9.9 demonstrates, in certain incidence circumstances, the metamaterial configuration exhibits nearly wideband and even perfect transmission – the property that is only slightly affected upon a small increase in μ-value. Finally, the obtained results indicate that the stated kind of graphene-over-graphite metamaterial can be used as filters in the visible and NIR regimes.

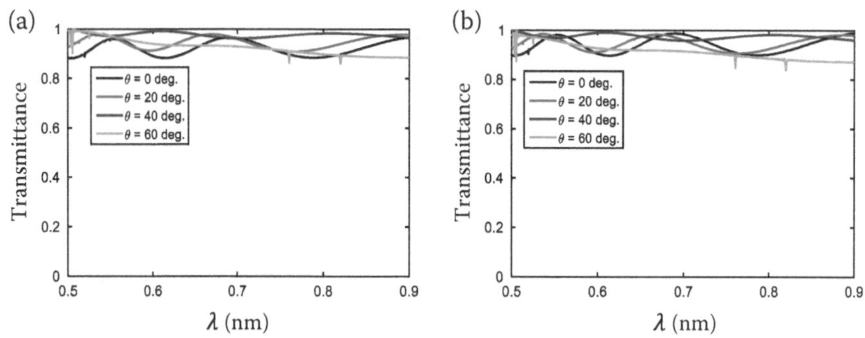

FIGURE 9.9 Transmittance vs. λ plots for (a) 0.5 eV, and (b) 1.5 eV graphene potentials under varying θ.

9.5 PLASMON-INDUCED TRANSPARENCY IN GRAPHENE-BASED METAMATERIALS

The basics of plasmonics and the propagation of SPP waves were discussed in the beginning of Section 9.4. Within the context, the phenomenon of localized surface plasmon (LSP) is interesting due to important applications, such as enhanced Raman scattering and many others [81,82]. Within the context, the interaction between two or more plasmonic particles would generate plasmonic modes, which can be either super-radiant (resulting in the radiative mode having a low quality factor) or sub-radiant (thereby giving rise to dark modes having a relatively large quality factor), essentially depending on the coupling of incident light with the plasmonic mode [83]. Interestingly, such plasmonic modes can be utilized in tuning the resonance frequency [84].

The concept of dark mode with a metastable level in an atomic system originates the phenomenon of electromagnetically induced transparency (EIT) in a medium [26,85,86], which possesses many potential applications, such as slow-light control, bio-chemical sensing, filtering, switching and nonlinear devices [87–89]. It is basically a quantum effect caused by the interference between coherent transitions excited in an atomic system by incidence EM fields. EIT can be exploited for the cancellation of absorption of EM wave propagating through metamaterials at a desired frequency, which would otherwise be non-transparent [90–93] – this is called plasmon-induced transparency (PIT) [26]. As such, PIT is the plasmonic analogue of EIT. It involves coupling between two distinct resonators in a metamaterial unit cell, wherein one of the resonators is super-radiative (yielding a relatively broad resonance feature with low Q-factor owing to a strong coupling to the incident field), called the "bright" resonator, and the other one is sub-radiant (showing a much sharper resonance linewidth with relatively large Q-factor), called the "dark" resonator. The dark modes usually cannot directly couple with the external field, but can be excited by the local field caused by the bright mode via near-field coupling. The destructive interference between these strongly coupled bright and dark resonator modes gives rise to a well-defined narrow-band transparency window. The phenomenon of PIT remains prudent in engineering metamaterial-based configurations because the geometrical parameters of these can be altered in order to achieve PIT tuning. Some of the notable applications are realizing slow-light propagation [27] and improvements in non-linear effect [94].

As described earlier, graphene is attractive in photoelectric modulation [95,96] – the feature that could be exploited to design THz functional devices and left-handed materials [97–100], modulators and switches [101–104]. In the context of PIT, several hybrid graphene-metal metamaterial structures have been reported to generate PIT effects in the THz regime [105–107]. For instance, Yee et al. [108] demonstrated a tunable PIT by using nanoengineered graphene-based metamaterial, and Shi et al. and He et al. [29,30] discussed achieving PIT in the THz regime by using graphene complementary metamaterials. Hybrid metal–graphene metamaterials can also be used to control the amplitude of the PIT window [31].

In this stream, an attempt has been made to achieve PIT by using a metasurface comprising a fractal-like structure (made of specifically designed silver nanospheres

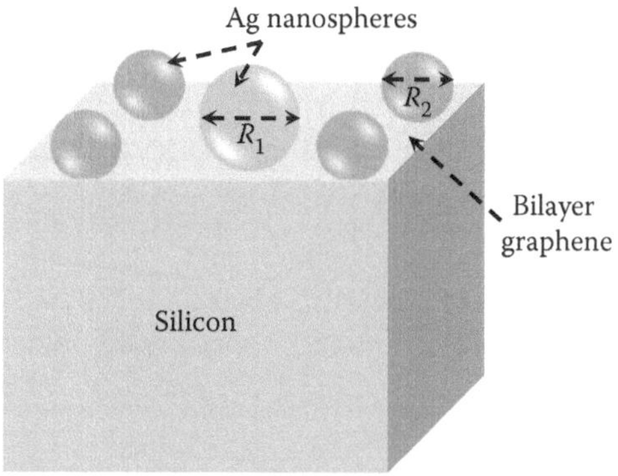

FIGURE 9.10 Schematic of the metamaterial structure.

of two different radii) at the top, as shown in the unit cell structure of Fig. 9.10 [109]. The schematic is basically a three-layer metasurface-graphene-dielectric configuration wherein the used silver nanospheres (at the top) are suitably arranged to attain a fractal design. Such a metamaterial has been studied for the TM-polarized incidence with an emphasis on the transmission characteristics under varying volume fraction elements (of Ag nanosphere), incidence obliquity and graphene chemical potential μ_c.

The effective permittivity of metasurface can be evaluated by exploiting the effective medium theory [110], according to which the parallel and perpendicular components are, respectively,

$$\epsilon_{\|}(\lambda) = \epsilon_d \frac{(1+f)\epsilon_m(\lambda) + (1-f)\epsilon_d}{(1-f)\epsilon_m(\lambda) + (1+f)\epsilon_d} \tag{9.10}$$

and

$$\epsilon_{\perp}(\lambda) = f\epsilon_m(\lambda) + (1-f)\epsilon_d. \tag{9.11}$$

In these equations, ϵ_m and ϵ_d are the permittivity of metallic and dielectric mediums, respectively. Also, f is the fill-factor of silver in the unit cell of the metasurface, the variation of which will essentially alter the dispersion characteristics of the metasurface. Notably, the value of ϵ_m can be obtained by using the Lorentz-Drude model [111].

Considering the fill-factor f to be 0.08 and 0.2, Fig. 9.11 illustrates the plots of the real ($\Re$; Fig. 9.11a) and imaginary ($\Im$; Fig. 9.11b) parts of effective permittivity against wavelength [109]. Notice that the $\Re(\epsilon_{\|})$-components remain negative, whereas the imaginary $\Im(\epsilon_{\|})$-components are positive; the values of both increase with the increase in f as well as λ under the same sign. Furthermore, both the $\Re(\epsilon_{\perp})$- and $\Im(\epsilon_{\perp})$-components become nearly vanishing corresponding to $f = 0.08$, thereby making the medium to be highly transparent in the used span of operating wavelength. Also, these permittivity components do not exhibit dispersion.

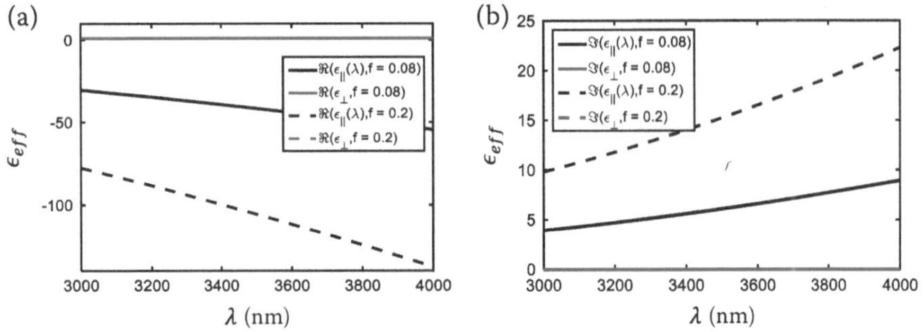

FIGURE 9.11 ϵ_{eff} vs. λ plots of the (a) real, and (b) imaginary parts.

The transmission spectra of the structure in Fig. 9.10 have also been investigated, as shown in Figs. 9.12, corresponding to R_1 = 500 nm and R_2 = 250 nm [109]. The bilayer graphene thickness has been taken to be 0.9 nm, and the unit cell has a surface area of 0.2 × 0.2 μm^2. Considering the normal as well as oblique incidence of the TM-polarized incidence waves, Fig. 9.12 illustrates the plots for three different μ_c-values, namely 0.1 eV, 1 eV and 2 eV. Herein, the presence of four PIT windows (corresponding to four bright-dark modes) with the peak transmissions positioned at nearly 3,050 nm, 3,150 nm, 3,600 nm and 3,950 nm wavelengths can be noticed. Also, the transmission windows become broader with increasing λ. However, relatively sharper transmission windows at low operating wavelengths essentially indicate a fairly large Q-factor.

As to the effect of graphene chemical potential, Fig. 9.12 determines that the increase in μ_c results in small red shifts in transmission peaks, and also the transmittance increases by a small amount. Furthermore, the increase in incidence obliquity causes reduced transmittance, thereby proving the use of normal incidence to be the best to attain strong light-matter interaction in order to achieve high transmittance with improved Q-factor, irrespective of the μ_c-value.

Based on these results, PIT can be achieved for the normal as well as oblique incidence of TM-polarized waves, which is evident from the sharp peaks in transmission spectra. Also, the case of normal incidence enhances transmittance. Further, the change in chemical potential does not significantly alter the PIT peak positions; instead, it merely contributes to a small increase in transparency. This feature of achieving transmittance through altering the obliquity of incidence waves and chemical potential of bilayer graphene triggers applications in sensing, filtering and absorption of radiation.

9.6 GRAPHENE-BASED ABSORBER

Metamaterial-based absorbers have been of key research importance due to exhibiting perfect absorption with certain kinds of configurations (demonstrating almost vanishing reflection) in particular frequency regimes [112–114]. Within this context, the propagation of SPP waves along the metal-dielectric boundary has been described above. However, the drawback remains due to material losses [115],

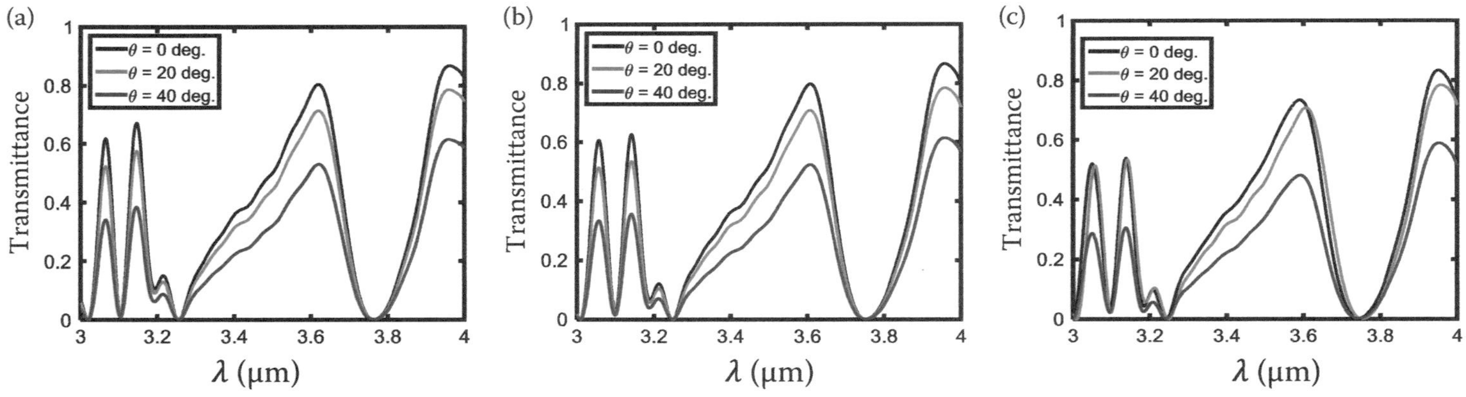

FIGURE 9.12 Transmittance vs. λ plots corresponding to different incidence obliquities under μ_c-values (a) 0.1 eV, (b) 1 eV and (c) 2 eV.

thereby affecting the functionality of metamaterial-based structures. Avoiding such demerit, graphene-supported metamaterials have been suitably attractive for optoelectronic applications. The use of graphene allows strongly confined SPP waves in graphene layers [4,65], which may be controlled through dynamical tuning of conductivity (of graphene). Also, graphene-based metasurfaces exhibit strong absorption characteristics [116,117].

9.6.1 Coned-graphene Metasurface Design

Among the varieties of graphene-based metamaterial absorbers reported in the literature, one of the unique designs would be the multilayered metasurface comprising coned graphene that forms a conductor-insulator-conductor (CIC) absorber [118]. Here, the metasurfaces are mounted at the top and bottom of a relatively thick SiO_2 dielectric substrate. In this absorber configuration, the dielectric medium is sandwiched between metasurfaces comprising arrays of coned graphite structures, as shown in Fig. 9.13. Within the context, the cascaded graphene structure ultimately yields a cone-shaped graphite medium. Also, different kinds of absorbers made of multilayered and/or cascaded graphene-based mediums have been reported [119,120].

It must be noted that the conventional metamaterial absorbers are made of three layers, among which the bottom layer is usually a highly reflecting medium (e.g., silver) that causes vanishing transmission of the incidence radiation, thereby resulting in trapping of energy within the dielectric section. However, in the present investigation, the bottom reflector has been replaced with the coned-graphite medium. As such, the incidence angle θ is important in order to determine the spectral characteristics. This is because of the LSPs, propagating along the interface, affect the near-field regions (of the medium interface). Furthermore, the variation in graphene Fermi energy E_f would govern the propagation of SPP waves over the effective contact area of coned-graphite layers with a SiO_2 substrate.

Following the schematic of Fig. 9.13, the absorber has the array of graphite cones as the metasurfaces above and below the SiO_2 dielectric medium of thickness t_d. Here, each metasurface unit cell contains a graphite cone [118]. Such an absorber

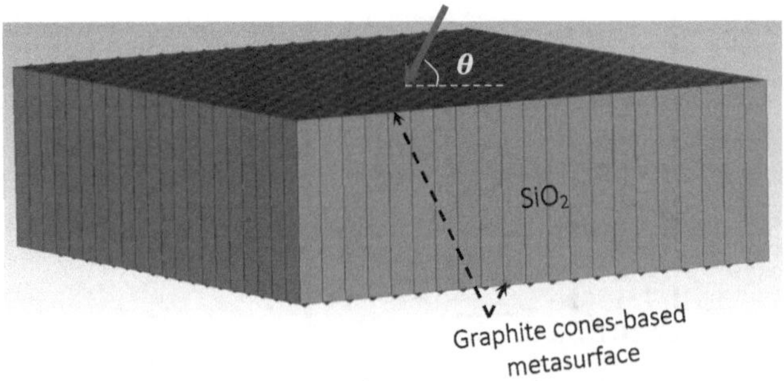

FIGURE 9.13 Schematic of coned-graphite metasurface-based absorber.

has been modeled as the unit cell configured along the z-direction, and the infinite arrays of graphite cones extend periodically along the x- and y-directions. The incidence waves fall upon the top metasurface at an angle θ, as shown in Fig. 9.13. The normal as well as oblique incidence of plane waves upon the metasurface provide the absorption spectra of the structure.

The absorption characteristics of this metamaterial can be evaluated by the expression $A(\omega) = 1 - R(\omega) - T(\omega)$. Here, R and T are, respectively, the reflection and transmission coefficients, with their respective values to be determined by the S-parameters as $R(\omega) = |S_{11}|^2$ and $T(\omega) = |S_{21}|^2$. As such, we consider the scattering matrices, and their relationship between the incidence and the outgoing waves at the input and the output ports, respectively.

Because the effective permittivity of metasurface is required to be evaluated before observing the absorption characteristics, the permittivity of graphene ε_g is

$$\varepsilon_g = 1 + \frac{i\sigma}{\varepsilon_o \omega t_g} \quad (9.12)$$

with σ being the conductivity of graphene, and ε_o, ω and t_g being the free-space permittivity, angular frequency and thickness of the graphene layer, respectively. The effective permittivity of the used metasurface can be evaluated by exploiting the Maxwell-Garnett theory [121], according to which

$$\varepsilon_{eff} = \varepsilon_d \frac{\varepsilon_g + 2\varepsilon_d + 2f(\varepsilon_g - \varepsilon_d)}{\varepsilon_g + 2\varepsilon_d - f(\varepsilon_g - \varepsilon_d)} \quad (9.13)$$

Here, ε_d and f are the dielectric permittivity and volume fraction element of graphene, respectively.

Fig. 9.14 shows wavelength dependence of effective permittivity of a coned-graphite-based metasurface [118]. Here, f has been considered as 0.31, and two

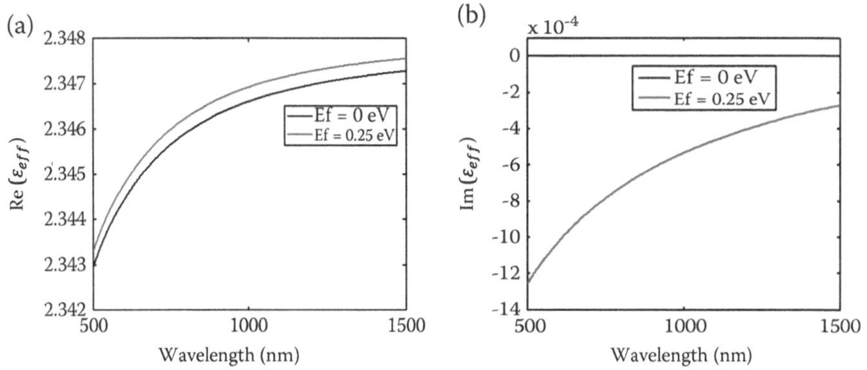

FIGURE 9.14 Wavelength dependence of the (a) real, and (b) imaginary parts of effective permittivity ε_{eff} of the coned-graphite metasurface.

different values of E_f as 0 eV and 0.25 eV. Figs. 9.14a and 9.14b, respectively, depict the plots of the real (Re) and imaginary (Im) parts of effective permittivity ε_{eff} against wavelength in the 500–1,500 nm span. These figures show that $\mathrm{Re}(\varepsilon_{eff})$ exhibits exponential increase with λ (Fig. 9.14a), though the increase remains faster corresponding to low values of wavelength. Also, the increase in E_f results in a little increase in ε_{eff}, which remains nearly uniform throughout the operating regime. Fig. 9.14b shows that, in the absence of bias (i.e., $E_f = 0$ eV), $\mathrm{Im}(\varepsilon_{eff})$ becomes vanishing. However, the use of $E_f = 0.25$ eV makes $\mathrm{Im}(\varepsilon_{eff})$ negative for the entire wavelength span; ε_{eff} is more negative corresponding to smaller wavelengths. As such, the metasurface under consideration shows the negative component of effective permittivity.

Figs. 9.15 and 9.16 illustrate the absorption spectra, as obtained against wavelength, for the E_f-values as 0 eV and 0.25 eV, respectively [118]. To plot these, each unit cell of metasurface has been considered as comprising one graphite cone and having its surface area as 130 × 130 nm^2. Also, the metasurface assumes 20 nm thickness, which is equal to the height of cones, and $f = 0.31$. Considering the varying incidence obliquity in the range of 0°–80°, the spectral plots are obtained for the TE- (Figs. 15a and 16a) and TM- (Figs. 9.15b and 9.16b) mode incidence excitations. Apart from these, the SiO_2 layer thickness has been taken as 1,000 nm.

It can be observed from these figures that multiple absorption peaks appear in the entire wavelength span, the amplitudes of which go on decreasing with increasing wavelength. It can be noticed that, for vanishing E_f, the TE-polarized incidence exhibits very high absorption (~98%) in the visible regime corresponding to grazing incidence with $\theta = 80°$ (Fig. 9.15a); the minimum absorption is attained for normal incidence. The oblique incidence exhibits large absorption in the visible regime; in the IR regime, however, the absorbance keeps on decreasing. The multiple absorption peaks in the operating wavelength span appear due to multiple reflections within the dielectric medium [118].

For the TM-polarized incidence excitation, corresponding to all incidence obliquities, the absorption is increased in the IR regime, as compared to what has been found before for the TE-polarized incidence (Fig. 9.15b). The case of TE-polarized grazing incidence ($\theta = 80°$) exhibits larger absorption in the visible regime, as compared to the TM-polarized grazing excitation, thereby indicating strong resonance conditions corresponding to the respective wavelengths.

Upon increasing E_f to 0.25 eV, we observe upon comparing the results in Fig. 9.16 with those in Fig. 9.15 (corresponding to 0 eV) that the absorption plots do not exhibit significant changes. As such, alterations in Fermi energy of multilayered graphene leave the absorption features almost unaffected, and therefore, the operation of the proposed CIC absorber under varying EM fields remains stable. Remember that the range of the operating wavelength remains greatly important. In the present case, however, the metasurface comprised of nano-coned graphite array does not exhibit significant alterations in spectral characteristics upon changing the chemical potential of graphene.

It can be seen from the above-mentioned results that such a doubly identical graphite-coned metasurface configuration can be used as absorber in the visible

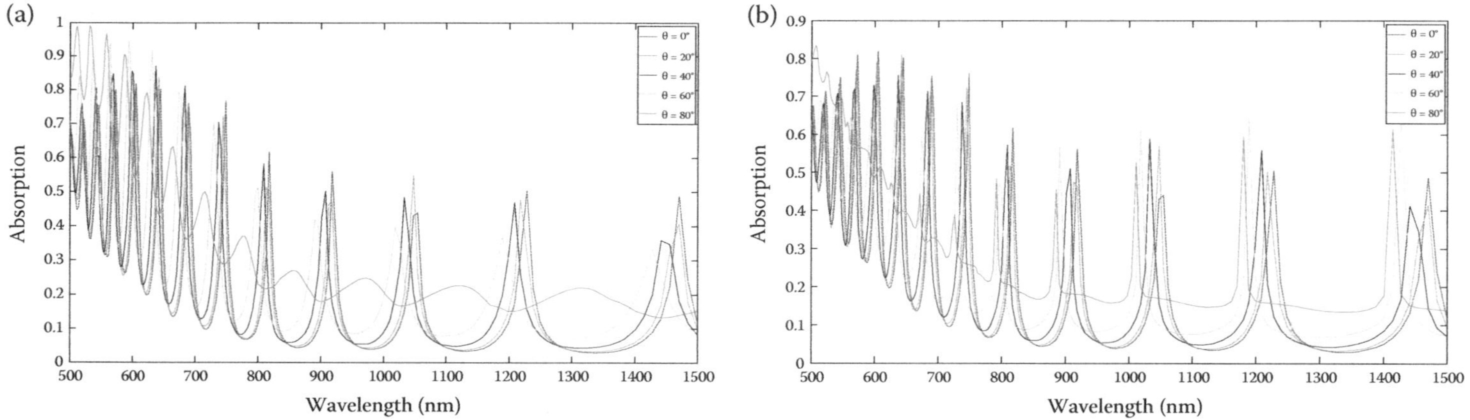

FIGURE 9.15 Plots of absorbance against wavelength λ corresponding to chemical potential $E_f = 0$ eV and in the situations of the (a) TE-, and (b) TM-polarized incidence excitations.

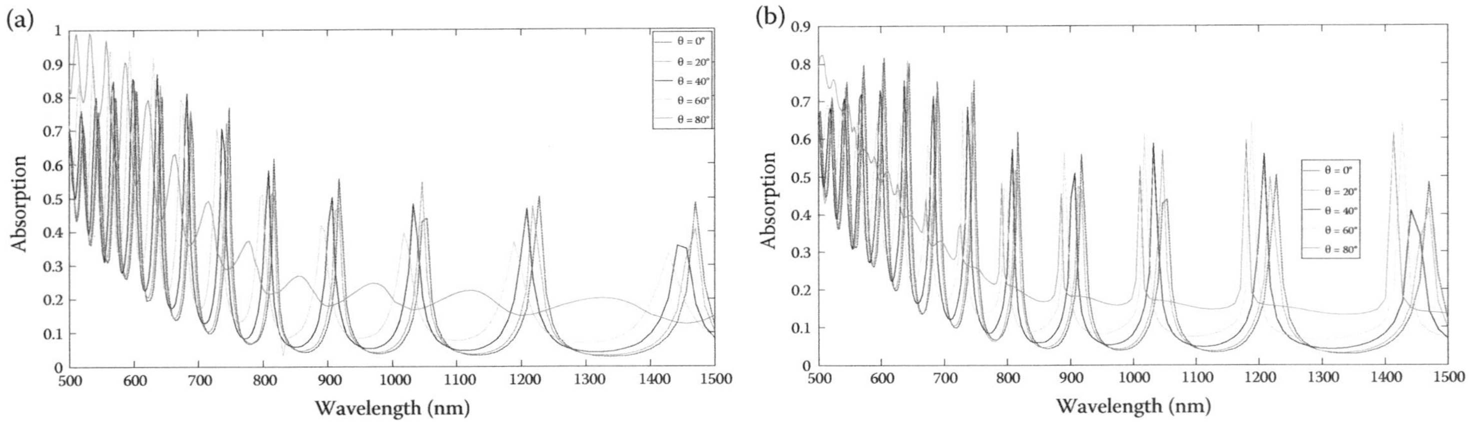

FIGURE 9.16 Plots of absorbance against wavelength λ corresponding to chemical potential $E_f = 0.25$ eV and in the situations of the (a) TE-, and (b) TM-polarized incidence excitations.

regime of EM spectrum. The absorber operates efficiently in the situation of grazing incidence of the TE-polarized incidence excitation. This is owing to strong resonance achieved in this case, which remains almost unaffected by small changes in graphene chemical potential.

9.6.2 Graphene Embedded Phase Change Mediums as Absorber

Phase change mediums (PCMs) have been attractive owing to possible usage in sensors and absorbers [122–124,125]. Some of the notable examples of such PCMs are vanadium dioxide (VO_2), strontium titanate ($SrTiO_3$), germanium-antimony-tellurium (GST), zirconium nitride (ZrN), etc. The use of these in forming metamaterials has been reported by many investigators for temperature and/or strain sensing applications.

The inherent property of phase change in PCMs remains the hallmark. For example, VO_2 is attractive owing to its constitutive and other conductive properties being temperature dependent [126]. Interestingly, VO_2 shows metallic characteristics when the temperature is equal to or above a certain value, and attains the dielectric phase when it is less than that [127–129]. Within the context, Inomata et al. [130] reported micro-structured VO_2 resonator-based sensors, wherein the resonant frequency varies linearly with temperature, and also, remains hysteresis-free. Yi et al. [131] developed dual-functional temperature sensor based on the coupling between VO_2 nanocrystal films and silver (Ag) nanoparticles. Dey et al. [132] reported VO_2 nanorods-based temperature sensor with efficient thermal conductivity over a temperature range of 20–100 °C. Baqir and Choudhury [133] studied temperature sensing characteristics of specially designed VO_2 metamaterial-based structure, wherein the metasurface comprises VO_2 nanodisks arranged in the form of array, and mounted on a periodic stack of SiO_2-Al_2O_3 planar structure. However, the major drawback of VO_2 medium is that the phase transition happens at a relatively low temperature (~68 °C) [134,135].

Among the PCMs, the VO_2-based systems are more useful to achieve perfect absorption in the IR wavelengths, whereas the GST-based (i.e., $Ge_2Se_2Te_5$) systems provide nearly perfect absorption in the IR and visible spans. Apart from this, unlike the VO_2, the phase of GST (amorphous or crystalline) remains stable in maintaining the primary state even in the absence of external control. Interestingly, the PCM-based devices can be tuned electrically [136], thermally [124], and/or optically [137,138].

Tunablility of structure is of great importance in designing active metamaterials and plasmonic absorbers [139,140]. Within the context, the properties of PCMs (such as the GST mediums)-based metamaterials can be tuned by suitably exploiting graphene layers in the structure, which would allow heating of the system. Notably, applying 200 ms electrical pulses of different amplitudes gradually decreases the resistivity of GST layer from high (the amorphous state) to low (the crystalline state) values, and the intermediate crystalline states can be configured by adjusting the amplitude of the excitation pulse [141].

Microheating systems have been important in designing miniaturized photonic devices [141]. Within the context, refs. [142–144] report techniques to exploit

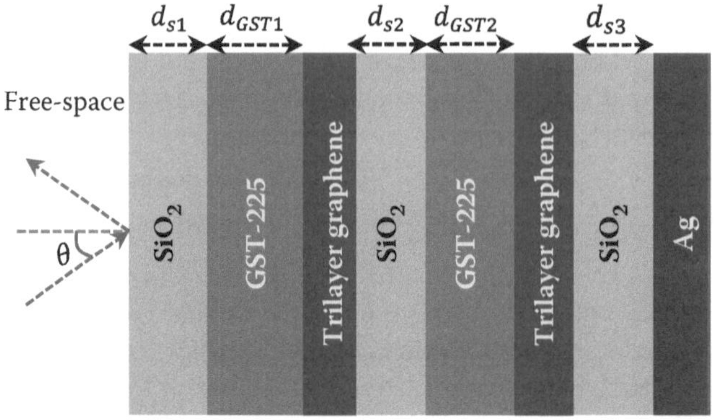

FIGURE 9.17 2D schematic of the GST-graphene-SiO_2 mediums in a multilayered planar configuration.

graphene as the microheater for GST mediums. In line with this, multilayer planar configuration has been investigated to achieve tunable absorption characteristics, wherein graphene layers act as microheaters for the embedded GST medium [34]. In particular, the structure comprises two graphene microheaters, each having tri-layer graphene thickness of ~1 nm. The absorption characteristics of the structure have been determined in the visible and IR regimes. The study includes the effect of two different phases of GST (amorphous and crystalline) on the absorption properties. Fig. 9.17 shows the schematic of the proposed structure, which is a planar multilayer configuration. Herein, the bottom Ag mirror layer is 50 nm thick, and there are 02 stacks of GST-graphene-SiO_2 mediums followed by a thin SiO_2 layer on top as the capping layer, in order to prevent the upper GST medium from evaporation and providing better confinement of heat.

In view of Fig. 9.17, the wavelength-dependence of the real (ϵ_r) and imaginary (ϵ_i) parts of permittivity, in the case of monolayer graphene (with chemical potentials μ_c as 0.5 eV and 1 eV) [34], have been shown in Fig. 9.18. It can be observed in this figure that the real part exhibits the positive and negative

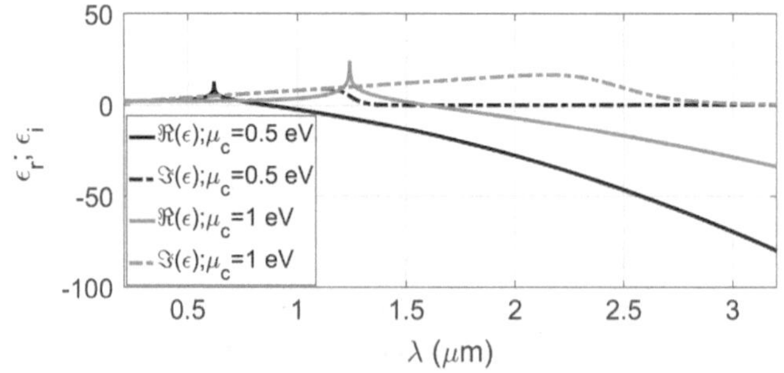

FIGURE 9.18 Wavelength dependence of monolayer graphene permittivity corresponding to the μ_c-values 0.5 eV and 1 eV.

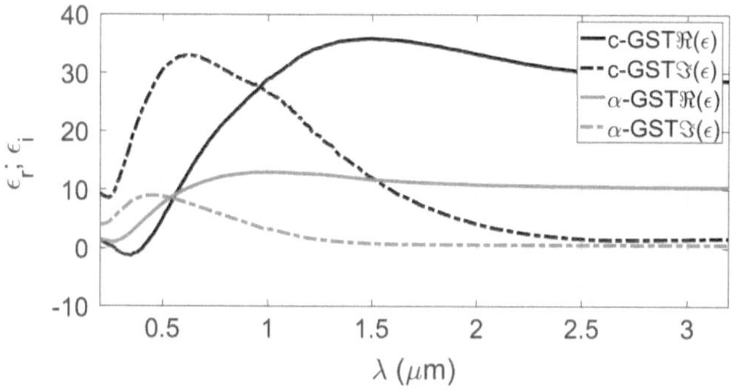

FIGURE 9.19 Wavelength dependence of GST permittivity in the amorphous (α) and crystalline (c) phases.

dependence on wavelength, whereas the imaginary part is positive only with altering wavelength. Further, the real parts of permittivity have sharp peaks, the positions of which, upon increasing μ_c, move toward longer wavelength. Though the imaginary components do not exhibit sharp peaks, the maxima of these also shift to longer wavelength with the increase in μ_c. As such, the graphene chemical potential would serve as the tuning parameter to alter the optical properties.

Fig. 9.19 shows the dispersion behaviour of GST in the amorphous (α) and crystalline (c) states [34]. To plot the real $\Re(\varepsilon)$ and imaginary $\Im(\varepsilon)$ components, the optical constants of mediums have been used in the context of refs. [145,146]. It can be observed that the permittivity dependence on wavelength is more prominent up to 2.5 μm in the case of c-phase of GST, whereas the α-phase exhibits it up to 1.5 μm only. Beyond these wavelengths, the permittivity values of the c- and α-phases become almost independent of wavelength.

As to the absorption characteristics, the influence of GST layers, in terms of their thickness values (i.e., d_{GST1} and d_{GST2}), on the spectral features has been observed. Fig. 9.20 illustrates such a spectral plot when the value of d_{GST1} is varied in the range of 90–150 nm, keeping the other parameters fixed [34]. For instance, Fig. 9.20 corresponds to the situation of d_{GST1} = 90 nm, d_{s1} = 100 nm, $d_{s2} = d_{s3}$ = 50 nm, d_g = 1 nm (i.e., a trilayer graphene), and μ_c = 0.1 eV. More specifically, this figure

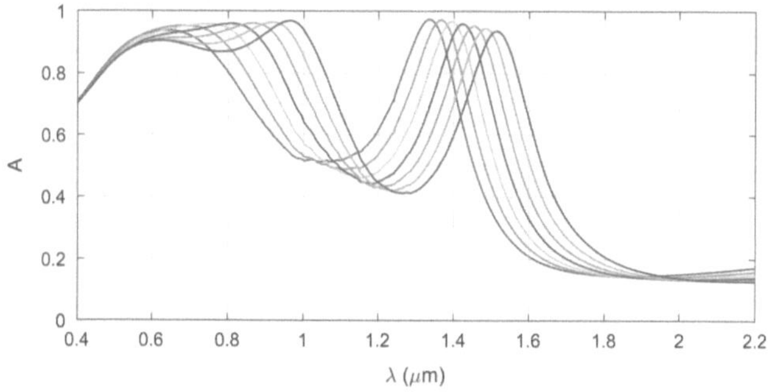

FIGURE 9.20 Plots of absorbance A vs. wavelength under varying d_{GST1}.

exhibits the wavelength dependence of absorbance A considering the TM-polarized incidence excitation impinging on the top SiO_2 surface normally (i.e., $\theta = 0°$).

It can be seen in Fig. 9.20 that the absorption spectra has two spans of wavelength in the visible (600–1000 nm) and IR (1300–1570 nm) regimes, wherein resonances have been observed. In these, the value of d_{GST1} has been varied from 90 nm to 150 nm in a step of 10 nm, and the second group of peaks shows shifts in d_{GST1} their positions to higher d_{GST1} wavelength with increasing d_{GST1}. It can be noticed that, in the visible span, significant amount of broadening of resonance absorption happens with ~99% absorption. Apart from this, red-shifts in absorption peak can be observed in the visible and IR regimes with increasing d_{GST1}.

Considering the effect of varying d_{GST2} on the absorption spectra, Fig. 9.21 shows the relevant plot, wherein values of d_{GST2} increase in a range of 50–110 nm in a step of 10 nm [34]; the positions of prominent peaks also shift to increased wavelength with the increase in d_{GST2}. In this plot, the parametric values have been as stated before, and $d_{GST1} = 130$ nm. It can be clearly observed in this figure that the lower GST layer thickness leaves significant impact on the absorption characteristics; in the IR regime, the resonance peaks are well-resolved now, whereas in the visible range, those remain almost intact. It is noticeable that the increase in d_{GST2} results in red-shifts of absorption peaks in the IR regime, and the shifts are linearly scaled for a larger GST layer thickness. It is worth mentioning that most of the environmental gas sensors operate in the mid-IR regime, and therefore, such a characteristic feature remains prudent for designing tunable optical gas sensors [147].

Being GST a phase change medium, which essentially depends on the temperature, the thermal tunability of the structure can be studied by manipulating the RI of GST layer(s) in both the α- and c-phases. The permittivity of GST for different ratios of crystallinity can be stated by the effective permittivity theory and Lorentz-Lorenz relations [148,149]; according to these

$$\frac{\epsilon_{eff}(\lambda) - 1}{\epsilon_{eff}(\lambda) + 2} = m\frac{\epsilon_c(\lambda) - 1}{\epsilon_c(\lambda) + 2} + (1 - m)\frac{\epsilon_a(\lambda) - 1}{\epsilon_a(\lambda) + 2} \tag{9.14}$$

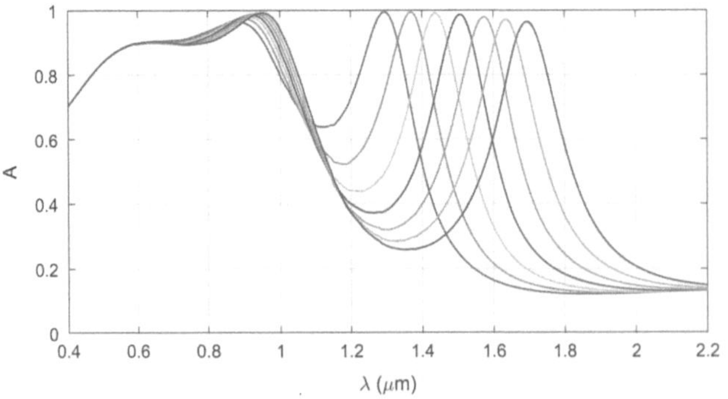

FIGURE 9.21 Plots of absorbance A vs. wavelength under varying d_{GST2}.

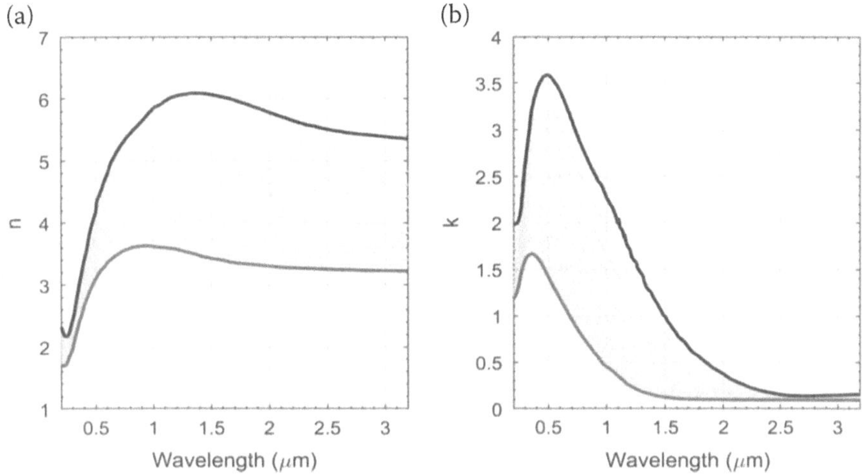

FIGURE 9.22 Plots of (a) RI n, and (b) extinction coefficient k vs. wavelength for $0 \leq m \leq 1$ – the α-state (lower solid line), the c-state (upper solid line), and intermediate states (gray dotted lines).

$$n_{eff}(\lambda) + jk_{eff}(\lambda) = \sqrt{\epsilon_{eff}(\lambda)} \tag{9.15}$$

with m being the ratio of crystallinity with its value varying from 0 (pure amorphous) to 1 (purely crystalline). In these equations, $\epsilon_{eff}(\lambda)$, $\epsilon_c(\lambda)$, and $\epsilon_a(\lambda)$ are the wavelength-dependent effective permittivity of GST medium, and its permittivity values in the c- and α-states., respectively. The wavelength dependence of GST layer RI is shown in Fig. 9.22. In view of the legend of this figure, it can be observed that with increasing m the RI gradually moves from the α-state (i.e., $m = 0$) toward the c-state (i.e., $m = 1$) [34].

As such, the kind of graphene embedded GST medium-based multilayer structure considered here can present high absorption in the visible and IR regimes; the positions of resonance peaks can be altered by varying the GST medium properties. The incorporation of two GST layers in the structure yields nearly perfect absorption. Herein, the graphene microheaters can be utilized to achieve GST phase transition, thereby introducing the property of tunability. Apart from this, the GST crystallization process adds additional tunability of the absorption spectra.

9.7 USE OF GRAPHENE IN SENSING

The use of sensors has been vital in almost every sector of human life that includes domestic, industries and the environment related measurements [150–152]. In this stream, the implementation of graphene has become significantly popular. Interestingly, graphene has reshaped the landscape of photonics technology. In particular, the IR and mid-IR response of graphene can be read through plasmonic characteristics, which can be dynamically tuned – the feature that has been greatly exploited in biosensing applications [153,154].

Pivoted to this, for example, graphene-supported tunable mid-IR biosensors were reported, which demonstrate potentials for quantitative protein detection and chemical-specific molecular identification [154]. In this work, a graphene layer has been synthesized by the chemical vapor deposition (CVD) technique, and transferred to a 280 nm thick native silica oxide of a silicon substrate. The investigation reported plasmonic features upon applying an electrostatic field across the SiO_2 layer through a bias voltage in the range of 0–120 V to dynamically control the chemical potential of graphene. Finally, the experiment could detect protein molecules, the primary material of life enabling most of the critical biological functions.

Monolayer graphene film-coated microfiber-based Bragg grating sensor has been reported to function as a gas sensor [155]. In the configuration, a fiber Bragg grating is first fixed on a MgF_2 crystal substrate with a low RI. After that, the graphene film is coated over the fiber exploiting the CVD method to form the device. While in operation, gas molecules are adsorbed by the graphene film, which results in alterations of the effective RI of fiber Bragg grating structure. This basically happens due to significant carrier movement of graphene. In reality, the device is conceptualized on the advantage of strong polar gas absorption and surface field enhancement of graphene film, which could detect ammonia (NH_3) gas of very low concentration ~0.2 ppm.

The above discussions indicate that patterned graphene structures have been of great importance in the application viewpoint when these are amalgamated with other suitably configured mediums. This could trigger the invention of many forms of tunable devices utilizing graphene-based metamaterials [156,157].

The use of graphene remains proper for heat spread in the neighboring mediums [158,159], which allows the implementation of PCMs along with graphene for certain sensing applications. Among the others, $SrTiO_3$ exhibits a high dielectric constant, which essentially depends on the change of ambient temperature [160]. As such, $SrTiO_3$ would be prudent in designing fairly rugged thermal sensors in the THz regime [161,162].

In this stream, a tunable dual-band metamaterial-supported thermal sensor has been investigated, which operates in the THz regime; Fig. 9.23 shows the schematic of the metasurface only having a pixelated pattern [163]. However, such a design of pixelated nanostructured mediums was conceptualized before, as reported

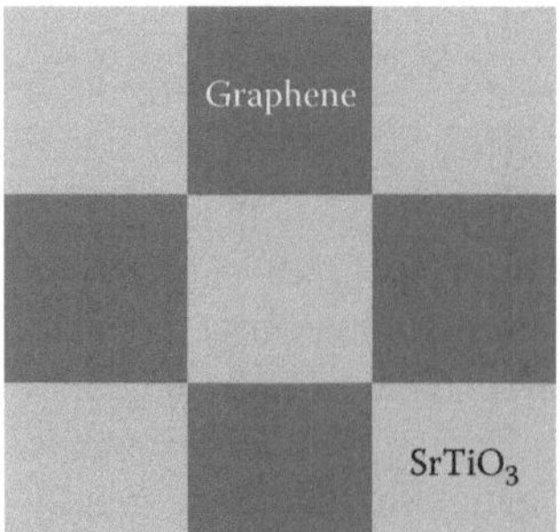

FIGURE 9.23 Top view of the pixelated metasurface unit cell.

elsewhere [164–166]. As can be seen, the metasurface consists of squared pixels of graphene that are symmetrically distributed over the surface of $SrTiO_3$ thin film. This yields a pattern of periodically arranged graphene-$SrTiO_3$ pixels, as the figure shows. The chemical potential μ of graphene can be varied for the tunability purpose, which will alter the constitutive properties of the metasutface, thereby varying the spectral characteristics of the sensor.

In the thermal sensor configuration, the metasurface of certain thickness (say h) is grown over a relatively thick (having a thickness of 500 nm) SiO_2 dielectric medium (not shown here) of the permittivity value as 2.1. The size of the pixels can vary. However, in this investigation, all pixels are square shaped. The transmission characteristics of such a structure have been evaluated by considering the area of each pixel (in the metasurface) as 50×50 nm^2 and the $SrTiO_3$ thin film thickness as 50 nm. Fig. 9.24 illustrates the plots of the real (Fig. 9.24a) and imaginary (Fig. 9.24b) parts of effective permittivity against the operating frequency (in the range of 1–6 THz) under μ as 0.4 eV. For this purpose, the permittivity of $SrTiO_3$ has been considered for two different values of ambient temperature, namely 250 K and 400 K. Also, the volume fraction element of graphene is considered to be 0.0043.

It can be seen in Fig. 9.24 that the real part exhibits the positive and negative dependence on frequency, whereas the imaginary part is positive valued only. With increasing frequency, the real part of effective permittivity becomes more positive. Further, the values of both the parallel and perpendicular components of the real part decrease with increasing temperature. Nevertheless, within the 2–6 THz frequency span, the effective permittivity is found to remain very high due to the presence of $SrTiO_3$, which possesses large relative permittivity over the used operating frequency range. Before 1.6 THz, the real component of effective permittivity is negative valued, and it becomes more negative as the operating frequency is reduced. Fig. 9.24 clearly determines the metasurface to be of the dielectric and Type I hyperbolic metamaterial kinds depending on the operating frequency.

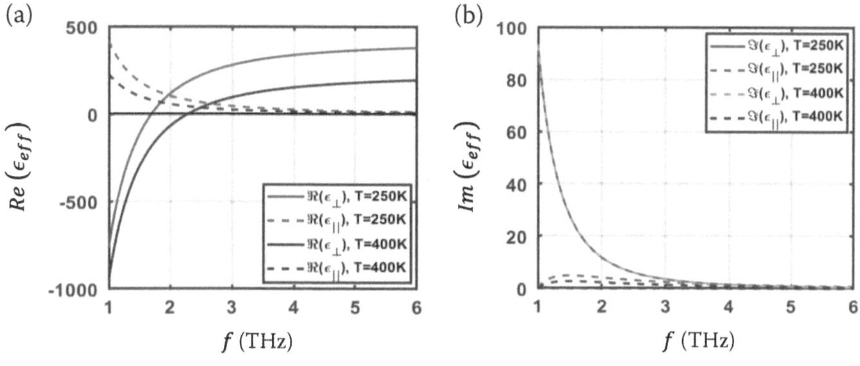

FIGURE 9.24 Dispersion characteristics of metasurface; (a) real part, and (b) imaginary part.

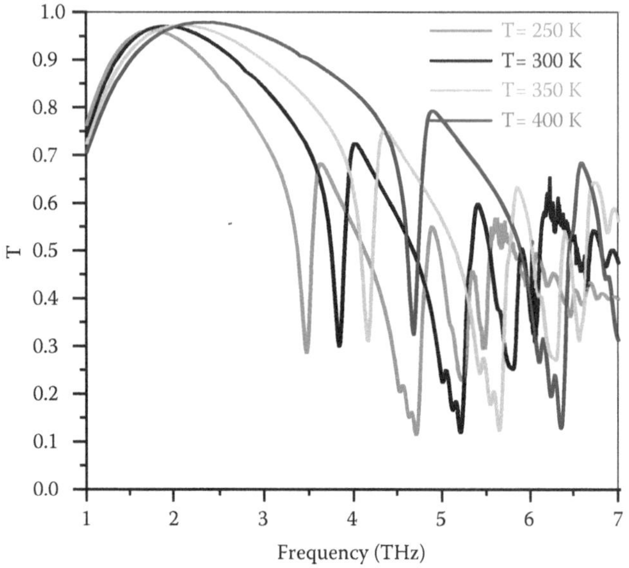

FIGURE 9.25 Transmission spectra of the metamaterial under varying temperature and $\mu = 0.4$ eV.

Fig. 9.25 shows the transmission spectra of metamaterial in the range of 1–7 THz under the ambient temperatures as 250 K, 300 K, 350 K and 400 K. Also, the graphene chemical potential has been considered as 0.4 eV. In this figure, corresponding to each value of ambient temperature, the structure exhibits 02 transmission dips. However, the increase in temperature by 50 K results in a shift of frequency minima by ~0.6 THz or more. This essentially demonstrates that the proposed design can be used for thermal sensing. It must be added here that Fig. 9.25 has been obtained under the normal incidence of waves. However, it has also been found that, upon altering the angle of incidence, positions of transmission minima do not shift; only the magnitude of transmission varies a little without affecting much the positions of transmission minima. As such, the investigated pixelated-graphene supported thermal sensor can be used in a fairly wide temperature range of 250–400 K. Indeed, alteration in graphene chemical potential introduces tunability to the structure.

9.8 SLOW-LIGHT STRUCTURES AND MODE FILTERING

We have explored various mechanisms, including the use of materials, to reduce the group velocity of light propagating in a medium. As discussed before, graphene enables slow-light propagation in plasmonic structures [23,24]. The cited references [167–169] demonstrate slow-light at telecom frequencies due to the surface plasmonic effect. Slowly varying surface plasmons in graphene- and silicon-based graded grating structures have also been reported [170]. In addition, the EIT in nanostructured graphene induces a significant amount of group delay with its magnitude larger than that offered by the metallic structure [29].

In the context of slow light in graphene-supported guides, the study of group velocity of a specially designed silica-clad slab waveguide structure having the core region comprising a gold nanolayer and a graphene monolayer, sandwiched between the core-clad interface, revealed the guide to behave as an optical filter in the visible regime [25].

Fig. 9.26 exhibits the schematic of the waveguide structure. Considering the propagation of time t-harmonic and the axis z-harmonic EM waves along the optical axis, the TE mode excitation provides the electric (E-) field components [171] as

$$E_{c\_y} = E_0 \begin{Bmatrix} \cos k_{\mathrm{Au}} x \\ \sin k_{\mathrm{Au}} x \end{Bmatrix} e^{j\omega t} |x| \leq d \tag{9.16a}$$

$$E_{cl\_y} = E_1 \left\{ e^{-k_{\mathrm{SiO_2}} x} \right\} e^{j\omega t} |x| \geq d \tag{9.16b}$$

in the core and clad sections, respectively. Here, E_0 and E_1 are the E-field amplitudes. Also, the magnetic (H-) field components corresponding to the TM mode in the core/clad sections will be

$$H_{c\_y} = H_0 \begin{Bmatrix} \cos k_{\mathrm{Au}} x \\ \sin k_{\mathrm{Au}} x \end{Bmatrix} e^{j\omega t} |x| \leq d \tag{9.17a}$$

$$H_{cl\_y} = H_1 \left\{ e^{-k_{\mathrm{SiO_2}} x} \right\} e^{j\omega t} |x| \geq d \tag{9.17b}$$

respectively, with H_0 and H_1 being the field amplitudes. In these equations, the subscripts c and cl in the left-hand sides represent the situations in the core and clad sections, respectively. In these equations, k_{Au} and $k_{\mathrm{SiO_2}}$ are the wave numbers of gold and silica mediums, respectively, which assume the forms

$$k_{\mathrm{Au}} = \sqrt{\varepsilon_{\mathrm{Au}}(\omega) k_0^2 - \beta^2} \tag{9.18a}$$

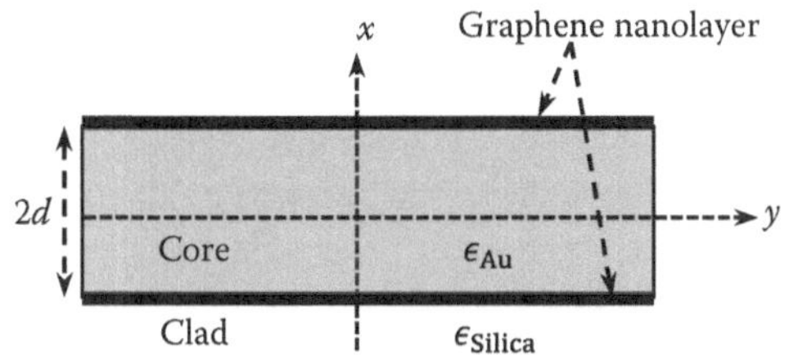

FIGURE 9.26 Schematic of the proposed graphene-based waveguide structure.

and

$$k_{\mathrm{SiO_2}} = \sqrt{\beta^2 - \varepsilon_{\mathrm{SiO_2}} k_0^2} \tag{9.18b}$$

Here, k_0 is the free-space wave number, and β is the propagation constant in medium. The real value of β yields propagation, whereas the imaginary part corresponds to absorption of waves (in the guide).

Now, using Maxwell's equations, and the continuity conditions of the tangential components of electric/magnetic fields at the core-clad interface, after some mathematical steps, one may derive the dispersion relations as

$$[f_1(\beta)]_{Even} = \tan(k_{\mathrm{Au}} d) - \frac{k_{\mathrm{SiO_2}} + j\omega\sigma_g(\omega)}{k_{\mathrm{Au}}} = 0 \tag{9.19a}$$

for the even-order TE modes, and

$$[f_1(\beta)]_{Odd} = \cot(k_{\mathrm{Au}} d) + \frac{k_{\mathrm{SiO_2}} + j\omega\sigma_g(\omega)}{k_{\mathrm{Au}}} = 0 \tag{9.19b}$$

for the odd-order TE modes [25]. Corresponding to the TM modes, the dispersion relations in the cases of even- and odd-order modes assume the forms

$$[f_2(\beta)]_e = \tan(k_{Au} d) - \frac{\varepsilon_{\mathrm{Au}}(\omega) k_{\mathrm{SiO_2}}}{\varepsilon_{\mathrm{SiO_2}} k_{\mathrm{Au}}(\omega)(1 + \sigma_g(\omega))} = 0 \tag{9.20a}$$

and

$$[f_2(\beta)]_o = \cot(k_{Au} d) + \frac{\varepsilon_{\mathrm{Au}}(\omega) k_{\mathrm{SiO_2}}}{\varepsilon_{\mathrm{SiO_2}} k_{\mathrm{Au}}(\omega)(1 + \sigma_g(\omega))} = 0 \tag{9.20b}$$

respectively [25]. In these equations, $\varepsilon_{\mathrm{Au}}$ and $\varepsilon_{\mathrm{SiO_2}}$ are the permittivity of gold and silica mediums, respectively, and $\sigma_g(\omega)$ is the frequency-dependent conductivity of graphene.

Interestingly, Eqs. (9.19) and (9.20) provide β-values corresponding to the TM mode only, whereas no solutions exist corresponding to the TE mode [25]. As such, one can infer that the proposed graphene-based waveguide configuration does not support the presence of TE mode. Apart from this, it has been found that the guide supports the sustainment of zero-order TM mode (i.e., the TM_0 mode) only.

Considering the operating wavelength to be 450 nm, Fig. 9.27 exhibits graphically the solutions corresponding to the TE and TM modes. Herein, Fig. 9.27a determines that both the even- and odd-order modes have similar dispersion characteristics. In Fig. 9.27b, the solid and dashed lines, respectively, depict the presence of even- and odd-order modes [25]. This figure shows the existence of

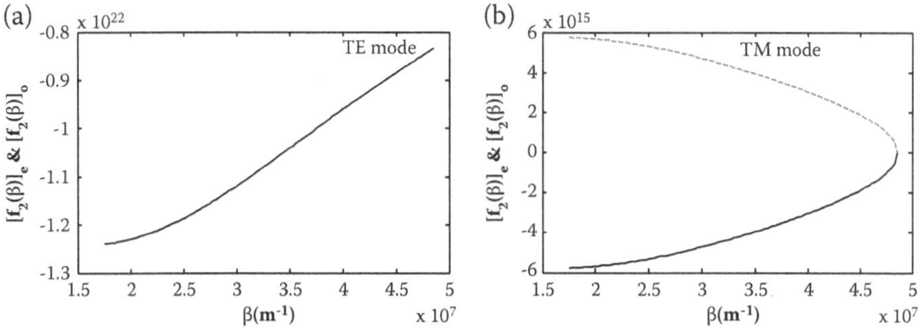

FIGURE 9.27 Solutions for the (a) TM, and (b) TE modes in the waveguide structure in Fig. 9.26.

only the zero-order root of β at its value $4.8 \times 10^7\ m^{-1}$, corresponding to the TM_0 mode in the entire visible spectral range. This clearly indicates that the proposed waveguide does not support the TE modes. In other words, it filters out the TE and the other higher-order TM modes, thereby affirming the proposed graphene-based structure to serve the purpose of mode filtering.

9.9 CONCLUSION

This chapter gave a glimpse of the importance of the graphene medium and its application in designing certain forms of metamaterial-inspired photonic components. In particular, we placed emphasis on filters, absorbers and sensors to be used in the THz, visible and IR regimes of the EM spectrum. In line with this, we touched upon a few relevant metamaterial designs and provided demonstrations of their usefulness in respective applications. In all these device configurations, graphene was primarily exploited to tune the spectral characteristics. In summary, the suggested graphene-embedded designs are attractive as they offer tunability. However, it remains more prudent to devise photonic components in the THz regime because they may more easily attain tunability.

ACKNOWLEDGMENTS

Discussions with M. Baqir, M. Ghasemi, M. Pourmand and E. Sheta on the various parts of this work are gratefully acknowledged.

REFERENCES

1. M. Dragoman and D. Dragoman, "Graphene-based quantum electronics," *Prog. Quantum Electron.*, vol. 33, pp. 165–214, 2009.
2. A. H. C. Neto, F. Guinea, N. M. R. Peres, K. S. Novoselov, and A. K. Geim, "The electronic properties of graphene," *Rev. Mod. Phys.*, vol. 81, pp. 109–162, 2009.
3. E. Stoye, "Graphene: Looking beyond the hype," *Sci. Amer.*, https://www.scientifcamerican.com/article/graphene-lookingbeyond-the-hype/, 2015.

4. A. N. Grigorenko, M. Polini, and K. S. Novoselov, "Graphene plasmonics," *Nature Photon.*, vol. 6, pp. 749–758, 2012.
5. Z.-Q. Ye, B.-Y. Cao, W.-J. Yao, T. Feng, and X. Ruan, "Spectral phonon thermal properties in graphene nanoribbons," *Carbon*, vol. 93, pp. 915–923, 2015.
6. G. Zhou, C. Cen, S. Wang, M. Deng, and O. V. Prezhdo, "Electron–phonon scattering is much weaker in carbon nanotubes than in graphene nanoribbons," *J. Phys. Chem. Lett.*, vol. 10, pp. 7179–7187, 2019.
7. C. Cao, J. Zhang, X. Wen, S. L. Dodson, N. T. Dao, L. M. Wong, S. Wang, S. Li, A. H. Phan, and Q. Xiong, "Metamaterials-based label-free nanosensor for conformation and affinity biosensing," *ACS Nano*, vol. 7, pp. 7583–7591, 2013.
8. Z. Chen, B. Guo, Y. Yang, and C. Cheng, "Metamaterials-based enhanced energy harvesting: A review," *Physica B.*, vol. 438, pp. 1–8, 2014.
9. F. Ma, Y.-S. Lin, X. Zhang, and C. Lee, "Tunable multiband terahertz metamaterials using a reconfigurable electric split-ring resonator array," *Light: Science & Appl.*, vol. 3, Article e171, 2014.
10. M. A. Baqir, M. Ghasemi, P. K. Choudhury, and B. Y. Majlis, "Design and analysis of nanostructured subwavelength metamaterial absorber operating in the UV and visible spectral range," *J. Electromagn. Waves and Appl.*, vol. 29, pp. 2408–2419, 2015.
11. M. Ghasemi, P. K. Choudhury, and A. Dehzangi, "Nanoengineered thin films of copper for the optical monitoring of urine – A comparative study of the helical and columnar nanostructures," *J. Electromagn. Waves and Appl.*, vol. 29, pp. 2321–2329, 2015.
12. M. Ghasemi, M. A. Baqir, and P. K. Choudhury, "On the metasurface based comb filters," *IEEE Photon. Technol. Lett.*, vol. 28, pp. 1100–1103, 2016.
13. M. Ghasemi, P. K. Choudhury, M. A. Baqir, M. A. Mohamed, A. R. M. Zain, and B. Y. Majlis, Metamaterial absorber comprised of chromium-gold nanorods-based columnar thin films, *J. Nanophoton.*, vol. 11, Article 043505, 2017.
14. M. A. Baqir and P. K. Choudhury, "Hyperbolic metamaterial-based UV absorber", *IEEE Phot. Technol. Lett.*, vol. 29, pp. 1548–1551, 2017.
15. M. Ghasemi and P. K. Choudhury, "Nanostructured concentric gold ring resonator-based metasurface filter device," *Optik*, vol. 127, pp. 9932–9936, 2016.
16. J. Yang, Z. Zhu, J. Zhang, C. Guo, W. Xu, K. Liu, X. Yuan, and S. Qin, "Broadband terahertz absorber based on multi-band continuous plasmon resonances in geometrically gradient dielectric-loaded graphene plasmon structure," *Sci. Repts.*, vol. 8, Article 3239, 2018.
17. P. Fu, F. Liu, G. J. Ren, F. Su, D. Li, and J. Q. Yao, "A broadband metamaterial absorber based on multi-layer graphene in the terahertz region," *Opt. Commun.*, vol. 417, pp. 62–66, 2018.
18. M. Huang, Y. Cheng, Z. Cheng, H. Chen, X. Mao, and R. Gong, "Based on graphene tunable dual-band terahertz metamaterial absorber with wide-angle," *Opt. Commun.*, vol. 415, pp. 194–201, 2018.
19. F. Yan, L. Li, R. Wang, H. Tian, J. Liu, J. Liu, F. Tian, and J. Zhang, "Ultrasensitive tunable terahertz sensor with graphene plasmonic grating," *J. Lightwave Technol.*, vol. 37, pp. 1103–1112, 2019.
20. X. Liu, G. Liu, P. Tang, G. Fu, G. Du, Q. Chen, and Z. Liu, "Quantitatively optical and electrical-adjusting high-performance switch by graphene plasmonic perfect absorbers," *Carbon*, vol. 140, pp. 362–367, 2018.
21. J. Wang, X. Ying, D. He, C. Li, S. Guo, H. Peng, L. Liu, Y. Jiang, J. Xu, and Z. Liu, "Enhanced absorption of graphene with variable bandwidth in quarter-wavelength cavities," *AIP Adv.*, vol. 8, Article 125301, 2018.

22. B. Liu, W. Yu, Z. Yan, C. Tang, J. Chen, P. Gu, Z. Liu, and Z. Huang "Ultra-narrowband light absorption enhancement of monolayer graphene from waveguide mode," *Opt. Express*, vol. 28, pp. 24908–24917, 2020.
23. R. Hao, J. M. Jin, X. Peng, and E. Li, "Dynamic control of wideband slow wave in graphene based waveguides," *Opt. Lett.*, vol. 39, pp. 3094–3097, 2014.
24. R. Hao, X.-L. Peng, E.-P. Li, Y. Xu, J.-M. Jin, X.-M. Zhang, and H.-S. Chen, "Improved slow light capacity in graphene-based waveguide," *Sci. Repts.*, vol. 5, Article 15335, 2015.
25. M. A. Baqir and P. K. Choudhury, "Graphene-based slab waveguide for slow-light propagation and mode filtering," *J. Electromagn. Waves and Appl.*, vol. 31, pp. 2055–2063, 2017.
26. S. Zhang, D. A. Genov, Y. Wang, M. Liu, and X. Zhang, "Plasmon-induced transparency in metamaterials," *Phys. Rev. Lett.*, vol. 101, Article 047401, 2008.
27. Z. Chai, X. Hu, Y. Zhu, S. Sun, H. Yang, and Q. Gong, "Ultracompact chip-integrated electromagnetically induced transparency in a single plasmonic composite nanocavity," *Adv. Opt. Mat.*, vol. 2, pp. 320–325, 2014.
28. Y. Ling, L. Huang, W. Hong, T. Liu, J. Luan, W. Liu, J. Lai, and H. Li, "Polarization-controlled dynamically switchable plasmon-induced transparency in plasmonic metamaterial," *Nanoscale*, vol. 10, pp. 19517–19523, 2018.
29. X. Shi, D. Han, Y. Dai, Z. Yu, Y. Sun, H. Chen, X. Liu, and J. Zi, "Plasmonic analog of electromagnetically induced transparency in nanostructure graphene," *Opt. Express*, vol. 21, pp. 28438–28443, 2013.
30. X. He, F. Liu, F. Lin, and W. Shi, "Graphene patterns supported terahertz tunable plasmon induced transparency," *Opt. Express*, vol. 26, pp. 9931–9944, 2018.
31. X. Yan, T. Wang, S. Xiao, T. Liu, H. Hou, L. Cheng, and X. Jiang, "Dynamically controllable plasmon induced transparency based on hybrid metal-graphene metamaterials," *Sci. Repts.*, vol. 7, Article 3917, 2017.
32. Y. Wang, H. Mi, Q. Zheng, Z. Ma, and S. Gong, "Graphene/phase change material nanocomposites: light-driven, reversible electrical resistivity regulation via form-stable phase transitions," *ACS Appl. Mater. Interfaces*, vol. 7, pp. 2641–2647, 2015.
33. A. Allahbakhsh and M. Arjmand, "Graphene-based phase change composites for energy harvesting and storage: state of the art and future prospects," *Carbon*, vol. 148, pp. 441–480, 2019.
34. M. Pourmand, P. K. Choudhury, and M. A. Mohamed, "Tunable absorber embedded with GST mediums and trilayer graphene strip microheaters," *Sci. Repts.*, vol. 11, Article 3603, 2021.
35. C. N. Santos, F. Joucken, D. De Sousa Meneses, P. Echegut, et al., "Terahertz and mid-infrared reflectance of epitaxial graphene," *Sci. Repts.*, vol. 6, Article 24301, 2016.
36. M. Wang and E.-H. Yang, "THz applications of 2D materials: graphene and beyond," *Nano-Struct. Nano Objects*, vol. 15, pp. 107–113, 2018.
37. D. A. Bandurin, D. Svintsov, I. Gayduchenko, S. G. Xu, et al., "Resonant terahertz detection using graphene plasmons," *Nature Commun.*, vol. 9, Article 5392, 2018.
38. L. A. Falkovsky and S. S. Pershoguba, "Optical far-infrared properties of a graphene monolayer and multilayer," *Phys. Rev. B.*, vol. 76, Article 153410, 2007.
39. S. Song, Q. Chen, L. Jin, and F. Sun, "Great light absorption enhancement in a graphene photodetector integrated with a metamaterial perfect absorber," *Nanoscale*, vol. 5, pp. 9615–9619, 2013.
40. Quan Li, L. Cong, R. Singh, N. Xu, et al., "Monolayer graphene sensing enabled by the strong Fano-resonant metasurface," *Nanoscale*, vol. 8, pp. 17278–17284, 2016.

41. J. Chen, S. Chen, P. Gu, Z. Yan, et al., "Electrically modulating and switching infrared absorption of monolayer graphene in metamaterials," *Carbon*, vol. 162, pp. 187–194, 2020.
42. H. Lin, B. C. P. Sturmberg, K.-T. Lin, Y. Yang, et al., "A 90-nm-thick graphene metamaterial for strong and extremely broadband absorption of unpolarized light," *Nature Photon.*, vol. 13, pp. 270–276, 2019.
43. Y. Yang, H. Lin, B. Y. Zhang, Y. Zhang, et al., "Graphene-based multilayered metamaterials with phototunable architecture for on-chip photonic devices," *ACS Photon.*, vol. 6, pp. 1033–1040, 2019.
44. A. Pizzi, G. Rosolen, L. J. Wong, R. Ischebeck, et al., "Graphene metamaterials for intense, tunable, and compact extreme ultraviolet and X-Ray sources," *Adv. Sci.*, vol. 7, Article 1901609, 2020.
45. S.-H. Lee, J.-H. Choe, C. Kim, S. Bae, et al., "Graphene assisted terahertz metamaterials for sensitive bio-sensing," *Sensors and Actuat. B: Chemical*, vol. 310, Article 127841, 2020.
46. B. Wang, X. Zhang, F. García-Vidal, X. Yuan, and J. Teng, "Strong coupling of surface plasmon polaritons in monolayer graphene sheet arrays," *Phys. Rev. Lett.*, vol. 109, Article 073901, 2012.
47. B. Zhao, J. M. Zhao, and Z. M. Zhang, "Enhancement of near-infrared absorption in graphene with metal gratings," *Appl. Phys. Lett.*, vol. 105, Article 031905, 2014.
48. A. Vakil and N. Engheta, "Transformation optics using graphene," *Science*, vol. 332, pp. 1291–1294, 2011.
49. T. Stauber, G. Santos, and F. Abajo, "Extraordinary absorption of decorated undoped graphene," *Phys. Rev. Lett.*, vol. 112, Article 077401, 2014.
50. G. W. Hanson, "Dyadic Green's functions and guided surface waves for a surface conductivity model of graphene," *J. Appl. Phys.*, vol. 103, Article 064302, 2008.
51. M. E. Morote, J. S. G. D. Gómez-Di, and J. P. Carrier, "Sinusoidally modulated graphene leaky-wave antenna for electronic beam scanning at THz," *IEEE Trans. Terahertz Sci. Technol.*, vol. 4, pp. 116–122, 2014.
52. Z. Liu, K. Suenaga, P. J. F. Harris, and S. Iijima, "Open and closed edges of graphene layers," *Phys. Rev. Lett.*, vol. 102, Article 015501, 2009.
53. K. Yan, H. Peng, Y. Zhou, H. Li, and Z. Liu, "Formation of bilayer bernal graphene: layer-by-layer epitaxy via chemical vapor deposition," *Nano Lett.*, vol. 11, pp. 1106–1110, 2011.
54. I.-T. Lin, J.-M. Liu, K.-Y. Shi, P.-S. Tseng, et al., "Terahertz optical properties of multilayer graphene: experimental observation of strong dependence on stacking arrangements and misorientation angles," *Phys. Rev. B.*, vol. 86, Article 235446, 2012.
55. S. Biabanifard, M. Biabanifard, S. Asgari, S. Asadi, and M. C. E. Yagoub, "Tunable ultra-wideband terahertz absorber based on graphene disks and ribbons," *Opt. Commun.*, vol. 427, pp. 418–425, 2018.
56. M. Biabanifard, A. Arsanjani, M. S. Abrishamian, and D. Abbott, "Tunable terahertz graphene-based absorber design method based on a circuit model approach," *IEEE Access*, vol. 8, pp. 70343–70354, 2020.
57. A. Najafi, M. Soltani, I. Chaharmahali, and S. Biabanifard, "Reliable design of THz absorbers based on graphene patterns: exploiting genetic algorithm," *Optik*, vol. 203, Article 163924, 2020.
58. X. F. Fan, W. T. Zheng, V. Chihaia, Z. X. Shen, and J. L. Kuo, "Interaction between graphene and the surface of SiO_2," *J. Phys.: Cond. Mat.*, vol. 24, Article 305004, 2012.

59. X. H. Deng, J. T. Liu, J. Yuan, T. B. Wang, and N. H. Liu, "Tunable THz absorption in graphene-based heterostructures," *Opt. Express*, vol. 22, pp. 30177–30183, 2014.
60. G. Kedawat, S. Srivastava, V. K. Jain, P. Kumar, et al., "Fabrication of artificially stacked ultrathin ZnS/MgF_2 multilayer dielectric optical filters," *ACS Appl. Mat. Inter.*, vol. 5, pp. 4872–4877, 2013.
61. S. Shi, C. Chen, and D. W. Prather, "Plane-wave expansion method for calculating band structure of photonic crystal slabs with perfectly matched layers," *J. Opt. Soc. Am. A.*, vol. 21, pp. 1769–1775, 2004.
62. Y.-J. Gao, H.-W. Yang, and G.-B. Wang, "A research on the electromagnetic properties of plasma photonic crystal based on the symplectic finite-difference time-domain method," *Optik*, vol. 127, pp. 1838–1841, 2016.
63. A. Madani and S. R. Entezar, "Surface polaritons of one-dimensional photonic crystals containing graphene monolayers," *Superlatt. and Microstruct.*, vol. 75, pp. 692–700, 2014.
64. Y. Li, L. Qi, J. Yu, Z. Chen, Y. Yao, and X. Liu, "One-dimensional multiband terahertz graphene photonic crystal filters," *Opt. Mater. Express*, vol. 7, pp. 1228–1239, 2017.
65. F. H. L. Koppens, D. E. Chang, and F. J. Garcia de Abajo, "Graphene plasmonics: a platform for strong light-matter interactions," *Nano Lett.*, vol. 11, pp. 3370–3377, 2011.
66. N. Ouchani, D. Bria, B. Djafari-Rouhani, and A. Nougaoui, "Defect modes in one-dimensional anisotropic photonic crystal," *J. Appl. Phys.*, vol. 106, Article 113107, 2009.
67. N. Ouchani, A. El-Moussaouy, H. Aynaou, Y. El-Hassouani, E. H. El-Boudouti, and B. Djafari-Rouhani, "Optical transmission properties of an anisotropic defect cavity in one-dimensional photonic crystal," *Phys. Lett. A.*, vol. 382, pp. 231–240, 2018.
68. M. Pourmand and P. K. Choudhury, "Wideband THz filtering by graphene-over-dielectric periodic structures with and without MgF_2 defect layer," *IEEE Access*, vol. 8, pp. 137385–137394, 2020.
69. H. Butt, Q. Dai, P. Farah, T. Butler, et al., "Metamaterial high pass filter based on periodic wire arrays of multiwalled carbon nanotubes," *Appl. Phys. Lett.*, vol. 97, Article 163102, 2010.
70. Q. Zhang, Q. Ma, S. Yan, F. Wu, X. He, and J. Jiang, "Tunable terahertz absorption in graphene-based metamaterial," *Opt. Commun.*, vol. 353, pp. 70–75, 2015.
71. W. Su and B. Chen, "Graphene-based tunable terahertz filter with rectangular ring resonator containing double narrow gaps," *Pramana – J. Phys.*, vol. 89, Article 37, 2017.
72. M. A. Baqir, P. K. Choudhury, and M. J. Mughal, "Gold nanowires-based hyperbolic metamaterial multiband absorber operating in the visible and near-infrared regimes," *Plasmonics*, vol. 14, pp. 485–492, 2019.
73. F. Qin and H.-X. Peng, "Ferromagnetic microwires enabled multifunctional composite materials," *Prog. In Mater. Sci.*, vol. 58, pp. 183–259, 2013.
74. P. K. Choudhury, K. K. Dey, and S. Basu, "Micro- and nanoscale structures/systems and their applications in certain directions: A brief review," In S. M. Musa (Ed.), *Nanoscale Spectroscopy with Applications*, UK: CRC Press, 2013.
75. J. D. Joannopoulos, P. R. Villeneuve, and S. Fan, "Photonic crystals: putting a new twist on light," *Nature*, vol. 386, pp. 143–149, 1997.
76. A. Pander, K. Takano, A. Hatta, M. Nakajima, and H. Furuta, "The influence of the inner structure of CNT forest metamaterials in the infrared regime," *Diamond and Related Mater.*, vol. 80, pp. 99–107, 2017.

77. M. Danaeifar, N. Granpayeh, A. Mohammadi, and A. Setayesh, "Graphene-based tunable terahertz and infrared band-pass filter," *Appl. Opt.*, vol. 52, pp. E68–E72, 2013.
78. H. Zhuang, F. Kong, K. Li, and S. Sheng, "Plasmonic bandpass filter based on graphene nanoribbon," *Appl. Opt.*, vol. 52, pp. 2558–2564, 2015.
79. B. Shi, W. Cai, X. Zhang, Y. Xiang, et al., "Tunable band-stop filters for graphene plasmons based on periodically modulated graphene," *Sci. Repts.*, vol. 6, Article 26796, 2016.
80. M. A. Baqir, P. K. Choudhury, T. Fatima, and A.-B. M. A. Ibrahim, "Graphene-over-graphite-based metamaterial structure as optical filter in the visible regime," *Optik*, vol. 180, pp. 832–839, 2019.
81. T. Ha, T. Enderle, D. F. Ogletree, D. S. Chemla, et al., "Probing the interaction between two single molecules: fluorescence resonance energy transfer between a single donor and a single acceptor," *Proc. Natl. Acad. Sci.*, vol. 93, pp. 6264–6268, 1996.
82. S. Nie and S. R. Emory, "Probing single molecules and single nanoparticles by surface-enhanced Raman scattering," *Science*, vol. 275, pp. 1102–1106, 1997.
83. M. I. Stockman, S. V. Faleev, and D. J. Bergman, "Localization versus delocalization of surface plasmons in nanosystems: Can one state have both characteristics?," *Phys. Rev. Lett.*, vol. 87, p. 167401, 2001.
84. E. Prodan, C. Radloff, N. J. Halas, and P. Nordlander, "A hybridization model for the plasmon response of complex nanostructures," *Science*, vol. 302, pp. 419–422, 2003.
85. J. Chen, W. Fan, T. Zhang, C. Tang, et al., "Engineering the magnetic plasmon resonances of metamaterials for high-quality sensing," *Opt. Express*, vol. 25, pp. 3675–3681, 2017.
86. J. Chen, H. Nie, C. Peng, S. Qi, et al., "Enhancing the magnetic plasmon resonance of three-dimensional optical metamaterials via strong coupling for high-sensitivity sensing," *J. Lightwave Technol.*, vol. 36, pp. 3481–3485, 2018.
87. J. Zhang, G. Hernandez, and Y. Zhu, "Slow light with cavity electromagnetically induced transparency," *Opt. Lett.*, vol. 33, pp. 46–48, 2008.
88. R. W. Boyd, "Material slow light and structural slow light: Similarities and differences for nonlinear optics," *J. Opt. Soc. Am. B.*, vol. 28, pp. A38–A44, 2011.
89. V. M. Acosta, K. Jensen, C. Santori, D. Budker, and R. G. Beausoleil, "Electromagnetically induced transparency in a diamond spin ensemble enables all-optical electromagnetic field sensing," *Phys. Rev. Lett.*, vol. 110, Article 213605, 2013.
90. J. Gu, R. Singh, X. Liu, X. Zhang, et al., "Active control of electromagnetically induced transparency analogue in terahertz metamaterials," *Nature Commun.*, vol. 3, Article 1151, 2012.
91. Y. Francescato, V. Giannini, and S. A. Maier, "Plasmonic systems unveiled by Fano resonances," *ACS Nano*, vol. 6, pp. 1830–1838, 2012.
92. X. Liu, J. Gu, R. Singh, Y. Ma, et al., "Electromagnetically induced transparency in terahertz plasmonic metamaterials via dual excitation pathways of the dark mode," *Appl. Phys. Lett.*, vol. 100, Article 131101, 2012.
93. W. Cao, R. Singh, I. A. I. Al-Naib, M. He, et al., "Low-loss ultra-high-Q dark mode plasmonic Fano metamaterials," *Opt. Lett.*, vol. 37, Article 3366, 2012.
94. H. Schmidt and A. Imamoglu, "Giant Kerr nonlinearities obtained by electromagnetically induced transparency," *Opt. Lett.*, vol. 21, pp. 1936–1938, 1996.
95. X. Zhou, X. Ling, H. Luo, and S. Wen, "Identifying graphene layers via spin hall effect of light," *Appl. Phys. Lett.*, vol. 101, Article 251602, 2012.

96. I. O. Zolotovskii, Y. S. Dadoenkova, S. G. Moiseev, A. S. Kadochkin, et al., "Plasmon-polariton distributed-feedback laser pumped by a fast drift current in graphene," *Phys. Rev. A.*, vol. 97, Article 053828, 2018.
97. B. Xu, C. Gu, Z. Li, and Z. Niu, "A novel structure for tunable terahertz absorber based on graphene," *Opt. Express*, vol. 21, pp. 23803–23811, 2013.
98. K. Yang, S. Liu, S. Arezoomandan, A. Nahata, and B. Sensale-Rodriguez, "Graphene-based tunable metamaterial terahertz filters," *Appl. Phys. Lett.*, vol. 105, Article 093105, 2014.
99. W. Zhu, F. Xiao, M. Kang, D. Sikdar, and M. Premaratne, "Tunable terahertz left-handed metamaterial based on multi-layer graphene-dielectric composite," *Appl. Phys. Lett.*, vol. 104, Article 051902, 2014.
100. X. He, "Tunable terahertz graphene metamaterials," *Carbon*, vol. 82, pp. 229–237, 2015.
101. S. H. Lee, M. Choi, T.-T. Kim, S. Lee, et al., "Switching terahertz waves with gate-controlled active graphene metamaterials," *Nature Mater.*, vol. 11, pp. 936–941, 2012.
102. F. Valmorra, G. Scalari, C. Maissen, W. Fu, et al., "Low-bias active control of terahertz waves by coupling large-area CVD graphene to a terahertz metamaterial," *Nano Lett.*, vol. 13, pp. 3193–3198, 2013.
103. R. Degl'Innocenti, D. S. Jessop, Y. D. Shah, J. Sibik, et al., "Low-bias terahertz amplitude modulator based on split-ring resonators and graphene," *ACS Nano*, vol. 8, pp. 2548–2554, 2014.
104. P. Q. Liu, I. J. Luxmoore, S. A. Mikhailov, N. A. Savostianova, et al., "Highly tunable hybrid metamaterials employing split-ring resonators strongly coupled to graphene surface plasmons," *Nature Commun.*, vol. 6, Article 8969, 2015.
105. M. Amin, M. Farhat, and H. Bağci, "A dynamically reconfigurable Fano metamaterial through graphene tuning for switching and sensing applications," *Sci. Repts.*, vol. 3, Article 2105, 2013.
106. Y. Zhang, T. Li, B. Zeng, H. Zhang, et al., "A graphene based tunable terahertz sensor with double Fano resonances," *Nanoscale*, vol. 7, Article 12682, 2015.
107. X. Zhao, C. Yuan, W. Lv, S. Xu, and J. Yao, "Plasmon-induced transparency in metamaterial based on graphene and split-ring resonators," *IEEE Photon. Technol. Lett.*, vol. 27, Article 1321, 2015.
108. K.-J. Yee, J.-H. Kim, M. H. Jung, B. H. Hong, and K.-J. Kong, "Ultrafast modulation of optical transitions in monolayer and multilayer graphene," *Carbon*, vol. 49, pp. 4781–4785, 2011.
109. M. A. Baqir, A. Farmani, P. K. Choudhury, T. Younas, et al., "Tunable plasmon induced transparency in graphene and hyperbolic metamaterial-based structure," *IEEE Photon. J.*, vol. 11, Article 4601510, 2019.
110. S. Berthier and J. Lafait, "Effective medium theory: mathematical determination of the physical solution for the dielectric constant," *Opt. Commun.*, vol. 33, pp. 303–306, 1980.
111. A. D. Rakić, A. B. Djurišić, J. M. Elazar, and M. L. Majewski, "Optical properties of metallic films for vertical-cavity optoelectronic devices," *Appl. Opt.*, vol. 37, pp. 5271–5283, 1998.
112. M. A. Baqir and P. K. Choudhury, "Toward filtering aspects of silver nanowire-based hyperbolic metamaterial," *Plasmonics*, vol. 13, pp. 2015–2020, 2018.
113. M. A. Baqir and P. K. Choudhury, "Design of hyperbolic metamaterial-based absorber comprised of Ti nanoshperes," *IEEE Photon. Technol. Lett.*, vol. 31, pp. 735–738, 2019.

114. K. V. Sreekanth, P. Mahalakshmi, S. Han, M. S. Mani Rajan, P. K. Choudhury, and R. Si, "Brewster mode-enhanced sensing with hyperbolic metamaterial," *Adv. Opt. Mater.*, vol. 7, Article 1900680, 2019.
115. A. N. Tzonev, G. G. Tsutsumanova, and S. C. Russev "Conditions for loss compensation of surface plasmon polaritons in a metal layer surrounded by an active dielectric: an analytic solution," *Opt. Quantum Electron.*, vol. 52, Article 313, 2020.
116. Y. Fan, C. Guo, Z. Zhu, W. Xu, et al., "Monolayer-graphene-based broadband and wide-angle perfect absorption structures in the near infrared," *Sci. Repts.*, vol. 8, Article 13709, 2018.
117. S. A. Nulli, M. S. Ukhtary, and R. Saito, "Significant enhancement of light absorption in undoped graphene using dielectric multilayer system," *Appl. Phys. Lett.*, vol. 112, Article 073101, 2018.
118. M. Ghasemi, P. K. Choudhury, and M. A. Baqir, "On the double nano-coned graphene metasurface-based multiband CIC absorber," *Plasmonics*, vol. 14, Article 1189, 2019.
119. I. S. Nefedov, C. A. Valagiannopoulos, and L. A. Melnikov,"Perfect absorption in graphene multilayers," *J. Opt.*, vol. 15, Article 114003, 2013.
120. Y. C. Zhang, Y. T. Fang, and Q. Cai, "Broad band absorber based on cascaded metamaterial layers including graphene," *Waves in Random and Complex Med.*, vol. 28, pp. 287–299, 2018.
121. O. Levy and D. Stroud, "Maxwell Garnett theory for mixtures of anisotropic inclusions: application to conducting polymers," *Phys. Rev. B.*, vol. 56, pp. 8035–8046, 1997.
122. A. Tittl, A. K. U. Michel, M. Schäferling, X. Yin, et al., "A switchable mid-infrared plasmonic perfect absorber with multispectral thermal imaging capability," *Adv. Mater.*, vol. 27, pp. 4597–5603, 2015.
123. M. Wuttig, H. Bhaskaran, and T. Taubner, "Phase-change materials for non-volatile photonic applications," *Nature Photon.*, vol. 11, pp. 465–476, 2017.
124. K. V. Sreekanth, S. Han, and R. Singh, "$Ge_2Sb_2Te_5$-based tunable perfect absorber cavity with phase singularity at visible frequencies," *Adv. Mater.*, vol. 30, Article 1706696, 2018.
125. S. Abdollahramezani, O. Hemmatyar, H. Taghinejad, A. Krasnok, et al., "Tunable nanophotonics enabled by chalcogenide phase-change materials," *Nanophotonics*, vol. 9, pp. 1189–1241, 2020.
126. J. B. Li and J. S. Li, "Research on a temperature controlled terahertz wave absorber based on VO_2 media," *Electron. Comp. and Mat.*, vol. 100, pp. 66–69, 2014.
127. H. Kocer, S. Butun, E. Palacios, Z. Liu, et al., "Intensity tunable infrared broadband absorbers based on VO_2 phase transition using planar layered thin films," *Sci. Repts.*, vol. 5, Article 13384, 2015.
128. F. Lu, Q. Tan, Y. Ji, Q. Guo, et al., "A novel metamaterial inspired high-temperature microwave sensor in harsh environments," *Sensors*, vol. 18, Article 2879, 2018.
129. R. Zhang, Q. S. Fu, C. Y. Yin, C. L. Li, et al., "Understanding of metal-insulator transition in VO_2 based on experimental and theoretical investigations of magnetic features," *Sci. Repts.*, vol. 8, Article 17093, 2018.
130. N. Inomata, L. Pan, M. Toda, and T. Ono, "Microfabricated vanadium oxide resonant thermal sensor with high temperature coefficient of resonant frequency," *Proc. 29th Intl. Conf. on Micro Electro Mechanical Systems (MEMS)*, Shanghai, China, pp. 1042–1045, 2016.

131. M. Yi, C. Lu, Y. Gong, Z. Qi, and Y. Cui, "Dual-functional sensor based on switchable plasmonic structure of VO_2 nano-crystal films and Ag nanoparticles" *Opt. Express*, vol. 22, pp. 29627–29635, 2014.
132. K. K. Dey, D. Bhatnagar, A. K. Srivastava, M. Wan, et al., "VO_2 nanorods for efficient performance in thermal fluids and sensors," *Nanoscale*, vol. 7, pp. 6159–6172, 2015.
133. M. A. Baqir and P. K. Choudhury, "On the VO_2 metasurface-based temperature sensor," *J. Opt. Soc. Am. B: Opt. Phys.*, vol. 36, pp. F123–F130, 2019.
134. Y. Chen, X. Li, Y. Sonnefraud, A. I. Fernández-Domínguez, et al., "Engineering the phase front of light with phase-change material based planar lenses," *Sci. Repts.*, vol. 5, Article 8660, 2015.
135. B. Trevon, I. Kim, and J. Rho, "Moth-eye shaped on-demand broadband and switchable perfect absorbers based on vanadium dioxide," *Sci. Repts.*, vol. 10, Article 4522, 2020.
136. P. Hosseini, C. Wright, and H. Bhaskaran, "An optoelectronic framework enabled by low-dimensional phase-change films," *Nature*, vol. 511, pp. 206–211, 2014.
137. F. F. Schlich, P. Zalden, A. M. Lindenberg, and R. Spolenak, "Color switching with enhanced optical contrast in ultrathin phase-change materials and semiconductors induced by femtosecond laser pulses," *ACS Photon.*, vol. 2, pp. 178–182, 2015.
138. M. Rudé, V. Mkhitaryan, A. E. Cetin, T. Miller, et al., "Ultrafast and broadband tuning of resonant optical nanostructures using phase-change materials," *Adv. Opt. Mater.*, vol. 4, Article 201600079, 2016.
139. Z. M. Liu, Y. Li, J. Zhang, Y. Q. Huang, et al., "A tunable metamaterial absorber based on VO_2/W multilayer structure," *IEEE Photon. Technol. Lett.*, vol. 29, pp. 1967–1970, 2017.
140. A. Chen and Z. Song, "Tunable isotropic absorber with phase change material VO_2," *IEEE Trans. Nanotechnol.*, vol. 19, pp. 197–200, 2020.
141. H. Taghinejad, S. Abdollahramezani, A. A. Eftekhar, T. Fan, et al., "ITO-based microheaters for reversible multi-stage switching of phase-change materials: Towards miniaturized beyond-binary reconfigurable integrated photonics," *Opt. Express*, vol. 29, pp. 20449–20462, 2021.
142. F. Giubileo and A. D. Bartolomeo, "The role of contact resistance in graphene field-effect devices," *Prog. in Surf. Sci.*, vol. 92, pp. 143–175, 2017.
143. K. Huang, J. Liu, L. Tan, J. Zuo, and L. Fu, "Ultrahigh temperature graphene molecular heater," *Adv. Mater. Interf.*, vol. 5, Article 1701299, 2018.
144. F. Luo, F. Yansong, P. Gang, X. Shuigang et al., "Graphene thermal emitter with enhanced Joule heating and localized light emission in air," *ACS Photon.*, vol. 6, pp. 2117–2125, 2019.
145. B. Gholipour, J. Zhang, K. F. MacDonald, D. W. Hewak, and N. I. Zheludev, "An all-optical, non-volatile, bidirectional, phase-change meta-switch," *Adv. Mater.*, vol. 25, pp. 3050–3054, 2013.
146. J. Wang, L. Yang, Z. Hu, W. He, and G. Zheng, "Analysis of graphene-based multilayer comb-like absorption system based on multiple waveguide theory," *IEEE Photon. Technol. Lett.*, vol. 31, pp. 561–564, 2019.
147. A. Lochbaum, A. Dorodnyy, U. Koch, S. M. Koepfli, et al., "Compact mid-infrared gas sensing enabled by an all-metamaterial design," *Nano Lett.*, vol. 20, pp. 4169–4176, 2020.
148. C. Ríos, Y. Zhang, S. Deckoff-Jones, H. Li, J. B. Chou et al., "Reversible switching of optical phase change materials using graphene microheaters," *2019 Conf. on Lasers and Electro-Optics (CLEO)*, vol. 1–2, 2019.

149. A. Forouzmand and H. Mosallaei, "Dynamic beam control via Mie-resonance based phase-change metasurface: A theoretical investigation," *Opt. Express*, vol. 26, pp. 17948–17963, 2018.
150. M. S. A. Rahman, S. C. Mukhopadhyay, P.-L. Yu, J. Goicoechea, et al., "Detection of bacterial endotoxin in food: New planar interdigital sensors based approach," *J. Food Eng.* vol. 114, pp. 346–360, 2013.
151. A. Nag, A. I. Zia, X. Li, S. C. Mukhopadhyay, and J. Kosel, "Novel sensing approachfor LPG leakage detection—Part II: Effects of particle size, composition, and coating layer thickness," *IEEE Sensors J.*, vol. 16, pp. 1088–1094, 2016.
152. A. Nag, S. C. Mukhopadhyay, and J. Kosel, "Flexible carbon nanotubenanocomposite sensor for multiple physiological parameter monitoring," *Sensors and Actuat. A Phys.*, vol. 251, pp. 148–155, 2016.
153. P. R. Griffiths and J. A. De Haseth, *Fourier Transform Infrared Spectrometry*, New York: Wiley, 2007.
154. D. Rodrigo, O. Limaj, D. Janner, D. Etezadi, et al. "Mid-infrared plasmonic biosensing with graphene," *Science*, vol. 349, pp. 165–168, 2015.
155. Y. Wu, B. Yao, A. Zhang, Y. Rao, et al. "Graphene-coated microfiber Bragg grating for high-sensitivity gas sensing," *Opt. Lett.*, vol. 39, pp. 1235–1237, 2014.
156. M. Amin, M. Farhat, and H. Bagci, "An ultra-broadband multilayered graphene absorber," *Opt. Express*, vol. 21, pp. 29938–29948, 2013.
157. Y. Cai, Y. Guo, Y. Zhou, X. Huang, et al., "Tunable dual-band terahertz absorber with all-dielectric configuration based on graphene," *Opt. Express*, vol. 28, pp. 31524–31534, 2020.
158. V. B. Svetovoy, P. J. van Zwol, and J. Chevrier, "Plasmon enhanced nearfield radiative heat transfer for graphene covered dielectrics," *Phys. Rev. B.*, vol. 85, Article 155418, 2012.
159. O. Ilic, M. Jablan, J. D. Joannopoulos, I. Celanovic, et al., "Near-field thermal radiation transfer controlled by plasmons in graphene," *Phys. Rev. B.*, vol. 85, Article 155422, 2012.
160. W. Li and Y. Cheng, "Dual-band tunable terahertz perfect metamaterial absorber based on strontium titanate (STO) resonator structure," *Opt. Commun.*, vol. 462, Article 125265, 2020.
161. B. Dong, H. Ma, J. Wang, P. Shi, et al., "Thermally tunable THz metamaterial frequency selective surface based on BST thin film," *J. Phys. D: Appl. Phys.*, vol. 52, Article 045301, 2019.
162. X. He, F. Lin, F. Liu, and W. Shi, "Tunable strontium titanate terahertz all-dielectric metamaterials," *J. Phys. D: Appl. Phys.*, vol. 53, Article 155105, 2020.
163. E. M. Sheta, P. K. Choudhury, and A.-B. M. A. Ibrahim, "Pixelated graphene-strontium titanate metamaterial supported tunable dual-band temperature sensor," *Opt. Mater.*, vol. 117, Article 111197, 2021.
164. P. Kumar, A. Lakhtakia, and P. K. Jain, "Graphene pixel-based polarization-insensitive metasurface for almost perfect and wideband terahertz absorption," *J. Opt. Soc. Am. B*, vol. 36, pp. F84–F88, 2019.
165. P. Kumar, A. Lakhtakia, and P. K. Jain, "Tricontrollable pixelated metasurface for absorbing terahertz radiation," *Appl. Opt.*, vol. 58, pp. 9614–9623, 2019.
166. P. Kumar, A. Lakhtakia, and P. K. Jain, "Tricontrollable pixelated metasurface for stopband for terahertz radiation," *J. Electromagn. Waves and Appl.*, vol. 34, pp. 2065–2078, 2020.
167. M. Sandtke and L. Kuipers, "Slow guided surface plasmons at telecom frequencies," *Nature Photon.*, vol. 1, pp. 573–576, 2007.

168. L. Yang, C. Min, and G. Veronis, "Guided subwavelength slow-light mode supported by a plasmonic waveguide system," *Opt. Lett.*, vol. 35, pp. 4184–4186, 2010.
169. Y. Huang, C. Min, and G. Veronis, "Subwavelength slow-light waveguides based on a plasmonic analogue of electromagnetically induced transparencies," *Appl. Phys. Lett.*, vol. 99, Article 143117.
170. H. Lu, C. Zheng, Q. Zhang, X. Liu, et al., "Graphene-based active slow surface plasmon polaritons," *Sci. Repts.*, vol. 5, Article 8443, 2015.
171. K. Okamoto, *Fundamental of Optical Waveguides*, London: Academic Press, 2000.

10 Asymmetric Split-H Based Metasurfaces for Identification of Organic Molecules

Ili F. Mohamad Ali Nasri[1,2,3],
Ifeoma G. Mbomson[1], *Marc Sorel*[1],
Nigel P. Johnson[1], *Caroline Gauchotte-Lindsay*[2],
and Richard M. De La Rue[1]

[1]School of Engineering, University of Glasgow, Glasgow, United Kingdom

[2]Environmental Engineering, Division of Infrastructure and Environment, School of Engineering, University of Glasgow, Glasgow, United Kingdom

[3]University Kuala Lumpur Malaysian Institute of Aviation Technology (UniKL MIAT), Jalan Jenderam Hulu, Jenderam Hulu, Selangor, Malaysia

10.1 INTRODUCTION

A plasmonic metamaterial is one that, through interaction with electromagnetic radiation, produces artificial optical properties based on surface plasmon effects. The resonant oscillation of free electron gas in metals formed by the interactions of incident light with dielectrics and metals, known as surface plasmon resonance, is currently being used widely for biosensor applications. Sensing applications benefit from the tunability features of plasmonic resonance produced in the many metamaterials used for various vibrational modes in the region of IR spectroscopy [1–8]. The term *metasurface* is now widely used for metamaterials that are created as metallic patterns on the surface of substrates via the lithographic processes that are typical of planar fabrication technology. There are many molecular vibrations that occur in the mid-IR spectral region [9].

Organic molecules are complex chemical compounds of which many are found in living systems. Many of the chemical bond vibrations in these molecules exhibit stretching band spectra that are in the IR, typically at wavelengths in the range between 2.5 μm and 12.00 μm [4,5,10,11]. Many organic compounds, for example, 17β-estradiol (also known as E2), consist largely of atoms, such as hydrogen and

DOI: 10.1201/9781003050162-10

carbon, that form strong molecular bonds. Significant characteristics for identifying the molecular bonds spectrally are through vibrational intensity and the positions of particular spectral features [12]. In this chapter, a numerically designed and fabricated metallic structure-based metasurface has been used to detect the significant features of C–H bonds in the E2 analyte, which is a type of human hormone responsible for the development of sex organs – with a molecular mass of 272.4 g/mol [13,14]. Furthermore, this hormone can be found naturally in the environment through human excreta or through some anthropogenic activities [15]. The high demand for E2 in clinical analyses and in determining the performance of human systems, as well as its occurrence in water treatment, has motivated interdisciplinary research groups to investigate and report on various sensor techniques [13,16,17]. We report on asymmetric split-H (ASH) structure-based metasurfaces that use plasmon resonances for identification and enhancement of the molecular vibrations of C–H bonds in E2.

10.2 THE ASYMMETRIC SPLIT-H STRUCTURE

Several forms of metallic metasurfaces have been reported in the literature, for example, surfaces based on arrays of split-ring resonators, U-shaped, nanorod, chiral, dipole and metasurface holographic movies [18–22]. In 2014, Yuan et al. studied asymmetric H-shaped structures that had no gap in the bar of the H-pattern [23]. Our group has pointed out that the plasmon resonance behavior produced by the basic structure can be improved by introducing a small gap in the bar of the H-structure. Basic metamaterial structures that include a nanometer-scale gap and multiple sharp edges have been demonstrated to form hot spots that can produce enhanced electric (*E*-) fields at or near resonance [24–26]. The *E*-field enhancement from the hot spots can greatly exceed the strength of the incident field [1,7]. In this chapter, the individual asymmetric split-H (ASH) elements in the arrays have both a 50 nm gap and multiple sharp edges of the arms, as shown in Fig. 10.1a,b. Fig. 10.1a also shows a normal incidence radiation on the plane of ASH arrays, with the direction of propagation constant (*K*) orthogonal to the periodic arrangement of a_x-

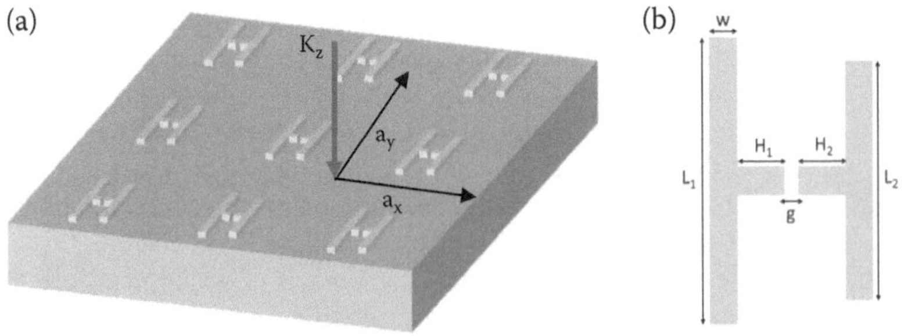

FIGURE 10.1 Schematic diagram of (a) gold ASH arrays deposited on a substrate, showing the periodic arrangements along the a_x- and a_y- axes, with the incident light propagation along the *z*-axis, and (b) detailed schematic of a single ASH structure.

and a_y- axes of the ASH structures. An appropriate optimization of the design and periodic arrangement of arrays has been demonstrated to exhibit high sensitivity [7,24–26]. The ASH design is composed of two conventional gold (Au) dipole nanostructures, separated by a nanometre-scale gap and two asymmetric vertical arms (Fig. 10.1). Metasurfaces based on the arrays of asymmetric split-ring resonators (ASRRs) have been introduced because of their *trapped-mode* resonance feature, as well as exhibition of a double-reflection *Fano*-resonance [27]. The combined presence of both a trapped mode and two *Fano*-resonances is particularly relevant for the potential use of these structures in sensing applications. The arm lengths of our ASH design are tuned to produce plasmon resonance peaks within the mid-infrared region.

10.3 NUMERICAL MODELING AND NANOFABRICATION

10.3.1 Numerical Modeling

Three-dimensional finite-difference time-domain (FDTD) software was used for the modeling and numerical simulation of the ASH arrays. The simulations enable optimization of the geometry – for example, the nanometre-scale gap dimensions, sharp edges, arm lengths and the periodic spacing arrangements in arrays. The FDTD method used covers a wide range of wavelengths in a single simulation run. The material properties were included in the simulations by applying the Drude model for the optical properties of a gold-film structure deposited on top of a fused silica substrate, with material data from Palik embedded in the software [28]. The basics of the FDTD time-stepping relationship show that the updated value of E-field in time is dependent on the stored value of E-field and on the numerical curl of the local distribution of the magnetic (H-) field in space [29]. Similarly, the updated value of H-field in time is dependent on the stored value of H-field and on the numerical curl of the local distribution of E-field in space.

Our simulations have used periodic boundary conditions along the x- and y-axes, with equal periodicity, to simulate an infinite array in both directions. A perfectly matched layer (PML) was used for the z-axis, over the wavelength range from 2 μm to 8 μm. The ASH array structures were excited by a plane wave propagating along the z-direction orthogonal to the E- and H-fields components of the x- and y- axes, as shown in Fig. 10.1a. A transmission monitor was placed between the substrates and nanostructures – and the reflection monitor was placed above the gold nanostructures.

Three different split-H nanostructures – the symmetric split-H (SSH), the asymmetric split-H (ASH) and the double asymmetric split-H (D-ASH) – were modeled for comparison with the same gap size throughout. Fig. 10.2 shows the transmittance and reflectance spectra for both the orthogonal polarizations. The E_X polarization applies with the E-field across the gap in the bar of H-shaped structure, while the E_Y polarization applies with the electric field parallel to the vertical dipole. The SSH structure (Fig. 10.2a) has identical vertical arm lengths. When L_1 and L_2 are of the value 1.6 μm, the gap (g) is 50 nm and the horizontal bar (H) is 0.35 μm, the single resonance peak appears in both the polarizations at a

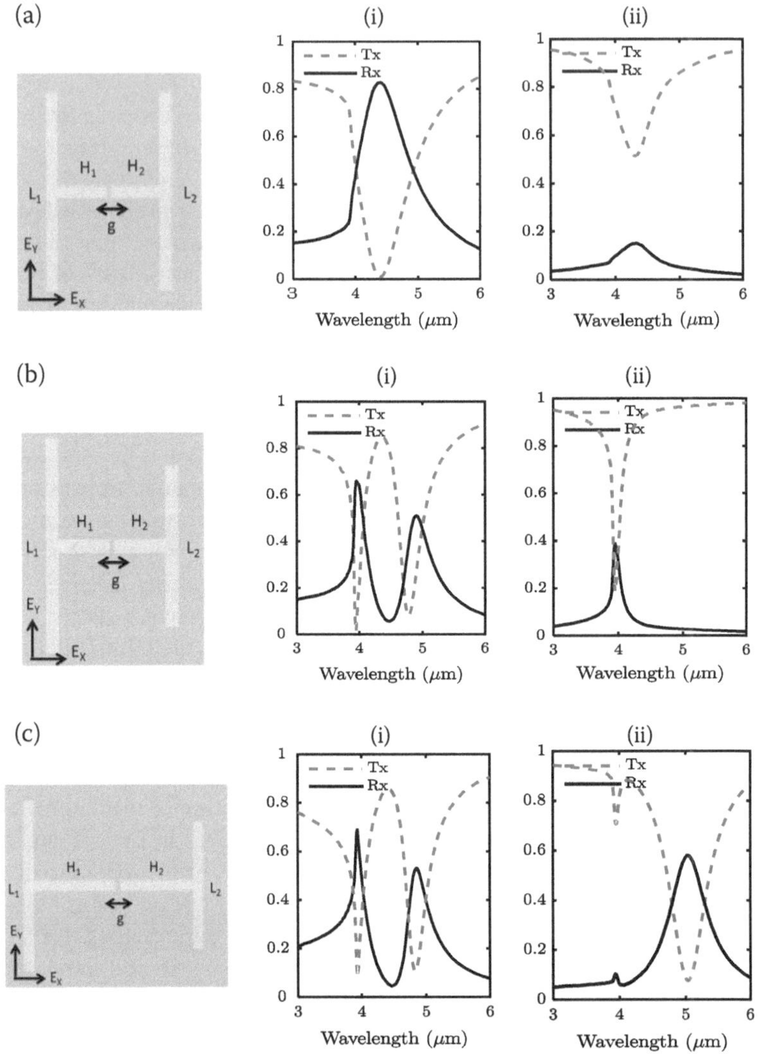

FIGURE 10.2 Schematic diagrams with transmittance and reflectance spectra from simulations of (a) SSH, (b) ASH and (c) D-ASH structures. The transmittance and reflectance spectra shown are for a 2.8 μm period arrangement in both (i) E_Y- and (ii) E_X- polarizations with reference to the vertical and horizontal arms.

wavelength of 4.40 μm. This behavior is due to the symmetric vertical arm length producing a single resonance peak. The nanostructures modeled were then modified to the ASH form (Fig. 10.2b) by changing the size of the vertical arm length L_2, keeping the arm length L_1 the same. By adjusting the arm length to introduce asymmetry into the structure, two separate vertical resonance peaks appear at wavelengths of 3.8 μm and 4.8 μm for the E_Y-polarized excitation.

Between the two resonances, the reflection trough at 4.40 μm with a Q-factor value of 5 can also be identified as exhibiting trapped mode characteristics.

Lastly, the D-ASH structures (Fig. 10.2c) were varied, for all the different arm lengths, in the vertical and horizontal directions. The obtained results show two separate resonance peaks produced in both polarizations. The asymmetric horizontal bar is H_1 = 0.8 μm and H_2 = 0.7 μm, and the sizes of L_1 and L_2 remain constant. The measurements show that, under the E_X-polarized excitation, double resonances are exhibited at the wavelengths of 3.8 μm and 5.00 μm, while under the E_Y-polarized excitation, resonances occur at the same wavelengths as for an ASH at 3.8 μm and 4.8 μm. Changing the two different arm lengths in the structure not only produces two distinct resonance peaks, but it also tunes the positions of the resonance peaks [30].

An additional simulation was performed to understand the coupling effect between the dipoles within an ASH with separate arm lengths, as shown in Fig. 10.3. The equal arm lengths within the individual ASH structure produce a single resonance that is similar to the resonance hybridization model that has been explored by Giessen et al. [31,32] and explained in other work [33,34]. The plasmonic dipole-like modes of the arm lengths L_1 (Fig. 10.3a) and L_2 (Fig. 10.3b) are excited at distinct frequencies, ω_1 and ω_2, respectively ($\hbar\omega$ represents the photon energy). The resonances occur at the respective wavelengths of 4.8 μm (dashed line) for the longer arm and 4.2 μm (dotted line) for the shorter arm, which correspond to the plasmonic eigenmodes $|\omega_1\rangle$ and $|\omega_2\rangle$. The interaction between these two modes results in new coupled modes as hybrid combinations of the plasmonic eigenmodes: a mode with the frequency $|\omega_+\rangle$ that corresponds to a

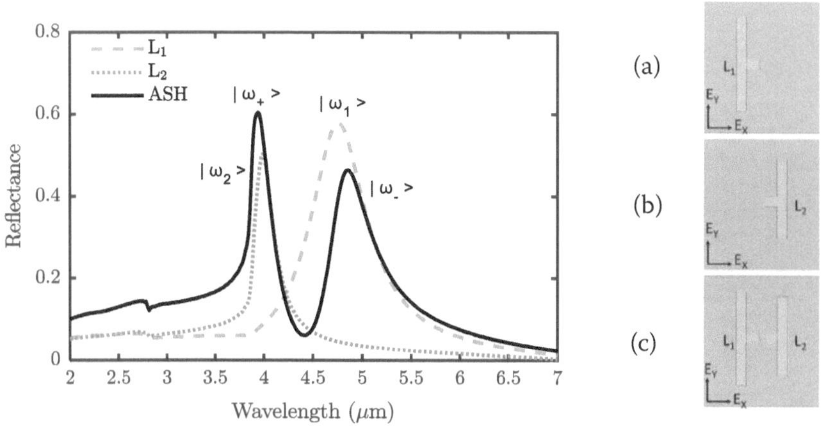

FIGURE 10.3 Schematic and simulation results for the component geometries of ASH structures. Illustration of structures with (a) the arm length L_1 of the ASH producing the resonance indicated by the green dashed line, (b) the arm length L_2 of the ASH segment producing the resonance indicated by the orange dotted line, and (c) the complete ASH producing the doubly resonant behavior indicated by the black line.

"symmetric" plasmonic mode of the combined structure (a higher energy mode), and a mode with frequency $|\omega_{-}\rangle$ that corresponds to an "anti-symmetric" plasmonic mode of the combined structure (a lower energy mode). The reflection dip at a wavelength of 4.4 μm is due to resonance hybridization between the two arms of ASHs that are positioned close to each other and produce two resonances: one at a wavelength of 3.8 μm and the other at a wavelength of 4.8 μm (solid black line).

Another illustration of the plasmon hybridization is shown in the schematic energy band diagram in Fig. 10.4. The hybrid resonances show a red shift for the arm length L_1 from 4.8 μm to 5.0 μm and a blue shift for the arm length L_2 from 4.2 μm to 3.8 μm. The coupling between the two vertical arm lengths L_1 and L_2 leads to the symmetric and asymmetric resonances $|\omega_{+}\rangle$ and $|\omega_{-}\rangle$. The resonance is red-shifted due to a reduced energy level (increasing wavelength). It is observable that the resonance $|\omega_1\rangle$ (dashed line) amplitude is much higher than the amplitude of $|\omega_2\rangle$ (dotted line), as demonstrated in Fig. 10.3. Because of the resonance hybridization, the amplitude of the symmetric resonance $|\omega_{+}\rangle$ is much higher than that of the asymmetric resonance $|\omega_{-}\rangle$. Furthermore, the resonance linewidth of the asymmetric resonance $|\omega_{-}\rangle$ is narrower than the linewidth of the longer arm resonance $|\omega_1\rangle$.

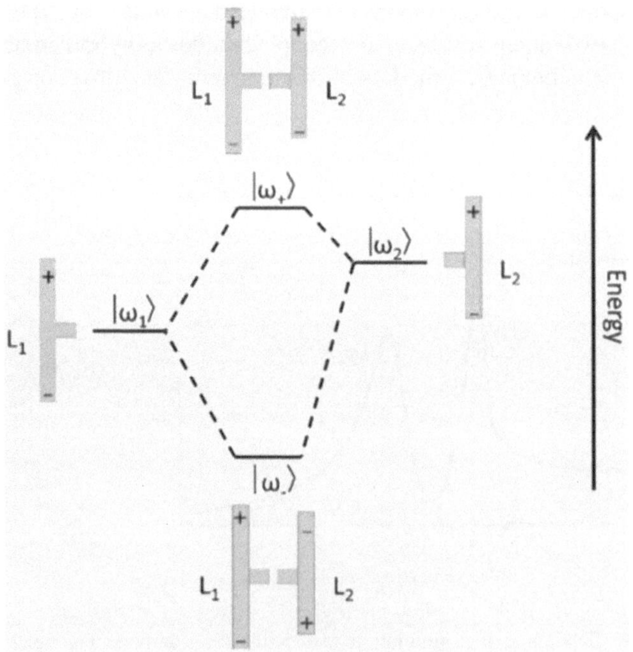

FIGURE 10.4 Energy level diagram showing schematically the resonance hybridization that occurs in the ASH structure. The coupling effect between the two dipole arms with lengths L_1 and L_2 produces two new modes, the higher energy "symmetric" mode and the lower energy "anti-symmetric" mode.

10.3.2 Nanofabrication on Fused Silica and Zinc Selenide Substrates

The arrays of ASHs were initially designed using the L-edit software, with the arm dimensions L_1 and L_2 being 1.1 µm and 0.9 µm, respectively, while the arms of the SSH structures in other arrays were chosen to be 1 µm, with the pattern widths (w) being 100 nm. Produced reflectance resonance peaks in the wavelength ranged from 2 µm to 8 µm to cover the region required for application. The thermal and optical properties of fused silica are superior to those of the other types of glass due to purity and simple composition. As a result, fused silica is widely used in laboratory equipment, sensor devices and in dielectric and semiconductor device fabrication. Applications of this material mostly exploit its wide transparency range within the electromagnetic spectrum, ranging from the ultraviolet, through the visible spectrum and some way into the IR regime. The ASH arrays were patterned on fused silica and single crystal zinc selenide substrates using electron beam lithography (EBL). Thorough cleaning of the substrates was performed using acetone, methanol and IPA, and then a bilayer of polymethyl methacrylate (PMMA) was spun on. The lift-off process was used after metallization of the pattern with 50 nm gold (Au) and 10 nm of titanium (Ti) for adhesion purposes. The same nanofabrication procedures were used as detailed in Mbomson and Mohamad Ali Nasri [14,26]. The used zinc selenide and fused silica substrates had thickness dimensions of 1 mm and 750 µm, respectively, and larger plates were then cleaved to nominally square areas of 12 mm^2. The benefits of using different substrate materials in relation to sensor applications are discussed in Section 10.4.2 of this chapter.

10.4 RESULTS AND DISCUSSION

The reflectance spectra of light from the fabricated gold ASH arrays have been used to characterize the plasmon resonance properties of the resulting metasurfaces. We have used the Fourier-transform infrared spectroscopy (FTIR) technique in the reflectance mode to observe the spectra. The mercury cadmium telluride (MCT) detector of the FTIR equipment was cooled with liquid nitrogen and the signal intensity was checked. The fabricated ASH sample was placed on the stage and positioned in the optical path. After the sample area was defined, an uncontaminated gold mirror was used for the reference measurement prior to measuring the ASH arrays. The experimental work focused on the production of improved distinct plasmon resonance peaks by varying the gap in the ASH bar and the periodic arrangement of the array. In this chapter, we present the results of measurements on the arrays of both the SSH and ASH arrays with 50 nm gaps and 2 µm periodic spacing along both the x-and the y-axes, as shown in the micrographs of Fig. 10.5.

10.4.1 Impact of Horizontal and Vertical Spacing of the Periodic Arrangement and Varying Gap of ASH Arrays

Work on the impact of spacing in periodically arranged dual-polarized ASH metasurface arrays has been described in Mbomson [35], including both the numerical simulations and FTIR measurements on the fabricated D-ASH arrays.

(a)

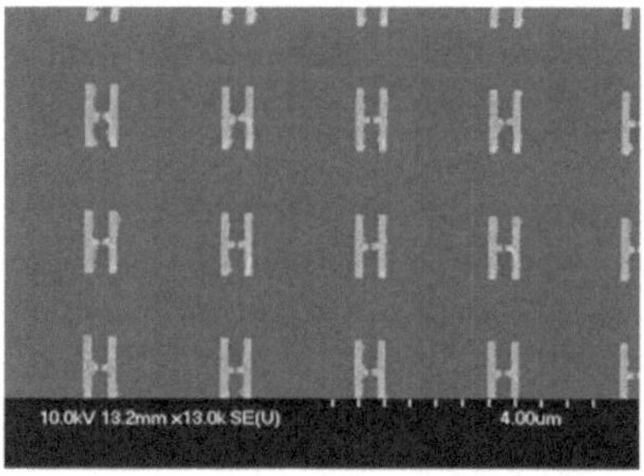

(b)

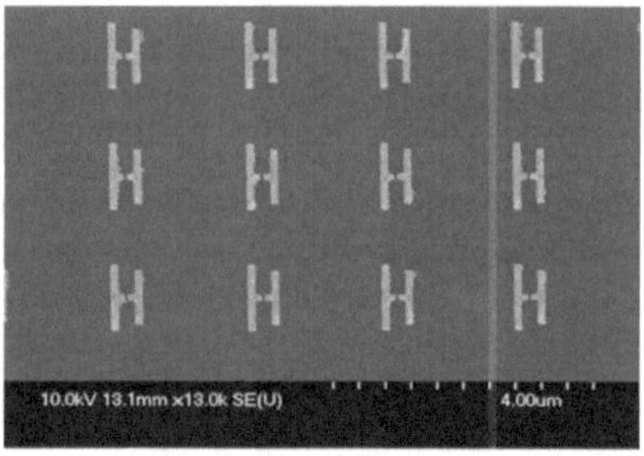

FIGURE 10.5 SEM image of (a) SSH, and (b) ASH arrays fabricated on fused silica substrate.

The asymmetric feature of ASH nanostructure was orthogonally designed in both the *x*- and *y*-axes of the electric mode. A *Q*-factor value of 26 was achieved both experimentally and in numerical simulation, through increase of the period of ASH array metasurface to a sufficiently large value [35]. This *Q*-factor value is comparably high to other work on metal film-based metasurfaces described in the literature [36]. The asymmetric nanostructures produce three distinct plasmonic resonance peaks that are strongly dependent on the polarization state of the incident electromagnetic waves. By applying a sufficiently large periodic spacing in the dual-polarization dependent asymmetric H-shaped (ASH) nanostructure array, a high *Q*-factor value can be achieved in the MIR region.

Variation of the gaps (g) in the bar-sections of dual-polarized ASH (D-ASH) arrays was also carried out (Fig. 10.2b). The sizes of the horizontal (*H*) bar section of the D-ASH structure were maintained constant at 0.875 μm and 0.675 μm (the structure formed by the gap in the horizontal bar-section), and the respective vertical (*L*) dipoles were of the lengths 1.6 μm and 1.4 μm. However, the gap size between the horizontal bars section, shown in Fig. 10.6, was varied from no gap up to 200 nm. The E_X polarization gave a variation in resonance peak position when the gap sizes were varied, as shown in Fig. 10.6b. When the ASH gap is closed, i.e., the gap is 0 nm, the total arm length of the horizontal bar is equal to 1.55 μm, and a single resonance appears at a wavelength of 7.5 μm due to the changes in the length of the horizontal bar section. The single resonance is attributed to the excitation of the surface plasmon along the vertical dipole being parallel to the E-field, as shown in Fig. 10.7. The peak positions are affected by the length, as discussed earlier in Section 10.3.1. When the gap was increased to 10 nm, the resonance was blue-shifted and two plasmon resonances were produced.

Fig. 10.7 shows the magnitude of E-plots of the *z*-component for D-ASH structures with varying gap over the range from 0 nm to 200 nm. The E-field was only along the horizontal asymmetric bar because the incident light was set to be polarized along the *x*-axis. The E-field monitor was placed in the center of the D-ASH structure, and the E-field maxima appear at the ends of the vertical dipoles or the horizontal bar. The dark blue color shows the minimum E-field intensity and dark red color shows the maximum E-field. It must be noted that the maximum brightness values on the two color scales are different in Fig. 10.7. When the gap in

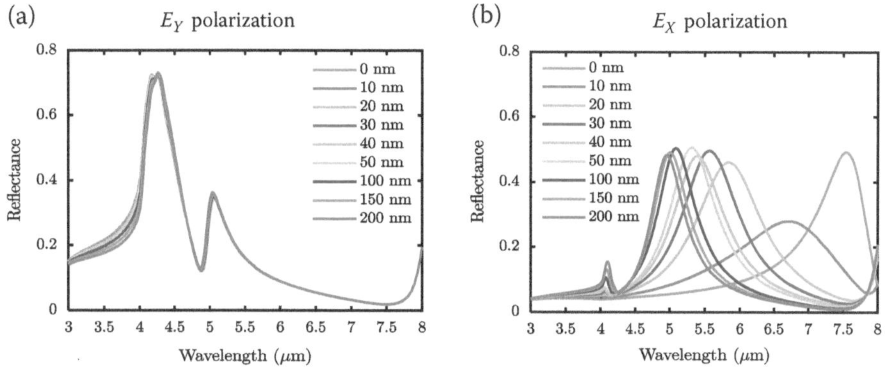

FIGURE 10.6 Effect on varying the gap sizes from 0 *nm* to 200 *nm* on the reflection spectra for (a) E_Y polarization, and (b) E_X polarization.

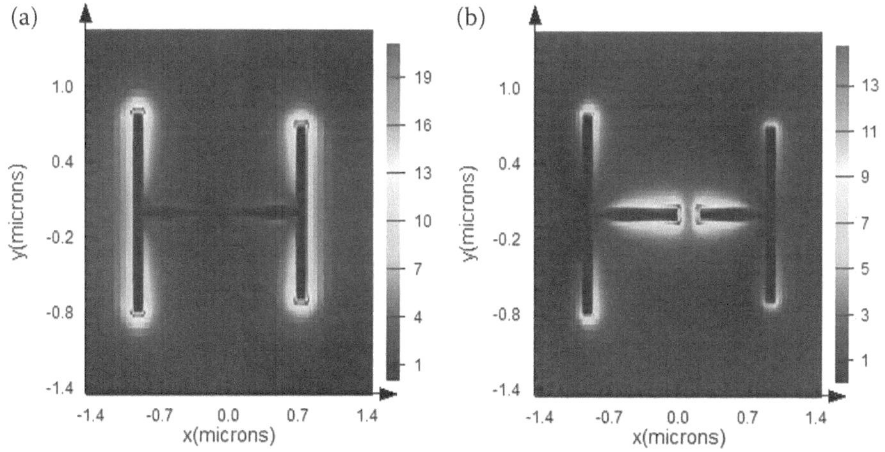

FIGURE 10.7 E-field plot for plasmonic resonance at a wavelength (a) 7.5 μm when the D-ASH has a 0 nm gap (i.e., no gap) in the bar section, and the plasmonic resonance at a wavelength (b) 4.28 μm with a 200 nm gap, as shown in reflectance spectra Fig. 4.2b.

the horizontal bar is 0 nm, large E-fields appear at the ends of the vertical arms, compared to the situation for 200 nm gaps, where the E-field maxima occur in the area around the gap. These areas form "hot spots", where the strong E-field, associated with the trapping of optical energy, produces enhanced absorption and resonance shifting in the presence of bio-material.

10.4.2 Effect of Different Substrates

The response of metamaterial nanostructures is not solely dependent on the size and geometric parameters of the patterned metallization, but also on the choice of the substrate material used. The criteria for choosing a suitable substrate include, in particular, the transmission range in the IR regime [28,37]. Generally, optical glasses have a refractive index (n) range between 1.4 and 2. One possible alternative substrate, single crystal zinc selenide, is transparent with a light-yellow colour – and has advantages for use with aqueous solutions. This is because it has a wide transmission wavelength range, from 0.6 μm up to 21 μm. In comparison, fused silica (SiO_2) substrates have a useful operating range from around 0.2 μm to 4.0 μm. Fused silica substrates, due to their low refractive index and transmittance range up to 4.0 μm, can be used in realizing metamaterial structures and metasurfaces with plasmonic resonances at shorter wavelengths (in the range from the visible through the near IR to shorter wavelengths in the mid IR regime). Both the substrate materials are highly transparent and insoluble in water, in contrast to other dielectric optical materials, such as calcium fluoride (CaF_2).

The ASH nanostructures were modeled with patterned gold metallization on the top of fused silica (Fig. 10.8) and single crystal zinc selenide (Fig. 10.9) substrates. The ASH structures were modeled using the same values on the two different planar

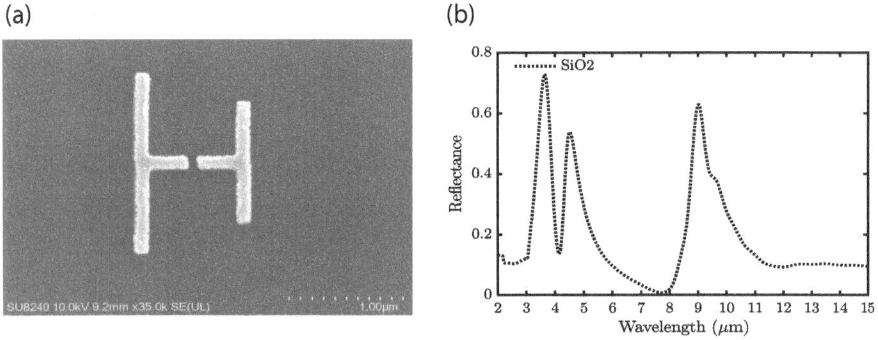

FIGURE 10.8 (a) SEM image of ASH on fused silica substrate. (b) Simulation results of ASH on fused silica substrate with the incident light polarized parallel with the asymmetric arms with lengths L_1 = 1.6 μm and L_2 = 1.2 μm.

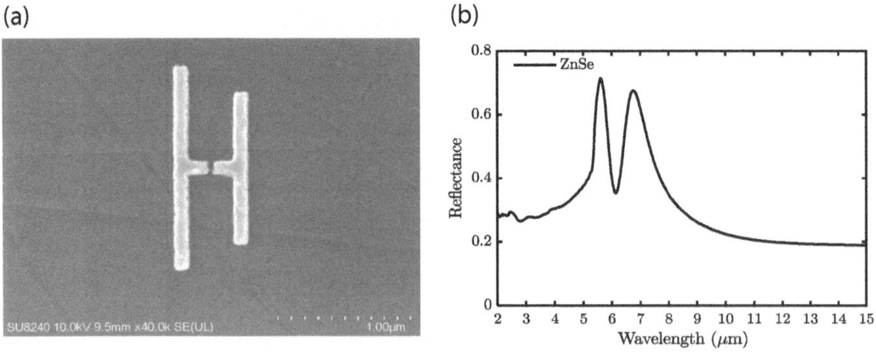

FIGURE 10.9 (a) SEM image of ASH on zinc selenide substrate. (b) Simulation results of ASH on zinc selenide substrate with the incident light polarized parallel with the asymmetric arms with lengths L_1 = 1.6 μm and L_2 = 1.2 μm.

substrates with the arm lengths L_1 and L_2 being 1.6 μm and 1.2 μm, respectively. The center-to-center of gap spacing in the ASH array was 1.1 μm along both the *x*- and *y*-axes. The refractive index $n = 2.43$ of the zinc selenide substrate was taken from the data sheet supplied by the Crystran company – and the data for fused silica were obtained from Palik [28,37]. Because zinc selenide substrates have a higher refractive index than that of fused silica, the ASH structure resonances shift toward longer wavelengths, producing plasmonic resonances at 5.5 μm and 6.8 μm. Fig. 10.8b shows a resonance peak with large absorption (approximately 0.4), as well as large reflectance at 9 μm, due to the intrinsic longer wavelength IR properties of the fused silica substrate. Fig. 10.8b also shows the corresponding plasmonic resonances at wavelengths of 3.8 μm and 4.8 μm. Although the transmission range and refractive index of zinc selenide are larger, the reflection loss increases by 30%, as shown in Fig. 10.9b. The designed dimensions of ASH structures on the zinc selenide substrates have been reduced to a smaller size to obtain resonances at shorter wavelengths, and thereby match with the targeted wavelengths.

10.4.3 Variation of the ASH Arm-length on ZnSe Substrate

In previous work, we focused on designing and developing the ASH structure on a fused silica substrate [7,35]. Although the array metasurface resonances can be manipulated to some extent by changing the periodic arrangement, in particular, the horizontal and vertical spacing of the individual elements, the overall size and the arm lengths in the ASH structure also have a direct effect on the resonances. In the present design, the periodic arrangement was set constant, and the arm-lengths of ASH structures on a ZnSe substrate were varied to produce resonance peaks at the targeted wavelengths. The periodic arrangement was set with a spacing of 1.1 μm along both the *x*- and *y*-axes. The ASH structures on ZnSe substrates are smaller than the sizes of corresponding ASH structures on fused silica substrates, as explained in Section 10.4.2, due to the ZnSe substrates having a higher refractive index than the fused silica substrates.

In order to optimize and demonstrate the coupling of ASH the structures, simulations were performed by changing the arm length L_2 and keeping a constant arm length L_1, or vice versa. The arm length L_1 was varied at 0.7 μm and 0.75 μm, while the arm length L_2 was varied from 0.5 μm to 0.6 μm, in order to exhibit a plasmonic resonance in the wavelength range between 2 μm and 4 μm. As the arm length is varied, the two resonance peaks are modified due to the presence of reflection dip. Table 10.1 shows the simulation and experimental results that are obtained by varying the arm length L_2. The resonance wavelength exhibits a red-shift, and linearly increases as the arm length increases, as shown in Fig. 10.10a, b. When the structures are less asymmetric, the resonance becomes broader and the amplitude of resonance peak at a wavelength of 3.5 μm is reduced. The value of *Q*-factor has been calculated; the average *Q*-factor is 8 for the arm length L_1 and 6 for arm length L_2. The reflection dip yields the *Q*-factor of 12.

To investigate the sensing capabilities of ASH structures fabricated on ZnSe substrates, the effect of a thin (50 nm) layer of different refractive index

TABLE 10.1
Simulation and Experimental Results of Varying the Arm Length L_2 with 0.5 μm, 0.55 μm, 0.6 μm and 0.65 μm. The Arm Length L_1 remains constant. The other parameters remain unchanged. All figures were labeled with ASH (L_1, L_2) μm.

Experiment	Simulation
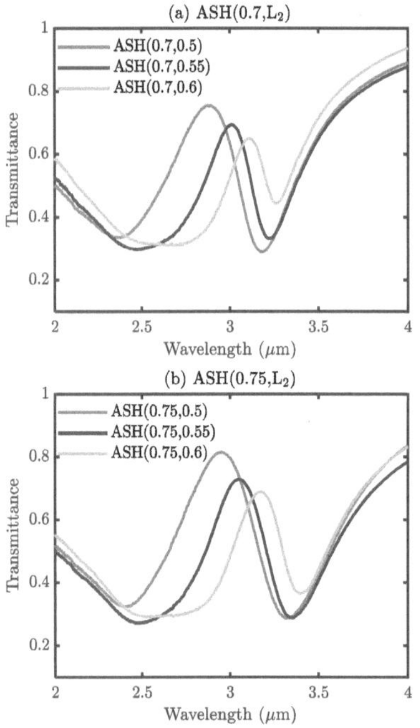	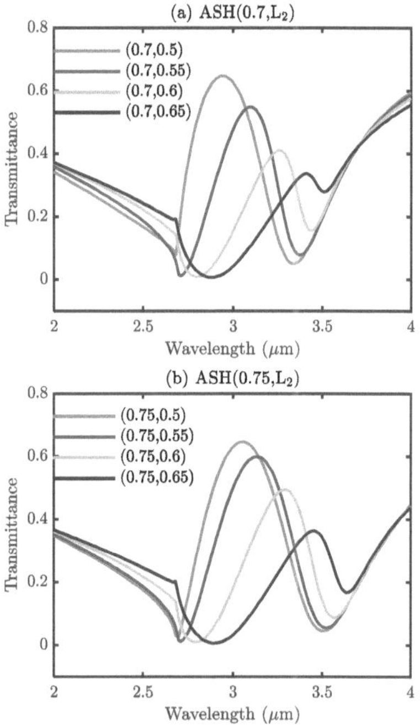

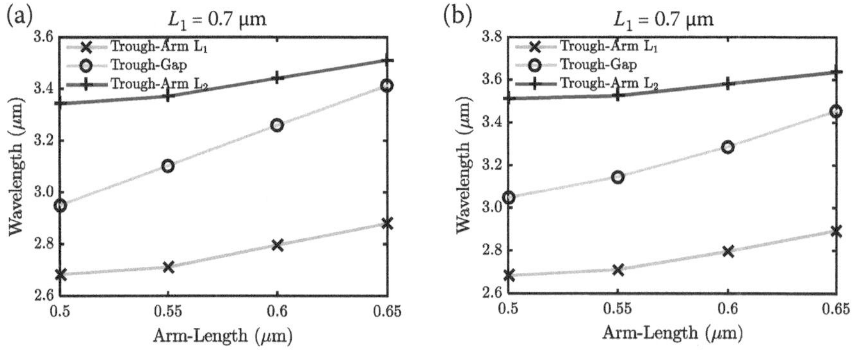

FIGURE 10.10 The resonance wavelength as a function of the arm length L_1 (0.7 μm and 0.75 μm). The arm length (L_1, L_2 and the gap g) increases as the resonance peak wavelength position increases.

dielectric material covering the metasurface was simulated (Fig. 10.11a). The value of refractive index was varied from 1.20 to 1.59, through the refractive index of water at n = 1.33 and that of PMMA at n = 1.46 [38]. The sensitivity was calculated from the refractive index change after the deposition of a thin layer of PMMA. The resonance of the bare ASH was red-shifted – and the resonance wavelength increased close to linearly with arm length, corresponding to a sensitivity value of 760 nm/RIU (Fig. 10.11c). Hence, the resonance wavelengths shifted by around 350 nm from their initial positions for the bare ASH structures.

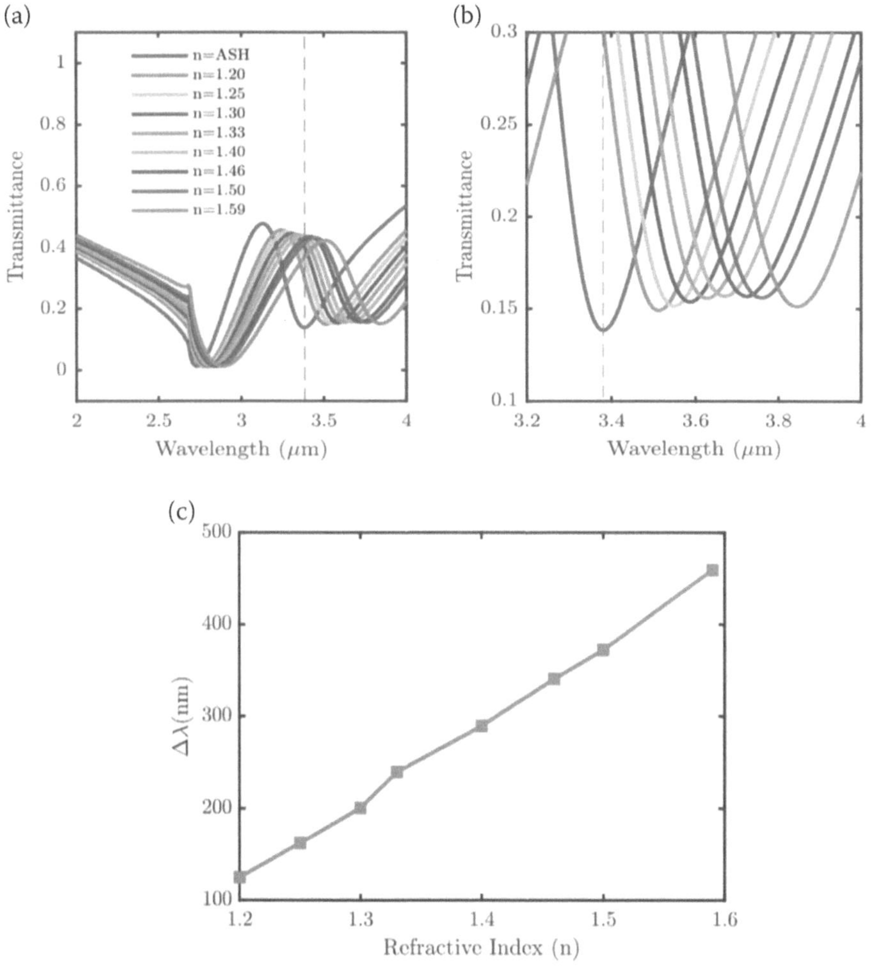

FIGURE 10.11 Simulated transmittance spectra of a 50 nm thick layer with different refractive indexes on an ASH-based metasurface modeled on ZnSe substrates. (b) Expanded view of the simulated transmittance spectra of (a). (c) Resonance shift (Δλ nm) versus refractive index (n) plot of the deposited layer for the ASH yields a linear relationship.

10.5 SENSING TECHNIQUES

A number of different sensing techniques have been employed by different researchers for assaying of organic molecules [39–43]. These methods, for example, screen-printed carbon electrodes (SPCEs) [39], gas chromatography-mass spectrometry (GC-MS) [40], gold electrode surfaces used for the detection of estradiol via under-potential deposition (UDP) [41] and aptamer-based optical fiber detection of 17β-estradiol [42], have demonstrated only limited sensitivity – while the laboratory-based experimental work involved remains challenging. The development of surface enhanced infrared absorption (SEIRA) technique, which often uses plasmon resonances produced by nanostructures or nanoparticles [43], has been shown to be effective for the specific assays of 17β-estradiol (E2) [44,45]. In our recent work, we emphasized the value of identification and interpretation of *Fano* resonances associated with the bond vibrations of E2 and their characteristic signatures [7]. Other researchers worked on the detection of E2 using a variety of methods with emphases on, for example, the interaction between E2 antigen and its receptor, as well as the use of molecular imprinted polyurethane [46,47].

In this chapter, we described the application of plasmon resonance sensor techniques based on the double resonance peaks produced by our designed ASH structures to evaluate the vibrational resonances present in E2. Articles on vibrational resonances of C-H bonds with other types of organic molecule, for example, polydimethylsiloxane (PDMS), 1-octadecanthiol (ODT) and poly-methyl-methacrylate (PMMA), that have also shown molecular bond stretch vibrations in the mid-IR region, have been published [4,5,8]. We carried out numerical simulations and experiments on E2 to study molecular bond vibrations that exist in the analyte. E2 is an example of organic molecule that exhibits vibrational resonances due to the presence of both the C=C and C–H bonds. These two bonds are, among the other molecular bonds, present in the range between 2 μm and 8 μm of the mid-IR spectrum. The C–H bond has very strong vibrational resonances that are observed over the range between 3.31 μm and 3.55 μm – with double resonance peaks, as shown in Fig. 10.12.

The reflectance spectra shown in Fig. 10.12 were obtained from the FTIR measurements and provide a comparison between the sensitivity values for the deposition of equal amounts of E2 at a concentration of 37 μmole/ml on the surface of fabricated SSH and ASH arrays. The sensitivity calculations for ASH structures were obtained from the red shift of resonance peak from 3.57 μm to 3.94 μm (a change in wavelength ($\Delta\lambda$) of 370 nm; (Fig. 10.12b.)) For SSH arrays, the red shift was from 3.23 μm to 3.37 μm with a $\Delta\lambda$ of 140 nm (Fig. 10.12a). The change in refractive index, Δn of 0.46 was calculated from $n_{E2} = 1.46$ and $n_{air} = 1$. From these results, the sensitivity values of 804 nm/RIU and 340 nm/RIU were obtained for the ASH and SSH arrays, respectively. Hence, the sensitivity value obtained for an SSH array metasurface is approximately 41% of that for an ASH array metasurface. From these results, we propose that an optimized ASH-based metasurface has promising features for application in high sensitivity bio-sensor devices, as we have shown in this section and in Mbomson et al. [7].

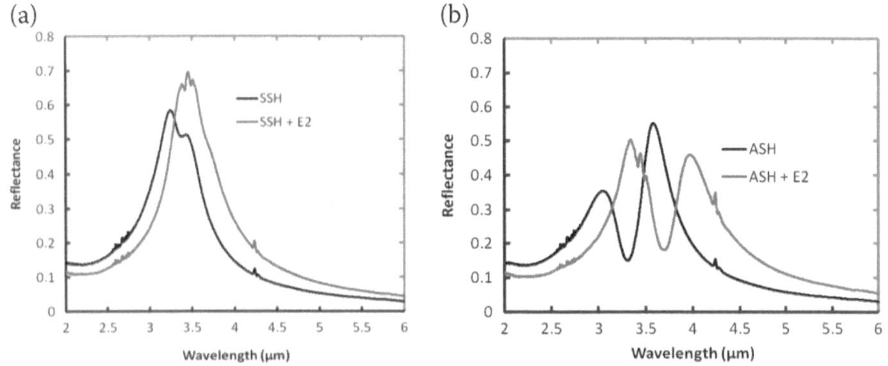

FIGURE 10.12 Reflectance spectra showing the red shift in the resonance peaks from an initial wavelength(s) of (a) 3.23 μm for the SSH structures. *Note:* The small dip on the plot of (a) at 3.3 μm is possibly due to imperfections in the fabrication process. (b) Initial wavelengths of 3.05 μm and 3.57 μm for the two peaks of ASH structures. The other spectral features are from water vapor at 2.85 μm, CO_2 at 4.20 μm and the C–H resonance stretch at 3.31 μm to 3.55 μm.

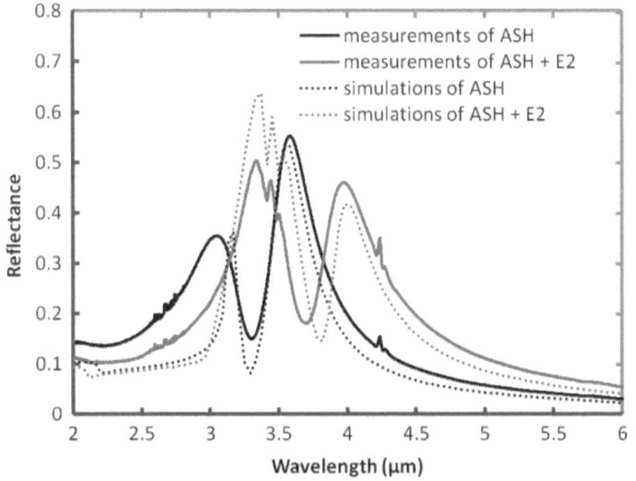

FIGURE 10.13 Reflectance spectra from measurement and simulation plots from ASH with E2 showing red-shifted resonance peaks due to refractive index change in the ASH surroundings – for three different samples. Ripples at 2.8 μm are from H_2O vapor, C–H vibrational resonance bond stretch from 3.31 μm to 3.55 μm and CO_2 at 4.2 μm.

The presence of C–H bond in E2 was evaluated, as shown in the measurement and simulation reflectance spectra of Fig. 10.13. The figure summarizes the work presented in this chapter by using plasmonic resonance peaks to demonstrate red shifts caused by positive changes in the refractive index of the medium above the metasurface array. Numerical modeling leads to the identification of

Fano-type resonance associated with E2, formed by the superposition of Lorentz resonance on a slowly varying dielectric medium background to produce wavelength-dependent negative and positive changes in the refractive index of analyte layer. For modeling, the background relative permittivity ($\mathcal{E}$ = 2.13) of the analyte was obtained from the expression $\mathcal{E} = (n_{E2})^2$, where n (= 1.46) was used as the refractive index of the analyte. The Lorentz permittivity ($\mathcal{E}_L$ = 0.0012) was used, together with the Lorentz linewidth (δ_L) of approximately 4×10^{12} rad/s in the numerical simulation, to provide a resonance peak closely matched with the experimental measurements. The vibrational resonances of C–H bonds are also enhanced by the gold ASH array metasurface, as can be observed in the figure – and has been calculated in the cited references [7,14,27]. The characteristic behavior of the ASH-based metasurface as a sensor device with high sensitivity is shown in Figs. 10.12 and 10.13.

10.6 CONCLUSIONS AND FUTURE WORK

The use of ASH arrays as sensor metasurfaces has been investigated – and future work will continue to provide a sensitive sensor platform for specific molecule detection. With low-cost synthesis, aptamers can be used as an alternative to antibodies [48]. Aptamers are single-stranded deoxyribonucleic acid (DNA) molecules and were isolated by Kim et al. [17]. They can be used as bio-recognition elements to bind specifically to targeted E2 molecules of (E2). The DNA aptamer strand used was 76-mer (basic molecular repetition units) long and had a molecular weight of 23 *kDa*. The 5' end of this aptamer was modified with a thiol linkage for sulfur gold (S-Au) bonding on the gold surface, and the other 3' end was kept label free. The 6-mercapto-1-hexanol (MCH) was used as a blocking agent to prevent non-specific binding on the gold ASH surface [11]. The aptamer was attached to the sensor surface and MCH was used to block the free spaces on the sensor surface and to capture E2 only.

Arrays of ASH metamaterial structures with multiple sharp edges and a 50 nm gap have been designed and fabricated, and then used to realize metasurfaces for assays of 17β-estradiol (E2). The parameters of ASH nanostructure arrays, such as the periodic arrangement of array, the effect of changing the substrate, and also variation of the vertical arm lengths L_1 and L_2 have been described in this chapter. All of these parameters affect the resonance wavelengths. The asymmetric structures give double resonance peaks with a dip that is controlled by the coupling between the two arms. Metasurfaces based on the arrays of asymmetric metamaterial structures have been shown to be well suited for organic sensing applications, because of their tunable resonance characteristics that can be affected by their overall geometric size, gap size, periodic arrangement, high Q-factor resonance and also the surrounding dielectric environment.

D-ASH nanostructures were reported to produce a Q-factor of 26 in the x-polarization, which is comparable with the highest values reported for metallic nanostructures in the mid-IR regime [35]. The double resonance peaks produced by the ASH structures can be used to enhance the sensitivity for two different molecular bonds – and also, enable dual polarization mode operation. Such ASH structures can be used to find close matches with the molecular resonances of target organic compounds – and by the use of surface enhanced IR spectroscopy (SEIRA spectroscopy) as a detection tool for mid-IR sensing. Plasmon resonances produced in ASH arrays have enhanced the vibrational resonances of C–H bond by a factor of 2.62×10^5. The limit of detection obtained from numerical simulation is 0.15 femtomole [14] for E2 with a molecular weight of 272.4 g/mol. A high sensitivity biosensor has been achieved with calculated value greater than 2700 nm/RIU at 3.59 μm and figure of merit (FOM) of 9 [14]. The resonance of C–H bond stretch vibration obtained using Lorentzian modeling closely matches the experimental results, as shown in Fig. 10.13. Plasmonic metamaterial-based sensor structures are effective for detection and analysis of the presence of vibrational resonances in molecular analytes, for example, 17*β*-estradiol.

ACKNOWLEDGMENT

This research work was funded by Majlis Amanah Rakyat (MARA) Malaysia and J.J. Mbomson Education Foundation. The authors IFMAN and IGM would also like to acknowledge the facilities and staff of the James Watt Nanofabrication Centre (JWNC) for their support during fabrication of the structures.

REFERENCES

1. I. M. Pryce, Y. A. Kelaita, K. Aydin, and H. A. Atwater, "Compliant metamaterials for resonantly enhanced infrared absorption spectroscopy and refractive index sensing," *Nano Lett.*, vol. 5, pp. 8167–8174, 2011.
2. W. Kubo and S. Fujikawa, "Au double nanopillars with nanogap for plasmonic sensor," *Nano Lett.*, vol. 11, pp. 8–15, 2011.
3. L. La Spada, F. Bilotti, and L. Vegni, "Metamaterial-based sensor design working in infrared frequency range," *Prog. In Electromagn. Res.*, vol. 34, pp. 205–223, 2011.
4. B. Lahiri, S. G. McMeekin, R. M. De La Rue, and N. P. Johnson, "Enhanced Fano resonance of organic material films deposited on arrays of asymmetric split-ring resonators (A-SRRs)," *Opt. Express*, vol. 21, pp. 9343–9352, 2013.
5. I. G. Mbomson, S. G. McMeekin, B. Lahiri, R. M. De La Rue, and N. P. Johnson, "Gold asymmetric split ring resonators (A-SRRs) for nano sensing of estradiol," *Proc. SPIE. 9125, Metamaterial IX*, vol. 9125, 2014.
6. J. Wu, C. Zhou, J. Yu, H. Cao, S. Li, and W. Jia, "Design of infrared surface plasmon resonance sensors based on graphene ribbon arrays," *Opt. and Laser Technol.*, vol. 59, pp. 99–103, 2014.
7. I. G. Mbomson, S. Tabor, B. Lahiri, G. Sharp, S. G. McMeekin, R. M. De La Rue, and N. P. Johnson, "Asymmetric split H-Shape nanoantennas for molecular sensing," *Biomed. Opt. Express.*, vol. 8, pp. 395–406, 2017.
8. J. Paul, S. G. McMeekin, R. M. De La Rue, and N. P. Johnson, "AFM imaging and plasmonic detection of organic thin-films deposited on nanoantenna arrays," *Sensors and Actuators A: Physical*, vol. 279, pp. 36–45, 2018.
9. D. Rodrigo, O. Limaj, D. Janner, D. Etezadi, F. J. D. De Abajo, V. Pruneri, and H. Altug, "Mid-infrared plasmonic biosensing with graphene," *Science*, vol. 349, no. 6244, pp. 165–168, 2015.
10. J. Terwisscha van Scheltinga, N. F. W. Ligterink, A. C. A. Boogert, E. F. Van Dishoeck, and H. Linnartz, "Infrared spectra of complex organic molecules in

astronomically relevant ice matrices – I. Acetaldehyde, ethanol, and dimethyl ether," *A&A.*, vol. 611, no. A35, p. 43, 2018.

11. A. Brognara, I. F. Mohamad Ali Nasri, B. R. Bricchi, A. Li Bassi, C. Gauchotte-Lindsay, M. Ghidelli, and N. Lidgi-Guigui, "Highly sensitive detection of estradiol by a SERS sensor based on TiO_2 covered with gold nanoparticles," *Beilstein J. Nanotechnol.*, vol. 11, no. 1, pp. 1026–1035, 2020.
12. D. Enders and A. Pucci, "Surface enhanced infrared absorption of octadecanethiol on wet-chemically prepared Au nanoparticle films," *Appl. Phys. Lett.*, vol. 88, p. 184104, 2006.
13. S. R. Staden, A. L. Gugoasa, B. Calenic, and J. Legler, "Pattern recognition of estradiol, testosterone and dihydrotestosterone in children's saliva samples using stochastic microsensors." *Sci. Repts.*, vol. 4, p. 5579, 2014.
14. I. G. Mbomson, "Mid-infrared photonic sensors based on metamaterial structures," PhD thesis, University of Glasgow, 2016.
15. Y. Zhang, J. L. Zhou, and B. Ning, "Photodegradation of estrone and 17b-estradiol in water," *Science Direct*, vol. 1, pp. 19–26, 2007.
16. L. M. Brown, L. Gent, K. Davis, and D. J. Clegg, "Metabolic impact of sex hormones on obesity," *Brain Res.*, vol. 1350, pp. 77–85, 2010.
17. Y. S. Kim, H. S. Jung, T. Matsuura, H. Y. Lee, T. Kawai, and M. B. Gu, "Electrochemical detection of 17β-estradiol using DNA aptamer immobilized gold electrode chip," *Biosens. Bioelectron.*, vol. 22, pp. 2525–2531, 2007.
18. L. La Spada, F. Bilotti, and L. Vegni, "Metamaterial-based sensor design working in infrared frequency range," *Prog. In Electromagn. Res. B*, vol. 34, pp. 205–223, 2011.
19. J. Yao, Z. Liu, Y. Liu, Y. Wang, C. Sun, G. Bartal, A. M. Stacy, and X. Zhang, "Optical negative refraction in bulk metamaterials of nanowires," *Science*, vol. 321, no. 5891, p. 930, 2008.
20. B. Wang, J. Zhou, T. Koschny, M. Kafesaki, and C. M. Soukoulis, "Chiral metamaterials: Simulations and experiments," *J. Opt. A: Pure and Appl. Opt.*, vol. 11, p. 114003, 2009.
21. I. Sersic, M. Frimmer, E. Verhagen, and A. F. Koenderink, "Electric and magnetic dipole coupling in near-infrared split-ring metamaterial arrays," *Phys. Rev. Lett.*, vol. 103, p. 213902, 2009.
22. H. Vilhena, S. G. McMeekin, A. S. Holmes-Smith, and N. P. Johnson, "Optimization of dipole structures for detection of organic compounds." *Proc. of SPIE*, vol. 9502, pp. 950212- 1, 2015.
23. B. Yuan, W. Zhou, and J. Wang, "Novel H-shaped plasmon nanoresonators for efficient dual-band SERS and optical sensing applications." *J. Opt.*, vol. 16, p. 105013, 2014.
24. P. Pavaskar, J. Theiss, and S. B. Cronin, "Plasmonic hot spots: nanogap enhancement vs. focusing effects from surrounding nanoparticles," *Opt. Express*, vol. 20, no. 13, pp. 14656–14662, 2012.
25. Zi.-Q. Cheng, F. Nan, D.-J. Yang, Y.-T. Zhong, L. Ma, Z.-H. Hao, L. Zhou, and Q.-Q. Wang, "Plasmonic nanorod arrays of a two-segment dimer and a coaxial cable with 1 nm gap for large field confinement and enhancement," *Nanoscale*, vol. 7, pp. 1463–1470, 2015.
26. I. F. Mohamad Ali Nasri, "Optical sensors based on asymmetric plasmonic nanostructures for environmental monitoring," PhD thesis, University of Glasgow, 2019.
27. V. A. Fedotov, M. Rose, S. L. Prosvirnin, N. Papasimakis, and N. I. Zheludev, "Sharp trapped-mode resonances in planar metamaterials with a broken structural symmetry," *Phys. Rev. Lett.*, vol. 99, p. 147401, 2007.

28. E. D. Palik, *Palik, Handbook of Optical Constants of Solids*. 2 Vols., 1991.
29. C. Schuster, A. Christ, and W. Fichtner, "Review of FDTD time-stepping schemes for efficient simulation of electric conductive media," *Microw. Opt. Technol. Lett.*, vol. 25, pp. 16–21, 2000.
30. S. Linden, "Magnetic response of metamaterials at 100 Terahertz," *Science*, vol. 306, no. 5700, p. 1351, 2004.
31. H. Guo, N. Liu, L. Fu, T. P. Meyrath, T. Zentgraf, H. Schweizer, and H. Giessen, "Resonance hybridization in double split-ring resonator metamaterials," *Opt. Express*, vol. 15, no. 19, p. 12095, 2007.
32. L. Na, G. Hongcang, F. Liwei, H. Schweizer, S. Kaiser, and H. Giessen, "Electromagnetic resonances in single and double split-ring resonator metamaterials in the near infrared spectral region," *Physica Status Solidi (B) Basic Res.*, vol. 244, no. 4, pp. 1251–1255, 2007.
33. B. Lahiri, S. G. McMeekin, R. M. De La Rue, and N. P. Johnson, "Resonance hybridization in nanoantenna arrays based on asymmetric split-ring resonators," *Appl. Phys. Lett.*, vol. 98, no. 15, p. 153116, 2011.
34. I. G. Mbomson, R. M. De La Rue, S. McMeekin, and N. P. Johnson, "Hybrid Asymmetric Nanostructures for the Mid-Infrared," *Intl. Conf. on UK-China Emerging Technologies (UCET)*, Glasgow, United Kingdom, pp. 1–4, 2020.
35. I. G. Mbomson, I. F. Mohamad Ali Nasri, R. M. De La Rue, and N. P. Johnson, Dual polarization operation of nanostructure arrays in the mid-infrared. *Appl. Phys. Lett.*, vol. 112, no. 7, p. 073105, 2018.
36. O. Limaj, S. Lupi, F. Mattioli, R. Leoni, and M. Ortolani, Midinfrared surface plasmon sensor based on a substrateless metal mesh. *Appl. Phys. Lett.*, vol. 98, no. 9, pp. 091902- 1, 2011.
37. Zinc Selenide, ZnSe, [Accessed: 20-Nov-2018]. https://www.crystran.co.uk/optical-materials/zinc-selenide-znse, 2007.
38. B. Lahiri, A. Z. Khokhar, R. M. De La Rue, S. G. McMeekin, and N. P. Johnson, "Asymmetric split ring resonators for optical sensing of organic materials," *Opt. Express*, vol. 17, no. 2, p. 1107, 2009.
39. P.-T. Katalin, B. Darius, and P. Laszlo, "Simultaneous measurement of 17β-estradiol, 17α-estradiol and estrone by GC–esotope eilution MS–MS," *Chromatographia*, vol. 71, pp. 311–315, 2010.
40. Y. Yang and E. P. C. Lai, "Optimization of molecularly imprinted polymer method for rapid screening of 17β-estradiol in water by fluorescence quenching." *Int. J Anal. Chem.*, vol. 214747, pp. 1–8, 2011.
41. X. Q. Liu, X. H. Wang, J. M. Zhang, H. Q. Feng, X. Q. Liu, and D. K. Y. Wong, "Detection of estradiol at an electrochemical immunosensor with a Cu UPD vertical bar DTBP-Protein G scaffold," *Biosens. Bioelectron.*, vol. 35, pp. 56–62, 2012.
42. N. Yildirim, F. Long, C. Gao, M. He, H.-Cg. Shi, and A. Z. Gu, "Aptamer-based optical biosensor for rapid and sensitive detection of 17β-estradiol in water samples," *Environ. Sci. Technol.*, vol. 46, pp. 3288–3294, 2012.
43. J. Liu, W. Bai, S. Niu, C. Zhu, S. Yang, and A. Chen, "Highly sensitive colorimetric detection of 17β-estradiol using split DNA aptamers immobilized on unmodified gold nanoparticles," *Sci. Repts.*, vol. 4, p. 7571, 2014.
44. Q. Zhang, Y. Wang, A. Mateescu, K. Sergelen, A. Kibrom, U. Jonas, T. Wei, and J. Dostalek, "Biosensor based on hydrogel optical waveguide spectroscopy for the detection of 17β-estradiol," *Talanta*, vol. 104, pp. 149–154, 2013.
45. W.-W. Zhang, Y.-C. Chen, Z.-F. Luo, J.-Y. Wang, and D.-Y. Ma, "Analysis of 17 beta-estradiol from sewage in coastal marine environment by surface plasmon resonance technique," *Chem. Res. Chin.*, vol. 23, pp. 404–407, 2007.

46. U. Latif, J. Qian, S. Can, and F. L. Dickert, "Biomimetic receptors for bioanalyte detection by quartz crystal microbalances – From molecules to cells," *Sensors*, vol. 14, pp. 23419–23438, 2014.
47. Y. Dai and C. C. Liu, "Detection of 17 β-estradiol in environmental samples and for health care using a single-use, cost-effective biosensor based on differential pulse voltammetry (DPV)," *Biosensors*, vol. 7, no. 2, p. 15, 2017.
48. https://en.wikipedia.org/wiki/Aptamer.

11 Acoustic Spoof Surface Waves Control in Corrugated Surfaces and Their Applications

Norbert Cselyuszka[1], *Nikolina Jankovic*[2], *Andrea Alu*[3,4], *and Vesna Bengin*[2]
[1]Silicon Austria Labs, Sensor Systems, Microsystem Technologies, Villach, Austria
[2]BioSense Institute-Research Institute for Information Technologies in Biosystems, University of Novi Sad, Novi Sad, Serbia
[3]Photonics Initiative, Advanced Science Research Center, City University of New York, New York, NY 10031, USA
[4]Physics Program, Graduate Center, City University of New York, New York, USA

11.1 INTRODUCTION

Studies have reported on the acoustic analogues of surface plasmon polaritons, which arise at the boundaries between highly conductive and dielectric media due to the coupling of light with free electron oscillations [1–7]. The principal challenge is the apparent nonexistence of an acoustic analogue of noble metals with acousto-plasmonic responses. However, spoof plasmons [8–16] can arise along patterned conducting grooved surfaces without requiring plasmonic responses. A corresponding acoustic phenomenon may, therefore, arise in the form of an acoustic surface wave at the boundary between a fluid and a hard-grooved surface [17–19]. Understanding of this analogy opens up avenues in the field of controlled propagation of surface acoustic waves and their applications.

In contrast to conventional acoustic surface waves, in which sound propagates on the surface of solids and the properties are closely dependent on the material parameters [20], acoustic spoof surface waves (ASSW) propagate along the interface between the background medium and a corrugated rigid plate, due to coupling between the oscillation inside the individual holes and the mutual near-field coupling between the adjacent holes. The geometrical parameters of corrugations provide a large degree of freedom to control the dispersion properties, which has led to a number of applications, such as collimation of sound [21–23], extraordinary

DOI: 10.1201/9781003050162-11

acoustic transmission [24–26], acoustic rainbow trapping [27–29], acoustic gas sensing [30], acoustic sensing enhancement [31], acoustic wave focusing [32,33], beam forming [34], beam steering [35], temperature-controlled acoustic surface waves [36] and surface acoustic waveguides [37].

In this chapter, acoustic surface waves propagating at the boundary between a fluid and a hard-grooved surface are discussed, along with their applications. Section 11.2 provides an in-depth theoretical analysis of the dispersion relation of ASSW and shows how ASSW can be controlled by tailoring geometrical parameters and the parameters of the surrounding medium. Owing to the peculiar nature of ASSW, there is a wide range of potential applications, including sound trapping and collimation, extraordinary energy transmission, enhanced sensing and acoustic lensing, which are discussed in Section 11.3. A new type of acoustic waveguide based on two corrugated surfaces separated by a fluid gap is presented in Section 11.3.7, as well as its possible application for acoustic delay lines, modulators and sensors.

11.2 THEORETICAL BACKGROUND

Surface acoustic waves have been extensively investigated in different forms, depending on the nature of the media forming the interface that supports them [38]. Since ASSWs occur between a corrugated surface and a background medium, the analysis of acoustic surface waves occurring on a fluid-fluid interface is of particular interest for understanding the behavior of ASSWs. To begin with, we consider the condition for the existence of a surface wave at the fluid–fluid interface, which is given by the equation [21]

$$\frac{k_z^I}{\rho^I} + \frac{k_z^{II}}{\rho^{II}} = 0 \tag{11.1}$$

where k^{I}_z and k^{II}_z are the inverse of the decay lengths of the acoustic surface waves in the mediums I and II, respectively, and ρ^{I} and ρ^{II} are the corresponding mass densities. Since fluids with negative mass densities do not exist in nature, it is clear that the surface waves cannot be realized in such a configuration. An alternative solution to achieve surface wave propagation was proposed in the form of a periodically corrugated surface. Its typical configuration is shown in Fig. 11.1, where a surface wave travels along the x-direction, i.e., at the interface of a one-dimensional (1D) array of grooves with a background medium. The array is characterized by the geometrical parameters, d, a and h, referring to the period, width and depth of the grooves, respectively.

The dispersion relation for the supported surface waves can be derived using an effective medium approximation, under the assumption that the period of grooves d is much smaller than the guided wavelength $\lambda \gg d$. In this regime, the periodic corrugated surface can be treated as an equivalent anisotropic homogeneous medium characterized by an effective density tensor. Assuming symmetry and infinite extent in the y-direction, the relevant components of mass density and bulk

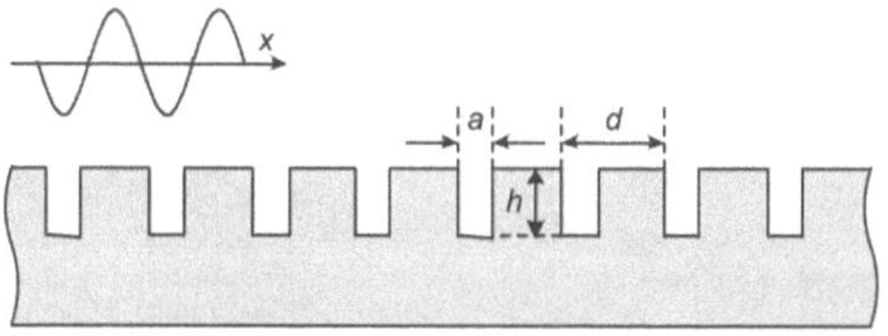

FIGURE 11.1 A typical grooved surface that supports propagation of acoustic surface waves.

modulus tensors for the corrugated surface are labeled as ρ_y^j, ρ_x^j and κ, respectively, where j takes the values I and II, denoting the top (fluid) and bottom layer, respectively. For the fluid-filled corrugated structure, the effective material parameters can be defined as $\rho_y^{II} = (d/a)\rho_{fluid}$, $\rho_x^{III} \longrightarrow \infty$, whilst $\kappa_{\text{eff}} = (d/a)\kappa_{\text{air}}$ [15]. In the top layer, the effective parameters are equal to the material parameters of fluid, which we consider to be air: $\rho^{II} = \rho_{air} = 1.21$ kg/m^3 and $\kappa_{\text{air}} = 1.41 \times 10^5$ Pa. Since the corrugated surface is assumed to be made of a perfectly rigid body, the wave is purely evanescent in the z-direction and propagates in the x-direction. The dispersion relation for the acoustic surface wave can be expressed as [17,28,36,37]

$$k_x(\omega) = k_0\sqrt{1 + \left(\frac{a}{d}\right)^2 (\tan k_0 h)^2} \qquad (11.2)$$

where ω is the angular frequency and k_0 is the wave number in medium I. Fig. 11.2 shows the dispersion relation of the acoustic waves propagating in air (k_0) and of the acoustic surface waves supported at the air/grooved surface interface, with geometrical parameters, d = 5 mm, a = 1 mm and h = 24 mm (k_x).

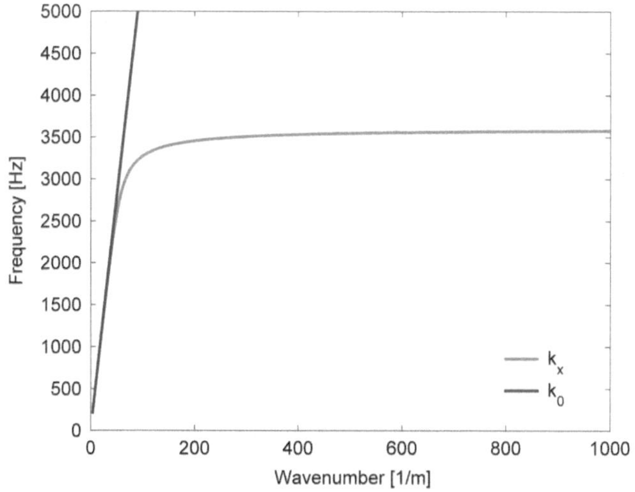

FIGURE 11.2 Dispersion diagram of the acoustic surface wave on a grooved surface.

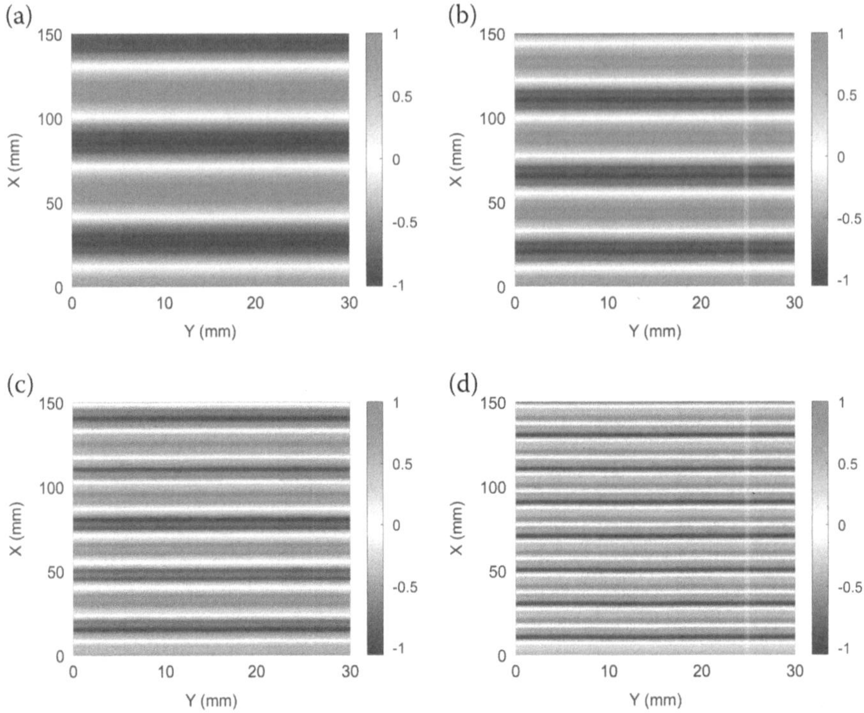

FIGURE 11.3 Acoustic pressure distribution indicating the dependence of wavelength on groove geometry variation (operating frequency f_0 = 3.4 kHz a) h = 21 mm, b) h = 23 mm, c) h = 24 mm and d) h = 24.5 mm).

One can note that the two dispersion relations exhibit similar properties at low frequencies. However, as frequency increases, strong dispersion emerges for the ASSW, and the effective wave number asymptotically approaches infinity, causing a stop-band to appear. Similar to surface plasmon polaritons (SPPs) in optics, at the certain frequency f_{ap}, the wave number tends to infinity, and f_{ap} depends on the geometrical parameters of the grooved surface as

$$f_{ap} = \frac{c_0}{4h} \tag{11.3}$$

These results indicate that the ASSW can be controlled and manipulated by varying the surface geometry. This is illustrated in Fig. 11.3, where the acoustic pressure field distribution over a grooved surface is shown for four different values of groove depth. One can note that, for the defined operational frequency, the wavelength of the surface acoustic wave varies significantly.

The dispersive nature, and the fact that the dispersion mainly depends on the geometrical parameters of the grooved surface, allows for arbitrary manipulation of the ASSW properties. Consequently, this opens up room for a number of potential applications, which are presented in the following section.

11.3 APPLICATIONS

Owing to their strong dispersion, acoustic surface waves on grooved surfaces can be used in many applications, such an acoustic wave trapping, extraordinary transmission and wave propagation control, some of which are presented in the following subsections.

11.3.1 Slowing Down Acoustic Surface Waves on a Grooved Surface

Slowing down optical waves with resonating photonic structures introduces controllable optical delays and allows temporary storage of light in all-optical memories and switches [28]. Similarly, acoustic surface wave propagation along a periodically grooved rigid surface surrounded by air can be engineered by geometrical means. These highly localized acoustic surface waves give rise to strong acoustic field confinement along the grooved rigid surface, whereas the slowing down of sound can eventually reduce the group velocity to zero [28].

In Devaux et al. [26], a metamaterial was designed, consisting of an array of grooves with constant periodicity and graded depths perforated on a rigid plate, to control the acoustic propagation (Fig. 11.4).

The group velocity of the acoustic surface wave in such a configuration can be expressed as

$$v_g = \frac{c_0}{\sqrt{1 + \frac{k_0}{k_x}(\tan k_0 h)^2} + k_0 h \left(\frac{k_0}{k_x}\right) \frac{(\tan(k_0 h))^2 + 1}{\sqrt{(\cot(k_0 h))^2 + \frac{k_0}{k_x}}}} \tag{11.4}$$

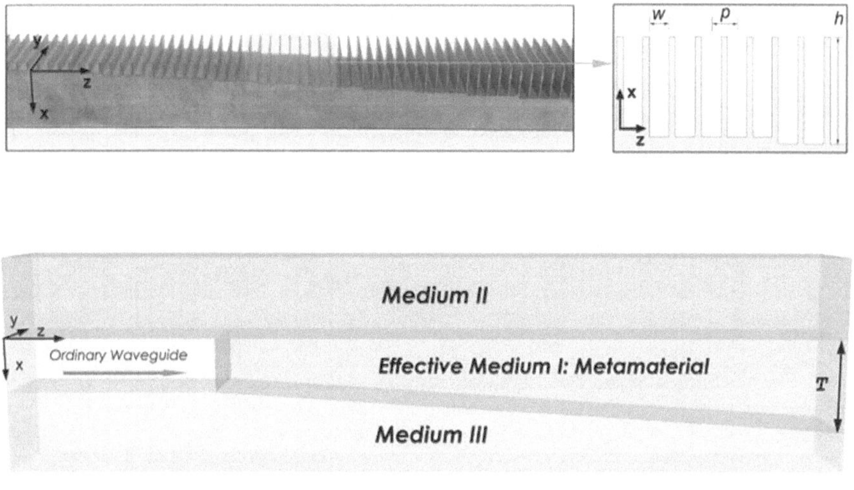

FIGURE 11.4 Metamaterial with graded depths perforated on a rigid plate and the effective medium model of the composite structure. (From Zhu et al. [28])

where c_0 is the speed of sound in air, k_0 is the wave number in air, h is the depth of grooves, and k_x is the surface wave number calculated from the dispersion relation Eq. (11.2). An interesting property of graded acoustic artificial structures, implied by the dispersion relation, is that they can control not only the propagation direction of the wave, but also the spatial distribution of its energy. This is illustrated in Fig. 11.5, which shows the group velocity along the graded metamaterial normalized with respect to the speed of sound in air. One can note that waves of different frequencies stop at different locations.

This is also confirmed in the experimental results for the sound intensity distribution on the surface (Fig. 11.6), where different frequencies slow down and stop at different positions. The stop positions depend on the frequency and the angle of linear change in the depth of grooves.

One should note that the same phenomena can be realized by linearly changing another geometrical surface parameter. Instead of the depth, for example, the width of grooves can be changed, as shown in Jia et al. [29]. Furthermore, the shape of the

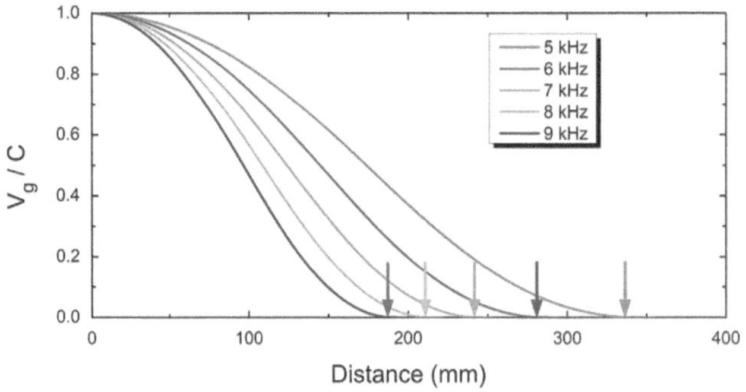

FIGURE 11.5 Calculated group velocities of acoustic waves (normalized to the speed of sound in air). (From Zhu et al. [28])

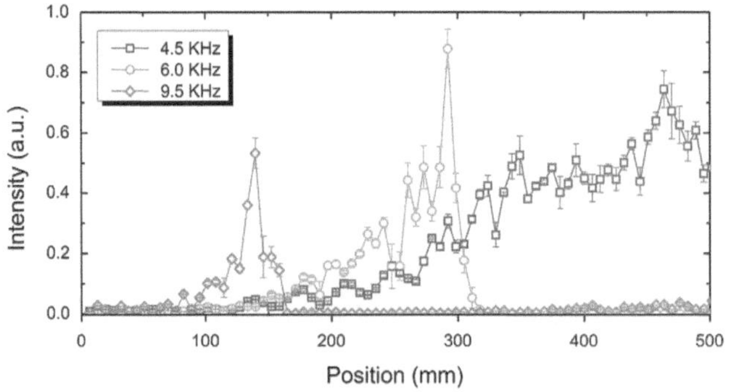

FIGURE 11.6 Experimental results of acoustic field intensity variation along the metamaterial. (From Zhu et al. [28])

graded surface does not have to be a plate in order to realize spatial frequency separation; it can be a cylindrical rod with gradually changing geometrical parameters, for example [27].

11.3.2 Extraordinary Transmission Assisted by Acoustic Surface Waves

Discovery of extraordinary optical transmission has opened a new line of research within optics with a number of applications [4]. Such peculiar transmission is achieved in a two-dimensional (2D) array of sub-wavelength holes in a metallic film, with the help of surface plasmon polaritons, and it has also inspired realization of another optical phenomenon – collimation of light in a single hole surrounded by a finite periodic array of indentations.

In analogy with the electromagnetic case, the main contributors necessary for the appearance of both enhanced transmission and collimation in acoustics are the acoustic surface waves [21]. Acoustic wave extraordinary transmission was first demonstrated in Christensen et al. [21], and to that end, a grooved surface, such as the one shown in Fig. 11.1, was used. Fig. 11.7 shows the normalized-to-area transmission spectrum across the single sub-wavelength slit for a normal incident acoustic plane wave propagating in air as well as the transmission spectrum in the case where the aperture is a slit of the same size but flanked by periodic grooves.

A resonant peak appearing at a wavelength close to the period of array clearly dominates the spectrum. For sound of that particular wavelength, the transmitted intensity is 70 times larger than the one impinging directly over a single slit opening. Namely, at the output side, grooves act as radiators to effectively couple acoustic surface waves to radiative waves in free space when the phases are matched [25]. Fig. 11.8 shows the experimental results of the full range intensity field pattern at the resonance frequency, which agrees with the simulation and calculation results, also shown in Fig. 11.8.

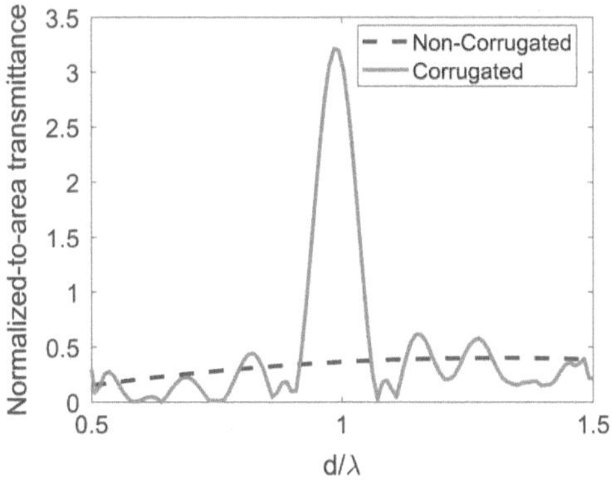

FIGURE 11.7 Transmission of sound through a single sub-wavelength slit, where d is the period of the grooves.

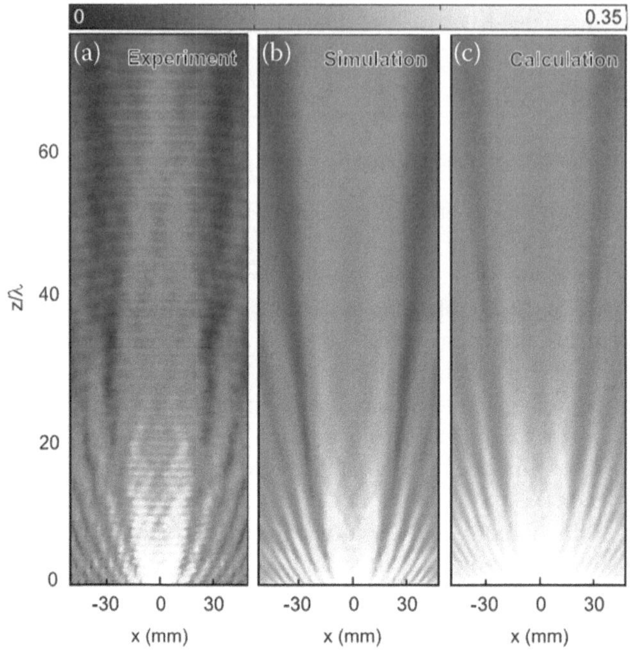

FIGURE 11.8 (a) Experimental, (b) simulated and (c) calculated results of full-range intensity field pattern in the x–z plane at the resonance frequency, showing the collimation effect. (From Yu et al. [25])

11.3.3 Acoustic Surface Wave Controlled by Temperature

In the previous section, we showed how the propagation properties of surface waves can be controlled by the geometrical parameters of corrugated surfaces. Nevertheless, the dispersion relation (Eq. (11.2)) indicates that the physical properties of the surrounding medium can also be used as a mechanism to control wave propagation. From the ideal gas state equation, the gas medium density can be obtained as a function of temperature, i.e.,

$$\rho_0(T) = \frac{p}{RT} \tag{11.5}$$

where R denotes the specific gas constant, and T is the temperature in Kelvin; thus, the temperature-dependent dispersion relation can be expressed as

$$k_x(T) = \omega\sqrt{\frac{1}{\kappa RT}}\sqrt{1 + \left(\frac{a}{d}\right)^2 \tan^2\left(h\omega\sqrt{\frac{1}{\kappa RT}}\right)} \tag{11.6}$$

where κ is the adiabatic constant. Fig. 11.9 shows the dispersion relations for different temperatures. One can note that the acoustic plasma frequency and the

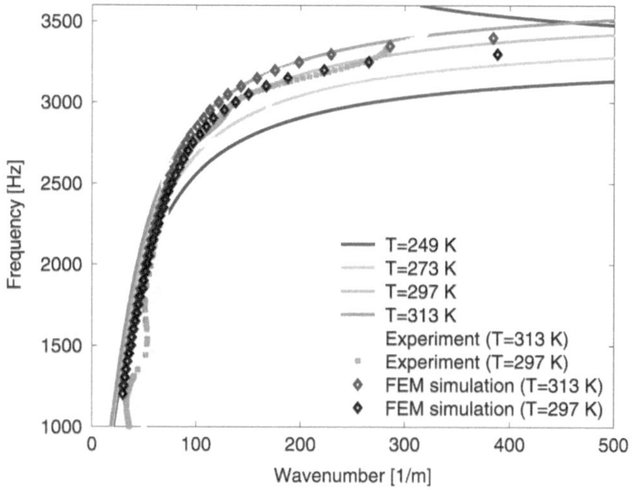

FIGURE 11.9 Experimental, simulated and theoretical dispersion curves at different temperatures. (From Cselyuszka et al. [36])

position of stop-band depends on the temperature, implying that the temperature is suitable for wave propagation control. The temperature dependence of wave number k_x, and the phase velocity for a defined frequency (3.4 kHz) and geometrical parameters are shown in Fig. 11.10.

While the phase velocity exhibits normal variation with temperature, the wave number exhibits somewhat different behavior, i.e., it rapidly increases at a specific temperature that corresponds to a critical value T_c,, whereas below T_c, no

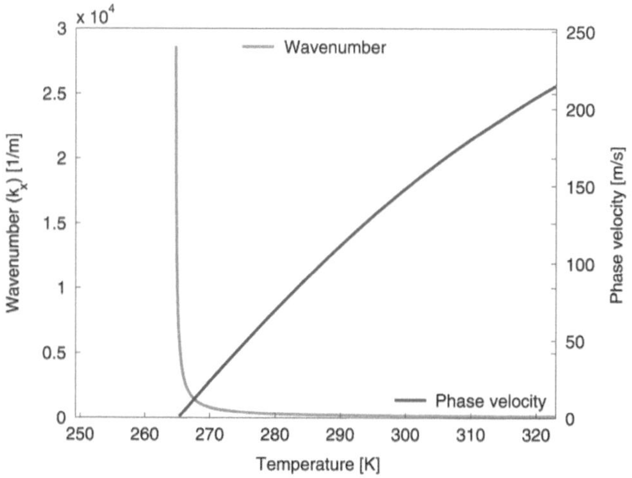

FIGURE 11.10 Temperature dependence of phase velocity and wave number at 3,400 Hz. ([From Cselyuszka et al. [36])

propagation can occur. The critical temperature T_c from Eq. (11.6), as a function of the operating frequency f and the depth of grooves h, is

$$T_c = \frac{16h^2 f^2}{\kappa R} \tag{11.7}$$

For the defined operating frequency $f_o = 3.4$, kHz and geometrical parameters $d = 8$ mm, $a = 5$ mm and $h = 24$ mm, $T_c = 264.9$ K. The discussion above does not hold in close vicinity T_c, where the wave number becomes too large and the effective medium concept is lost since the guided wavelength becomes comparable with the period d. The effective media concept can thus be applied as long as the guided wavelength is larger than approximately $4d$, i.e., $k_x < \pi/2d$.

11.3.4 Slowing Down Acoustic Surface Waves by Applying Temperature Gradient along the Wave Propagation/Spatial Spectral Separation

In the previous section, we showed that the wave slows down with increasing wave number, and the sound intensity also increases to a maximal value immediately before the surface wave stops. If the temperature of the surrounding medium decreases gradually along the propagation direction, the acoustic surface wave will slow down and stop when the temperature reaches the specific value T_c. In this manner, an acoustic wave of a given frequency can be trapped at any desired point along the surface, simply by tuning the temperature of the surrounding medium. This is illustrated in Fig. 11.11, where three arbitrary cases are shown, each obtained by using a different linear temperature gradient. The trapping temperature is

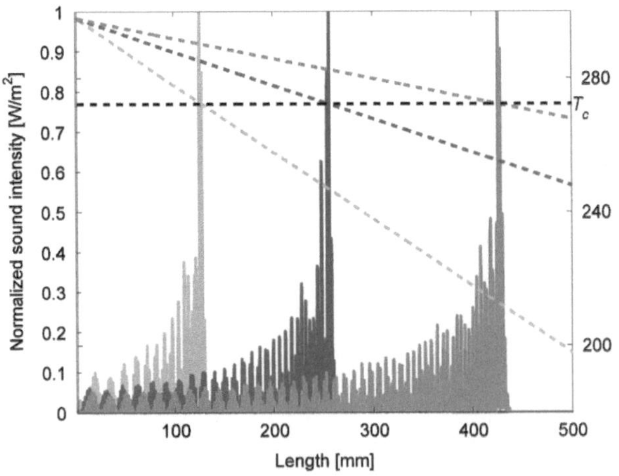

FIGURE 11.11 Acoustic surface wave trapping at arbitrary locations along the grooved medium is obtained by applying different temperature gradients (f = 3.4 kHz). (From Cselyuszka et al. [36])

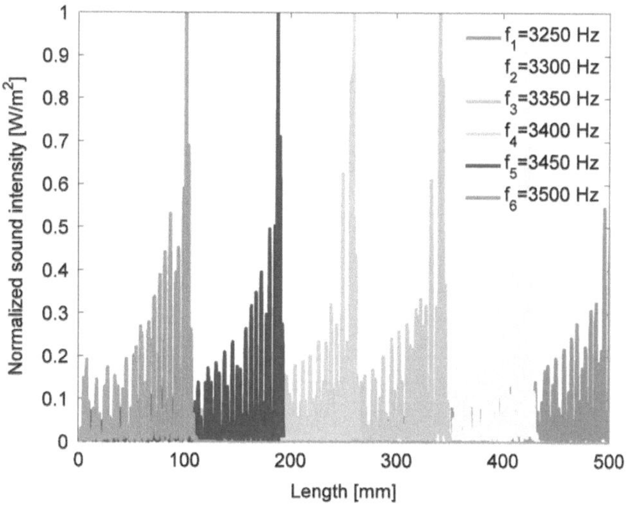

FIGURE 11.12 Spatial acoustic spectral analysis is obtained as acoustic waves of different frequencies are trapped at different positions along the surface. (From Cselyuszka et al. [36])

shown in inset with a black horizontal line. In each case, trapping occurs at the position along the grooved surface where the surrounding temperature reaches T_C.

Alternatively, for a given temperature gradient, the surface waves of different frequencies will be trapped at different points along the surface, as shown in Fig. 11.12. In this manner, spatial spectral analysis of acoustic surface waves can easily be realized.

11.3.5 Temperature-controlled Tunable Gradient Refractive Index (GRIN) Acoustic Medium

The temperature gradient can also be applied in a direction transverse to the propagation direction. In that manner, an acoustic gradient refractive index (GRIN) medium can be obtained [36]. The temperature profile can be easily changed without changing the geometrical properties of structure. So, the same structure can be used for different applications. Such a concept is illustrated in Fig. 11.13, where the middle of the acoustic surface waveguide has a finite width section built in, and its refractive index is tailored to be a gradient in transversal direction [36].

The temperature in the host waveguide is kept constant, while the temperature profile in the middle region can be set to achieve the desired refractive index profile function. The refractive index between the host waveguide and the GRIN lens can be calculated as the ratio between the wave numbers in two different media. With the linear refractive index profile in the GRIN medium, the incident plane wave can be bent by an arbitrary angle [29]. The bending angle is a function of the width and thickness of the lens and the maximal value of the refractive index. The temperature profile for the defined refractive index profile is calculated from the dispersion relation (Eq. (11.7)) and the two calculated temperature profiles, which correspond

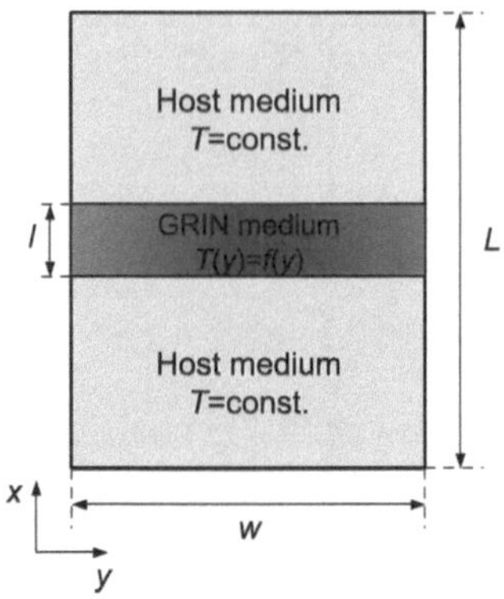

FIGURE 11.13 Illustration of the GRIN structure.

to the linear index profile along the lens, are shown in Fig. 11.14. Both profiles have a similar shape and the same maximum temperature, which corresponds to the host medium temperature, while the lowest temperatures differ by 10 K. The profiles correspond to the wave bending profiles with $\Theta_1 = 15°$ and $\Theta_2 = 30°$ bending angle. This is confirmed by the pressure field distributions shown in Fig. 11.15, where the incident plane wave is, indeed, bent by the defined angles $\Theta = 15°$ and $\Theta = 30°$ after the interaction with the GRIN medium.

Different propagation properties can be achieved by applying more complex refractive index profiles. For instance, the most common refractive index profile in the GRIN medium for focusing is the hyperbolic secant refractive index profile [39–41], which exhibits no distortions [42]. With different hyperbolic secant temperature profiles, the focal length can be varied. To demonstrate versatility and tunability of such profiles, Fig. 11.16 shows the acoustic intensity distribution on the same surface with two different hyperbolic secant temperature profiles. Whereas

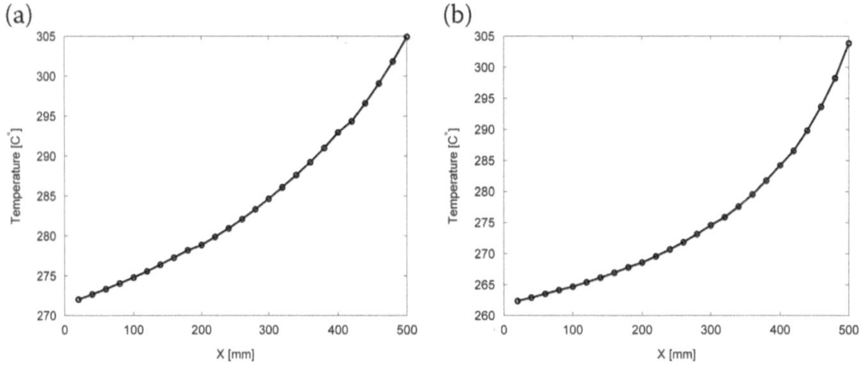

FIGURE 11.14 Temperature T_i [K] in the lens along the direction transverse to propagation direction for bending the acoustic surface wave by a) $\Theta = 15°$ and b) $\Theta = 30°$.

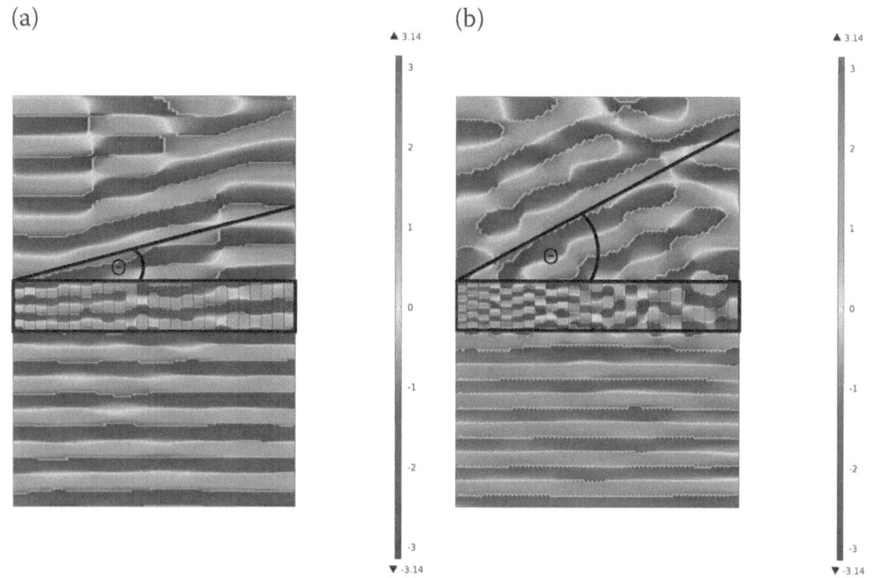

FIGURE 11.15 Acoustic pressure phase distribution on the surface a) Θ = 15° and b) Θ = 30°.

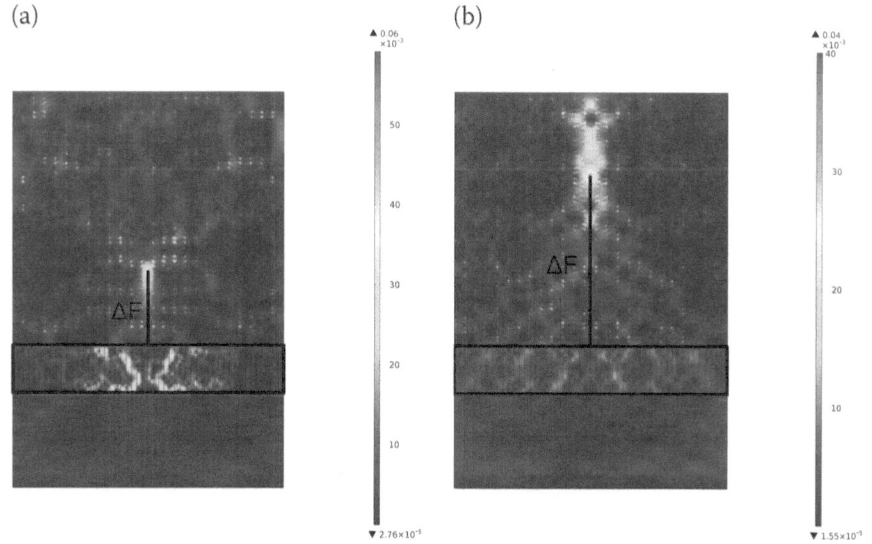

FIGURE 11.16 Distribution of sound intensity on the surface with (a) lower and (b) larger focal length.

in the first case the ASSWs are focused at 120 mm from the lens, in second case, the focal distance is three times larger.

11.3.6 Gas Sensing with ASSW/acoustic Mach-Zehnder Interferometer

Besides temperature, the wave number of ASSW is very sensitive to other physical properties of the background medium, which can be employed for sensing applications. Namely, gaseous fluid is characterized by its bulk modulus and mass density, which are a function of temperature and molar mass and which affect the acoustic dispersion relation. Thus, the dispersion relation and stop-bands of ASSW can be used to determine the properties of the surrounding medium.

This possibility was demonstrated in Cselyuszka et al. [43], in which an acoustic Mach-Zehnder interferometer [44] was employed to determine the properties of a binary gas mixture of air and CO. The ASSW-based Mach-Zehnder interferometer (aMZI) configuration is shown in Fig. 11.17. It consists of an input waveguide, after which an acoustic wave is split into two signals, i.e., into sensing and reference waves, and which are afterward combined at the output waveguide. The acoustic waveguides in all sections are based on corrugated surfaces that support the ASSW modes.

Due to the interference of the reference and sensing waves, the amplitude of the output wave depends on the relative phase difference between the two waves at the point where the waves are superimposed. In this manner, the wave interference, and consequently the phase difference, is converted into amplitude modulation. The phase difference can be calculated as

$$\Delta\Phi = (k_s - k_r)L \tag{11.8}$$

where k_R and k_S are the ASSW wave numbers in the referent and sensing arms, respectively, and L is the length of the sensing part.

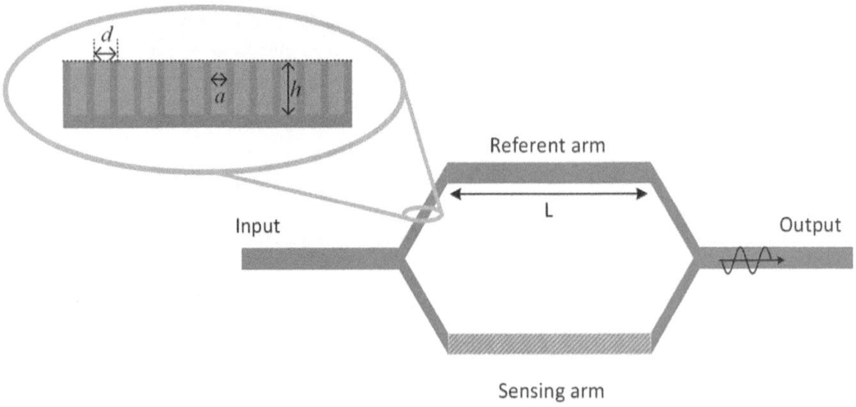

FIGURE 11.17 Structure of the aMZI based on ASSW.

The ASSW wave number can be calculated from the dispersion relation (Eq. (11.2)), while the acoustic wave number in the binary gas mixture can be expressed as

$$k_B = \frac{\omega}{\sqrt{\frac{w_1 c_{p1} + w_2 c_{p2}}{w_1 c_{V2} + w_2 c_{V1}} \left[\frac{w_1}{M_1} + \frac{w_2}{M_2} \right] RT}} \tag{11.9}$$

where ω is the angular frequency, R is the universal gas constant, T is temperature, w is the relative concentration of gas, c_v and c_p are the specific heat capacities of gas on constant velocity and constant pressure, respectively, and M is the molar mass of gas. The indices 1 and 2 in the subscript relate to gas 1 and gas 2 in the binary mixture. The wave numbers k_R and k_S are calculated using Eqs. (11.1) and (11.9), whereas w_2 is equal to zero in the case of k_r since it is related to the pure air, whereas in the case of k_s, the parameter w_2 is varied since k_s is related to the sensing arm.

To demonstrate the gas-sensing potential of aMZI, consider a scenario in which the aMZI is used for monitoring relative gas concentrations in the CO/air mixture. To that end, the background medium above the sensing arm is set to be the gas mixture CO/air with predefined relative concentration of CO, whereas the background gas above the rest of the structure is pure air [45]. The operating frequency and length L can be optimized to achieve the phase difference $\Delta\Phi$ in the range $0 < \Delta\Phi < \pi$.

The phase difference between the reference and sensing arms as a function of CO concentration in the gas mixture is shown in Fig. 11.18. It corresponds to the predefined CO concentration range from 0% to 1.5%, and it implies linear

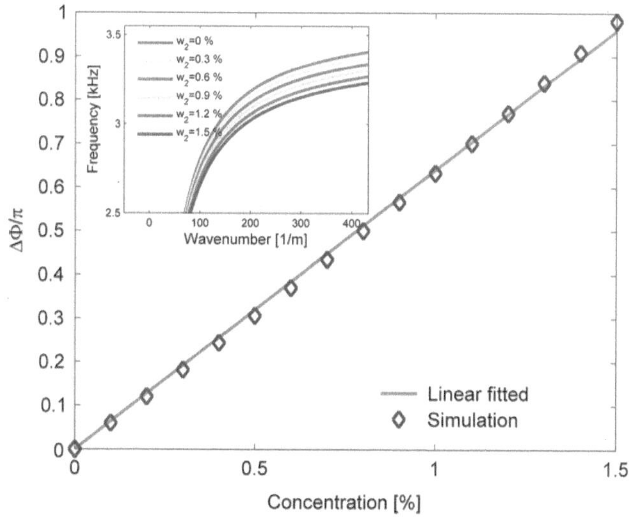

FIGURE 11.18 Normalized phase difference between the two branches of interferometer as a function of CO concentration in the gas mixture. The inset shows the dispersion variation due to concentration changes.

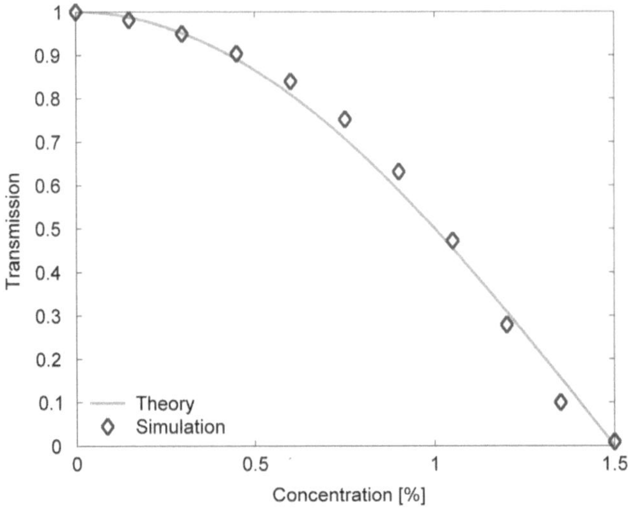

FIGURE 11.19 Transmission coefficient of aMZI as a function of CO concentration in binary gas mixture.

dependence and high sensitivity of the structure behavior to the small changes in the CO concentration. The simulated transmission coefficient of the aMZI as a function of CO concentration using the calculation method from Cselyuszka et al. [46] is in Fig. 11.19. The results are in excellent agreement with the theoretical predictions, and they also imply the high sensitivity of the structure to the CO concentration.

Acoustic pressure field distributions for different CO concentrations are shown in Fig. 11.20, and they confirm the operating principle of the acoustic Mach-Zehnder interferometer. For the CO concentration of 0%, the reference and sensing waves are in phase, whereas for the CO concentration of 1.5%, they are out of phase and thus cancel at the output.

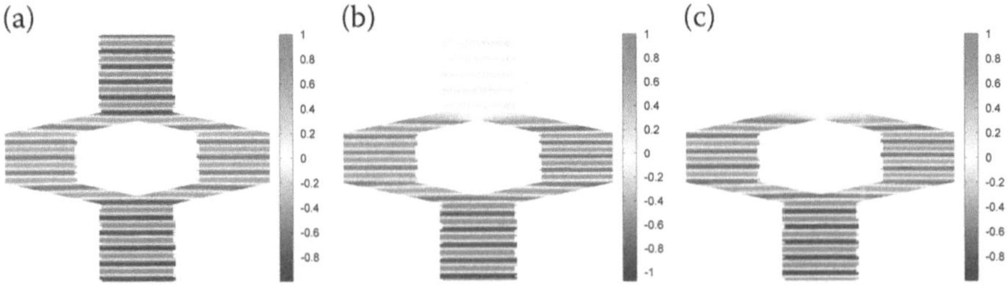

FIGURE 11.20 Acoustic pressure distribution for different CO concentrations in the gas mixture a) 0%, b) 0.9% and c) 1.5%.

11.3.7 Spoof-fluid-spoof Acoustic Waveguide and Its Applications for Sound Manipulation

Among the most common plasmonic waveguide structures, metal-insulator-metal (MIM) waveguides have received special attention for guiding SPPs [47–49]. The spoof plasmon analogue of MIM is the doubly corrugated waveguide, which comprises a spoof-insulator-spoof (SIS) waveguide [50,51]. In this section, the acoustic counterpart of the plasmonic MIM structure, based on the ASSW concept, is presented.

Two corrugated rigid surfaces separated by a gap filled with a fluid medium form the spoof-fluid-spoof (SFS) acoustic waveguide (Fig. 11.21). The parameters d, a and h stand for the period, width and depth of grooves, respectively, while g is the height of the fluid gap between the rigid corrugated surfaces. For the sake of simplicity and without losing generality, a lossless 2D case is considered, where the structure extends infinitely along the z-direction, and the acoustic wave propagates in x-y plane.

The dispersion relation of the SFS waveguide is derived using an effective medium approximation, and they can be written as

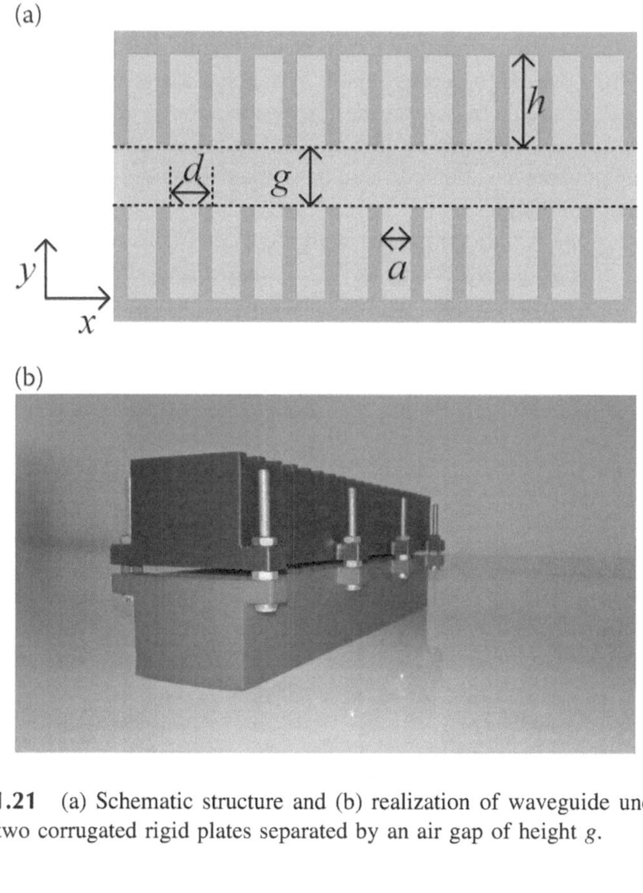

FIGURE 11.21 (a) Schematic structure and (b) realization of waveguide under analysis, formed by two corrugated rigid plates separated by an air gap of height g.

$$\frac{a}{d}k_0 \tan(hk_0) = \sqrt{\beta^2 - k_0^2} \tanh\left(\frac{g}{2}\sqrt{\beta^2 - k_0^2}\right) \tag{11.10}$$

$$\frac{a}{d}k_0 \tan(hk_0) = \sqrt{\beta^2 - k_0^2} \coth\left(\frac{g}{2}\sqrt{\beta^2 - k_0^2}\right) \tag{11.11}$$

where k_0 is the wave number in the fluid, and β is the propagation constant in the SFS waveguide. The two solutions imply two propagation modes – even and odd, i.e., symmetric and anti-symmetric modes in the SFS waveguide. The symmetric and anti-symmetric modes refer to the symmetry or anti-symmetry of the pressure and transverse velocity fields $p(y)$ and $v_y(y)$ as a function of y within the fluid region. When the air gap between the corrugated plates is large, the SFS waveguide modes split into two single ASSWs, and in the limit $g \longrightarrow \infty$, both Eqs. (11.10) and (11.11) become Eq. (11.2), which is the dispersion relation of a single ASSW. Fig. 11.22a,b shows the symmetric and anti-symmetric modes, respectively, for the SFS waveguide. The insets show the corresponding spatial distribution of the acoustic pressure within the waveguide for each mode.

The acoustic pressure field distribution inside the waveguide shows the expected characteristics for both modes – in the symmetric case, the field distribution along the y-direction is symmetrical, whereas the anti-symmetric mode exhibits anti-symmetric features. The symmetric mode is the dominant mode in the SFS waveguide, without a cut-off frequency, and it is weakly dispersive at low frequencies. However, as frequency increases, strong dispersion occurs, and the effective wave number asymptotically approaches infinity. Nevertheless, due to the finite size of the period $d > 0$, the propagation constant has a maximum at $\beta = \pi/d$, beyond which the mode hits a stop-band.

The second mode is anti-symmetric with a cut-off frequency, and it has a very flat dispersion, implying slow sound waves. A smaller gap induces a smaller group velocity for the symmetric mode, and the maximal frequency, which is supported in

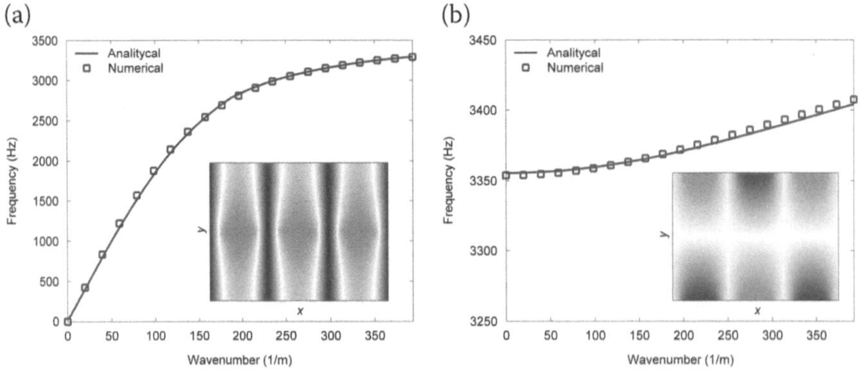

FIGURE 11.22 Dispersion curve of (a) symmetric and (b) anti-symmetric mode. The insets show the spatial pressure distribution inside the waveguide for the corresponding modes when $\beta = 100\ m^{-1}$. ([From Cselyuska et al. [37])

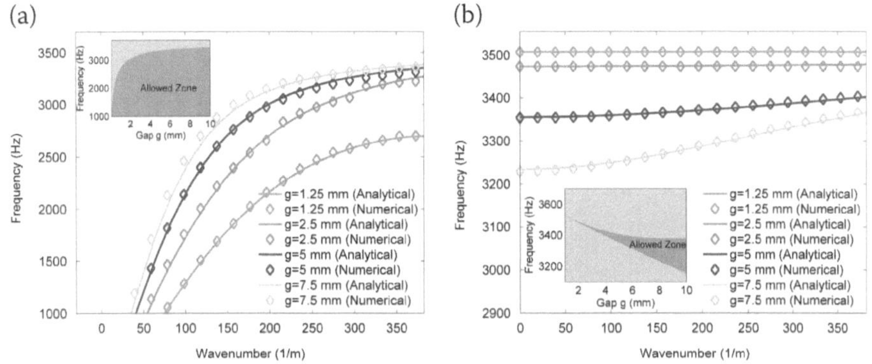

FIGURE 11.23 Dispersion curve of (a) symmetric and b) anti-symmetric mode for different values of gap size. The insets show the allowed spectral zones and forbidden spectral zones for each mode, as a function of the gap size. (From Cselyuska et al. [37])

the SFS waveguide, correspondingly decreases. Similarly, the anti-symmetric mode becomes even flatter as the gap height decreases, causing a corresponding narrowing of the operating frequency range.

Fig. 11.23a,b shows the dispersion properties of the symmetric and anti-symmetric modes with different gap thicknesses, whereas the insets show the allowed propagation zone for each mode as a function of the gap size. Such sensitivity and extreme slow-wave features, especially for the anti-symmetric mode, make the proposed acoustic waveguide a good candidate for acoustic wave modulation and sensing applications. While the anti-symmetric mode supports truly extreme dispersion features, in practice the symmetric mode may be more easily excited and detected.

Fig. 11.24 shows the measured wave number as a function of the gap height g, and it reveals that the dispersion significantly depends on the gap height, which is further confirmed by the inset showing the simulated values of the guided normalized wavelengths versus the height of the air gap.

The concept is further demonstrated in Fig. 11.25a–c, which shows the acoustic pressure field distribution for different values of g and the case of $g \longrightarrow \infty$.

The previous analysis shows that the model dispersion of SFS waveguide is very sensitive to the gap height, and this behavior can be expected if other geometrical and/or fluid parameters are varied. Also, this indicates a strong potential of the proposed structure for acoustic wave modulation or sensing. For instance, the dispersion and group velocity in the SFS waveguide can be dynamically changed by mechanically adjusting the gap g, creating a tunable delay line, while extreme sensing features can be achieved if the sensed analyte is the fluid in the gap. Moreover, the flat dispersion property of the antisymmetric mode of the proposed structure and its finite cut-off frequency can be exploited as a near-zero medium, which implies a number of possible applications, such as acoustic super coupling and extreme non-reciprocity [52,53].

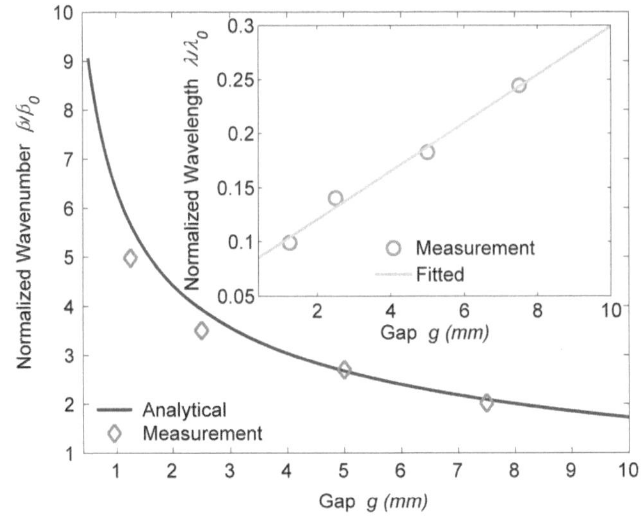

FIGURE 11.24 Analytical and measured wave number as a function of air gap height between the corrugated surfaces. Analytical and measured wavelengths as a function of the gap height are shown in the inset. (From Cselyuska et al. [37])

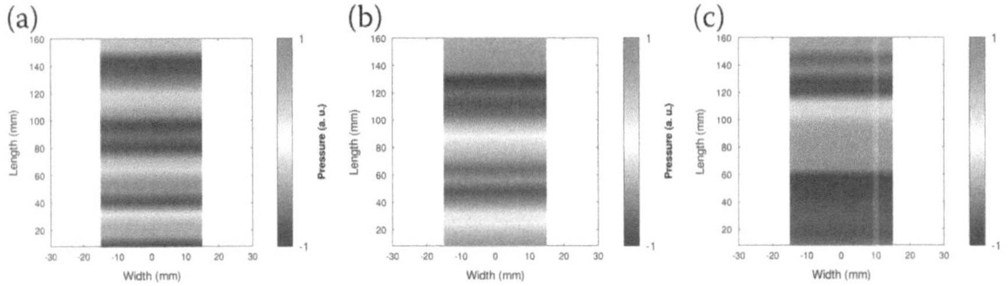

FIGURE 11.25 Measured acoustic pressure distribution along the structure at the operating frequency for different values of g: (a) g = 1.25 mm, (b) g = 5 mm and (c) single corrugate surface ($g \longrightarrow \infty$.) (From Cselyuska et al. [37])

11.4 CONCLUSION

In this chapter, the acoustic spoof surface wave propagation along the interface between the background fluid medium and a corrugated rigid plate was presented. We explained the theoretical background of the phenomenon and provided the dispersion relation for the acoustic wave propagation on the surface. We presented the versatility and potential of the phenomenon through applications, such as slowing down the surface acoustic waves, extraordinary transmission, propagation control by temperature, spatial spectral analysis of wideband acoustic waves, gas sensing and steering of acoustic spoof surface waves. Also, a new type of acoustic spoof-fluid-spoof waveguide was presented, which is an analogue of the metal-insulator-metal plasmonic waveguide. We explained properties of the guided acoustic modes and their dispersion relations and showed that the group and phase velocities can be controlled by changing the waveguide geometrical parameters, thus enabling the application of this technology in acoustic delay lines, modulators and sensors.

ACKNOWLEDGMENTS

This work was supported in part by the National Science Foundation EFRI program and Simons Foundation, and by the European Union's Horizon 2020 research and innovation program under Grant Agreement 777714.

REFERENCES

1. J.-C. Weeber, A. Dereux, C. Girard, J. R. Krenn, and J.-P. Goudonnet, "Plasmon polaritons of metallic nanowires for controlling submicron propagation of light," *Phy. Rev. B*, vol. 60, p. 9061, 1999.
2. H. Ditlbacher J.-P. Krenn, G. Schider, A. Leitner, and F. R. Aussenegg, "Two-dimensional optics with surface plasmon polaritons," *Appl. Phys. Lett.*, vol. 81, p. 1762, 2002.
3. W. L. Barnes, A. Dereux, and T. W. Ebbesen, "Surface plasmon subwavelength optics," *Nature* vol. 14, p. 824, 2003.
4. T. W. Ebbesen, H. J. Lezec, H. F. Ghaemi, T. Thio, and P. A. Wolff, "Extraordinary optical transmission through sub-wavelength hole arrays," *Nature*, vol. 391, p. 667, 1998.
5. C. Genet and T. W. Ebbesen, "Light in tiny holes," *Nature*, vol. 445, p. 39, 2007.
6. M. I. Stockman, "Nanofocusing of optical energy in tapered plasmonic waveguides," *Phys. Rev. Lett.*, vol. 93, p. 137404, 2004.
7. A. Rusina M. Durach, K. A. Nelson, and M. I. Stockman, "Nanoconcentration of terahertz radiation in plasmonic waveguides," *Opt. Exp.*, vol. 16, p. 18576, 2008.
8. J. B. Pendry, L. Martin-Moreno, and F. J. Garcia-Vidal, "Mimicking surface plasmons with structured surfaces," *Science*, vol. 305, p. 847, 2004.
9. A. P. Hibbins, B. R. Evans, and J. R. Sambles, "Experimental verification of designer surface plasmons," *Science*, vol. 308, p. 670, 2005.
10. N. Yu, Q. J. Wang, M. A. Kats, et al., "Designer spoof surface plasmon structures collimate terahertz laser beams," *Nat. Mat.*, vol. 9, p. 730, 2010.
11. A. I. Fernandez-Domguez, L. Martin-Moreno, F. J. Garcia-Vidal, S. R. Andrews, and S. A. Maier, "Spoof surface plasmon polariton modes propagating along periodically corrugated wires," *IEEE J. Sel. Top. Q. Elect.*, vol. 14, p. 1515, 2008.
12. Y. J. Zhou and T. J. Cu, "Broadband slow-wave systems of subwavelength thickness excited by a metal wire," *Appl. Phys. Lett.*, vol. 99, p. 101906, 2011.
13. D. Martin-Cano, M. L. Nesterov, A. I. Fernandez-Dominguez, F. J. Garcia-Vidal, L. Martin-Moreno, and E. Moreno, "Domino plasmons for subwavelength terahertz circuitry," *Opt. Exp.*, vol. 18, p. 754, 2010.
14. Y. G. Ma, L. Lan, S. M. Zhong, and C. K. Ong, "Experimental demonstration of subwavelength domino plasmon devices for compact high frequency circuit," *Opt. Exp.*, vol. 19, p. 21189, 2011.
15. F. J. Garcia-Vidal, L. Martin-Moreno, and J. B. Pendry, "Surfaces with holes in them: new plasmonic metamaterials," *J. Opt. A: Pure Appl. Opt.*, vol. 7, p. 97, 2005.
16. Z. Ruana and M. Qiub, "Slow electromagnetic wave guided in subwavelength region along one-dimensional periodically structured metal surface," *Appl. Phys. Lett.*, vol. 90, p. 201906, 2007.

17. L. Kelders J. F. Allard, and W. Lauriks, "Ultrasonic surface waves above rectangular-groove gratings," *J. Acoust. Soc. Am.*, vol. 103, p. 2730, 1998.
18. L. Kelders, W. Lauriks, and J. F. Allard, "Surface waves above thin porous layers saturated by air at ultrasonic frequencies," *J. Acoust. Soc. Am.*, vol. 104, p. 882, 1998.
19. Z. He, H. Jia, C. Qiu, et al., "Nonleaky surface acoustic wave s on a textured rigid surface," *Phys. Rev. B*, vol. 83, p. 132101, 2011.
20. L. Rayleigh, "On waves propagated along the plane surface of an elastic solid," *Proc. Lond. Math. Soc.*, vol. 17, p. 4, 1885.
21. J. Christensen, A. I. Fernandez-Dominguez, F. L. Perez, L. Martin-Moreno, and F. J. Garcia-Vidal, "Collimation of sound assisted by acoustic surface waves," *Nat. Phys.*, vol. 3, p. 851, 2007.
22. Z. He, H. Jia, C. Qiu, et al., "Nonleaky surface acoustic waves on a textured rigid surface," *Phys. Rev. B*, vol. 83, p. 132101, 2011.
23. N. Korozlu, O. A. Kaya, A. Cicek, and B. Ulug, "Self-collimation and slow-sound effect of spoof surface acoustic waves," *J. App. Phys.*, vol. 125, p. 074901, 2019.
24. J. Christensen, L. Martín-Moreno, and F. J. García-Vidal, "Enhanced acoustical transmission and beaming effect through a single aperture," *Phys. Rev. B*, vol. 81, p. 174104, 2010.
25. Z. Yu M.-H. Lu, L. Feng, et al., "Acoustic surface evanescent wave and its dominant contribution to extraordinary acoustic transmission and collimation of sound," *Phys. Rev. Lett.*, vol. 104, p. 164301, 2010.
26. T. Devaux, H. Tozawa, P. H. Otsuka, et al., "Giant extraordinary transmission of acoustic waves through a nanowire," *Sci. Adv.*, vol. 6, p. 8507, 2020.
27. J. J. Christensen, P. A. Huidobro, L. Martín-Moreno, and F. J. García-Vidal, "Confining and slowing airborne sound with a corrugated metawire," *Appl. Phys. Lett.*, vol. 93, p. 083502, 2008.
28. J. Zhu, Y. Chen, X. Zhu, et al., "Acoustic rainbow trapping," *Sci. Rep.*, vol. 3, p. 1728, 2013.
29. H. Jia, M. Lu, X. Ni, M. Bao, and X. Li, "Spatial separation of spoof surface acoustic waves on the graded groove grating," *J. Appl. Phys.*, vol. 116, p. 124504, 2014.
30. A. Ciceka, Y. Arslan, D. Trak, et al., "Gas sensing through evanescent coupling of spoof surface acoustic waves," *Sens. Act. B: Chem.*, vol. 288, p. 259, 2019.
31. Y. Chen, H. H. Liu, M. Reilly, H. Bae, and M. Yu, "Enhanced acoustic sensing through wave compression and pressure amplification in anisotropic metamaterials," *Nat. Commun.*, vol. 5, p. 5247, 2014.
32. Y. Ye, M. Ke, Y. Li, T. Wang, and Z. Liu, "Focusing of spoof surface-acoustic-waves by a gradient-index structure," *J. Appl. Phys.*, vol. 114, p. 154504, 2013.
33. T. Liu, F. Chen, S. Liang, H. Gao, and J. Zhu, "Subwavelength sound focusing and imaging via gradient metasurface-enabled spoof surface acoustic wave modulation," *Phys. Rev. App.*, vol. 11, p. 034061, 2019.
34. S. R. Craig, J. H. Lee, and C. Shi, "Beamforming with transformation acoustics in anisotropic media," *Appl. Phys. Lett.*, vol. 117, p. 011907, 2020.
35. J. Kim, S. Park, Md. Anzan-Uz-Zaman, and K. Song, "Holographic acoustic admittance surface for acoustic beam steering," *Appl. Phys. Lett.*, vol. 115, p. 193501, 2019.
36. N. Cselyuszka, M. Secujski, N. Engheta, and V. Bengin, "Temperature-controlled acoustic surface waves," *New J. Phys.*, vol. 18, p. 103006, 2016.
37. N. Cselyuszka, A. Alu, and N. Jankovic, "Spoof-Fluid-Spoof acoustic waveguide and its applications for sound manipulation," *Phys. Rev. Appl.*, vol. 12, p. 054014, 2019.
38. P. Hess, "Surface acoustic waves in materials science," *Phys. Today*, vol. 55, p. 42, 2002.
39. B. I. Popa and S. A. Cummer, "Design and characterization of broadband acoustic composite metamaterials," *Phys. Rev. B*, vol. 80, p. 174303, 2009.

40. A. Climente, D. D. Torrent, and J. Sánchez-Dehesa, "Sound focusing by gradient index sonic lenses," *Appl. Phys. Lett*, vol. 97, p. 104103, 2010.
41. Sz-C. S. Lin and T. J. Huang, "Gradient-index phononic crystals," *Phys. Rev. B*, vol. 79, p. 094302, 2009.
42. C. Gomez-Reino, M. V. Perez, and C. Bao, *Gradient-Index Optics: Fundamentals and Applications*, New York, NY, USA: Springer, 2002.
43. N. Cselyuszka, A. Alu, and N. N. Jankovic, "Binary gas concentration sensing using acoustic Mach-Zehnder interferometer based on acoustic spoof surface waves," *Proc. of Metamaterials*, 16–19 September 2019.
44. K. P. Zetie, S. F. Adams, and R. M. Tocknell, "How does a Mach-Zehnder interferometer work," *Phys. Edu.*, vol. 35, p. 46, 2000.
45. B. E. Poling, J. M. Prausnitz, and J. P. O'Connell, *Properties of gases and liquids*, 5th ed., New York: McGraw-Hill Education, 2001.
46. N. Cselyuszka, M. Secujski, and V. Bengin, "Analysis of acoustic metamaterials – acoustic scattering matrix and extraction of effective parameters," *Proc. of Metamaterials*, 170–172, 2012.
47. E. N. Economou, "Surface plasmons in thin films," *Phys. Rev.*, vol. 182, p. 539, 1969.
48. V. R. Almeida, Q. Xu, C. A. Barrios, and M. Lipson, "Guiding and confining light in void nanostructure," *Opt. Lett.*, vol. 29, p. 1209, 2004.
49. J. A. Dionne, L. A. Sweatlock, H. A. Atwater, and A. A. Polman, "Plasmon slot waveguides: Towards chip-scale propagation with subwavelength-scale localization," *Phys. Rev. B*, vol. 73, p. 035407, 2006.
50. M. A. Kats, D. Woolf, R. Blanchard, N. Yu, and F. Capasso, "Spoof plasmon analogue of metal-insulator-metal waveguides," *Opt. Exp.*, vol. 19, p. 14860, 2011.
51. D. Woolf, M. A. Kats, and F. Capasso, "Spoof surface plasmon waveguide forces," *Opt. Lett.*, vol. 39, p. 517, 2014.
52. H. Esfahlani M. S. Byrne, M. McDermott, and A. Alù, "Acoustic supercoupling in a zero-compressibility waveguide," *Research*, vol. 2019, p. 2457870, 2019.
53. L. Quan, D. L. Sounas, and A. Alù, "Non-reciprocal Willis coupling in zero-index moving media," *Phys. Rev. Lett.*, vol. 123, p. 064301, 2019.

12 The Principle of Miniaturization of Microwave Patch Antennas

Oleg Rybin
School of Radio Physics, Biomedical Electronics & Computer Systems, V.N. Karazin Kharkiv National University, Kharkiv, UKRAINE

12.1 INTRODUCTION

An antenna (or antenna array) is an essential element of any electromagnetic (EM) system. Wireless technology has been growing exponentially since the beginning of this century due to application domains such as e-health, internet of things, and smart buildings. This fast growth can continue if we develop smart antenna systems that can combat a widened spectrum with energy efficiency, at low cost, and with small size, and can operate in variable embeddings, such as chip packaging, the human body, matching networks, etc.

One of the possible candidates for compact wireless antenna systems is a patch antenna. These antennas are very popular printed aperture antennas due to their simple design, low cost, light mass, low profile and ease of integration with circuits and forming arrays.

The design features of patch antennas make it tempting to miniaturize their volume profile by means of an increase in the dielectric constant of the substrate. Indeed, the simplest design of a rectangular patch antenna is defined by the formulas [1,2]

$$\left.\begin{aligned} L &= \frac{c}{2f_r\sqrt{\varepsilon_{reff}}} - 0.824\cdot d\frac{(\varepsilon_{reff}+0.3)(W/d+0.264)}{(\varepsilon_{reff}-0.258)(W/d+0.8)}, \\ \varepsilon_{reff} &= \frac{\varepsilon_r+1}{2} + \frac{\varepsilon_r-1}{2}\left[1+12\frac{d}{W}\right]^{-1/2}, \\ W &= \frac{c}{2f_r}\sqrt{\frac{2}{1+\varepsilon_r}}, \\ L_a &= L + 6\cdot d, \quad W_a = W + 6\frac{\log 16}{\pi}d, \end{aligned}\right\} \tag{12.1}$$

DOI: 10.1201/9781003050162-12

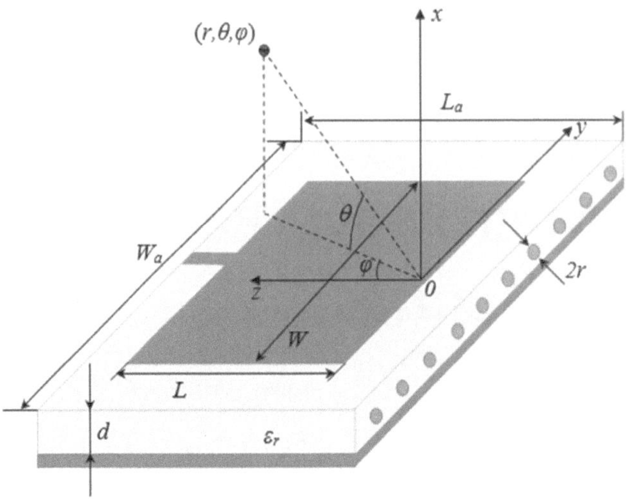

FIGURE 12.1 Patch antenna with composite substrate in spherical coordinate system.

where L is the length of patch, W is the width of patch, f_r is the resonance frequency of antenna, $\varepsilon_{\text{reff}}$ is the effective relative permittivity of antenna, c is the speed of light in vacuum, d is the thickness of patch, ε_r is the dielectric constant of substrate, L_a is the length of antenna, and W_a is the width of antenna (see Fig. 12.1).

Looking at Eqs. (12.1), one may state: the greater the value of relative permeability ε_r, the smaller the linear dimensions L_a and W_a of the antenna patch. If this conclusion is true, it leads to the idea of miniaturization of a patch antenna: an increase in relative permittivity of the antenna substrate leads to a decrease in the antenna volume profile. Several attempts to realize this idea were made at the end of the last century; see, for example, Mongia et al. and Colburn and Rahmat-Samii [3,4]. Although miniaturization was achieved in these studies, the performance of such antennas was considerably degraded. This was because of a strong capacitive coupling between the antenna patch and the antenna ground plane. In order to decrease the above coupling, it was proposed to replace a homogenous high permittivity dielectric by an inhomogeneous composite/metamaterial with the same value of constitutive parameters [5–7]. Such usability of composites/metamaterials is possible due to the idea of tuning the effective frequency-dependent relative permittivity and relative permeability of these artificial materials through individually adjusting the dimensions of the electric and magnetic resonant inclusions in the unit cell.

Initially, double negative metamaterials, metamaterials with low effective refractive index or mu-negative materials were used for miniaturizing the patch antennas; see, for example, the cited works [8–13]. Metamaterials with an increase in the effective refractive index are much less commonly used for creating the composite substrates, although the idea of replacing a homogenous high permittivity dielectric substrate with a composite/metamaterial substrate with a high value in the complex effective refractive index for miniaturization purposes is more logical.

Moreover, this idea was proposed long ago [14–16]. Thus, the utilization of composites/metamaterials in creating miniaturized patch antennas is still a very challenging problem. Nevertheless, creating an appropriate miniaturization concept can be considered a sufficient condition for design and fabrication of pilot samples of novel compact antennas.

This section is devoted to creating the concept of miniaturization of rectangular patch antennas on metal-dielectric composites/metamaterials with enhanced effective relative permittivity (i.e., $\mathrm{Re}(\varepsilon_r) > 1$). It is assumed that we consider only non-magnetic substrates (i.e., $\mathrm{Re}(\varepsilon_r) \approx 1$).

12.2 2D MODEL OF A RECTANGULAR PATCH ANTENNA WITH ONE-LAYER WIRE COMPOSITE/METAMATERIAL SUBSTRATE

12.2.1 Standard Design

According to the above hypothesis, tuning the electric properties of composite substrates enables one to improve the performance of rectangular patch antennas. The idea of using a dielectric matrix with a built-in wire grid as a composite arose when a relevant experiment was reported by Zouganelis et al. [17]. According to the results, putting a flat metal-dielectric composite sample (Fig. 12.2) on the top of patch antenna (Fig. 12.3) results in improved directivity of the antenna. The sample is presented as the parallelepiped dielectric matrix with a built-in wire grid. Moreover, the sample is located on the top of the antenna in such a way that its wires are exactly parallel to the antenna patch. This phenomenon was explained in short in Rybin and Shulga [18] using the ray-optic effect based on Snell's law. Later, in Rybin [19], this theory was developed for the case of patch antennas with wire grids embedded into dielectric substrates. In this study, we present the full-wave analysis of patch antenna with a one-layer wire grid embedded into dielectric substrate (Fig. 12.1). The analysis is based on the approach of given surface current distribution [20,21].

Consider 2D geometry of the rectangular patch antenna with a composite substrate depicted in Fig. 12.1, with respect to a Cartesian coordinate system (Fig. 12.4). The perfect conducting patch of zero thickness is placed in the plane

FIGURE 12.2 Composite samples. Left: Sample with holes. Right: Same sample with iron wires inside the holes.

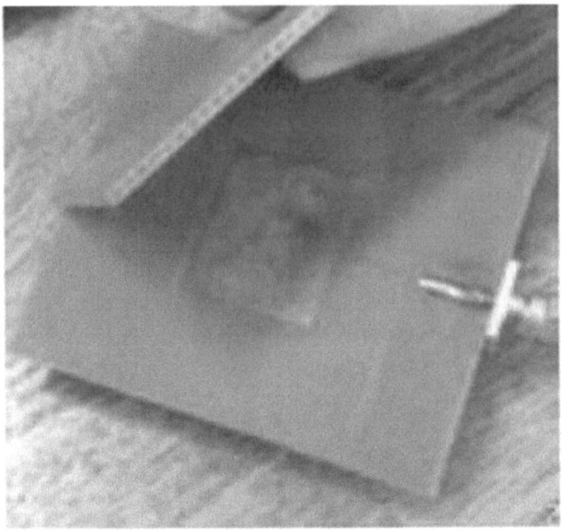

FIGURE 12.3 Composite sample located on the top of a patch antenna.

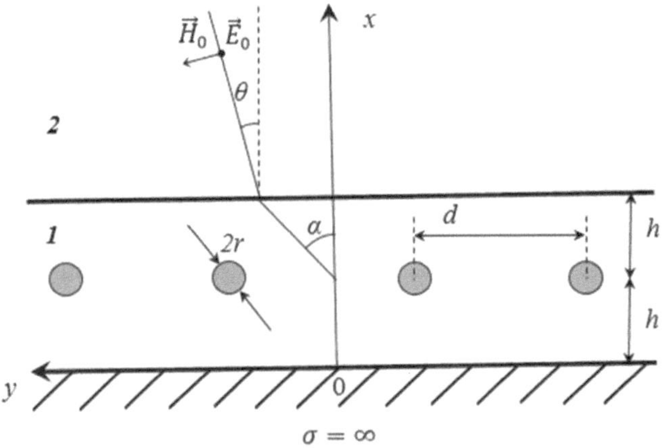

FIGURE 12.4 2D geometry of rectangular patch antenna on composite substrate with one-layer wire grids.

with $x = 2h$ and $-L/2 \leq z \leq L/2$, $-W/2 \leq y \leq W/2$, where W and L are the width and length of the patch, respectively. The non-magnetic wire grid is contained in the plane $x = h$ and is backed by a perfect conducting wall at $x = 0$. The grid is composed of an array of copper wires of circular cross-section. The radius of wires is r. The wires are parallel to the z-axis and spaced at a distance d between the centers. The substrate host dielectric material ($0 \leq x$ $2h$) has a relative permittivity ε_m and a relative permeability μ_r. The microwave approximation of the relative permittivity of wires ε_i is given by [22]

$$\varepsilon_i(\omega) = 1 + \sigma_i / i\omega\varepsilon_0 \tag{12.2}$$

where ε_0 is the permittivity of the vacuum, and σ_i is the conductivity of the wire material. The microwave approximation of the relative permeability of the wires μ_i is given by [23]

$$\mu_i(\omega) = \frac{2}{k_s r} \frac{J_1(k_s r)}{J_0(k_s r)} \tag{12.3}$$

where $k_s(\omega) = k_0\sqrt{\varepsilon_i(\omega)}$, $J_m(x)$ is the *mth* order Bessel function of a real variable x of the first kind, $J_m(x)$ is the wave number in the free space, and ω is the circular frequency.

Let a plane monochromatic EM wave of frequency ω, with the electric field component of magnitude E_0 be excited by a surface current j of the antenna patch in such a way that the electric intensity vector, parallel to the z-axis, is incident on the wire grid with angle of incidence θ, as shown in Fig. 12.4. The primary electric intensity vector is given by

$$\vec{E}^{in} = E^{in}(x, y)\cdot\vec{z_0} = E_0 e^{-ik_2(x \cos\theta + y \sin\theta)}\cdot\vec{z_0}, \tag{12.4}$$

where $k_2 = k_0$ is the wave number in the 2nd ($x > 2h$) space domains which is the free space, and z_o is the unit vector of the z-axis. The time factor $e^{i\omega t}$ is omitted throughout the study. Then, considering the results in Wait [24], neglecting the evanescent waves and under the conditions $r << d$, $r << h$, gives for the 1st ($0 < x < 2h$) space domain

$$\vec{E_1} = E_1(x, y)\cdot\vec{z_0} = E_0 e^{ik_1 y \sin\alpha}\left(e^{ik_1 x \cos\alpha} + Re^{-ik_1 x \cos\alpha}\right)\cdot\vec{z_0}, \tag{12.5}$$

and for the 2nd space domain

$$\vec{E_2} = E_2(x, y)\cdot\vec{z_0} = A\cdot e^{-ik_2 y \sin\theta} e^{-ik_2(x-2h)\cos\theta}\cdot\vec{z_0}, \tag{12.6}$$

Here [24]

$$R = -1 + \frac{(2Z_1/\cos\theta)\sin^2(k_1 h \cos\theta)}{(Z_1/2\cos\theta)(1 - e^{-2ik_1 h \cos\theta}) + Z_g} = -1 + G, \tag{12.7}$$

with $k_1 = \sqrt{\varepsilon_m \mu_m}\, k_0$ as the wave number in the substrate host dielectric of the 1st space domain, $Z_1 = \sqrt{\mu_0 \mu_m / \varepsilon_0 \varepsilon_m}$ as the impedance of host dielectric of the 1st space domain, and Z_g is the surface impedance of the wire grid given by [24]

$$Z_g = \frac{i\mu_0 \omega d}{2\pi}\left(\log\frac{d}{2\pi r} + \Delta\right) + (1 + i)\frac{d}{2\pi r}\sqrt{\frac{\mu_0 \omega}{2\sigma_i}}, \tag{12.8}$$

where μ_0 is the permeability of vacuum and the summand Δ is given by

$$\Delta = \frac{1}{2}\sum_{m=1}^{\infty}\left[\frac{1-e^{-4\pi(h/d)\sqrt{(m+(d/\lambda_1)\sin\theta)^2-(d/\lambda_1)^2}}}{\sqrt{(m+(d/\lambda_1)\sin\theta)^2-(d/\lambda_1)^2}}+\frac{1-e^{-4\pi(h/d)\sqrt{(m-(d/\lambda_1)\sin\theta)^2-(d/\lambda_1)^2}}}{\sqrt{(m-(d/\lambda_1)\sin\theta)^2-(d/\lambda_1)^2}}-\frac{2}{m}\right], \quad (12.9)$$

In this equation, λ_1 is the wavelength in the 1st space domain.

It is assumed throughout that $d << \lambda_1$ and $d = 2h$. That is why the factor Δ is negligible compared with $\log(d/2\pi r)$ and $\lim\limits_{d/\lambda_1\to 0}\Delta|_{d=2h} = e^{-2\pi}/(e^{-2\pi}-1)$. In this study, we neglect the evanescent waves. That is why we are not able to evaluate the antenna performance parameters associated with the near-field distribution.

The approximations for the magnetic intensity vector in the 1st and 2nd space domains are to be defined using Faraday's law. The appropriate straightforward algebra gives

$$\vec{H_1}(x, y) = -\frac{E_0}{Z_1}[E^{1+}(x, y)\cdot \sin\alpha\cdot\vec{x_0} - E^{1-}(x, y)\cdot\cos\alpha\cdot\vec{y_0}] \quad (12.10)$$

$$\vec{H_2}(x, y) = -\frac{E_2(x, y)}{Z_2}[\sin\theta\cdot\vec{x_0} + \cos\theta\cdot\vec{y_0}] \quad (12.11)$$

where

$$E^{1\pm}(x, y) = e^{ik_1 y\sin\theta}\left(e^{ik_1 x\cos\theta} \pm Re^{-ik_1 x\cos\theta}\right) \quad (12.12)$$

Let us present the current flow on the patch in the form

$$\vec{j} = j\cdot\delta(x-2h)\cdot\vec{z_0} = -j_0\cdot\delta(x-2h)\cdot e^{-ik_2 y\sin\theta}\cdot\vec{z_0} \quad (12.13)$$

where j_0 (= const.) is to be predetermined.

The unknowns E_0, A and α can be calculated by imposing the boundary conditions for the tangential electric and magnetic fields to be continuous at the interface $x = 2h$, i.e.,

$$\vec{E_1}|_{x=2h} = \vec{E_2}|_{x=2h} \quad (12.14)$$

$$[\vec{x_0}\times(\vec{H_2}-\vec{H_1})]|_{x=2h} = j\cdot\vec{z_0}, \quad (12.15)$$

where x_0 is the unit vectorvector of the x-axis. Substituting Eqs. (12.5)–(12.13) into Eqs. (12.14) and (12.15) finally gives

$$E_0 = \frac{j_0 Z_2 Z_1}{Z_1 \cos\theta \cdot E^{1+}(2h, 0) + Z_2 \cos\alpha \cdot E^{1-}(2h, 0)}, \tag{12.16}$$

$$E_0 = \frac{j_0 Z_2 Z_1}{Z_1 \cos\theta \cdot E^{1+}(2h, 0) + Z_2 \cos\alpha \cdot E^{1-}(2h, 0)}, \tag{12.17}$$

where $\cos\alpha = \sqrt{1 - (\sin^2\theta/\varepsilon_m \mu_m)}$, and $Z_2 = Z_0$ is the impedance of host dielectric of the 22nd space domain.

The efficiency of the antenna is to be evaluated by the formula [1]

$$e = \frac{P_{rad}}{P_T} = \frac{\mathrm{Re}(P_r)}{P_T} \tag{12.18}$$

where P_{rad} is the total radiated power, and P_r is the complex input power of the antenna, which can be determined following Rybin et al. [21] (it is assumed that Im (P_r) is due to energy that is mostly stored in the vicinity of the wire grid and is just partially stored in the vicinity of antenna patch) as

$$P_r = -\frac{1}{2}\int_S \vec{E_2} \cdot \vec{j}^{\,*} ds = \frac{j_0 S}{2} A, \tag{12.19}$$

and P_T is the power delivered to the antenna terminal from generator via a transmission line with $S = w \times L$ as the patch area. P_T is given by [1,25]

$$P_T = P_{rad} + P_{ohm}. \tag{12.20}$$

Here, P_{ohm} is the power dissipated due to ohmic losses, given by

$$P_{ohm} = \frac{1}{2}\int_S |\vec{j}\,|^2 R_p ds \tag{12.21}$$

with R_p as the ohmic resistance of the antenna patch. Throughout the study, we consider a perfectly thin copper patch along with $R_p = R_s S$, where $R_s = \sqrt{\omega\mu_0/2\sigma_i}$ is the surface resistance of a highly conductive non-magnetic metal in the microwave frequency range [26].

Substituting Eqs. (12.19) and (12.21) in Eq. (12.20) finally gives

$$P_T = \frac{j_0 S}{2} \cdot \mathrm{Re}(A) + \frac{j_0^2 S^2}{2} \sqrt{\frac{\omega \mu_0}{2\sigma_i}}. \tag{12.22}$$

Substituting Eqs. (12.19) and (12.22) in Eq. (12.18) finally yields

$$e = \left(1 + \frac{j_0 S}{\mathrm{Re}(A)} \sqrt{\frac{\omega \mu_0}{2\sigma_i}}\right)^{-1}. \tag{12.23}$$

To calculate both the normalized gain and directivity, the radiated far-field representation in the polar coordinate system is required. According to Balanis [1], we have

$$\left.\begin{aligned} E_\theta &= ik_0 \frac{e^{-ik_0 r}}{2\pi r} \frac{1 + \cos\theta}{2} f(\theta, \varphi) \sin\varphi, \\ E_\varphi &= ik_0 \frac{e^{-ik_0 r}}{2\pi r} \frac{1 + \cos\theta}{2} f(\theta, \varphi) \cos\varphi, \end{aligned}\right\} \tag{12.24}$$

where the expression for the radiation pattern $f(\theta, \varphi)$ is to be evaluated as follows:

$$\begin{aligned} f(\theta, \varphi) &= \int_S E_2(2h, y) \cdot e^{ik_{2y}y + ik_{2z}z} dydz = \\ &= 4|A| \cdot S \cdot \sin c\left(\frac{L}{\lambda_2}(\cos\varphi - 1)\sin\theta\right) \cdot \sin c\left(\frac{L}{\lambda_2}\sin\theta \sin\varphi\right) \cdot \sin c\left(\frac{W}{\lambda_2}\sin\theta\cos\varphi\right) \end{aligned} \tag{12.25}$$

where $\lambda_2 = \lambda_0$ is the wavelength in the 2nd space domain, and $\sin c(x) = \sin(\pi x)/\pi x$. Then, the expression for the radiation density is given by

$$U(\theta, \varphi) = \frac{k_0^2}{8\pi^2 Z_0} \frac{(1 + \cos\theta)^2}{4} |f(\theta, \varphi)|^2, \tag{12.26}$$

and the expression for the directive gain is given as

$$\begin{aligned} D(\theta, \varphi) &= 4\pi \frac{U(\theta, \varphi)}{\mathrm{Re}(P_r)} = \frac{(2k_0 S|A|)^2}{\pi Z_0} \frac{(1 + \cos\theta)^2}{\mathrm{Re}(A)} \cdot \\ &\cdot \sin c^2\left(\frac{L}{\lambda_0}(\cos\varphi - 1)\sin\theta\right) \cdot \sin c^2\left(\frac{L}{\lambda_0}\sin\theta\sin\varphi\right) \cdot \sin c^2\left(\frac{W}{\lambda_0}\sin\theta\cos\varphi\right). \end{aligned} \tag{12.27}$$

As can be seen from Eq. (12.27), the maximum value of directive gain is achievable for $\theta = 0^0$, and its value, notably, the directivity, is given as

$$D_{\max} = \frac{8k_0^2}{\pi Z_0} \frac{(S|A|_{\theta=0})^2}{\mathrm{Re}(A)|_{\theta=0}}. \tag{12.28}$$

At the same time, the maximum gain can be written as

$$G_{\max} = eD_{\max} = \frac{8k_0^2}{\pi Z_0}\frac{(S\,|A|_{\theta=0})^2}{\mathrm{Re}(A)|_{\theta=0}}\left(1 + \frac{j_0 S}{\mathrm{Re}(A)|_{\theta=0}}\sqrt{\frac{\omega\mu_0}{2\sigma_i}}\right)^{-1}. \tag{12.29}$$

Taking into account Eq. (12.25), we are able to derive the expression for the normalized gain in the form [1]

$$\begin{aligned} g(\theta,\ \varphi) &= \\ &= \frac{(1+\cos\theta)^2}{4}\frac{|f(\theta,\varphi)|^2}{|f(\theta,\varphi)|^2_{\max}} = \frac{(1+\cos\theta)^2}{4}\sin c^2\left(\frac{L}{\lambda_0}(\cos\varphi - 1)\sin\theta\right) \\ &\cdot \sin c^2\left(\frac{W}{\lambda_0}\sin\theta\cos\varphi\right). \end{aligned} \tag{12.30}$$

Now, the expressions for the normalized gain along two principle planes are given as follows. For the H-plane, we set $\varphi = 0^0$, which gives $E_\theta = 0$ in Eq. (12.30), i.e.,

$$g_H(\theta) = \lim_{\varphi\to 0} g(\varphi,\ \theta) = \frac{(1+\cos\theta)^2}{4}\sin c^2\left(\frac{W}{\lambda_0}\sin\theta\right), \tag{12.31}$$

and, for the E-plane, we set $\varphi = 90^0$, which gives $E_\varphi = 0$ in 12.30, i.e.,

$$g_E(\theta) = \lim_{\varphi\to 90^0} g(\varphi,\ \theta) = \frac{(1+\cos\theta)^2}{4}\sin c^2\left(\frac{L}{\lambda_0}\sin\theta\right). \tag{12.32}$$

Throughout this study, the assessment of antenna performance is mostly based on the modeling of the normalized gain along the two principle planes as well as the estimation of both the efficiency and maximum power gain. However, the evaluation of the near-field intensity will also be performed.

Let us create a simple design of rectangular patch antenna with a dielectric substrate. It is assumed that the substrate will be created from RT-duroid-5880 ($\varepsilon_r = 2.2$). The antenna operates at the resonant frequency $f_r = 7.3$ GHz. We also assume that $h = 0.0004$ m and $d = 2h$. Then, using 12.1 finally gives $L = 0.0132$ m, $W = 0.016$ m.

The dimensions of antenna with composite substrate is the same as the dimension of antenna with dielectric substrate. Let us evaluate the material parameters of the composite substrate (i.e., ε_m, μ_m, r), assuming that copper wires with circular cross-section are chosen as inclusions. In order to do that, we take into account that the above-mentioned use of composites/metamaterials is possible. This is due to the idea of tuning the effective frequency-dependent relative permittivity and relative permeability of these artificial materials by individually adjusting the dimensions of the electric and magnetic resonant inclusions in the unit cell. That is why the values of ε_m, μ_m, r can be evaluated using the effective medium theory (EMT) [27]. The appropriate calculations finally give $\varepsilon_m \approx 2$, $\mu_m \approx 1$, $r = 0.00009$ m.

In this study, we use finite-difference time-domain (FDTD) simulations of the S11-spectra, the far-field patterns in the *E*- and *H*-planes, the efficiency and power gain for four patch antenna models, and the intensity of near-field around them. Thus, four design models are considered in the study. The first model considers the FDTD simulations for patch antenna with dielectric substrate; let us call it the standard homogenous numerical model/design. The second model corresponds to the FDTD simulations for patch antenna with composite substrate; we call it the standard inhomogeneous numerical model/design. The third model assumes the FDTD simulations for the miniaturized patch antenna with composite substrate obtained and discussed in the next paragraph; we call it the miniaturized inhomogeneous numerical model. Finally, the fourth model considers the analytical simulations using Eqs. (12.22)–(12.32) for patch antenna with composite substrate; we call it the standard inhomogeneous analytical model/design.

We now use the above-given classification of antennas throughout the study. The commercial FDTD software PLANC FDTD (Ver. 6.2, created by the Information and Mathematical Science Laboratory Inc.) is used in this study for making numerical simulations. In fact, the above-obtained values of r, h, d, ε_m, μ_m, f_r, L and W are being used in the parameter settings of software for performing the above-mentioned FTFD numerical simulations for antenna performance.

The results of the FDTD simulations of the S11-spectra for all the considered antennas are shown in Fig. 12.5. Throughout this study, we are going to call any first peak of the S11-parameter spectrum as the main peak. As one can observe from the figure, the first peak in the S11-spectrum of each numerical model is shifted from the "theoretical" resonant frequency (7.3 GHz). It occurs due to the fringe phenomenon on the patch edges and the spatial dispersion in the case of inhomogeneous models. Indeed, the numerical model with dielectric substrate

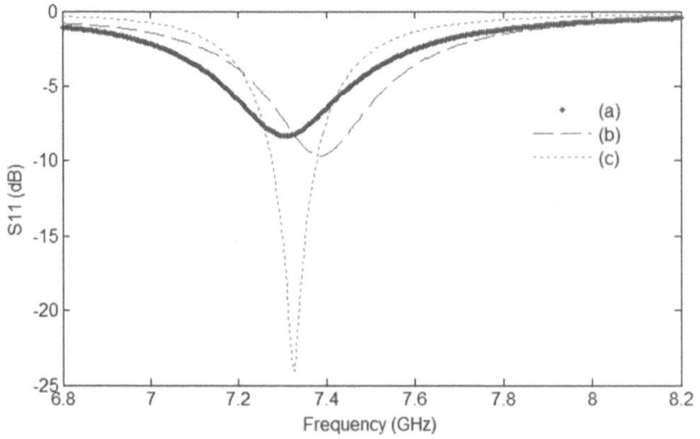

FIGURE 12.5 Simulated S11-spectra for (a) standard homogenous numerical model, (b) standard inhomogeneous numerical model and (c) miniaturized inhomogeneous numerical model.

resonates at 7.307 GHz and with composite substrate, resonates at 7.385 GHz. Note that the proposed analytical model does not take into account the fringe phenomenon on the edges of antenna patch. Thus, by replacing a homogenous high permittivity dielectric substrate with an inhomogeneous metal-dielectric composite/metamaterial with the same values of both the effective relative permittivity and permeability leads to the redistribution of the reactive part of load antenna impedance between patch and metal inclusions. It results in increasing the energy stored in the EM field radiated by the antenna due to a "more effective" use of the magnetic component of the radiated EM field compared to the case of the antenna with dielectric substrate. We can also observe from Fig. 12.5 that the antenna with composite substrate is better impedance-matched than the antenna with dielectric substrate.

The far-field patterns in the *H*- and *E*-planes for all the antenna models evaluated at their own resonant frequencies are presented in Fig. 12.6 and Fig. 12.7. As seen from these figures, no significant differences are observed between the analytical far-field patterns and the numerical ones except for the back lobes. The back lobes are, in fact, almost absent in the far-field patterns of analytical model. This is due to considering the antenna model infinite in the direction of *z*-axis.

The analytical modeling (using Eqs. (12.23) and (12.29)) and numerical FDTD calculations (using the commercial software) have shown: (i) the maximum power gain of standard homogenous numerical model is 4.6 dB, and it is found at $\theta = -2^0$; (ii) the maximum power gain of standard inhomogeneous analytical model is 4.84

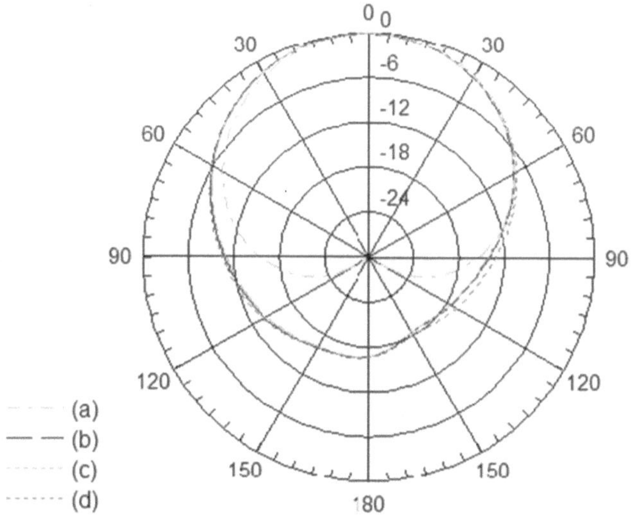

FIGURE 12.6 Far-field patterns in the *H*-plane (in dB) of (a) standard inhomogeneous analytical model, (b) standard homogenous numerical model, (c) standard inhomogeneous numerical model and (d) miniaturized inhomogeneous numerical model evaluated at the resonant frequencies.

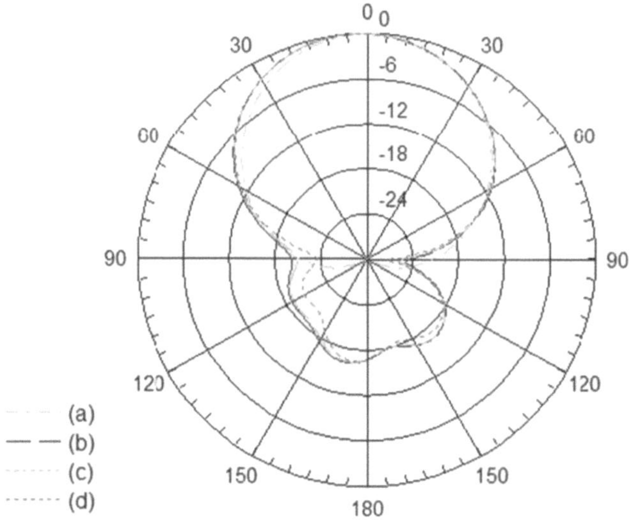

FIGURE 12.7 Far-field patterns in the *E*-plane (in dB) of (a) standard inhomogeneous analytical model, (b) standard homogenous numerical model, (c) standard inhomogeneous numerical model and (d) miniaturized inhomogeneous numerical model evaluated at the resonant frequencies.

dB, and it is found at $\theta = 0^0$; (iii) the maximum power gain of standard inhomogeneous numerical model is 5.01 dB, and it is found at $\theta = -2^0$. Also, it is found that the appropriate efficiency value is 40.72% for the case of standard homogeneous numerical model, 42.76% for the case of standard inhomogeneous analytical model, and 43.98% for the case of standard inhomogeneous numerical model. These results are in good agreement with the main idea of the study presented in the introductory section.

In this study, the FDTD simulations for near-fields around the antenna in two mutually perpendicular directions have been performed. The appropriate field distributions are depicted in Figs. 12.8 and 12.9. As can be observed from these figures, the implementation of composite/metamaterial substrates results into considerable decrease in the intensity of near-fields around the antennas. This result is especially important with a viewpoint of creating gadgets because of the possibility of a considerable decrease in the negative impact of EM irradiation on humans and living objects. Moreover, as shown in Rybin and Shulga [28], such antennas can be basically used for creating novel transponders.

Thus, taking into account the above performance analysis of all the considered antenna models, we can suppose that the proposed analytical model of rectangular patch antenna can be used for designing patch antennas of a new generation.

12.2.2 Miniaturized Design

Let us find a miniaturized design of antenna for the given above values of d, ε_r, μ_r, and f_r, if it is possible. Taking into account numerous FDTD simulations, we were

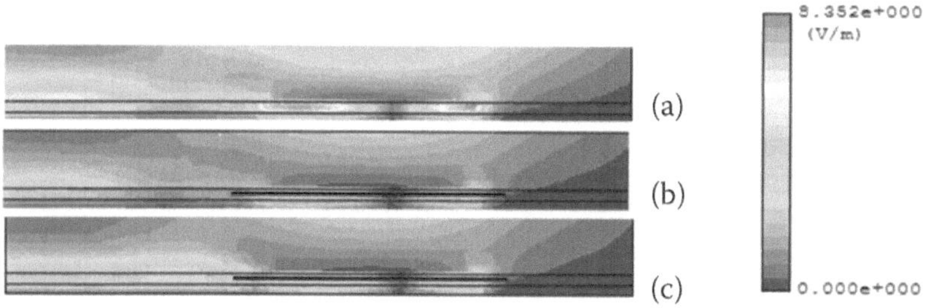

FIGURE 12.8 Near-field calculated in the cross-section $y = 0$ m of patch antenna for (a) standard homogenous numerical model at the frequency 7.307 GHz, (b) standard inhomogeneous numerical model at the frequency 7.383 GHz and (c) miniaturized inhomogeneous numerical model at the frequency 7.326 GHz.

able to obtain the appropriate antenna design with $L = 0.0117$ m and $W = 0.0156$ m. Throughout the study, we will call the appropriate antenna design the miniaturized inhomogeneous numerical model/design. The evaluated resonant frequency is 7.418 GHz. It is important to mention that the miniaturized design theoretically gives the most impedance-matched antenna compared to all the previously considered antenna designs in the study. Moreover, no significant differences are observed between the far-field patterns of the miniaturized antenna design and that of the other antenna designs presented at their own resonant frequencies except the back lobes (Figs. 12.6 and 12.7).

The FDTD simulations have shown that the maximum power gain of the standard homogenous numerical model at the resonant frequency is 6.06 dB, and it is found at $\theta = -2^0$, while the appropriate efficiency value is reached 50.93%. At the same time, the intensity of near-field around the miniaturized patch antenna is less than that of the standard patch antenna on a dielectric substrate (Figs. 12.8 and 12.9).

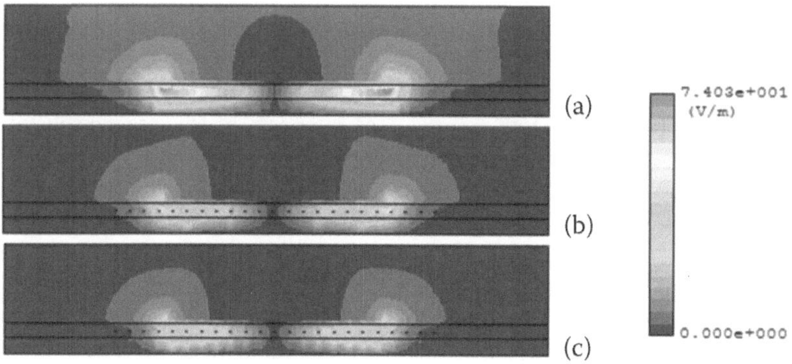

FIGURE 12.9 Near-field calculated in the middle cross-section along the x-axis of patch antenna for (a) standard homogenous numerical model at the frequency 7.307 GHz, (b) standard inhomogeneous numerical model at the frequency 7.383 GHz and (c) miniaturized inhomogeneous numerical model at the frequency 7.326 GHz.

In order to evaluate the advantage of the miniaturized design in the volume profile with regard to the analytical design, let us use the volume profile miniaturization gain factor (VPMGF) introduced in Rybin and Shulga [29]:

$$VPMGF = \left(1 - \frac{W_{\min} \cdot L_{\min}}{W \cdot L}\right) \cdot 100\%, \tag{12.33}$$

where L_{min} and W_{min} are the length and width of the patch of the miniaturized design, respectively.

Substituting the appropriate dimensional values in Eq. (12.33) gives the value of VPMGF to be 13.4%. In fact, VPMGF can reach larger values for large real part of the effective relative permeability of composite/metamaterial substrates.

The last miniaturized design has shown that the miniaturization of rectangular patch antennas on enhanced non-resonant composite/metamaterial substrates (Re $(\varepsilon_r) > 1$, $\mathrm{Re}(\mu_r) \geq 1$) is possible in principle. Moreover, its quantitative estimation can be evaluated by using 12.33. However, 12.33 says nothing about how to calculate the miniaturized values of L_{min} and W_{min}. In other words, the appropriate miniaturization concept for rectangular patch antennas is required. The concept will be able to bring breakthroughs in the domain of compact antennas with extra capabilities of reconfigurable and wearable antenna systems with regard to band tuning operating, beam steering operating, and so on.

12.2.3 Miniaturization Concept for Antenna with One-layer Wire Composite/metamaterial Substrate

We found above that the volume profile of a rectangular patch antenna on substrate created on basis of enhanced non-resonant composite/metamaterial is being decreased by increasing the value of real part of complex effective relative permittivity of substrate. In this way, it is logical to answer the question: to what extent can we increase the value of real part of complex effective relative permittivity of substrate in order to miniaturize the antenna volume profile? In order to answer the question, we have to solve the minimization problem of volume profile of a rectangular patch antenna with variable/parameters: the resonant frequency $f_r = \omega/2\pi$ of the antenna (or its wavelength $\lambda_r = c/f_r$), the thickness of the antenna substrate d, and the effective relative permittivity ε_r of the antenna substrate.

Taking into account that the antenna part of volume covered by the patch is equal to $V = S{\cdot}d$, the above-mentioned miniaturization problem for the antenna volume profile is reduced to the non-linear problem of mathematical programming with the objective function presented by the maximum gain of patch antenna G_{max} as a function of S and λ_r (or $f_r = c/\lambda_r$). Here, the thickness of substrate d and ε_r is already the given parameters. Thus, the mentioned miniaturization problem is defined as follows:

$$\left.\begin{aligned} &\max G_{\max}(S, \lambda_r); \\ &4d < \frac{\lambda_r}{\sqrt{\varepsilon_m \mu_m}} \leq 10d, \end{aligned}\right\} \tag{12.34}$$

where the last two inequalities define the condition for the composite/metamaterial to be non-resonant (semi-resonant, in fact) [28]. It means that the resonance phenomena related to wave diffraction on the lattice of wires are not taken into account while the resonances inside the wires are taken into account.

Note, the maximization in Eq. (12.34) can give us the optimal value of the antenna patch area S, but it cannot give us the appropriate values of antenna patch length L and the antenna patch width W. In order to be able to obtain the values of L and W together with the optimal value of S, let us subdivide the above-mentioned optimization problem into two other problems. The first one is the following maximization problem:

$$\left.\begin{aligned} &\max G_{\max}(S); \\ &S > 0, \end{aligned}\right\} \tag{12.35}$$

where the maximum gain $G_{\max}$ is a function of S, while ε_r, d and λ_r (or f_r) are the given parameters. The second problem is with regard to the system

$$\left.\begin{aligned} &S_{\min}(\varepsilon_r, d) = L(\varepsilon_r, d)\cdot W(\varepsilon_r, d), \\ &4d < \frac{\lambda_r}{\sqrt{\varepsilon_m \mu_m}} \le 10d, \end{aligned}\right\} \tag{12.36}$$

where the patch area S_{sol} is the solution of the problem in Eq. (12.35). Here, λ_r (or f_r) is the given parameter.

Substituting Eq. (12.29) into 12.35 and solving the resulting problem by means of the minimization method of classical calculus finally gives the expression of the minimal antenna patch area in closed form:

$$S_{\min} = \frac{C - 2B}{2BC}, \tag{12.37}$$

where $B = \frac{8k_0^2}{\pi Z_0} \frac{|A|^2_{\theta=0}}{\text{Re}(A)|_{\theta=0}}$ and $C = \frac{j_0}{\text{Re}(A)|_{\theta=0}} \sqrt{\frac{\omega\mu_0}{2\sigma_i}}$.

The appropriate calculations finally give $S_{min} = 0.18252 \times 10^{-3}$ m^2.

Substituting Eq. (12.37) and 12.1 into the equality of system, Eq. (12.36) finally gives the following transcendental equality:

$$\frac{j_0 \pi Z_0 \sqrt{\omega\mu_0 / 2\sigma_i} - 16k_0^2 |A|^2_{\theta=0}}{16k_0^2 j_0 \pi Z_0 \sqrt{\omega\mu_0 / 2\sigma_i}\, |A|^2_{\theta=0}} \cdot \text{Re}(A)|_{\theta=0} = \frac{1}{2}\lambda_r^2 \left((1 + \varepsilon_r)^2 - \frac{1 - \varepsilon_r^2}{\sqrt{1 + 12\sqrt{2}\,(d/\lambda_r)\sqrt{1+\varepsilon_r}}} \right)^{-\frac{1}{2}} - \\ - \frac{0.824}{\sqrt{2}} \lambda_r d \left(\frac{\varepsilon_r - 1 + (1.6 + \varepsilon_r)\sqrt{1 + 12\sqrt{2}\,(d/\lambda_r)\sqrt{1+\varepsilon_r}}}{\varepsilon_r - 1 + (0.484 + \varepsilon_r)\sqrt{1 + 12\sqrt{2}\,(d/\lambda_r)\sqrt{1+\varepsilon_r}}} \right) \frac{1 + 0.264 \cdot \sqrt{2}\,(d/\lambda_r)\sqrt{1+\varepsilon_r}}{\sqrt{1+\varepsilon_r} + 0.8 \cdot \sqrt{2}\,(d/\lambda_r)(1+\varepsilon_r)}. \tag{12.38}$$

Equation (12.38) must be solved with respect to any of the variables: εr, λr (or fr) and d. The remaining two variables are defined as already-given parameters in this case. Considering that we have already set the variables ε_r and d in advance, and the variable always "floats" in our numerical simulations, then 12.38 will be solved with respect to the variable f_r. The found value of the variable f_r and the already-given parameter of ε_r can be used to obtain minimal values of L and W using 12.1:

$$\left.\begin{aligned} W_{\min} &= \frac{c}{2f_{rm}}\sqrt{\frac{2}{1+\varepsilon_r}}, \\ L_{\min} &= \frac{S_{\min}}{W_{\min}}, \end{aligned}\right\} \tag{12.39}$$

where f_{rm} is the solution of 12.38. As can be observed, this equation has no analytical solution in closed form and can only be solved numerically using, for example, the corresponding functions of the MatLab software. The appropriate solution gives f_{rm} = 7.481 GHz, W_{min} = 0.01585 m, L_{min} = 0.01152 m, that are in relatively good agreement with the numerical results that give f_{rm} = 7.418 GHz, W_{min} = 0.0156 m and L_{min} = 0.0117 m.

Despite the fact that 12.38 was obtained for the specific structure of substrate, it can also be used for a relatively arbitrary metal-dielectric composite filling the substrate matrix – we just need to be able to calculate the amplitude of the electric intensity vector component of the EM wave in a far region, which is, in fact, the main excited wave. Once again, we would remind that the approximation in 12.29 has been obtained by not taking into account the fringe phenomenon on the edges of the antenna patch. That is why the near-field distribution can actually be more complex. As a result, the $S11$-spectrum of the antennas with composite substrates can contain additional peaks of interest except for the peak of the main wave. Let us give a simple example. For this purpose, we consider the $S11$-spectrum of the miniaturized inhomogeneous numerical model in the frequency range of 5–20 GHz (Fig. 12.10). As may be observed from the figure, the second, third, fourth and fifth peaks situated at the frequencies 9.556 GHz, 11.29 GHz, 13.72 GHz, 14.45 GHz, 15.93 GHz, 16.84 GHz and 18.57 GHz, correspondingly. These peaks are of interest with a viewpoint of the possibility for the miniaturized antenna to be a reconfigurable one in frequency functionality.

FDTD simulations for the far-field patterns, efficiency and power gain at the above-mentioned frequency peaks have shown that the peak located at the frequency 16.84 GHz may be of some interest. Indeed, comparing the far-field patterns of the last peak with the far-field patterns of the main (first) peak (see Figs. 12.11 and 12.12), we can assume that miniaturized patch antennas on composite/metamaterial substrates can be multi-band and even multi-directional. This is especially evident from 3D radiation patterns depicted in Fig. 12.13. The maximum power gain at the frequency 16.84 GHz is equal to 6.97 dB, and it is found at $\theta = -10^0$. The appropriate efficiency value is 71.73%. This result is quite interesting from the point of view of designing reconfigurable compact antenna systems in frequency functionality and polarization.

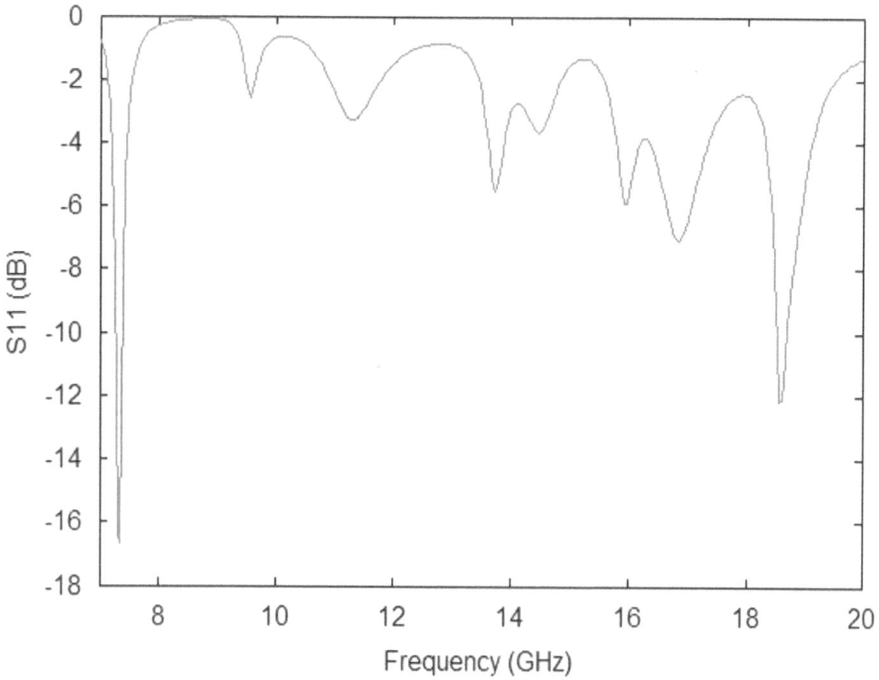

FIGURE 12.10 *S*11-spectrum of the miniaturized inhomogeneous numerical model.

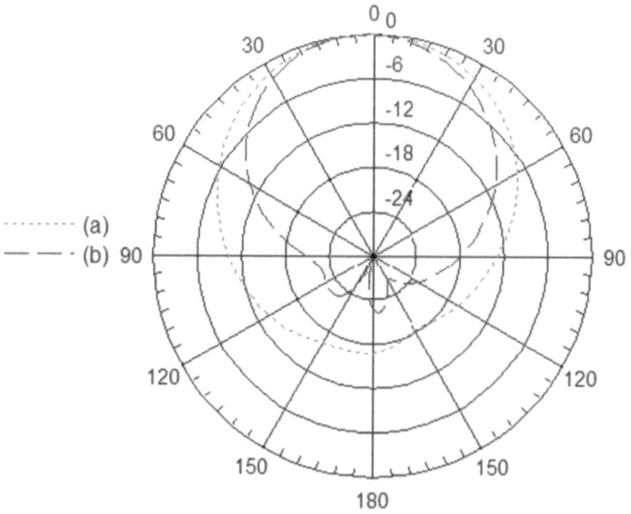

FIGURE 12.11 Far-field patterns in the *H*-plane of miniaturized inhomogeneous numerical model evaluated at (a) the main peak frequency (7.383 GHz) and (b) 16.84 GHz.

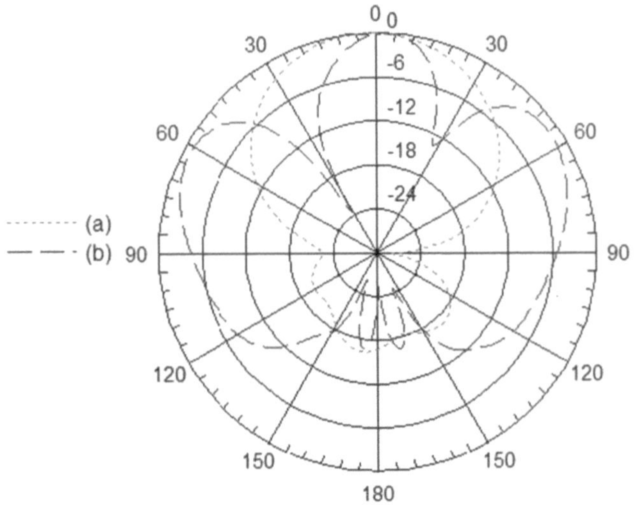

FIGURE 12.12 Far-field patterns in the *E*-plane (in dB) of miniaturized numerical model evaluated at (a) the main peak frequency (7.383 GHz) and (b) 16.84 GHz.

Taking into account the case of standard numerical model with composite substrate, it is logical to conjecture that the intensity of near-field distribution around the miniaturized antenna at the frequency 16.84 GHz is quite low. Indeed, according to the FDTD simulations performed for near-field distributions (regarding patch antenna) in two mutually perpendicular directions, their intensity is very low (Figs. 12.14 and 12.15). It says much for the above-mentioned idea for using the additional resonant frequencies of patch antennas on composite/metamaterial substrates. The appropriate FDTD simulations have also shown that the same qualitative result takes place for the standard inhomogeneous model. This enables one to draw a general conclusion about the multi-directional and multi-band nature of patch antennas on composite/ metamaterial substrates. Such a design can avoid the use of multiple antennas, for instance, by integrating receiving and transmitting

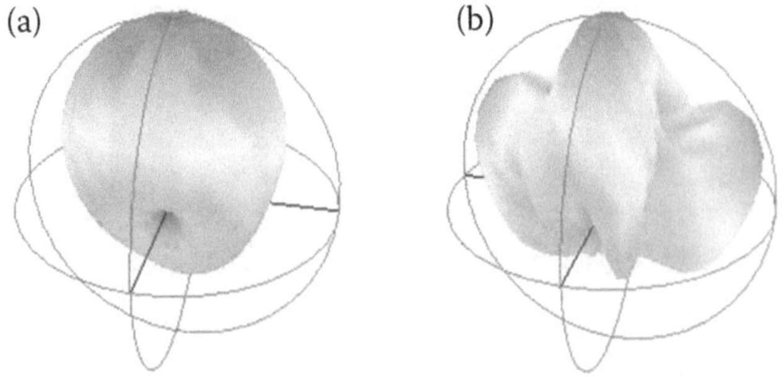

FIGURE 12.13 3D far-field patterns of the miniaturized numerical antenna model created at (a) the main peak frequency, and (b) 16.84 GHz.

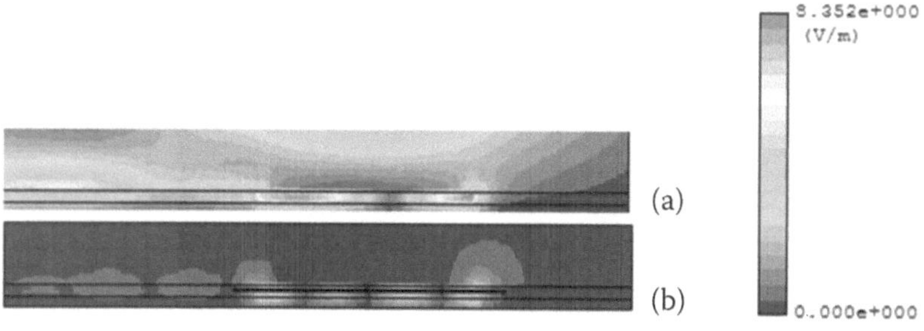

FIGURE 12.14 Near-field calculated in the cross-section $y = 0$ m for (a) standard homogenous numerical model calculated at the frequency 7.307 GHz, and (b) the miniaturized numerical antenna model calculated at 16.84 GHz.

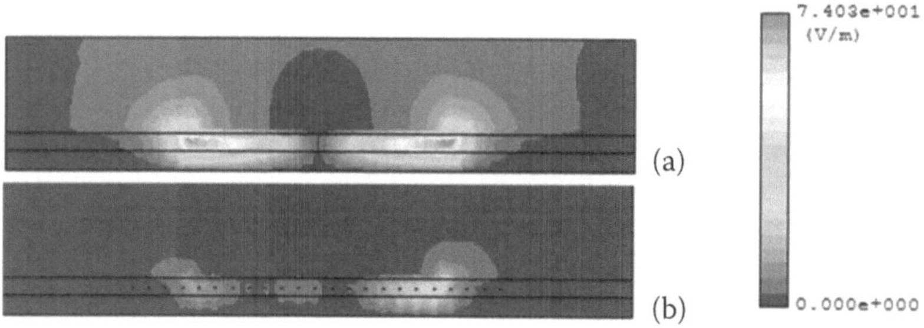

FIGURE 12.15 Near-field calculated in the middle cross-section along the x-axis of patch antenna for (a) standard homogenous numerical model calculated at the frequency 7.307 GHz, and (b) the miniaturized numerical antenna model calculated at 16.84 GHz.

functions into the same communication systems or the same antennas operating in the GSM 1,800 MHz and WLAN 2.4 GHz application bands.

12.3 FAR-FIELD FOCUSING FOR RECTANGULAR PATCH ANTENNAS WITH COMPOSITE/METAMATERIAL SUBSTRATES

12.3.1 Fabry-Perot Approach for a Patch Antenna with Superstrate

The idea of using superstrates to improve the performance of patch antenna is not new [30–32], but still attracts the attention of researchers working on designing compact antenna systems [33–35] and the near-field phase transforming structures developed for directive radio-frequency (RF) front-end antennas for applications like point-to-point communication and satellite communication [36]. Moreover, a hard material superstrate can play the role of protection to antenna from environmental hazards.

Note that most of the above-mentioned publications deal with non-magnetic substrates. Thus, in this study, we will consider ferrite superstrates. The substrate

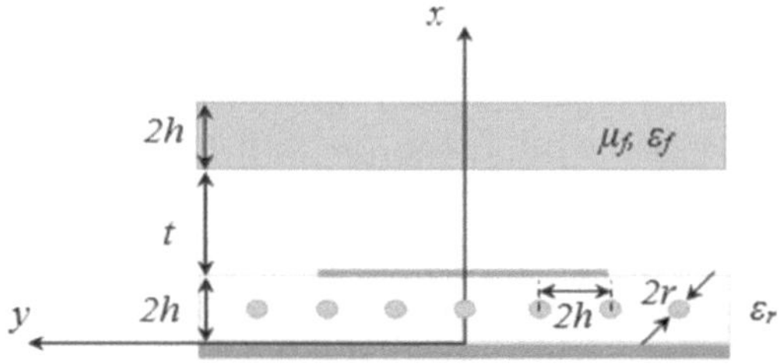

FIGURE 12.16 Metamaterial patch antenna with magnetic superstrate.

positioning is going to be defined in terms of Fabry-Perot cavity theory, which will be developed appropriately.

Let us consider the miniaturized patch antenna on composite substrate designed in the previous paragraph. At the same time, we assume that the antenna is covered by a superstrate that has the same geometry as the substrate (Fig. 12.16), where ε_f is the relative permittivity of superstrate and μ_f is its relative permeability. It is also assumed that the superstrate is situated at a distance t from the antenna patch. The expression for evaluating the value of t will be derived later. In order to do that, we are going to consider the antenna patch surface as the lower surface of the Fabry-Perot cavity. At the same time, we are going to consider the lower surface of the superstrate as the upper surface of the cavity. Then, the upper component of the cavity is accepted as a layer of thickness, whereas the lower surface of the cavity is accepted as a half-space with the refractive index $n_{\text{eff}} = (\varepsilon_{\text{eff}})^{1/2}$, where ε_{eff} is defined by the second formula in 12.1.

Let us assume that $R = |R|e^{i\delta_R}$ is the full reflection coefficient of superstrate, $T = |T|e^{i\delta_T}$ is the full transmission coefficient of superstrate, and $\rho_l = |\rho_l|e^{i\delta_l}$ is the Fresnel reflection coefficient of the lower surface of the cavity. The component of electric intensity vector E in the far-zone consists of the vector sum of partial rays of the electric field component, as a result of the reflections from the cavity surfaces (Fig. 12.17). Then, using the approach of work in Boutayeb et al. [37], it is logical to write down the expression for the component of the electric intensity vector:

$$E(\theta) = E_0 \cdot D(\theta) \cdot T \cdot \sum_{n=0}^{\infty} (|R| \cdot |\rho_l| \cdot e^{-i\Phi})^n, \tag{12.40}$$

where $\Phi = 2\frac{2\pi}{\lambda_r} t \cdot \cos\theta - \delta_R - \delta_l$, $D(\theta)$ is the electric field pattern, and λ_r is the resonant wavelength of the patch antenna.

The sum in the right side of Eq. (12.40) is the infinite geometric progression. That is why 12.40 can finally be rewritten in the form

$$E(\theta) = \frac{E_0 \cdot D(\theta) \cdot T}{1 - |R||\rho_l|e^{-i\Phi}}, \tag{12.41}$$

where, as shown in Wan et al. [38],

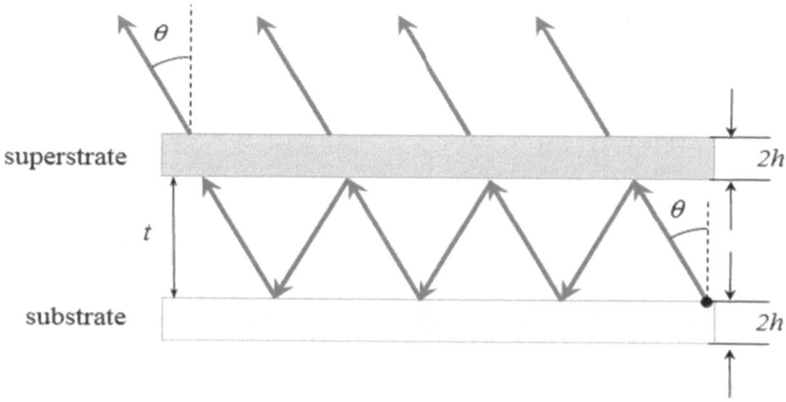

FIGURE 12.17 Ray distribution between the superstrate and free space.

$$\left.\begin{aligned} R &= \rho_l + \frac{t_l \tilde{t}_l \tilde{\rho}_l \exp\left\{-i\frac{2\pi}{\lambda_r} l \cdot \cos\theta\right\}}{1 - \tilde{\rho}_l^{\,2} \exp\left\{-i\frac{4\pi}{\lambda_r} l \cdot \cos\theta\right\}}, \\ T &= t_l \tilde{t}_l + \frac{t_l \tilde{t}_l \tilde{\rho}_l^{\,2} \exp\left\{-i\frac{4\pi}{\lambda_r} l \cdot \cos\theta\right\}}{1 - \tilde{\rho}_l^{\,2} \exp\left\{-i\frac{4\pi}{\lambda_r} l \cdot \cos\theta\right\}}, \end{aligned}\right\} \tag{12.42}$$

where the Fresnel coefficients ρ_l, t_l, $\tilde{\rho}_u$, $\tilde{t}_u$ are defined by

$$\left.\begin{aligned} \rho_l &= \frac{\varepsilon_{reff} \cdot \cos\theta - \sqrt{\varepsilon_{reff} - \sin^2\theta}}{\varepsilon_{reff} \cdot \cos\alpha + \sqrt{\varepsilon_{reff} - \sin^2\alpha}}, \quad & t_l &= \frac{2 \cdot \sqrt{\varepsilon_{reff}} \cdot \cos\theta}{\varepsilon_{reff} \cdot \cos\alpha + \sqrt{\varepsilon_{reff} - \sin^2\theta}}, \\ \tilde{\rho}_l &= \frac{\sqrt{\varepsilon_{reff} - \sin^2\theta} - \cos\theta}{\sqrt{\varepsilon_{reff} - \sin^2\theta} + \cos\alpha}, \quad & \tilde{t}_l &= \frac{\sqrt{\varepsilon_{reff}}\sqrt{\varepsilon_{reff} - \sin^2\theta}}{\sqrt{\varepsilon_{reff} - \sin^2\theta} + \cos\theta}. \end{aligned}\right\} \tag{12.43}$$

Taking into account Eq. (12.41), the power pattern $P_E(\theta)$ can be written in the form

$$P_E(\theta) = \frac{|T|^2}{1 - 2|R|\left|\rho_l\right| \cdot \cos\Phi + |R|^2\left|\rho_l\right|^2} |D(\theta)|^2. \tag{12.44}$$

As seen from Eq. (12.44), the power pattern achieves its maximum value in the forward reflection direction ($\theta = 0°$) when $\cos\varphi = 1$. The last conclusions lead to the formula for evaluating the optimal value of t as a function of parameters of the patch antenna and superstrate:

$$t = \frac{\lambda_r}{4\pi}(\delta_R + \delta_l). \tag{12.45}$$

Eq. (12.45) provides the maximum possible value of the power pattern $P_E(\theta)$.

12.3.2 Main Relations for Evaluating the Performance of Patch Antenna with Metamaterial Substrate and Ferrite Superstrate

Taking into account Eq. (12.44), the radiation density of the antenna with superstrate can be defined by

$$U(\theta, \varphi) = \frac{k_0^2}{8\pi^2 Z_0} \frac{(1 + \cos\theta)^2}{4} \frac{|T|^2}{1 - 2|R||\rho_l|\cdot \cos\Phi + |R|^2|\rho_l|^2} |f(\theta, \varphi)|^2. \tag{12.46}$$

where θ and φ are the angles of the spherical coordinate system (see Fig. 12.1), and $f(\theta, \varphi)$ is the radiation pattern without superstrate, given by Eq. (12.25). Then, the radiation pattern of the antenna with superstrate is given by

$$f_{sup}(\theta, \varphi) = 4S\frac{A\cdot T}{1 - |R||\rho_l|} \sin c\left(\frac{L}{\lambda_r}(\cos\varphi - 1)\sin\theta\right)\cdot \sin c\left(\frac{L}{\lambda_r}\sin\theta\sin\varphi\right) \cdot \sin c\left(\frac{W}{\lambda_r}\sin\theta\cos\varphi\right) \tag{12.47}$$

Using Eqs. (12.18) and (12.45) enables one to derive the expression for the efficiency of antenna under consideration in the form

$$e = \frac{|T|\cdot \mathrm{Re}(A)}{1 - |R||\rho_l|}\left(1 + \frac{j_0 S}{\mathrm{Re}(A)}\sqrt{\frac{\omega\mu_0}{2\sigma_i}}\right)^{-1}. \tag{12.48}$$

The expression for the directive gain is given by

$$D(\theta, \varphi) = \frac{(2k_0 S|A|)^2}{\pi Z_0} \frac{(1 + \cos\theta)^2}{\mathrm{Re}(A)} \frac{|T|^2}{1 - 2|R||\rho_l| + |R|^2|\rho_l|^2} \times$$
$$\times \sin c^2\left(\frac{L}{\lambda_r}(\cos\varphi - 1)\sin\theta\right)\cdot \sin c^2\left(\frac{L}{\lambda_r}\sin\theta\sin\varphi\right)\cdot \sin c^2\left(\frac{W}{\lambda_r}\sin\theta\cos\varphi\right). \tag{12.49}$$

As one can observe from Eq. (12.49), the maxima of directive gain is achievable at $\theta = 0°$. Then, the maxima of gain is given by

$$G_{max} = eD_{max} = \frac{8k_0^2}{\pi Z_0} \frac{|T|^3 \mathrm{Re}(A)|_{\theta=0}}{(1 - |R||\rho_l|)^3} \frac{S^2|A|^2}{\mathrm{Re}(A)|_{\theta=0} + j_0 S\sqrt{\frac{\omega\mu_0}{2\sigma_i}}}. \tag{12.50}$$

Taking into account Eqs. (12.45) and (12.46), we derive the expression for the normalized gain in the form

$$g(\theta,\varphi)=\frac{(1+\cos\theta)^2}{4}\frac{|f_{\sup}(\theta,\varphi)|^2}{|f_{\sup}(\theta,\varphi)|^2_{\max}}=\frac{(1+\cos\theta)^2}{4}\frac{|T|^2\cdot(\mathrm{Re}(A))^2}{(1-|R|\,|\rho_l|)^2}\left[\frac{(1-|R|\,|\rho_l|)^2}{|T|^2\cdot(\mathrm{Re}(A))^2}\right]\Bigg|_{\theta=0^0}$$
$$\times\sin c^2\left(\frac{L}{\lambda_r}(\cos\varphi-1)\sin\theta\right)\cdot\sin c^2\left(\frac{W}{\lambda_r}\sin\theta\cos\varphi\right)\cdot\sin c^2\left(\frac{W}{\lambda_r}\sin\theta\right).\tag{12.51}$$

The expressions for the normalized gain along the two principle planes are given as follows. For the H-plane, we set $\varphi = 0°$ in Eq. (12.51):

$$g_H(\theta)=\lim_{\varphi\to0^0}g(\varphi,\theta)=\frac{(1+\cos\theta)^2}{4}\frac{|T|^2\cdot(\mathrm{Re}(A))^2}{(1-|R||\rho_l|)^2}\left[\frac{(1-|R||\rho_l|)^2}{|T|^2\cdot(\mathrm{Re}(A))^2}\right]\Bigg|_{\theta=0^0}$$
$$\cdot\sin c^2\left(\frac{W}{\lambda_r}\sin\theta\right)\tag{12.52}$$

and for the E-plane, we set $\varphi_0 = 0°$ in Eq. (12.50):

$$g_E(\theta)=\lim_{\varphi\to90^0}g(\varphi,\theta)=\frac{(1+\cos\theta)^2}{4}\frac{|T|^2\cdot(\mathrm{Re}(A))^2}{(1-|R||\rho_l|)^2}\left[\frac{(1-|R||\rho_l|)^2}{|T|^2\cdot(\mathrm{Re}(A))^2}\right]\Bigg|_{\theta=0^0}$$
$$\cdot\sin c^2\left(\frac{L}{\lambda_r}\sin\theta\right).\tag{12.53}$$

Let us make the analysis of performance of the antennas on composite substrates of the previous section with ferrite superstrate. In order to do that, we initially define the constitutive parameters of the antenna. In order to be consistent with the results of Wan et al. [38], we are going to use the following values of the superstrate: $\varepsilon_f =$ 9.66 – 0.174i and μ_f = 1.2–0.974i. It means $MgO*Fe_2O_3$ is chosen as the superstrate material. It is worth mentioning that its linear dimensions are exactly the same as the linear dimensions of superstrate. The values of other parameters, except the value of t, are chosen from the previous paragraph. The calculations using Eq. (12.44) give t = 0.0004 m.

The results of FDTD simulations regarding the S11-spectra of antenna designs with ferrite superstrate are shown in Fig. 12.18. As can be observed from the figure, these antenna designs are well impedance-matched, especially the miniaturized antenna design. Moreover, the miniaturized design has a broader beamwidth as opposed to the case of no superstrate (Fig. 12.4). At the same time, the resonant frequency of the miniaturized design with ferrite superstrate is the most blue-shifted with regard to all the above considered numerical antenna designs (it resonates at 7.404 GHz), whereas the standard design with ferrite superstrate resonates at 7.312 GHz. This frequency is the closest to the theoretical resonant frequency (7.3 GHz).

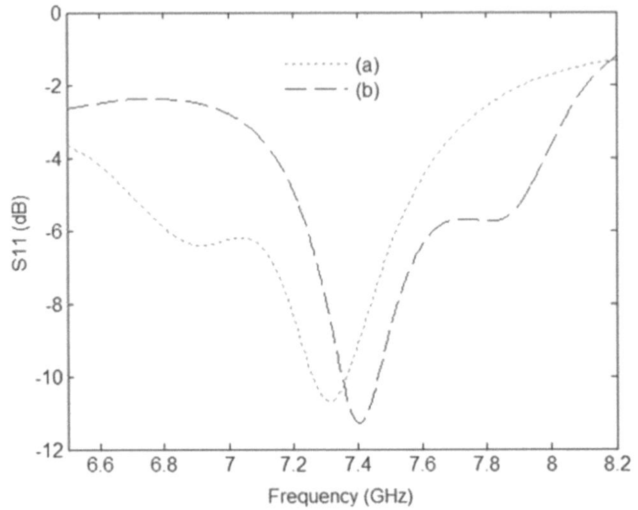

FIGURE 12.18 Simulated $S11$-spectra for (a) standard inhomogeneous numerical model with ferrite superstrate, and (b) standard miniaturized numerical model with ferrite superstrate.

The results of FDTD simulations of far-field patterns for all the antenna designs with ferrite superstrates are presented in Figs. 12.19 and 12.20. As seen from these figures, no significant differences are observed between the far-field patterns of the last two antenna designs (with ferrite superstrates) and all the previous simulated ones (without ferrite superstrates) in the H-plane, while worse coincidences of the

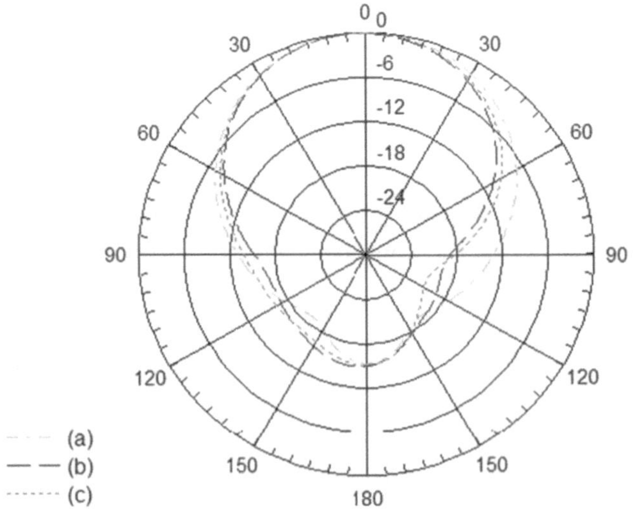

FIGURE 12.19 Far-field patterns in the H-plane (in dB) of (a) standard inhomogeneous analytical model with ferrite superstrate, (b) standard inhomogeneous numerical model with ferrite superstrate, and (c) miniaturized inhomogeneous numerical model with ferrite superstrate evaluated at the resonant frequencies.

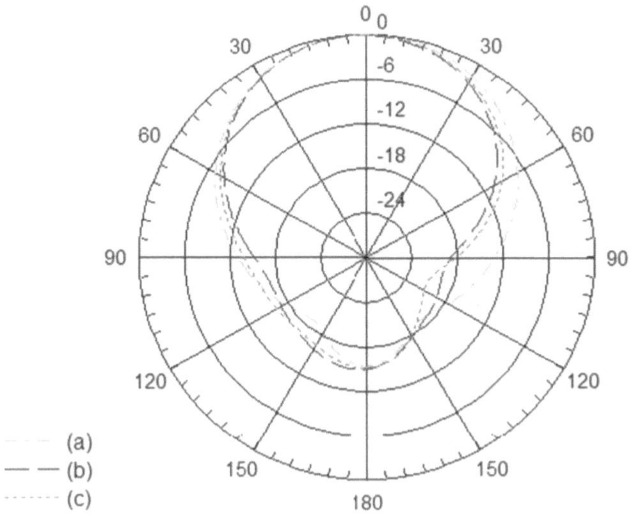

FIGURE 12.20 Far-field patterns in the *E*-plane (in dB) of (a) standard inhomogeneous analytical model with ferrite superstrate, (b) standard inhomogeneous numerical model with ferrite superstrate, and (c) miniaturized inhomogeneous numerical model with ferrite superstrate evaluated at the resonant frequencies.

far-field patterns are observed in the *E*-plane. As can be observed from Figs. 12.19 and 12.20, the analytical model has back lobes, which was not the case in the absence of ferrite superstrate. This is because the superstrate has finite dimensions.

Right now, we are going to show that the presence of ferrite superstrate can be energetically advantageous. Indeed, the appropriate FDTD simulations and analytical calculations have shown: (i) the maximum power gain of the standard inhomogeneous analytical model with ferrite superstrate is 6.01 dB, and it is found at $\theta = 0^0$; (ii) the maximum power gain of the standard inhomogeneous numerical model is 5.92 dB, and it is found at $\theta = -2^0$; (iii) the maximum power gain of the miniaturized inhomogeneous numerical model is 6.59 dB, and it is found at $\theta = 0^0$. Also, it has been found that the appropriate efficiency value is 49.37% for the case of the standard inhomogeneous analytical model, 46.01% for the case of the standard inhomogeneous numerical model, and 54.34% for the case of the miniaturized inhomogeneous numerical model. These results are in good agreement with the main idea of Wan et al. [38]: using a ferrite superstrate can improve the performance of the patch antenna. It takes place because of two mechanisms: (1) the mechanism of Fabry-Perot resonant cavity, which is already very well-known; and (2) increased energy stored in the magnetic component of the EM field radiated by the antenna due to the tendency of the magnetic moment of ferromagnetic superstrate to be aligned with the magnetic component of the EM field produced by the antenna and to reinforce it by virtue of its own magnetic fields. Note that the ferrite superstrate can serve as one of the antenna housing panels.

The FDTD simulations of near-fields for both the antenna designs with ferrite superstrates in two mutually perpendicular directions have shown that these fields

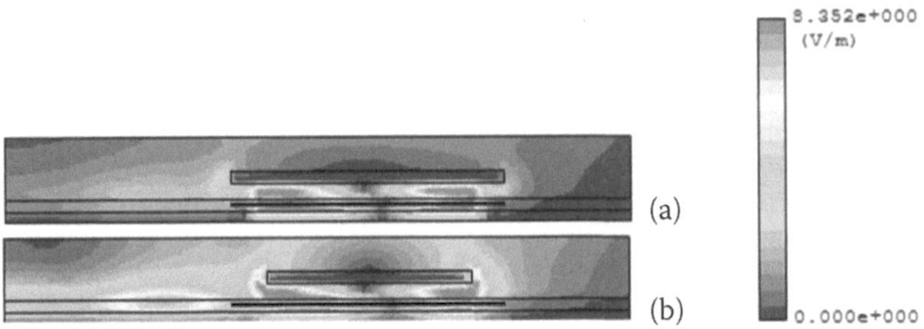

FIGURE 12.21 Near-field calculated in the cross-section $y = 0$ m of patch antenna with ferrite superstrate for (a) standard inhomogeneous numerical model at the frequency 7.322 GHz, and (b) miniaturized inhomogeneous numerical model at the frequency 7.404 GHz.

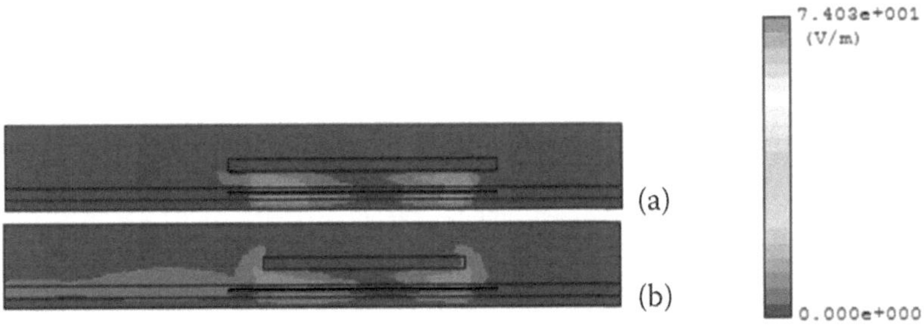

FIGURE 12.22 Near-field calculated in the middle cross-section along the x-axis of patch antenna with ferrite superstrate for (a) standard inhomogeneous numerical model at the frequency 7.322 GHz, and (b) miniaturized inhomogeneous numerical model at the frequency 7.404 GHz.

are mostly concentrated under the superstrate, while the intensity of near-fields is considerably less than the intensity of near-fields of the initial patch antenna with dielectric substrate (Figs. 12.21 and 12.22). This, in turn, means that such antennas can be used for gadgets.

12.4 2D MODEL OF RECTANGULAR PATCH ANTENNA WITH TWO-LAYER WIRE COMPOSITE/METAMATERIAL SUBSTRATE

12.4.1 The Main Analytical Relations

Let us complicate the model of our antenna compared to the one considered in Section 12.2: consider the model of Section 12.2 with two layers of non-magnetic metallic wires symmetrically embedded into the host dielectric substrate (Fig. 12.23).

First of all, let us neglect the evanescent waves as done in Section 12.2 regarding the case of the one-layer wire composite substrate. It is also supposed that $r \ll d = h$. Then, considering the double-layer grid of antenna in free space as a four-terminal network with its own S-matrix (Fig. 12.24) enables to write

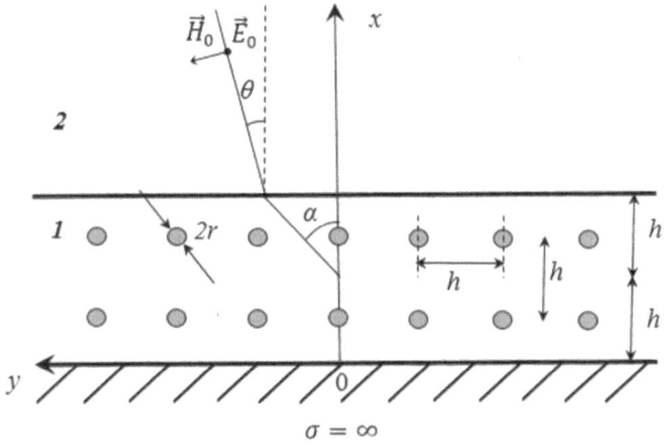

FIGURE 12.23 2D geometry of rectangular patch antenna with one-layer wire composite substrate.

FIGURE 12.24 Representing the double-layer grid as a four-terminal network.

$$\begin{bmatrix} F_I \\ L_{II} \end{bmatrix} = S \cdot \begin{bmatrix} F_I \\ L_{II} \end{bmatrix}, \tag{12.54}$$

where the scattering matrix S is symmetrical in our case, i.e.,

$$S = \begin{bmatrix} S_{11} & S_{12} \\ S_{21} & S_{22} \end{bmatrix} = \begin{bmatrix} \Gamma & K \\ K & \Gamma \end{bmatrix}, \tag{12.55}$$

where Γ and K are the Fresnel reflection and transmission coefficients, respectively, at the left and right plane boundaries of the layer.

Let us derive the elements of matrix S. According to the results of Wait [24],

$$\left.\begin{aligned} L_{II} &= E_0 e^{ik_0(h\cos\theta + y\sin\theta)}, \\ F_I &= -E_0 e^{-ik_0(h\cos\theta - y\sin\theta)}, \\ F_{II} &= E_0 G e^{-ik_0(h\cos\theta - y\sin\theta)}, \\ L_I &= -E_0 G e^{-ik_0(h\cos\theta - y\sin\theta)}, \end{aligned}\right\} \tag{12.56}$$

where G is defined by Eq. (12.7).

Substituting Eqs. (12.55) into Eq. (12.54) and resolving the resulting equations with regard to Γ and K finally gives

$$\left.\begin{aligned} \mathrm{K} &= \frac{F_I F_{II} - L_I L_{II}}{F_I^2 - L_{II}^2}, \\ \Gamma &= \frac{F_I L_I - F_{II} L_{II}}{F_I^2 - L_{II}^2}. \end{aligned}\right\} \tag{12.57}$$

Substituting Eqs. (12.56) into Eqs. (12.57) gives

$$\left.\begin{aligned} \mathrm{K} &= \frac{G}{2} i \frac{1 + e^{-2ik_0 h \cos\theta}}{\sin(2k_0 h \cos\theta)}, \\ \Gamma &= \frac{G}{2} i \frac{1 - e^{-2ik_0 h \cos\theta}}{\sin(2k_0 h \cos\theta)}. \end{aligned}\right\} \tag{12.58}$$

The expressions of full transmission T_f and reflection R_f coefficients are very well known

$$\left.\begin{aligned} T_f &= -\frac{\mathrm{K}^2 \cos\alpha \cos\theta \cdot e^{2ik_r \cos\alpha}}{1 - \Gamma^2 \cos\theta \cos\alpha \cdot e^{2ik_r \cos\alpha}}, \\ R_f &= \frac{\Gamma(\cos\theta - \cos\alpha) \cdot e^{2ik_r \cos\alpha}}{1 - \Gamma^2 \cos\theta \cos\alpha \cdot e^{2ik_r \cos\alpha}}, \end{aligned}\right\} \tag{12.59}$$

where $k_r = k_0\sqrt{\varepsilon_r \mu_r}$; ε_r and μ_r, respectively, are the effective relative permittivity and permeability of the substrate composite. Note that, in order to be absolutely accurate, we must replace in all expressions of Section 12.2 the constitutive parameters of the index m with the same parameters of index c. However, the accuracy of all analytical results still remains acceptable even without the above replacement. This is caused by the fact that, in all the obtained approximations, only a small value of metal volume fractions ($\pi r^2/d^2 < 0.25$) is considered.

Taking into account that the antenna substrate is backed by a perfect conducting wall ($x = 0$), the approximation of the electric intensity vector in the 2nd space domain is given by

$$\vec{E}_2 = E_2(x, y)\cdot\vec{z}_0 = R_{tot}\cdot e^{-ik_0 y \sin\theta} e^{-ik_0(x-2h)\cos\theta}\cdot\vec{z}_0, \tag{12.60}$$

where $R_{tot} = R_f - T_f^2$.

In order to be able to analytically evaluate the antenna performance, the boundary-value problem, similar to that of solved in Section 12.2 can be solved with regard to the antenna presented in Fig. 12.23. The approximation for the electric intensity vector in the substrate domain ($0 < x < 2h$) is given by

$$\vec{E}_1 = E_1(x, y)\cdot\vec{z}_0 = \left[C_1^+ e^{ik_1(x\cos\alpha + y\sin\alpha)} + C_1^- e^{-ik_1(x\cos\alpha - y\sin\alpha)}\right]\cdot\vec{z}_0. \tag{12.61}$$

The approximations for the magnetic intensity vector in the 1st and 2nd space domains are to be defined through Faraday's law. The appropriate straightforward algebra finally gives

$$\vec{H}_1(x, y) = -\frac{1}{Z_1}[F^{1+}(x, y)\cdot \sin\alpha\cdot\vec{x_0} - F^{1-}(x, y)\cdot \cos\alpha\cdot\vec{y_0}], \quad (12.62)$$

$$\vec{H}_2(x, y) = -\frac{E_2(x, y)}{Z_0}[\sin\theta\cdot\vec{x_0} + \cos\theta\cdot\vec{y_0}]. \quad (12.63)$$

where

$$F^{1\pm}(x, y) = e^{ik_r y\sin\theta}\left(C_1^+ e^{ik_r x\cos\theta} \pm C_1^- e^{-ik_r x\cos\theta}\right), \quad (12.64)$$

Let us present the current flows on the patch in the form

$$\vec{j} = j\cdot\delta(x - 2h)\cdot\vec{z_0} = -j_0\cdot\delta(x - 2h)\cdot e^{-ik_r y\sin\theta}\cdot\vec{z_0}, \quad (12.65)$$

where j_0 = cont. is already given. The unknowns $C_1^\pm$ and α can be calculated by imposing the boundary conditions that the tangential electric field is continuous at the interface, i.e., $x = 2h$:

$$\vec{E}_1|_{x=2h} = \vec{E}_2|_{x=2h}, \quad (12.66)$$

and the boundary conditions for the tangential magnetic field at the interface $x = 2h$:

$$[\vec{x_0} \times (\vec{H}_2 - \vec{H}_1)]|_{x=2h} = j\cdot\vec{z_0}, \quad (12.67)$$

where x_0 is the unit vector of the x-axis. Indeed, substituting Eqs. (12.59), (12.60) and (12.64) into Eqs. (12.65)–(12.66), finally gives the equations with regard to the variables $C_1^\pm$:

$$\left.\begin{aligned} &R_{tot} = F^{1+}(2h, 0), \\ &Z_c\cos\theta\cdot F^{1+}(2h, 0) + Z_0\cos\alpha\cdot F^{1-}(2h, 0) = j_0 Z_0 Z_c. \end{aligned}\right\} \quad (12.68)$$

where $\cos\alpha = \sqrt{1 - (\sin^2\theta/\varepsilon_r\mu_r)}$, $Z_r = \sqrt{\mu_0\mu_r/\varepsilon_0\varepsilon_r}$ is the impedance of the composite material. Solving Eq. (12.68) with regard to the variables $C_1^\pm$ finally gives:

$$\left.\begin{aligned} C_1^+ &= \frac{e^{-2ik_r h \cos\theta}}{2Z_0 \cos\alpha}(j_0 Z_0 Z_r - R_{tot}[Z_r \cos\theta - Z_0 \cos\alpha]), \\ C_1^- &= \frac{e^{-2ik_r h \cos\theta}}{2Z_0 \cos\alpha}(R_{tot}[Z_r \cos\theta + Z_0 \cos\alpha] - j_0 Z_0 Z_r). \end{aligned}\right\} \tag{12.69}$$

As seen from the results in the current section and the results in Section 12.2.1, the field structure of antenna depicted in Fig. 12.2 is coincident with that depicted in Fig. 12.21. It enables one to generalize some results of the theory in Section 12.2.1 to the case of antenna depicted in Fig. 12.23 without repeating the tedious algebra of Section 12.2.1 with regard to deriving the main approximations describing the performance of this antenna. Thus, the approximations for the main parameters describing the performance of the antenna depicted in Fig. 12.23 are determined as follows:

The efficiency of the antenna is given by

$$e = \left(1 + \frac{j_0 S}{\mathrm{Re}(R_{tot})}\sqrt{\frac{\omega\mu_0}{2\sigma_i}}\right)^{-1}. \tag{12.70}$$

The radiation pattern of the antenna is given by

$$f(\theta, \varphi) = 4|R_{tot}|\cdot S\cdot \sin c\left(\frac{L}{\lambda_0}(\cos\varphi - 1)\sin\theta\right)\cdot \sin c\left(\frac{L}{\lambda_0}\sin\theta\sin\varphi\right) \cdot \sin c\left(\frac{W}{\lambda_0}\sin\theta\cos\varphi\right). \tag{12.71}$$

The directive gain of the antenna is given by

$$D(\theta, \varphi) = \frac{(2k_0 S|R_{tot}|)^2}{\pi Z_0}\frac{(1+\cos\theta)^2}{\mathrm{Re}(A)}\cdot \sin c^2\left(\frac{L}{\lambda_0}(\cos\varphi - 1)\sin\theta\right)\cdot \sin c^2\left(\frac{L}{\lambda_0}\sin\theta\sin\varphi\right)\cdot \sin c^2\left(\frac{W}{\lambda_0}\sin\theta\cos\varphi\right). \tag{12.72}$$

The maximum value of the directive gain of the antenna is achievable for $\theta = 0°$, and the directivity is given by

$$D_{\max} = \frac{8k_0^2}{\pi Z_0}\frac{(S|R_{tot}|_{\theta=0})^2}{\mathrm{Re}(R_{tot})|_{\theta=0}}. \tag{12.73}$$

The maximum gain of the antenna is given by

$$G_{\max} = \frac{8k_0^2}{\pi Z_0}\frac{(S|R_{tot}|_{\theta=0})^2}{\mathrm{Re}(R_{tot})|_{\theta=0}}\left(1 + \frac{j_0 S}{\mathrm{Re}(R_{tot})|_{\theta=0}}\sqrt{\frac{\omega\mu_0}{2\sigma_i}}\right)^{-1}. \tag{12.74}$$

It is obvious from Eqs. (12.69)–(12.73) that the expressions for the normalized gain of the antenna and its normalized gain along the two principle planes are given by 12.30 and Eqs. (12.31) and (12.32), respectively.

In order to make both the analytical and numerical analysis of antenna performance, we evaluate the values of the radius of inclusions r using the free-space method [39]. The appropriate calculations give: r = 0.00045 m. It is interesting to mention that using the formulas in Rybin and Shulga [27] gives almost the same results. Thus, it is tempting to conclude: if the values of the metal volume fractions are the same, the values of the effective parameters will also be the same. The last conclusion is valid only for small values of the metal volume fraction ($\pi r^2/d^2 < 0.25$). This is because the dependence of the values of effective parameters on the metal volume fraction is non-linear (the dependence seems almost linear for low values of the metal volume fraction) [27,28].

The miniaturized design for the antenna depicted in Fig. 12.23 can be found from the algorithm based on Eqs. (12.36)–(12.39) if the coefficient A is replaced with the coefficient R_{tot}. Running the appropriate procedure finally gives almost the same results for ε_r, L, and W with a given value of the substrate thickness. Indeed, the theoretical calculations for the miniaturized antenna design (S_{min} = 0.00017628555 m^2) give: f_{rm} = 7.531 GHz, W_{min} = 0.01574 m, and L_{min} = 0.0112 m. Last results are in good agreement with the results of numerical experiment: f_{rm} = 7.512 GHz, W_{min} = 0.0156 m, L_{min} = 0.0113 m. Notice that f_r = 7.512 GHz for the standard numerical design.

Putting the values of L_{min}, W_{min}, L, and W into 12.33 gives: $VPMGF$ = 16.5%, that is a slightly better value of the volume profile miniaturization gain factor compared with the case of one-layer wire grid substrates. At the same time, blue shifts of the main frequencies are stronger for the case of two-layer wire grid substrates (Fig. 12.25). This is mostly caused by two reasons: (i) a more complicated distribution of the reactive part of the load antenna impedance around the wires for the case of two-layer wire grid substrates, and (ii) relatively clear diamagnetic behavior of the antenna substrate for the case of two-layer wire grid substrates (in fact, Re(μ_r) ≈ 0.95 for the case of the two-layer wire grid substrate and Re(μ_r) ≈ 0.98 for the case of one-layer wire grid substrates, whereas 12.1 is created for Re(μ_r) = 1).

The results of FDTD simulations of the far-field patterns for the antenna designs with two-layer wire grid superstrates are presented in Figs. 12.26–12.27.

As seen from Figs. 12.26 and 12.27, no significant differences are observed between the far-field patterns of the numerical antenna designs with one-layer wire grid superstrates and the far-field patterns of the antenna designs with two-layer wire grid superstrates.

The analytical modeling (using Eqs. (12.69) and (12.73)) and numerical FDTD calculations have shown: (i) the maximum power gain of standard inhomogeneous analytical model is 5.17 dB, and it is found at $\theta = -1^0$; (ii) the maximum power gain of standard inhomogeneous numerical model is 5.26 dB, and it is found at $\theta = -1^0$; (iii) the maximum power gain of miniaturized inhomogeneous numerical model is 5.81 dB, and it is found at $\theta = -1^0$. At the same time, it has been found that the

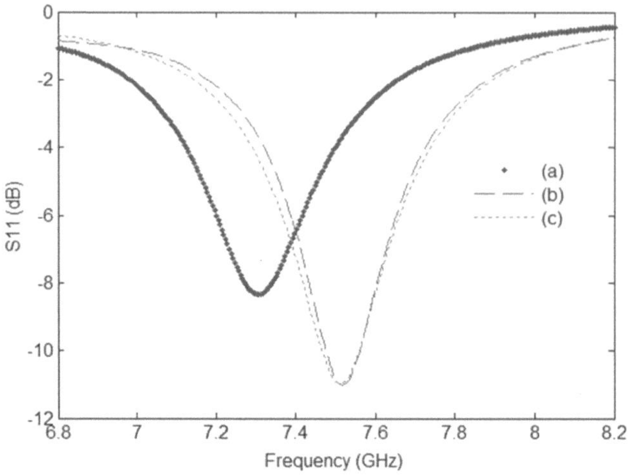

FIGURE 12.25 Simulated $S11$-spectra for (a) standard homogenous numerical model, (b) standard inhomogeneous numerical model and (c) miniaturized inhomogeneous numerical model.

appropriate efficiency value is 40.72% for the case of the standard inhomogeneous analytical model, 44.82% for the case of the standard inhomogeneous numerical model, and 49.36% for the case of the miniaturized inhomogeneous numerical model. Moreover, according to the results of appropriate FDTD simulations, we can

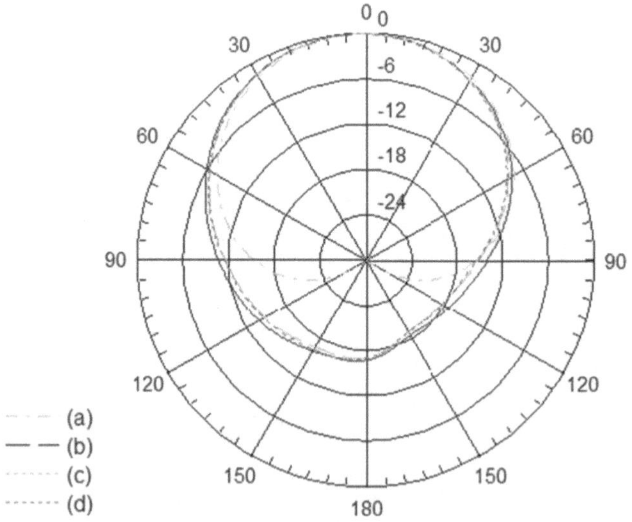

FIGURE 12.26 Far-field patterns in the H-plane (in dB) of (a) standard inhomogeneous analytical model, (b) standard homogenous numerical model, (c) standard inhomogeneous numerical model and (d) miniaturized inhomogeneous numerical model evaluated at the resonant frequencies.

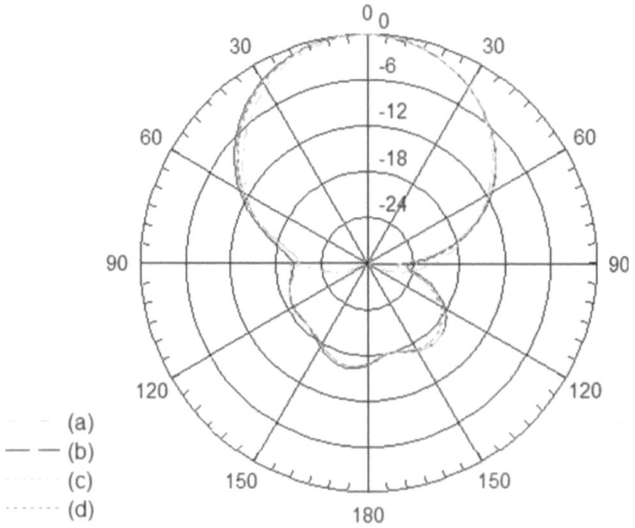

FIGURE 12.27 Far-field patterns in the *E*-plane (in dB) of (a) standard inhomogeneous analytical model, (b) standard homogenous numerical model, (c) standard inhomogeneous numerical model and (d) miniaturized inhomogeneous numerical model evaluated at the resonant frequencies.

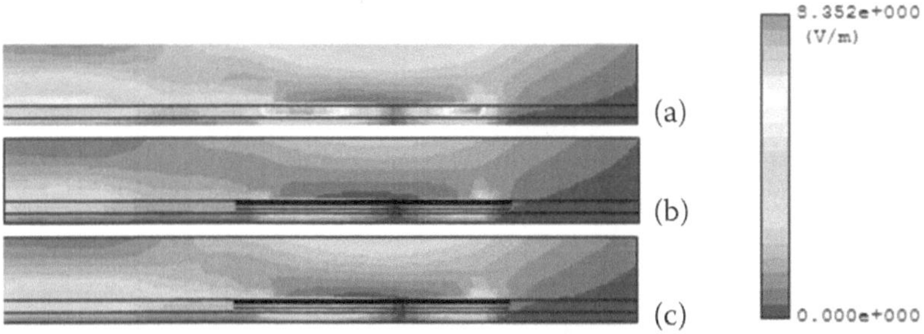

FIGURE 12.28 Near-field calculated in the cross-section $y = 0$ m of patch antenna for (a) standard homogenous numerical model at the frequency 7.307 GHz, (b) standard inhomogeneous numerical model at the frequency 7.518 GHz, and (c) miniaturized inhomogeneous numerical model at the frequency 7.515 GHz.

conclude that the intensity of near-fields for the case of antennas with two-layer wire grid substrates (Figs. 12.27 and 12.28) is quite comparable to the intensity of the near-fields for the case of two-layer wire grid substrates (Figs. 12.8 and 12.9).

Table 12.1 compares the parameters of all the considered antenna designs. Analyzing the contents of Table 12.1 enables us conclude that: (i) the performance of miniaturized design is greater than the performance of appropriate design; (ii) using the ferrite superstrates enables one to considerably improve the performance of patch antennas; (iii) using the composite metal-dielectric

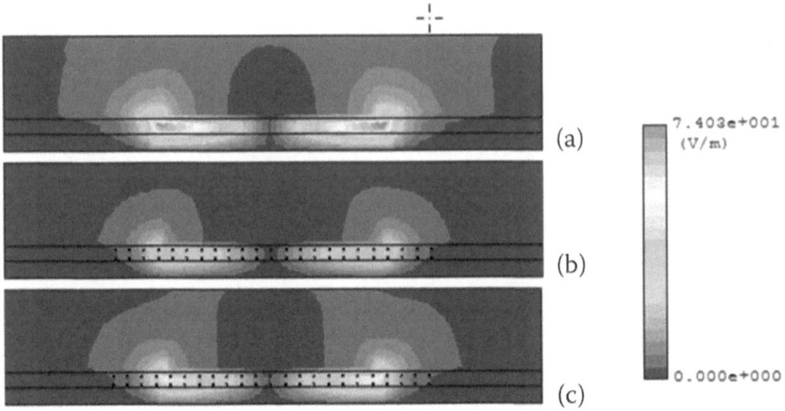

FIGURE 12.29 Near-field calculated in the middle cross-section along the x-axis of patch antenna for (a) standard homogenous numerical model at the frequency 7.307 GHz, (b) standard inhomogeneous numerical model at the frequency 7.518 GHz, and (c) miniaturized inhomogeneous numerical model at the frequency 7.515 GHz.

TABLE 12.1
Main Characteristics of Antennas

Design	Frequency (GHz)	E (%)	D_{max} (dB)	*VPMGF* (%)	Substrate
Standard numerical without superstrate	7.307	40.72	4.6	–	Dielectric
Standard analytical without superstrate	7.3	42.96	5.05	–	Composite with one-layer wire grid
Standard numerical without superstrate	7.385	43.98	5.01	–	Composite with one-layer wire grid
Miniaturized numerical without superstrate	7.418	50.93	6.06	13.4	Composite with one-layer wire grid
Standard numerical with superstrate	7.322	46.1	5.92	–	Composite with one-layer wire grid
Standard analytical with superstrate	7.3	49.37	6.01	–	Composite with one-layer wire grid
Miniaturized numerical with superstrate	7.404	54.34	6.59	13.4	Composite with one-layer wire grid
Standard numerical without superstrate	7.518	44.82	5.26	–	Composite with two-layer wire grid
Standard analytical without superstrate	7.3	40.72	5.17	–	Composite with two-layer wire grid
Miniaturized numerical without superstrate	7.515	49.36	5.81	16.5	Composite with two-layer wire grid

substrates can make patch antennas multi-band and multi-directional; (iv) adding the wire layers to the composite substrates enables one to slightly decrease the volume profile of patch antennas; (v) using composite substrates with multi-layer wire grids blue-shifts the main resonant frequency of patch antennas compared to composite substrates with one-layer wire grids; and (vi) adding the wire layers to the composite substrates does not improve the performance of patch antennas. The last two conclusions are the reason why patch antennas with multi-layer wire grids embedded into dielectric substrates with ferrite superstrate are not being considered in this study.

12.5 CONCLUSION

Theoretical foundations for designing microwave patch antennas with one- and two-layer wire grids embedded in the dielectric substrates are presented in this chapter as well as the principle of miniaturization of such antennas. Such antennas have improved performance compared to patch antennas on dielectric substrates with the same numerical values of constitutive parameters. The first type of antennas are reconfigurable in frequency functionality and polarization. Moreover, the performance of such antennas can be improved by using ferrite superstrates. However, an increase in the number of layers of wire grids in the antenna substrates does not lead to further improvement in performance, but instead to an essential blue shift in the resonance frequency. The proposed antennas with composite substrates and ferrite superstrates can be used in designing novel antennas for gadgets and transponders.

ACKNOWLEDGMENTS

The author would like to acknowledge his teacher, Dr. Georgios Zouganelis, and Prof. Dr. N. N. Gorobets for his advice during the writing of the chapter. Also, he acknowledges Ms. Julia (Iuliia) Rybin for her valuable help in preparing the manuscript.

REFERENCES

1. C. A. Balanis, *Antenna Theory: Analysis and Design*. 3rd edition. Hoboken, NJ, USA: John Wiley & Sons, Inc., p. 1072, 2005.
2. H. A. Wheeler, "Transmission-line properties of parallel strips separated by a dielectric sheet," *IEEE Trans. Microw. Theory Tech.*, vol. MTT-13, no. 16, pp. 172–185, 1965.
3. R. K. Mongia, A. Ittipiboon, and M. Cuhaci, "Low profile dielectric resonator antennas using a very high permittivity material," *Electron. Lett.*, vol. 30, pp. 1362–1363, 1994.
4. J. S. Colburn and Y. Rahmat-Samii, "Patch antennas externally perforated high dielectric constant substrates," *IEEE Trans. Antennas and Propagat.*, vol. 47, no. 12, pp. 1785–1794, 1999.
5. S. Encoh, G. Taybe, P. Sabouroux, and N. Guerin, "A metamaterial for directive emission," *Phys. Rev. Lett.*, vol. 89, no. 21, Article 213902, 2002.

6. K. Buell, H. Mosallaei, and K. Sarabandi, "A substrate for small patch antennas providing tunable miniaturization factors," *IEEE Trans. Microw. Theory Tech.*, vol. 54, no. 1, pp. 135–146, 2006.
7. P. M. T. Ikonen, K. N. Rozanov, A. V. Osipov, P. Alitalo, and S. A. Tretyakov, "Magneto-dielectric substrates in antenna miniaturization: Potential and limitations," *IEEE Trans. Antennas and Propagat.*, vol. 54, no. 11, pp. 3391–3399, 2006.
8. G. V. Eleftheriades and R. Islam, "Miniaturized microwave components and antennas using negative-refractive-index transmission-line (NRI-TL)," *Metamaterials*, vol. 1, no. 2, pp. 53–61, 2007.
9. F. Bilotti, A. Alu, and L. Vegni, "Design of miniaturized metamaterial patch antennas with μ-negative loading," *IEEE Trans. Antennas and Propagat.*, vol. 56, no. 6, pp. 1640–1647, 2008.
10. H. A. Jang, D. O. Kim, and C. Y. Kim, "Size reduction of patch antenna array using CSRRs loaded ground plane," *Proc. of 31st Prog. in Electromagn. Res. Symp. (PIERS'2012)*, Kuala Lumur (Malaysia), pp. 1487–1489, 27–30 March, 2012.
11. S. K. Patel and Y. P. Kosta, "Size reduction in microstrip based meandered radiating structure using artificial substrate," *Int. J. Appl. Electrom.*, vol. 43, no. 2, pp. 207–216, 2013.
12. Y. Liu, X. Guo, S. Gu, and X. Zhao, "Zero index metamaterial for designing high-gain patch antenna," *Int. J. Antennas Propagat.*, vol. 2013, Article ID 215681, pp. 1–12, 2013.
13. M. Rahimi, F. B. Zarrabi, R. Ahmadian, Z. Mansouri, and A. Keshtkar, "Miniaturization of antenna for wireless application with difference metamaterial structures," *Prog. in Electromagn. Res.*, vol. 145, pp. 19–29, 2014.
14. R. C. Hansen and M. Burke, "Antennas with magneto-dielectrics," *Microw. Opt. Technol. Lett.*, vol. 26, no. 2, pp. 75–78, 2000.
15. H. Mosallaei and K. Sarabandi, "Magneto-dielectrics in electromagnetics: Concept and applications," *IEEE Trans. Antennas and Propagat.*, vol. 52, no. 6, pp. 1558–1567, 2004.
16. G. Lovat, P. Burghignoli, F. Capolino, and D. R. Jackson, "Combinations of low/high permittivity and/or permeability substrates for highly directive planar metamaterial antennas," *IET Microw. Antennas Propagat.*, vol. 1, no. 1, pp. 177–183, 2007.
17. G. Zouganelis, F. Soma, O. Rybin, and H. Ohsato, "Study of ultra low index ($0 < \varepsilon < 1$) metamaterial using a patch antenna," *Proc. of 1st Asia-Oceania Ceramic Federation Conference (AOCF) convened in conjunction with 18th Fall Meeting of The Ceramic Society of Japan*, Osaka (Japan), vol. 18, pp. 148, 25–27 September, 2005.
18. O. Rybin and S. Shulga, "Feedback magnetization of ultra-low index irradiative structure," *Mod. Phys. Lett. B.*, vol. 29, no. 29, Article 1550179, 2015.
19. O. Rybin, "Microwave miniaturization concept for narrow band rectangular patch antenna structures," *Int. J. Appl. Electrom.*, vol. 48, no. 1, pp. 69–75, 2015.
20. S. Prosvirnin and Yu Nechaev, *Calculation of Microstrip Antennas in the Approximation of a Given Surface Current Distribution*, Voronezh City: Voronezh State University, pp. 112, 1992 (in Russian).
21. O. Rybin, V. Shulga, and S. Shulga, "The given surface current distribution model of a rectangular patch antenna with metamaterial-like substrate," *Results in Phys.*, vol. 15, Article 102573, 2019.
22. I. A. Deryugin and M. A. Sigal, "Frequency dependence of the magnetic permeability and dielectric between 500 and 35000 Mcs," *Sov. Phys. – Techn. Phys.*, vol. 6, no. 1, pp. 72–77, 1961.
23. L. D. Landau, E. M. Lifshitz, and L. P. Pitaevskii, *Electrodynamics of Continuous Media*, 2nd edition, Burlington: Elsevier, p. 460, 2008.

24. J. R. Wait, "Reflection from a wire grid parallel to a conducting plane," *Can. J. Phys.*, vol. 32, no. 9, pp. 571–579, 1954.
25. P. Perlmutter, S. Shtrikman, and D. Treves, "Electric surface current model for the analysis of microstrip antennas with application to rectangular elements," *IEEE Trans. Antennas and Propagat.*, vol. AP-33, no. 3, pp. 301–311, 1985.
26. D. K. Cheng. *Field and Wave Electromagnetics*, 2nd edition, Reading, MA, USA: Eddison-Wesley Publ. Co., Inc., p. 709, 1983.
27. O. Rybin and S. Shulga, "Revised homogenization for two-component metamaterial with non-magnetic metallic cylindrical inclusions", *Appl. Phys. A*, vol. 125, no. 2, pp. 153–160, 2019.
28. O. Rybin and S. Shulga, "Generalized broad-band effective medium theory of two-component metamaterials including magnetic ones: A review," *J. Electromagnet. Wave*, vol. 34, no. 11, pp. 1513–1549, 2020.
29. O. Rybin and S. Shulga, "Utilization of double metal-dielectric composite substrates for microwave miniaturization of rectangular patch antennas," *J. Comput. Electron.*, vol. 15, no. 3, pp. 1023–1027, 2016.
30. D. D. Krishna, M. Gopikrishna, C. K. Aanandan, P. Mohanan, and K. Vasudevan, "Compact dual band slot loaded circular patch antenna with a superstrate," *Prog. in Electromagn. Res.*, vol. 83, pp. 245–255, 2008.
31. H. Vettikalladi, O. Lafond, and M. Himdi, "High-efficient and high-gain superstrate antenna for 60-GHz indoor communication," *IEEE Antennas Wireless Propagat. Lett.*, vol. 8, pp. 1422–1425, 2009.
32. H. Attia, L. Yousefi, O. Siddiqui,·and O. M. Ramahi, "Analytical formulation of the radiation field of printed antennas in the presence of artificial magnetic superstrates," *Appl. Phys. A*, vol. 103, no. 3, pp. 877–880, 2011.
33. E. K. I. Hamad and A. Abdelaziz, "Metamaterial superstrate microstrip patch antenna for 5G wireless communication based on the theory of characteristic modes," *J. Electr. Eng.*, vol. 70, no. 3, pp. 187–197, 2019.
34. B. Hasa and K. Raza, "Dual band slotted printed circular patch antenna with superstrate and EBG structure for 5G applications," *Mehran Univ. Research J. Eng. and Technol.*, vol. 38, no. 1, 2019, pp. 227–238.
35. Y. Al-Alem and A. A. Kishk, "Low-cost high gain superstrate antenna array for 5 G applications," *IEEE Antennas Wireless Propagat. Lett.*, vol. 19, no. 11, pp. 1920–1923, 2020.
36. T. Hayat, M. U. Afzal, A. Lalbakhsh, and K. P. Essel, "Additively Manufactured Perforated Superstrate to Improve Directive Radiation Characteristics of Electromagnetic Source," *IEEE Access*, vol. 7, pp. 153445–153452, 2019.
37. H. Boutayeb, K. Mahdjoubi, A. C. Tarot, and T. A. Denidni, "Directivity of an antenna embedded inside a Fabry-Perot cavity: analysis and design," *Microw. Opt. Technol. Lett.*, vol. 48, no. 1, pp. 12–17, 2006.
38. J. Wan, O. Rybin, and S. Shulga, "Far field focusing for a microwave patch antenna with composite substrate," *Results in Phys.*, vol. 8, pp. 971–976, 2018.
39. D. K. Ghodgaonkar, V. V. Varadan, and V. K. Varadan, "Free-space measurement of complex permittivity and complex permeability of magnetic materials at microwave frequencies," *IEEE Trans. Instrum. Meas.*, vol. 39, no. 2, pp. 387–394, 1990.

13 Review of Metamaterial-Assisted Vacuum Electron Devices

Raktim Guha[1], *Xin Wang*[2], *Amit K. Varshney*[3], *Zhaoyun Duan*[4], *Michael A. Shapiro*[5], *and B. N. Basu*[6]

[1]Academy of Scientific and Innovative Research (AcSIR), Vacuum Electron Devices Development Group, CSIR-Central Electronics Engineering Research Institute (CEERI), Pilani, Rajasthan, India

[2]University of Electronic Science and Technology of China (UESTC), Chengdu, Sichuan, China

[3]Department of Electronics and Communication Engineering (ECE), Sir J. C. Bose School of Engineering, Supreme Knowledge Foundation Group of Institutions (SKFGI), Mankundu, West Bengal, India

[4]University of Electronic Science and Technology of China (UESTC), Chengdu, Sichuan, China

[5]Gyrotron Research Group, Plasma Science and Fusion Center (PSFC), Massachusetts Institute of Technology, Cambridge, Massachusetts, USA

[6]Sir J.C. Bose School of Engineering, Supreme Knowledge Foundation Group of Institutions (SKFGI), Mankundu, West Bengal, India

13.1 INTRODUCTION

Engheta and Ziolkowski in a book, edited by them, wrote: "*To the best of our knowledge, the first attempt to explore the concept of 'artificial' material appears to trace back to the late part of the nineteenth century when in 1898 Jagadis Chunder Bose conducted the first microwave experiment on twisted structures – geometries that were essentially artificial chiral elements by today's terminology*" [1,2]. Since then considerable efforts have been made to explore artificial materials, and one such material is the metamaterial (MTM) – a periodic, macroscopic structure – which can be formed by embedding inclusions and material components in a host medium. MTMs can be engineered to have some electromagnetic properties of the materials altered to something 'meta', meaning 'beyond' what can be found in

DOI: 10.1201/9781003050162-13

nature, such as negative refractive index, reversed Doppler effect, and reversed Cherenkov effect [3–6].

The objective of the present chapter is to bring out the advantages of MTM assistance in improving the performance of the structure interaction impedance (*K*) of vacuum electron devices (VEDs), and the device output power, gain, efficiency, etc., remembering that VEDs enjoy more power producing and handling capabilities than their conventional solid-state device counterparts [7,8]. The chapter reviews the global efforts to showcase this point, while being aware that the conventional VEDs, such as travelling-wave tubes (TWTs) [9,10], klystrons [9,10] and backward-wave oscillators (BWOs) [10] have their limitations in terms of RF output power, gain, bandwidth, overall efficiency, design complexity at higher frequencies, etc., as typified in Table 13.1.

In a VED, the kinetic or potential energy of an electron beam is transferred to electromagnetic waves supported by a material medium, which may be termed as

TABLE 13.1
Performance comparison of typical conventional VEDs [7–10]

Parameters	TWTs	Klystrons	BWOs
Interaction impedance (*K*)	Low (~10–40 Ω using helix slow-wave structures (SWSs); 400–600 Ω using coupled-cavity SWSs, at the operating frequency band)	–	Medium (~20–300 Ω, at the operating frequency band)
Overall efficiency	Low-to-medium (~7–60%, varying due to different SWSs used, at different operating frequencies)	Medium-to-high (~30–70%, efficiency increasing with the number of depressed collector stages)	Medium (~40–60%)
RF output power	Low-to-medium (~40–500 W for CW mode, and 1–250 kW for pulsed mode)	Medium-to-high (~ 200 kW (average), and up to several GW for pulsed mode)	Medium-to-high (~250 kW–1.5 GW for pulsed mode)
Gain	Medium (~30–50 dB)	High (~40–60 dB)	-
Bandwidth	Low-to-medium (from ~few GHz to 2 octave with high design complexity)	Very narrow	Wide-range frequency operation implemented by varying only the beam voltage
Design complexity at high frequency	Very high	Very high	Medium
Multi-band operation	–	–	Dual-band operation

the beam-wave interaction structure. A material, treated as a medium for electromagnetic waves to interact with, can be classified in terms of the medium permittivity ε and permeability μ as: (i) double-positive (DPS) (Re(ε)> 0, Re(μ)>0) material, supporting forward-wave propagation (e.g., isotropic dielectrics); (ii) epsilon negative (ENG) (Re(ε)< 0, Re(μ)> 0) material, supporting evanescent wave-modes, e.g., electric plasmas: metal thin wires (microwave) frequencies: $\omega<\omega_{pe}$, where ω_{pe} represents the electric plasma frequency; (iii) double-negative (DNG) (Re(ε)< 0, Re(μ)< 0) material – which cannot be found in nature, but can be physically realized – supporting backward-wave propagation; and (iv) mu negative (MNG) (Re(ε)> 0, Re(μ)< 0) material, supporting evanescent wave-modes, e.g., ferrimagnetic materials and artificial resonant magnetic materials: magnetic plasmas at frequencies $\omega<\omega_{pm}$, where ω_{pm} represents the magnetic plasma frequency [1].

If appropriately engineered, an MTM can be designed to exhibit not only the characteristics of DNG material, but also those of DPS, MNG and ENG materials – the latter two being also termed as single-negative (SNG) materials [1,3]. The notable examples of MNG, ENG and DNG MTMs are the arrays of parallel metallic wires/rods/strips [11,12] and split-ring resonators (SRRs) [13], spiral resonators [14], labyrinth resonators (LR) [14], complementary split-ring resonators (CSRRs) [15,16], I-shaped MTMs [17], electric-LC resonator [18–20], S-shaped MTMs [21], cut-wire pairs [22], fishnet MTMs [23], thin wire-LR MTMs [24], and so on.

There has been a recent surge in the exploration of MTM assistance in VEDs and their interaction structures [4,5,16,16,16,25–46]), which have alleviated the shortcomings of conventional VEDs listed in Table 13.1. These have been reviewed here in Sections –13.3-13.5, after outlining, in Section 13.2, the approach of retrieving the parameters, namely the effective relative permeability and permittivity, of an MTM medium treating it as an effectively homogeneous structure. This helps in developing the understanding of the study encompassing electromagnetic analysis, simulation and experimental characterization of MTM-assisted VEDs and their interaction and RF coupling structures, and thus in appreciating the results of such study. Further, after reviewing challenging aspects and future scope of MTM-assisted VEDS in Section 13.6, the present review is concluded in Section 13.7 highlighting the major findings of the review.

13.2 MTM EFFECTIVE MEDIUM

Before diving into the 'magical' world of MTM-assisted VEDs, one needs to understand the 'magic' of MTMs by retrieving its effective medium or constitutive properties, namely the effective relative permittivity $\varepsilon_{r,\text{eff}}(f)$ [11] and effective relative permeability $\mu_{r,\text{eff}}(f)$ [13]. There are three approaches to the retrieval of the constitutive parameters of an MTM medium, which use (i) Lorentz and Drude models [11,13], (ii) refection/transmission or *S*-parameter method [12,21,47,48], and (iii) equivalent-circuit method [14], respectively. However, the first and the second approaches have been proved to be more popular than the third approach in the study of the effect of MTM assistance on VEDs. The readers are encouraged to see Chapter 2 of the present memoir for a brief overview of the first two of these approaches.

13.3 MTM-ASSISTED INTERACTION STRUCTURES OF MICROWAVE VEDS ANALYSED ONLY BY THEORY AND/OR SIMULATION

In this section, we review in brief the status of the study that is based on the electromagnetic analysis and simulation of MTM-assisted interaction structures of VEDs and underlying beam-wave interaction mechanisms, citing, however, only a few representing examples. The experimental aspects of this study have been taken up in Section 13.4.

13.3.1 MTM-Assisted Interaction Structures for Microwave Amplifiers

All over the world, extensive efforts have been made to incorporate the exotic properties of MTMs into the VEDs, and notably, to this end, some researchers combined the MTMs with conventional interaction structures [4,25–35]. Some other researchers have come out with novel MTM based SWSs [5,16,38–46]. In this regard, electromagnetic analysis or equivalent circuit analysis along with simulations, using commercial software's (CST, HFSS, MAGIC, COMSOL, etc.), have been carried out while making some simplifying assumptions on the constitutive parameters of MTMs (such as, the consideration of isotropic and homogeneous material properties instead anisotropic/bi-anisotropic and inhomogeneous properties, etc.) to make the analysis tractable and easily understandable.

13.3.1.1 Traveling-wave amplifiers

MTM loaded helix SWS

A helix supported by a number of dielectric rods ($\varepsilon_{r,\text{eff}}(f) = \varepsilon_r = 5.1, 6.5, 9.4$, etc. and $\mu_{r,\text{eff}}(f) = \mu_r = 1$ for conventional dielectric materials) in a metal envelope is an extensively used SWS of a TWT [49] (Fig. 13.1a). The structure has been analysed by smoothing out the discrete dielectric supports into a continuous dielectric tube of an equivalent relative permittivity ($\varepsilon'_{r,\text{eff}}(f) = \varepsilon'_{r,\text{eff}}$, $\mu'_{r,\text{eff}}(f) = \mu'_{r,\text{eff}} = 1$) (Fig. 13.1b) [49,50], the primed quantities, in this context, representing the parameters of the equivalent smoothed-out dielectric tube.

Purushothaman and Ghosh [27] investigated into an SWS consisting of a helix supported by discrete ENG, MNG, and DNG MTM rods in a metal envelope (Fig. 13.1a) with the help of electromagnetic field analysis in the tape-helix model that takes into account the axial harmonic effects of the helix turns; they however did not carry out any simulation study. In their study, they have assumed that the properties of the MTM rods are isotropic, homogeneous and frequency independent ($\varepsilon_{r,\text{eff}}(f) = \varepsilon_{r,\text{eff}}$, $\mu_{r,\text{eff}}(f) = \mu_{r,\text{eff}}$ and $\varepsilon'_{r,\text{eff}}(f) = \varepsilon'_{r,\text{eff}}$, $\mu'_{r,\text{eff}}(f) = \mu'_{r,\text{eff}}$). They have also considered the real values of ENG, MNG, and DNG such that $\varepsilon'_{r,\text{eff}}$ and $\mu'_{r,\text{eff}}$ both remain positive to support the forward wave propagation in the SWS. As far as the dispersion control at enhanced K of the SWS is concerned, they found that the ENG- and MNG-MTM helix-support rods are superior to the DNG-MTM rods. However, in view of the oversimplified assumptions made, their findings need to be carefully checked before taking them for the development of any relevant design concepts.

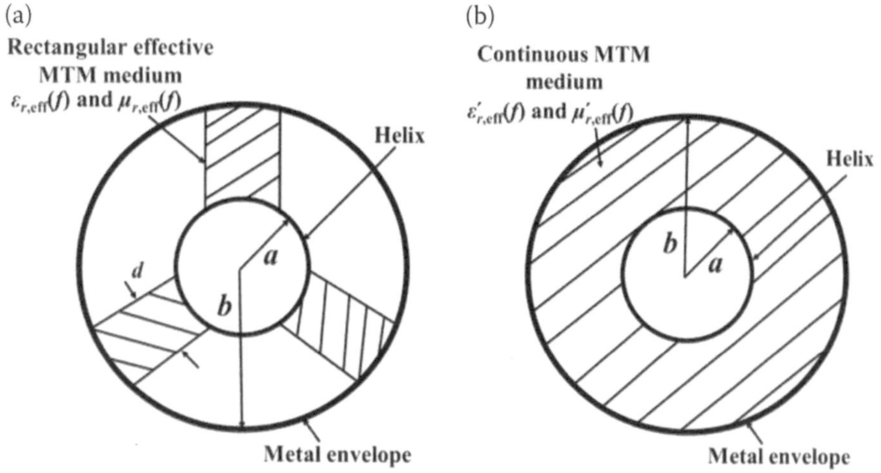

FIGURE 13.1 (a) Helix of radius a with discrete MTM support rods, typically of rectangular cross-section, characterized by the effective parameters, namely, the relative permittivity and the relative permeability $\varepsilon_{r,\text{eff}}(f)$ and $\mu_{r,\text{eff}}(f)$, respectively, enclosed in a metal envelope of radius b, and (b) its equivalent structure model consisting of a helix surrounded by an equivalent MTM tube, characterized by the equivalent effective parameters $\varepsilon'_{r,\text{eff}}$ and $\mu'_{r,\text{eff}}$, in a metal envelope.

It has been a conventional method to widen the bandwidth of a helix-TWT by controlling the helix dispersion with low K values by metal vane loading of the metal envelope though the method entails the risk of arcing in the tube, and though the method has the scope for further enhancement of the structure interaction impedance leading to the consequent enhancement of the device gain and efficiency [9,10,51]. Guha and Ghosh [30] have proposed and simulated, using CST Microwave Studio, a novel DPS-MTM loaded helix SWS in which I-shaped metallic strip printed on the both the azimuthal faces of the conventional dielectric (BeO, $\varepsilon_r = 6.5$) helix-support rods. The proposed structure exhibits the dispersion characteristics similar to those of the conventional vane loaded helix SWS with a higher K than the latter leading to the enhanced device gain and efficiency with relaxed possibility of arcing [30]. However, this proposed MTM loaded helix SWS needs further investigation into parameter retrieval and the validation of the simulated results against electromagnetic analysis.

MTM loaded folded-waveguide SWS

Conventional folded-waveguide (FW) SWSs for TWTs are extensively used for high frequency applications as transverse dimensions of the waveguide remain bigger than helix SWSs, for which they have better power handling capability. However, typically FW-SWSs exhibit extremely low K values at their wide operating bandwidth. In this regard, Rashidi and Behdad [32] studied ENG-MTM (with frequency independent effective medium property) inclusion in FW-SWS for TWTs (Fig. 13.2(a)) where, due to the ENG-MTM insertion, operating frequency gets

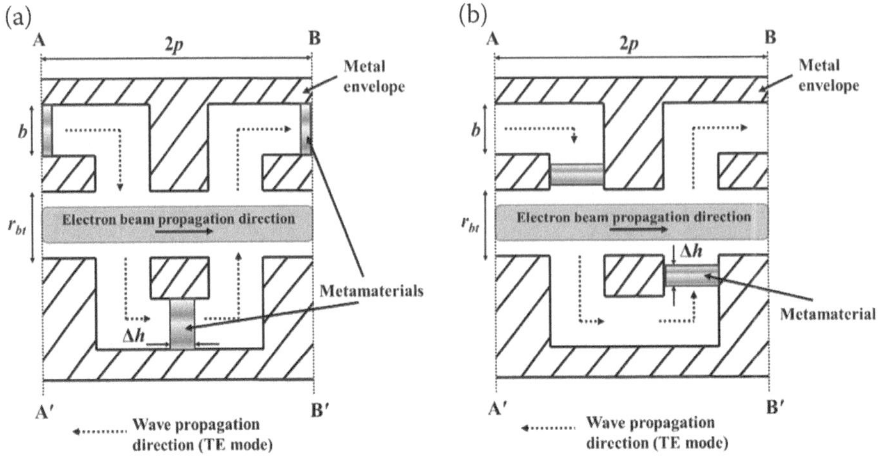

FIGURE 13.2 Unit cell of an FW loaded by (a) ENG-MTM, and (b) DNG-MTM inserts in a rectangular waveguide of height b with a beam tunnel diameter r_{bt}, showing also the axial periodicity p of the insert, and the thickness Δh of both the types of MTM loading.

up-shifted caused by the increase of phase velocity, obtained using CST Microwave Studio, owing to the introduction of ENG-MTM insert acting as a shunt inductance [32]. However, if the transverse dimensions of FW with an ENG-MTM insert are increased by a scale factor, this up-shifting of operating frequency can be prevented, and can be restored back to that corresponding to the case of FW without ENG-MTM loading, however, at the cost of the bandwidth. Interestingly, the larger transverse dimensions of the ENG-MTM loaded FW would allow for a larger beam tunnel diameter, and consequently, a larger beam current, larger beam power and a higher device RF output power for a given frequency of operation. This advantage of ENG-MTM assistance is of special significance in designing high-frequency TWTs, such as the millimetre-wave TWTs, for which, in the conventional design, the beam diameter and current become restricted due to the smaller transverse dimensions of the SWS. Further, added to this advantage of the ENG-MTM assistance is enhanced K of FW making it potentially employable as SWS of a high-gain, high-efficiency ENG-MTM-assisted FW-TWT [32]. The TWT designers face the difficulty of fabrication as well as thermal management in designing and developing the conventional TWTs in high-frequency, such as the millimetre-wave regime, due to their reduced transverse dimensions. Clearly, the ENG-MTM loading of FW-TWT has alleviated this problem by providing a scaled-up larger transverse dimensions as required for preventing the operating frequency band from shifting to higher frequencies. Further, a higher K of an ENG-MTM-loaded FW would enhance the gain and efficiency of an ENG-MTM-assisted FW-TWT [32].

In another version, the FW is loaded with DNG-MTM, which has been analytically studied by Tan and Seviour [34], leading to its dispersion relation involving the $\varepsilon_{r,\mathrm{eff}}(f)$ and $\mu_{r,\mathrm{eff}}(f)$ of the MTM given by Drude and Lorentz models, respectively. Further, the dispersion relation of an MTM-loaded FW can be used to plot its dispersion curve, which peculiarly deviates from that of an unloaded FW in that there appears the so-called turning point. Around this tuning point, the regime of simultaneous negative relative permeability and relative permittivity exists with a discontinuity in both the left and right branches of the dispersion curve, though only the right branch components are physically triggered with the forward electromagnetic wave [34].

As far as the DNG-MTM-loaded FW (Fig. 13.2b) is concerned, Tan and Seviour [33,34] deduced an expression for the increment of electromagnetic power ΔP transferred from the beam to electromagnetic wave in a loaded waveguide, in terms of the waveguide transverse dimensions, structure periodicity, propagation constant of an unloaded rectangular waveguide excited in the TE_{01} mode, loaded-propagation-constant, electron velocity, beam current and beam accelerating potential. They also studied the beam accelerating potential dependent variation of ΔP with frequency, the positive and negative values of ΔP representing the amplification of electromagnetic waves and inverse Cherenkov acceleration of the electron beam, respectively. Thus, they could find the frequency and beam accelerating potential for the amplification of electromagnetic waves for the maximum positive value of ΔP, as is necessary for the beam-excited DNG-MTM loaded FW to function as an FW-TWT [34]. It was also revealed by their work that, in a DNG-MTM loaded FW-TWT, the MTM parameters provide an additional control over the gain-frequency response, unlike in a conventional FW-TWT without MTM loading in which only the structure dimensions control such a response. Further, the dependency of frequency corresponding to the maximum power transfer from the electron beam to electromagnetic waves on the beam accelerating potential makes the DNG-MTM loaded FWT rather frequency tuneable. However, the limited frequency regime in which the MTM loaded FW exhibits the DNG property makes the DNG-MTM loaded FW-TWT essentially bandwidth-limited and dependent on the properties of the DNG-MTM used. The need of the hour is to reduce the inherent ohmic losses associated with the MTM and widen the DNG frequency regime of MTM to make the DNG-MTM loaded FW-TWT more useful a VED [33,34].

13.3.1.2 Resistive-wall amplifiers

The resistive-wall amplifier (RWA) is a growing-wave VED in which the gain in the device is obtained through the interaction between the charge of electron beam and that on the resistive metal wall that is induced by the beam such that the charge induced on the wall acts on the electron beam so as to cause larger and larger electron bunches to be formed, thereby resulting in the spatial growth of signal [52,53].

Subsequently, Rowe et al. [35,54] extended the study on RWA further by exploring the effect of MTM assistance on its performance. For the sake of simplicity of analysis, Rowe et al. [35] treated the MTM as isotropic, in order to get a quick insight into the behaviour of an MTM-assisted RWA, though later on in [54], they made the analysis more rigorous by treating the MTM as anisotropic.

Rowe et al. [35] proposed two different configurations of RWAs assisted by an MTM that was supposedly isotropic: (i) one using a lossless ENG-MTM as the support material for the resistive layer (Fig. 13.3), where region I is the electron beam

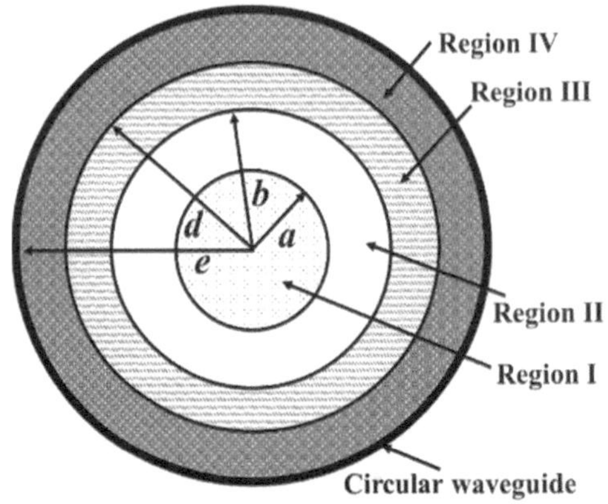

FIGURE 13.3 Cross-sectional views of an MTM-assisted RWA showing (i) beam region (Region I), (ii) vacuum region (Region II), (iii) resistive layer region (Region III), and (iv) lossless MTM region (Region IV). $r = a$ represents the beam radius; $r = b$ and $r = d$ represent the inner radii of the resistive layer and MTM regions, respectively; and $r = e$ represents the inner wall radius of the circular waveguide.

region, region II is a vacuum region, region III is a resistive layer region and region IV is a lossless MTM region, and (ii) yet another using a lossy ENG-MTM as both the resistive and support mediums integrated into a single layer, the regions III and IV being merged to a single region. The MTM exhibits here a plasma-like response with a positive permeability and a negative permittivity, responsible to maintain the inductive wall admittance, which, in turn, yield high gains of the device over relatively broad bandwidths with moderate beam filling factors [35]. For any given set of operating parameters (such as beam power, beam diameter, MTM layer thickness, etc.), they found that a negative permittivity value could be chosen such that it would yield a very large gain rate of the device corresponding to a purely inductive admittance at the beam edge. Further, they found that the gain rate increased with the beam filling factor as in a conventional RWA, though with somewhat larger value [35]. However, there is the difficulty of implementing felt in thin-coating a lossless ENG-MTM (Fig. 13.3), which has been subsequently overcome by the use of a lossy ENG-MTM [35].

Later on, in order to add practical relevance to the study, the simplified assumption of Rowe et al. [35] that the MTM behaves as isotropic was removed by them in [54] by treating the MTM as anisotropic in their study on ENG-MTM-assisted RWA. Therefore, in order to gain more practical insight into the ENG-MTM-assisted RWA, they investigated into (i) a periodic array of metallic rods to provide the ENG response along the length of the rods, with lumped element inductors, to reduce the effective plasma frequency of the structure, and (ii) a meander-line MTM, which also provides the ENG response along its periodicity with much lower effective plasma frequency than that of the array of metallic rods used as the lossy MTM liner of a rectangular waveguide [54]. Rowe et al. [54] through their analysis also found that the anisotropic MTM liner provided higher inductive wall admittance as well as an increased peak gain rate within a narrow bandwidth as compared to an isotropic MTM liner.

However, one of the practical challenges in the design of a wire array-based MTM-assisted RWA is to trade-off between the conflicting requirements with respect to the wire spacing. On the one hand, the continuous-medium-like response of an MTM is achievable only for wire spacing less than wavelength (sub-wavelength spacing) for which the effective plasma frequency should be much greater than the operating frequency. On the other hand, in order to obtain an appreciable device gain, the effective plasma frequency should be slightly (not very much) greater than the operating frequency [54]. However, the issue of meeting these conflicting requirements has been resolved by proposing a periodically spaced inductive meander-line structure for the MTM-assisted RWA (constructed in CST Particle Studio), which reduces the effective plasma frequency of the structure, while at the same time, maintains the sub-wavelength wire spacing [54,55].

13.3.1.3 Klystron amplifiers

Multi-beam klystron

The multi-beam klystron (MBK), which is a breakthrough in the development of high power, light weight, compact klystrons, uses several electron beams, each propagating in its own channel, and then interacting with the field of a common interaction structure [7–10,51]. One can attain high values of the total beam current and perveance of the entire multi-beam stream without increasing these parameters for the individual beams, thus, making it possible to increase the beam power, and hence, the RF power of the device by increasing the number of beams (in the device). However, the limitations of an MBK are: (i) reduction of cathode lifetime due to its smaller size and higher loading, (ii) non-availability of a multi-emitter cathode with higher emission density and longer life, (iii) voltage breakdown leading to the reduction of the reliability of the tube if a low-perveance, high-voltage beam is used for high device efficiency, and (iv) non-functionality of the tube if, on the contrary, a high-perveance, low-voltage beam is used that gives rise to space-charge repulsion, thereby preventing the required bunching of electrons in the tube [51].

In this background of the capabilities and limitations of MBKs, Galdetskiy [31] added a new dimension to the development of MBKs by proposing the MTM-assisted MBK cavity intended for enhancing the power capability of an MBK. Their design of MTM-assisted MBK was comprised of the beam channels, the MTM in the form of periodic inductive inserts between the beam channels, and the quasi-T wave-guiding interaction gap region of the rectangular cavity [31]. Further, Galdetskiy [31] has shown how the dispersion characteristics of MTM cavity get modified by the MTM filling of the cavity with the appearance of a cut-off frequency. Also, at a given frequency, with MTM filling, the wave number decreases, and therefore, the wavelength increases, thereby allowing for a larger diameter of the ferrule of cavity, which is limited by a half of wavelength. This, in turn, allows a greater number of electron beams to pass through the cavity and, consequently, a higher total beam current and larger output power from an MBK [31]. However, no parameter retrieval of the MTM and no validation of simulated results with electromagnetic analysis and/or experiments have been reported by them.

Extended interaction klystron

At low-frequency regions, transverse dimensions of the conventional klystron cavities are very large which makes the entire device extremely bulky and expensive when multi-cavity structures have been considered. Therefore, Wang et al. [56] have designed an all-metal MTM-assisted interaction cavity structure for klystron using CST Microwave Studio and Particle Studio, called as MTM extended interaction klystron (MEIK) to miniaturized the cavities as well as the entire device which also reduces the total cost of the device while targeting the wide application areas in the military and civilian fields, such as high power radars, particle accelerators, radio-astronomy, industrial heating, medical imaging, etc.

The MEIK cavities can be derived from the concept of CeSRR MTM based SWSs for BWOs which can support slot mode as well as cavity mode [57]. The cavity mode provides strong longitudinal electric field for energy conversion between EM wave and electron beam that makes the MEIK superior to its conventional counterpart. The MTM-assisted extended interaction resonant cavities (EIRCs), with a smaller inner cavity radius than their conventional counterparts, exhibit higher effective characteristic impedance, which means the MTM EIRCs provide stronger velocity modulation to the electron beam as well as a larger fundamental modulation current from a shorter drift tube, thereby reducing the whole length of the MEIK and achieving longitudinal and transversal miniaturization [56].

For a backward-wave amplifier based on a coaxial negative refractive index transmission-line MTM, one may follow Weiss and Grbic [58].

13.3.2 MTM-Assisted Interaction Structures for Microwave Oscillators

13.3.2.1 Backward-wave oscillators

MTM loaded helix SWS

Datta et al. [25] and Varshney et al. [26,28] carried out the equivalent circuit and electromagnetic field analysis, respectively, of a helix surrounded by a continuous DNG-MTM tube in a metal envelope (Fig. 13.1b), and obtained one and the same dispersion relation and identical dispersion characteristics. This DNG-MTM medium shows $\varepsilon'_{r,\mathrm{eff}}$ and $\mu'_{r,\mathrm{eff}}$ both exhibiting negative constant values to support the backward wave propagation in the SWS. Their analyses are, however, oversimplified, presumably to make the calculation tractable, by ignoring the frequency dependence and resonant behaviour of the MTM.

Subsequently, Guha et al. [29] carried out a more rigorous analysis by incorporating the frequency dependence of DNG-MTM parameters according to Lorentz and Drude models. They analysed a helix with discrete DNG-MTM support rods, defined by the frequency-dependent constitutive effective parameters $\varepsilon_{r,\mathrm{eff}}(f)$ and $\mu_{r,\mathrm{eff}}(f)$, symmetrically arranged around the helix in a metal envelope (Fig. 13.2a), by smoothing out the discrete supports into a continuous DNG-MTM tube of equivalent effective parameters $\varepsilon'_{r,\mathrm{eff}}(f)$ and $\mu'_{r,\mathrm{eff}}(f)$ (Fig. 13.1b). The succinct findings of these simple analyses by treating the helix by the sheath-helix model, with due care to validate the analyses against the CST simulation, are as follows [29]:

1. The structure supports a negative value of the structure K and wave group velocity corresponding to a backward power flow, thereby proving it to be a suitable interaction structure of a backward-wave amplifier or oscillator.
2. The magnitude of K of a DNG-MTM loaded helical structure is much higher than that of its conventional counterpart, suggesting that a device employing a DNG-MTM-assisted helix yields a higher gain and efficiency as well as a greater potential for miniaturization.
3. Out of all the structure parameters, it is the value of relative permittivity of DNG-MTM which is most effective in controlling the structure dispersion characteristics, and can provide nearly flat dispersion characteristics, as required for wideband gain–frequency characteristics of the device employing the structure.

Also, over this DNG-MTM frequency regime, Guha et al. [29] obtained the dispersion and K vs. frequency characteristics of a DNG-MTM loaded helical SWS, and validated them against simulation using the tool CST Studio Suite. Further, the capability of multi-band operation of the DNG-MTM loaded helical SWS was established, interestingly, using one and the same structure dimensions with respect to the helix radius, helix pitch angle and metal envelope radius. For this purpose, the band-specific values of frequency parameters taken in the analysis are: (i) magnetic plasma frequency f_{mp}, magnetic resonant frequency f_{m0} and collision frequency Γ_m (which accounts for the magnetic loss) occurring in the Lorentz model expression for $\mu_{r,\text{eff}}(f)$, and (ii) electric plasma frequency f_{ep} and collision frequency Γ_e (which represents the electronic dissipation) occurring in the Drude-model expression for $\varepsilon_{r,\text{eff}}(f)$ [29] (Fig. 13.4). This multi-band operation can be possible with the help of multi-band DNG-MTMs [21,22]. This multi-band DNG-MTM loaded helix SWS provides higher K values in comparison with the other MTM based SWS for BWOs without altering physical dimensions of the proposed SWS [29].

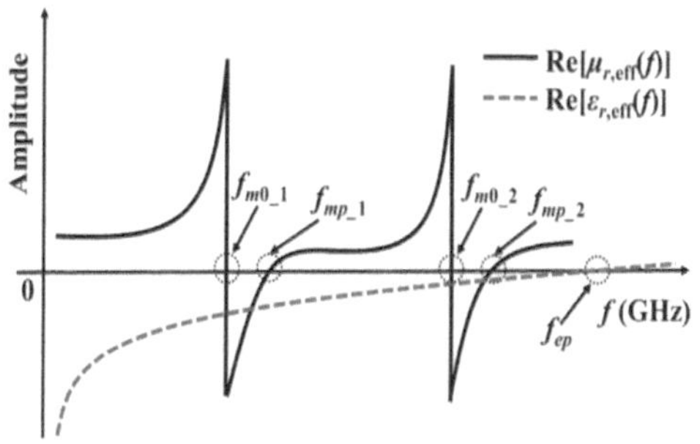

FIGURE 13.4 Effective medium responses of $\varepsilon_{r,\text{eff}}(f)$ using the Drude model (shown in broken line) and $\mu_{r,\text{eff}}(f)$ using the Lorentz model (shown in solid line).

Furthermore, Guha et al. [29] proposed a typical realistic design of a unit cell comprised of (i) rectangular SRRs on a dielectric substrate typically made of beryllia ($\varepsilon_r = 6.3$) of a given thickness printed on one of the radial edges of rectangular helix-support rod, and (ii) a metallic strip of a given thickness on the dielectric substrate on the other radial edge of the rod [12,14]. The retrieved parameters ($\varepsilon_{r,\text{eff}}(f)$ and $\mu_{r,\text{eff}}(f)$ using [12]) of the unit cell are found to be simultaneously negative as the DNG-MTM medium. Hence, they found the dispersion and K (negative) vs. frequency characteristics of the structure by both electromagnetic and simulation that agreed well, the negative value of K indicating a backward-wave interaction due to backward energy flow, and suggesting the potential application of the structure in backward-wave amplifier and oscillator.

CSRR and Below Cut-off Waveguide Based Combined SWS

A hollow rectangular/circular waveguide below its cut-off frequency behaves as the MNG or ENG MTM with stop-band responses depending on whether it is excited in the TM or TE mode. However, these stop-band responses will be transformed into pass-band responses, rather than an attenuating evanescent mode, as a backward-wave propagating mode in a waveguide to make the latter a DNG-MTM waveguide, when the waveguide is excited below cut-off either in a TM mode, and is loaded periodically with an ENG MTM, or in a TE mode and is loaded periodically with an MNG MTM [36–39,42,57,59–64].

Hummelt et al. [42,65] studied a DNG-MTM waveguide formed by stacking two sets of parallel CSRR MTM plates (Fig. 13.5a) periodically along wider wall of the rectangular waveguide which operates in the below cut-off frequency region (Fig. 13.5b). The potential use of the structure is in a BWO, in which a circular electron beam runs parallel to these two plates to interact with a TM_{11} mode supported by the structure. The advantage of such a waveguide is that it operates below cut-off that allows the transverse dimensions of the structure to be

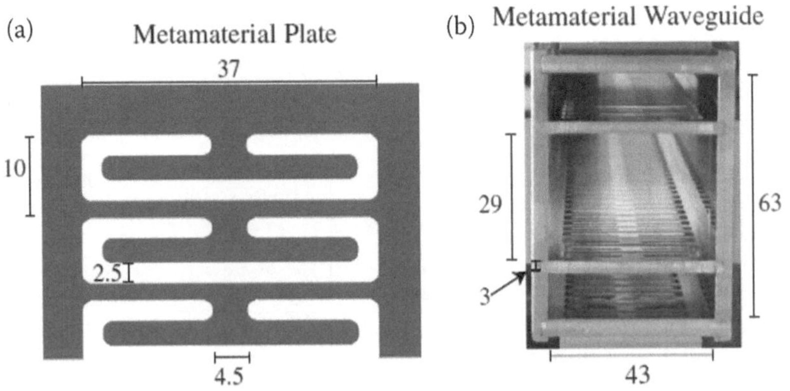

FIGURE 13.5 (a) One of the CSRR-MTM plates and (b) the assembled MTM-loaded waveguide at MIT (USA), the relevant dimensions being shown in mm.
[Source: Hummelt et al. [65]].

much smaller than a wavelength. Such an advantage is of relevance of the structure to the miniaturization of microwave VEDs, for instance, an S-band BWO [42]. The structure supports two negative index modes. Out of these two modes, one with relatively small values of the axial electric field ($E_z \approx 0$) at typically around 2 GHz is not expected to strongly interact with the electron beam, while another at around 2.6 GHz with a stronger TM mode-like axial electric field predominantly interacts with the beam at the point of intersection between the beam line and the dispersion characteristics, obtained using CST Microwave Studio, where, though the wave phase velocity is positive, the wave group velocity is negative corresponding to a negative slope of the dispersion characteristics, the magnitude of the group velocity being $\approx 0.075\ c$. The frequency corresponding to this point of intersection is found to be 2.65 GHz, which, thus, is the operating frequency of MTM-BWO employing the structure proposed by Hummelt et al. [42]. A notable observation is the auto-modulation instability of the CST-PIC simulated output power, which is indicated by the multi-frequency behaviour of the output spectra. However, with the adjustment in the values of length of the structure and beam current, such auto-modulation of the output power would disappear [42]. In another approach, Hummelt et al. [42] modelled the two plates of DNG-MTM waveguide as the two identical 'effective-medium' slabs to make an equivalent effective-medium BWO, with a view to make the CST-PIC simulation less computationally intensive and less time-consuming. In this approach, for the purpose of simulation of the equivalent effective-medium BWO, Hummelt et al. [42] took the typical values of structure length, beam radius, beam voltage, beam current, and beam confining magnetic field. They found that the simulated results with respect to the output power as a function of time have qualitatively agreed for a BWO in a CSRR-based DNG-MTM waveguide.

Wang et al. [57] developed an equivalent circuit model of an all-metal metamaterial SWS (MSWS), to theoretically predict the dispersion characteristics of the structure using the Curnow's formula [66], which agreed well with their HFSS-simulated results, following the concepts in [15]. The structure consists of a circular waveguide periodically loaded with CeSRRs. They calculated the resonant frequency of the CeSRR based on the simulated surface current distribution on it obtained using the eigen mode solver in CST Microwave Studio. Further, based on the finding of Esteban et al. [62] that a hollow waveguide excited in the TM mode operating below the cut-off frequency exhibits an effective MNG property, Wang et al. [57] took up a hollow circular waveguide of 20 mm radius excited in the TM_{01} mode that has the cut-off frequency of 5.74 GHz, and made an MSWS out of such a waveguide by periodically loading it with an array of CeSRRs. Hence, they found that MSWS so designed would behave as an effective DNG medium in the frequency range of 2.0–2.8 GHz providing a backward wave passband mode rather than an attenuating evanescent mode. Interestingly, MSWS exhibits a high value of its K, typically greater than 110 Ω over the frequency range of 2.1–2.6 GHz [57]. A high value of K of MSWS makes a VED employing the structure to have a high value of efficiency, which, in turn, makes the device have the potential for being miniaturized.

Shapiro et al. [46] have investiagted a negative-index metawaveguide (NIMW) composed of a stack of metal plates patterned by CSRRs as a coherent radiation source (extracting energy from the beam) or a particle accelerator (transfering the electromagnetic energy to the beam). The CSRR and TM mode of NIMW have exhibited negative $\varepsilon_{r,\text{eff}}(f)$ and negative $\mu_{r,\text{eff}}(f)$, respectively, simultaneously to support backward wave propagation.The highlights of the NIMW are: (i) the opposite sign of the group velocity and the beam velocity can result in an instability utilized in backward-wave oscillators (BWOs) or (for lower beam currents) backward-wave amplifiers (BWAs) in which the interaction has taken place between an electron beam and a spatial harmonic of the electromagnetic field in a periodic structure; (ii) low group velocity of the negative-index wave increases spatial gain while reducing the start oscillation current in a shorter interaction structure; (iii) the structure can be fabricated using standard planar fabrication technique unlike conventional BWOs where high-precision machining occurs at high frequencies; (iv) the NIMW is further facilitated by uniform spatial distribution of longitudinal electric field which is maximized at the beam's location as shown in Fig. 13.6a; (v) the output resonant frequency of the device can be controlled accurately and precisely while controlling the resonant frequency of the CSRR by electrical or optical tuning. On an interesting note, the CSRR metallic plates are placed parallel to the direction of the beam propagation, therefore there is no need to make holes in the CSRR for beam propagation, which reduces fabrication complexity at high frequencies. Shapiro et al. [46] analytically modelled the NIMW considering the bianisotropy of the structure and obtained the dispersion relation (Fig. 13.6b) using COMSOL Multiphysics software.

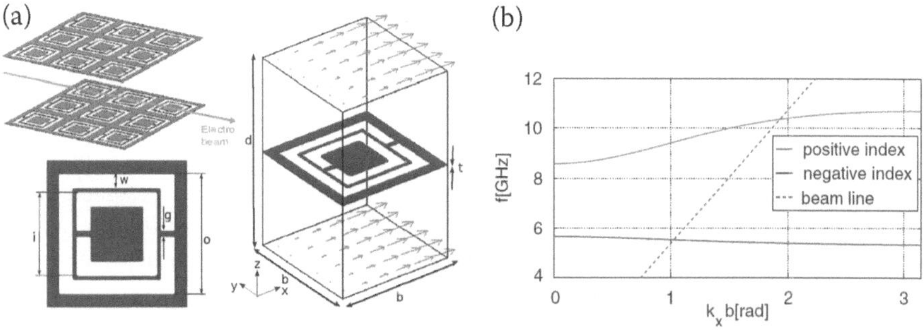

FIGURE 13.6 (a) Schematic of a negative-index metawaveguide (NIMW) composed of a stack of metal plates patterned by CSRRs typically in which $o = 6.6$ mm, $i = 4.6$ mm, $w = 0.8$ mm, $g = 0.3$ mm. An electron beam propagating in the x direction and interacting with the NIMW is also shown. Also shown is a single cell of an NIMW and the midplane electric fields (indicated by arrows) interacting with the electron beam, the other relevant typical dimensions being shown $d = 12.8$ mm, $b = 8$ mm, and $t = 0.05$ mm, so chosen for a frequency near $f_0 = 5$ GHz. (b) Dispersion characteristics for the lowest-order TM_1-like modes (shown in solid line), obtained using COMSOL, and the 'beam-mode line' (shown in broken line) defined by $\omega = k_x v_b$, where $v_b = 0.9c$.
[Source: Shapiro et al. [46]]

There are several other MTM derived SWSs for BWOs, such as thin wire MTM based SWS [39], split-ring based MTM SWS [67], complementary electric split-ring resonators [41], planar two parallel plates DNG-MTM SWS for multi-beam BWO [45] without using below cut-off waveguide, etc. The working principle is almost same for all the DNG-MTM waveguides of circular or square or rectangular cross-sections. They differ only by *K* values, operating frequencies, output powers, efficiencies, etc. Therefore, it is up to the readers that they may consider one of these SWSs or may proposed novel MTM-derived SWSs for potential application in BWOs.

13.3.2.2 Extended Interaction Oscillators

If one short-circuits both ends of a CeSRR-based SWS and at the same time increases the number of periods within the same length then the frequency gaps between the adjacent modes become smaller which can cause oscillations in the MEIK. These oscillations turn the MEIK into an MEIO (metamaterial extended interaction oscillator) [56]. The MEIO has several merits over conventional BWO such as (i) transverse dimension miniaturization of the cavities as well as the entire device, (ii) increased efficiency and gain of the device, and (iii) shortening of the length of the device, etc. The MEIO has been analysed theoretically and by simulation by Wang et al. [56].

13.3.3 Cherenkov Radiation Sources

Cherenkov radiation (CR) is emitted when charged particles move in a material with a speed faster than the phase velocity of light in that material. P. A. Cherenkov [68] observed CR in conventional materials (RH media) experimentally. After 3 years of this experiment, Frank and Tamm [69] explained the CR using macroscopic EM theory and after another 3 years Ginzburg [70] explained CR using quantum theory. Veselago [71] introduced double negative medium (DNM) where relative permittivity and relative permeability were negative simultaneously and stated that LH medium exhibited reverse CR (RCR), as one of the unusual properties of the medium, without obtaining a mathematical solution of RCR. The RCR signifies that the wave vector and the time-averaged Poynting vector are "anti-parallel" and that the wave has a negative group velocity while an electron beam passes very close to the surface of a DNG-MTM. DNG-MTMs exhibit the RCR property and, to the best of our knowledge, it was Duan et al. [72–75] to first ever demonstrate that, for the RCR in an 'anisotropic' DNG-MTM, the wave vector and the time-averaged Poynting vector are not exactly anti-parallel, contrary to what it would be in an isotropic DNG-MTM [76]. This is tantamount to stating that, in an anisotropic DNG-MTM, θ_{CR} is not exactly equal to θ_{sv}, where θ_{CR} is the angle between the wave vector $\bar{k}$ and the velocity vector $\bar{\upsilon}$ of the particle of charge q, and θ_{sv} is the angle between the time-averaged Poynting vector $<\bar{S}>$ and the charged particle velocity vector $\bar{\upsilon}$ [5,77].

Hummelt et al. [65] attributed the generation of output radiation from the beam-wave interaction in the waveguide to Cherenkov-cyclotron or anomalous Doppler instability, in the presence of a beam-confining magnetic field, which dominates over the Cherenkov instability. The output from beam-wave interaction in the waveguide is generated at a frequency equal to the Cherenkov frequency minus the cyclotron frequency [65]. Two distinct lowest order modes are supported by the MTM-loaded waveguide – one symmetric and another anti-symmetric – corresponding to zero on-axis transverse and zero on-axis axial electric fields, respectively, the dispersion characteristics of which, obtained by the CST simulation, are shown in Fig. 13.6. Interestingly, these waveguide modes are below cut-off TM modes in the 43-by-63-mm waveguide (Fig. 13.5b), which can only propagate in the waveguide due to the presence of MTM plates Hummelt et al. [65].

In order to appreciate the negative values of effective permeability and permittivity for modelling a MTM-loaded waveguide, Hummelt et al. [65] found it helpful to refer to the expressions $\mu_{eff} = 1-\omega_{c0}^2/\omega^2$ and $\varepsilon_{eff} = 1-\omega_p^2/(\omega^2-\omega_0^2)$ of the medium as the effective permeability and permittivity, respectively. Here, ω is the wave angular frequency, ω_{c0} is the waveguide cutoff angular frequency for TM modes, ω_p is the effective plasma frequency, and ω_0 is the resonant frequency of medium – the relevant quantities, in the present context of the structure, depicted in Fig. 13.5, being: $\omega_0/2\pi$~2.1 GHz, $\omega_{c0}/2\pi$~4.2 GHz and $\omega_p/2\pi$~1.7 GHz Hummelt et al. [65]. Further, the Cherenkov-cyclotron instability causing the radiation refers to the value $n = -1$ in the beam-mode cyclotron-wave dispersion relation $\omega = k_z v_z - n\omega_c/\gamma$, where v_z is the axial DC beam velocity, k_z is the wave number, γ is the Lorentz relativistic mass factor and n is the beam harmonic mode number, $n = 0$ referring to the ordinary Cherenkov instability and $n \neq 0$ to the Cherenkov-cyclotron instability, the special case $n = -1$ of which is the anomalous Doppler instability, which can lead to enhanced efficiency and output power from the beam-wave interaction in the waveguide Hummelt et al. [65]. Interestingly, at the operating point corresponding to the intersection between the beam-mode dispersion line and the waveguide-mode dispersion characteristics (plot), typically, around 2.4 GHz (Fig. 13.7), the slope of waveguide-mode dispersion plot is negative, thereby making the wave group velocity negative as well, which signifies a backward wave.

Duan et al. [78] considered a sheet electron beam (SEB) to transport along the $+z$ direction over a periodic array of CeSRR (Fig. 13.8a). In the HFSS-simulated dispersion characteristics of the structure in Fig. 13.8b, the frequency $f\,(= \beta_0 v_0/2\pi)$ is plotted against the phase advance $\phi\,(= \beta_0 p)$ of the propagating wave, p being the periodicity of the structure and β_0 the phase propagation constant. The SEB line is determined by the initial beam velocity v_0, which, in turn, is determined by the beam potential $V = (\gamma-1)mc^2/e$, where γ, c, e, and m are the relativistic factor, speed of light in vacuum, charge of an electron and rest mass of an electron, respectively. They took V = 160 kV and 99.7 kV as the beam voltages corresponding to the SEB lines 1 and 2, respectively (Fig. 13.8b). Further, the $n = 0$ space-harmonic of the fundamental TM-dominant mode of the waveguide – acting as a magnetic plasma – exhibits a left-handed behaviour or backward-wave dispersion, corresponding to a negative slope of the dispersion plot of waveguide in Fig. 13.8b, signifying a negative group velocity, n being the space-harmonic number. The synchronous condition requiring that the initial velocity v_0 should be slightly greater than the phase velocity of electromagnetic waves can be found

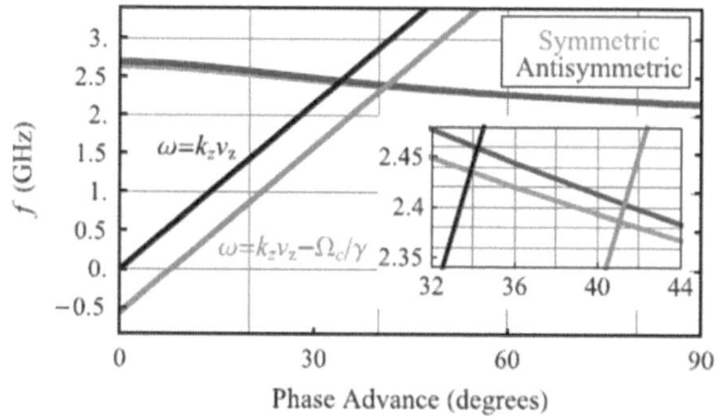

FIGURE 13.7 CST Microwave Studio simulated dispersion characteristics of the symmetric mode (shown in red) and anti-symmetric mode (shown in blue) of the MTM-loaded waveguide showing also the Cherenkov (shown in black) and anomalous Doppler (shown in green) beam-mode dispersion lines. The phase advance is $k_z p$, where k_z is the wave number with p as the ring periodicity of the CSRR-MTM plate, taking, typically, p = 10 mm (Fig. 13.5). Also, the beam energy is typically taken as 490 keV and the magnetic field as 400 G.
[Source: Hummelt et al. [65]].

from the point of intersection between the beam-mode SEB line and the dispersion characteristics of waveguide [78]. Further, Duan et al. [78] used the post-processing in HFSS to find K of a square waveguide loaded with CeSRRs over the typical frequency range of 2.83–3.05 GHz (Fig. 13.8c). For this purpose, they averaged the value of K over the 12 mm ×2 mm cross-section of the SEB bunch, the bottom surface of which was taken 0.5 mm above the surface of the layer of CeSRRs. Attributable to the large axial electric field intensity, the CeSRR-loaded waveguide has much higher K.

Tang et al. [64] proposed a square waveguide periodically loaded with an array of CeSRRs (Fig. 13.8a), with input and output couplers. The structure having the provision for a sheet electron beam to pass through it has potential applications in a Cherenkov oscillator – a VED based on Cherenkov radiation (CR) that is emitted by charged particles moving in a transparent medium with a velocity exceeding the phase velocity of light in the medium. Further, the HFSS simulation

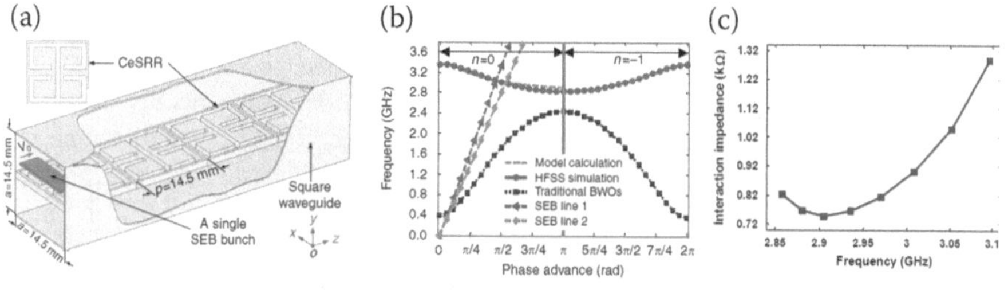

FIGURE 13.8 (a) Schematic of the square waveguide loaded with CeSRRs together with its (b) HFSS-simulated dispersion and (c) K vs. frequency characteristics.
[Source: Duan et al. [78]].

due to Tang et al. [64] indicated that the MTM waveguide supports both the $n = 0$ space-harmonic, fundamental mode and the $n = -1$ space-harmonic, first higher-order mode exhibiting the "backward" wave-mode properties with high K. This opens up the potential of the structure to be employed in a Cherenkov oscillator. Further, they also simulated the beam-wave interaction in the device using the CST-PIC solver and the highly efficient electromagnetic PIC-CHIPIC code. They found that the dual-band frequencies corresponding to the two modes 1 and 2, respectively, are generated continually for the duration of pulse, and that the frequencies eventually become stable after ~60 ns. Thus, they predicted the device peak output power of the mode 1 as ~111 kW at 2.832 GHz, and that of the mode 2 is ~47 kW at 4.512 GHz. Also, they found that the CST-PIC solver and the PIC-CHIPIC code are in good agreement in predicting the dependence of peak output power for both the modes and for predicting total electronic efficiency, both on the beam voltage Further, the predicted total electronic efficiency up to ~52% of the MTM Cherenkov oscillator is much higher than that obtainable from the conventional BWOs.

13.4 MTM-ASSISTED VEDS AS MICROWAVE SOURCES–FABRICATION AND EXPERIMENTAL CHARACTERIZATION

In Section 13.3, we have reviewed the investigations into MTM-assisted VEDs based on analysis and/or simulation. Now, this section includes a review on the fabrications and experimental measurement efforts in the development of MTM-assisted VEDs.

13.4.1 Backward-wave oscillators

Hummelt et al. [77] used a waveguide filled with MTM complementary split-ring elements as the SWS in their experimental MTM-BWO. The 352 mm long structure consists of two plates, each of 1 mm thickness with a separation of 36 mm between them, and each loaded with an array of CSRRs of 8 mm periodicity for S-band MTM-BWO which generates 40 W of power at 2.83 GHz, using a 500 keV, 80 A electron beam. This, to the best of the knowledge of the authors of Hummelt et al. [77], is the first ever experimental demonstration of microwave generation employing an MTM structure. The experimental setup of Hummelt et al. [77] shows the fully assembled device and relevant components under vacuum after high temperature bake-out. In their experiment, the microwave power is dumped into two loads after it is coupled out of the vacuum chamber using 60 dB couplers through two waveguide arms. Hummelt et al. [77] used a 6 GHz oscilloscope to measure the signal behaviour and a peak power meter to measure the microwave power of MTM-BWO developed. They measured the RF power at one of the microwave couplers, and also, measured the gun voltage and the collector current. It is also of interest to study the frequency tuning by the gun voltage for which the frequency is obtained from the Fourier transform of microwave signal recorded in the experiment with a 6 GHz oscilloscope. The frequency so obtained from the interpretation of experiment agreed well with that predicted by the HFSS

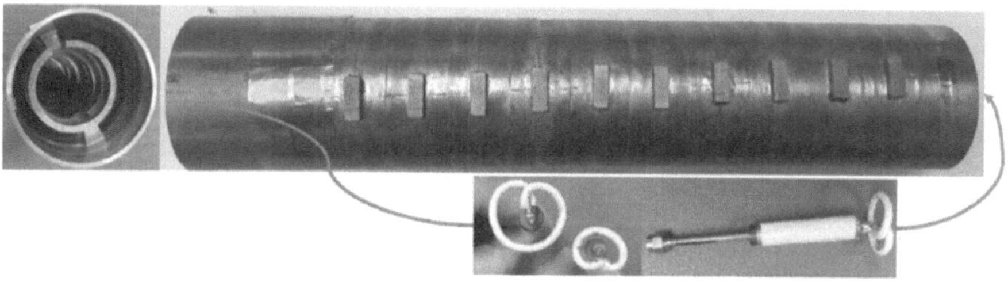

FIGURE 13.9 Experimental setup of MSWS with a two-mode launcher at University of New Mexico.
[Source: [67]]

simulation. Subsequent tests of a longer and slightly modified structure, as mentioned in Hummelt et al. [77], will produce over 1 MW of microwave power at 2.83 GHz.

Yurt et al. [67] used an MSWS for an O-type HPM microwave source – "O" standing for TPO (tubes à propagation des ondes). The MSWS, shown in Fig. 13.9, is comprised of periodically alternating, two oppositely oriented SRRs constituting a unit cell with broadside coupling, and also, 180^0 out of phase, which are coaxial with and connected to a metal envelope. Each split-ring is electrically connected to the output waveguide with a 30^0 conducting tab. Further, to be more specific, the SRRs provide the negative permeability of MSWS, while the metal tube – treated as a waveguide with its diameter taken such that the frequency of oscillation to be generated is below the cut-off frequency of a regular waveguide – provides the negative permittivity of MSWS [67]. Further, Yurt et al. [67] envisaged an annular electron beam, derived from an explosive emission annular cathode, to transmit through the MSWS and extract the excited radiation in the beam in an end-fire manner through the gaps between SRRs [67], via a conical horn section. A crude prototype has been made of the MSWS for doing cold test using network analyser. The geometry of the mode launcher consists of a loop antenna, which is fed coaxially. A pair of mode launchers is placed at both ends of the MSWS as a transmitter and receiver (Fig. 13.9) [67]. They performed typical transmission and reflection measurements of the tube loaded with rings using two mode launchers. Both experimental and simulation results are in good agreement. Yurt et al. [67] used 3D MAGIC PIC simulation code of the SINUS-6 accelerator of the University of Mexico (UNM) for the 'hot test' results of beam-wave interaction in MSWS. Thus, they used the UNM SINUS-6 voltage pulse, which has a 16 ns pulse length to predict the maximum radiated power of 260 MW at 1.4 GHz frequency from the beam transporting through the MSWS with the anode current of 5.3 kA for a beam confining magnetic field of 2 T.

13.4.2 Cherenkov radiation sources

RCR, already introduced in the beginning of sub-section 3.3, has been studied by theory, simulation and experiment, for instance, in [65,72–75,78]. One such study

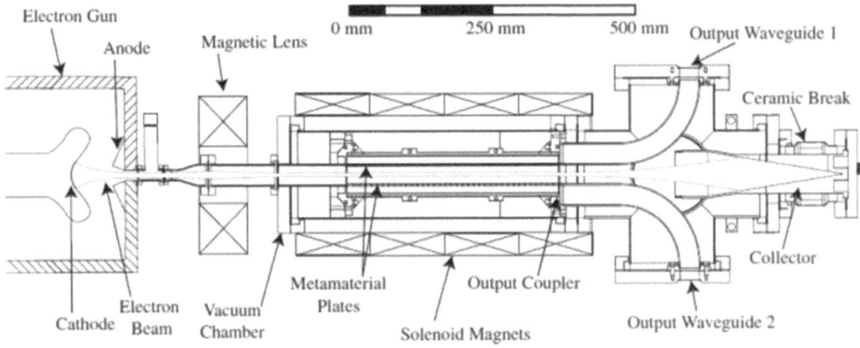

FIGURE 13.10 Schematic of the experimental setup at MIT (USA). [Source: [65]].

due to Duan et al. [78] concerns with the development of an all-metal square waveguide loaded with CeSRRs (Fig. 13.8a&b) and that of a compatible novel output coupler specially designed to extract the RCR signal output from the waveguide. This type of MTM exhibits a left-handed behavior that was also confirmed for the first time by them. In their experiment, they also observed that the radiation takes place in a direction predominantly opposite to the movement of a single sheet electron beam (SEB) bunch [78].

Hummelt et al. [65] presented an experimental test (at the MIT, USA) of the phenomenon of RCR using an intense electron beam transporting through an MTM-loaded waveguide between two copper MTM plates machined with CSRRs inserted in the waveguide (Fig. 13.5b). It is the first ever experimental demonstration of coherent microwave generation from a continuous electron beam interacting with an MTM structure. The experimental arrangement of Hummelt et al. [65] is shown in Fig. 13.10. An electrostatic-focused, 1 μs pulsed electron gun forms the electron beam. The backward-wave power generated in the MTM-loaded waveguide is reflected back from the gun end towards the collector end, and is coupled out through two output rectangular waveguides (WR284: 72.14 mm × 34.04 mm).

The experiment was carried out by Lu et al. [43] on the test bench of MIT (USA) as illustrated in Fig. 13.11, which shows the Pierce electron gun, magnetic lens, the MTM-R structure, solenoid magnet, collector, theand Arm 1 and Arm 2 WR284 waveguides for guiding microwave powers generated, besides the calibrated Bethe-hole directional couplers and RF loads on each of the two arms. The gun cathode is high-voltage-pulsed up to 490 kV by 1 μs long pulses forming a 490 kV, 84 A beam, which is confined by using a magnetic lens (giving adjustable fields up to 840 G) closer to the gun and a solenoid magnet (giving adjustable fields up to 1500 G) in the region farther from the gun and overlapping the MTM-R structure (Fig. 13.11). While the spent beam is dumped into the collector, the microwave powers generated in the MTM-R structure are guided through the Arm 1 and Arm 2 waveguides to the terminating RF loads of each of the arms. The two calibrated Bethe-hole directional couplers of coupling coefficients –64 dB and –61 dB, respectively,

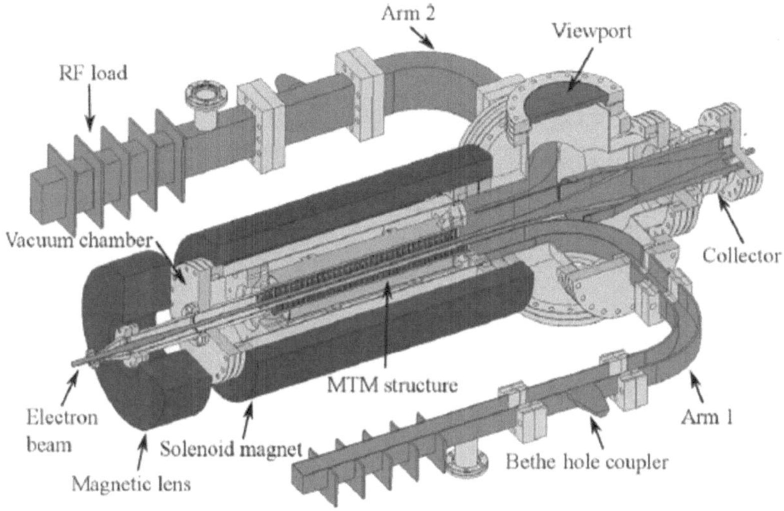

FIGURE 13.11 Measurement bench at MIT (USA) showing the Pierce electron gun, magnetic lens closer to the gun, the MTM-R structure, the solenoid magnet farther away from the gun and overlapping the MTM-R structure, collector, the Arm 1 and Arm 2 WR284 waveguides, the Bethe-hole directional coupler and the RF load on each of the arms.
[Source: Lu et al. [43]].

pick up small amounts of power from the Arm 1 and Arm 2 waveguides (Fig. 13.11). The couplers measure power in the operating mode at 2.4 GHz. A power meter and a fast oscilloscope were used to measure power and frequency, respectively [43].

In order to extensively explore the properties of RCR excited in a square waveguide loaded with CeSRRs, Duan et al. [78] developed an experimental setup as shown in Fig. 13.12a. An SEB, generated by the cathode and Tesla transformer (Fig. 13.12b), and confined by an axial magnetic field, is made to interact with the electromagnetic waves supported by the MTM-loaded waveguide. The two signals,

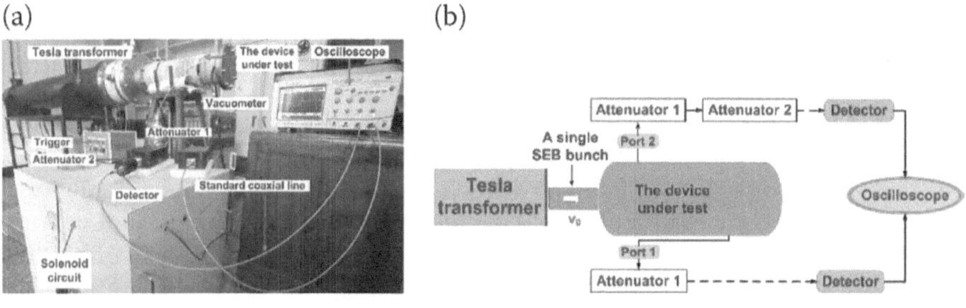

FIGURE 13.12 (a) Experimental platform at University of Electronic Science and Technology of China (UESTC) at Chengdu and (b) block diagram for the experimentation on the platform.
[Source: Duan et al. [78]].

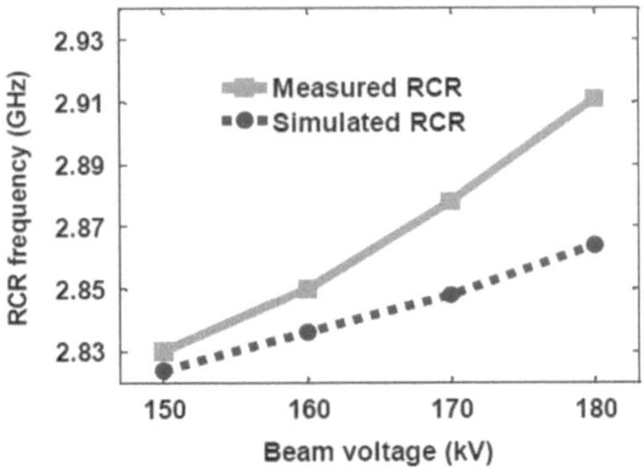

FIGURE 13.13 RCR frequency as a function of beam voltage.
[Source: Duan et al. [78]].

one due to the RCR and the other due to its unavoidable reflection, are picked up from the two ports 2 and 1 and attenuated by 63 dB and 53 dB, respectively, and then each of them is converted to the voltages by the two detectors, and finally read out by an oscilloscope (Fig. 13.12b). Further, there is a delay of 5.5 ns of signal at port 1 relative to that at port 2, which measures the propagation time for RCR signal unavoidably reflected from port 2 to port 1.

The study of the variation of RCR frequency with the beam voltage carried out by Duan et al. [78], as shown in Fig. 13.13, is a crucial support for the negative band corresponding to the backward-wave dispersion. Both the simulated and measured results in Fig. 13.12 show that the RCR frequency increases with beam voltage. This finding is consistent with the observation in Fig. 13.8b that the slope of beam-mode line increases with the beam voltage causing an increase in the operating RCR frequency where the beam-mode line intersects with the $n = 0$ space-harmonic mode of waveguide.

Further, one may investigates about the experimental setups of RWA and MEIK in [79] and [56], respectively.

13.5 MTM-ASSISTED CROSS-FIELD VEDS

13.5.1 Magnetron

Esfahani et al. [80–82] have addressed the well-known conflicting issue between the simultaneous increase of the quality factor and of the characteristic admittance and have increased efficiency and average output power of spatial harmonic magnetrons (SHM) by loading CSRR epsilon near zero (ENZ) MTM unit cell in the SWS of a SHM (Fig. 13.14a&b). They have used both HFSS and CST-PIC simulations to obtain retrieved parameters of CSRR MTM unit cell (Fig. 13.14c&d) and output power as well as efficiency, respectively. Esfahani [82] has fabricated a prototype of CSRR MTM loaded SHM and measured the output power and efficiency.

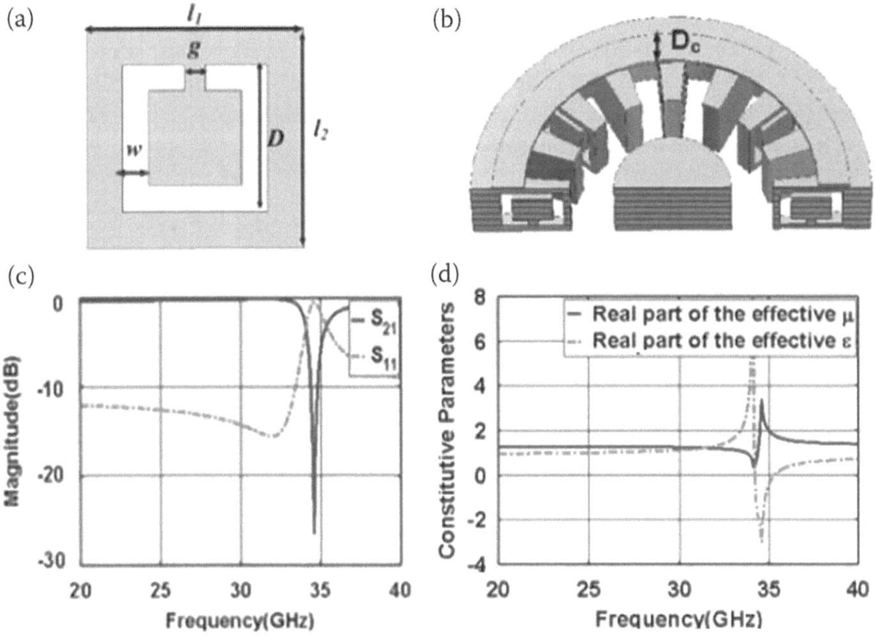

FIGURE 13.14 (a) A CSRR unit cell with the relevant dimensions: $l1$ = 1.6 mm, $l2$ = 1.65 mm, g = 0.2 mm, w = 0.25 mm, D = 1.4 mm; (b) CSRRR-loaded SWS SHM with the dimension D_c = 1 mm; (c) simulated scattering parameters of the CSRR unit-cell (obtained using ANSYS HFSS); (d) retrieved constitutive parameters of CSRR unit cell. [Source: Esfahani et al. [82]]

Andreev and Hendricks [83] have explored and developed new methods to increase the microwave output power by increasing anode current of multi-cavity magnetrons using the metal thin-wire (MTW) and the metal split-ring MTM-like structures as cold (non-thermionic) cathodes. For the multi-cavity magnetron, operating in one of the TE-like modes, the bulk MTM cathode may be designed in the form of array of rods, placed either radially (parallel to the transverse radial electric field) or longitudinally (parallel to the axial magnetic field). Therefore, by varying the lattice constant or periods of the array of rod MTMs, one can enhance the output power and efficiency of the magnetron [83].

13.5.2 Gyrotron

Gyrotrons are fast-wave VEDs that use an over-moded cylindrical cavity and higher order modes to generate high output power in the microwave and THz frequency regimes. However, these higher order output modes are not compatible with the receiver for direct use and it is very difficult to transmit them remotely. Therefore, it is necessary to transform the higher order modes into the Gaussian mode or lower order linearly polarized modes. Usually, Vlasov-type or

Denisov-type quasi-optical mode converter (QOC) is used in a transverse-output gyrotron, for instance, for fusion plasma heating application [84]. Further, Fu et al. [85] have proposed and simulated, using HFSS, a cross-shaped MTM array based QOC to convert the TE_{01} mode to a well-focused Gaussian mode with 95% scalar Gaussian mode content.

13.6 CHALLENGING ASPECTS AND FUTURE SCOPE OF MTM-ASSISTED VEDS

In Sections 13.3 through 13.5, we have briefly reviewed the global research and development trends in MTM-assisted VEDs, for the past one and half decades highlighting the merits of MTM-assisted VEDs over their conventional counterparts that do not accrue the advantage of such MTM assistance. However, the researchers, all over the world, have faced several challenges while developing MTM-assisted VEDs. Some of them are as follows:

i. Almost all MTMs are strongly resonant in nature and, therefore, MTM-assisted VEDs exhibit strong dispersion and, consequently, a narrow bandwidth.
ii. At the resonant frequency region, an MTM has large values of the imaginary parts of $\varepsilon_{r,\mathrm{eff}}(f)$ and $\mu_{r,\mathrm{eff}}(f)$, the latter representing the electric and magnetic losses, respectively. Therefore one needs to obtain the figure-of-merit of the MTM to determine the usable bandwidth without much loss.
iii. MTM-assisted SWSs exhibit relatively high magnitudes of RF phase velocity thereby warranting very high beam voltages for the required near-synchronous beam-wave interaction in the device. Concurrently, a very high voltage power supply is required to fulfil this requirement during hot-testing. This further increases the measurement/testing cost.
iv. Almost all MTM-assisted BWOs enjoy the advantage of device miniaturization at lower frequency bands over their conventional counterparts. Correspondingly, however, the beam tunnel diameter as well as the beam diameter also gets reduced, which, in turn, reduces the power handling capability of the device. Hoverer, this reduction of power is compensated for by higher K values and resulting higher device efficiencies at low frequency bands. However, yet there remains the necessity of addressing this point at high frequencies.
v. From the standpoint of the effective medium/sub-wavelength consideration, the physical dimensions of MTMs are rather very small in the operating wavelength regime. Therefore, in the sub-THz frequency regime, the fabrication of MTMs becomes extremely difficult.

The above challenges are the stepping stones for future progress in the development of MTM-assisted VEDs. As an incidental reference to the context [30], a proper investigation is required for the development of low-loss MTMs and MTM-assisted wide-band TWTs [30]. One or more highly efficient and accurate MTM fabrication techniques have to be invented for extremely high frequencies.

13.7 SUMMARY AND CONCLUSION

In parallel with the global efforts in improving the performance of microwave lenses, phase shifters, directional couplers, antennas, sensors, absorbers, system components for cloaking/radar cross-section reduction, etc., by MTM assistance, the efforts have also been made to improve the performance of VEDs by MTM assistance. In this review, a survey has been made on the status of worldwide investigations into the MTM-assisted beam-wave interaction structures and MTM-assisted VEDs. However, only a few representative examples have been brought within the purview of the review in the four categories of (i) whether or (ii) not they had been studied by electromagnetic analysis and/or simulation and, if so, (iii) whether or (iv) not any experimental appreciation had gone into such study. The study on the MTM-assisted interaction structures and VEDs/radiation sources reviewed in this chapter under these categories (i) through (iv), therefore, need to be elevated in the category level through research and development. In other words, if the structures and devices have been studied only by electromagnetic analysis, they need to be studied by simulation, too; and, if they have been only analysed and simulated, they need to be studied by the actual experimentation, too. Moreover, newer MTM-assisted interaction structures and devices need to be explored for higher frequencies and higher powers as well as for the size reduction or compactness. Interestingly, the MTM assistance can be used not only for the miniaturization of VEDs, but also for the enhancement of their sizes, as we have seen here in the case of an MTM-assisted MBK, in which the enhancement of cavity diameter, caused by MTM assistance, has allowed a greater number of electron beams to pass through the cavity and, consequently, for a higher total beam current and a higher RF output power from the device. Similarly, in another case, the enhancement of the transverse dimensions of FW has been implemented by MTM assistance to prevent the shifting of operating frequency of an ENG-MTM-loaded FW-TWT. Thus, there is much scope for further research and development in individual MTM-assisted VEDs and their interaction structures already investigated and reviewed in this chapter. As a typical example, with respect to the DNG-MTM-loaded FW-TWT, the need of the hour is to reduce the inherent Ohmic losses associated with MTM, and widen the DNG frequency regime in order to make the device more useful to the VED community. It is hoped that the present review of the status of worldwide development efforts in MTM-assisted VEDs would arouse interest among the global VED community, and help one to choose suitable problems in the area for further investigations.

A note to add is that MTM-assisted accelerators as well as MTM-assisted THz VEDs, which have been excluded from the present review due to the paucity of space, may be explored by interested readers. For MTM-assisted accelerators, they can refer to McGregor and Hock [16]; Antipov et al. [86]; McGregor and Hock [87]; Sharples and Letizia [88]; Lu et al. [89]; and Lu et al. [90], etc. Similarly, for MTM-assisted THz VEDs, they can refer to Starinshak and Wilson [91]; Duan et al. [92]; and Duan et al. [93], etc. Furthermore, some worth pursuing problems to draw the attention of the readers are: (i) wideband, compact MTM absorber to achieve gain flatness in meander line TWT [94]; (ii) lumped resistance MTM absorber to be used as a sever in FW-TWT [95]; and (iii) beam curling reduction in a sheet electron beam using all-metal MTM [96], etc. Similarly, closely related problems worth investigating into are the VEDs assisted by artificial materials other than MTMs: (i) electromagnetic bandgap meander line SWS for enhancing the structure *K* and reducing the structure attenuation [97]; (ii) Cherenkov oscillator based on photonic crystal waveguide [98]; (iii) electromagnetic wave shielding using a 2D array of photonic crystals in an FW-TWT [99]; (iv) photonic crystal based THz VEDs [100,101]; (v) excitation of surface plasmon polaritons (SPPs) [102]; and (vi) generation of spoof SPPs surface waves using electron beam [103], etc. It is believed that the foregoing emerging and exciting research areas have a tremendous research and development potential in the global VED community.

REFERENCES

1. N. Engheta and R. W. Ziolkowski, *Metamaterials: Physics and Engineering Explorations*, New Jersey: IEEE Press/John Wiley & Sons, Inc, 2006.
2. J. C. Bose, "On the rotation of plane of polarization of electric wave by a twisted structure," *Proc. Royal Soc.*, vol. 63, pp. 146–152, 1898.
3. C. Caloz, *Electromagnetic metamaterials: Transmission line theory and microwave applications: The Engineering Approach*, New Jersey: John Wiley & Sons, Inc., 2006.
4. N. Purushothaman, "Design of metamaterial-based interaction structures for microwave tubes," Ph. D thesis, AcSIR, CSIR-CEERI, Pilani, India, 2019.
5. Z. Duan, M. A. Shapiro, E. Schamiloglu, N. Behdad, Y. Gong, J. H. Booske, B. N. Basu, and R. J. Temkin, "Metamaterial-inspired vacuum electron devices and accelerators," *IEEE Trans. Electron Devices*, vol. 66, no. 1, pp. 207–218, 2019.
6. N. Engheta and R. W. Ziolkowski, "A positive future for double-negative metamaterials," *IEEE Trans. Microw. Theory Techn.*, vol. 53, no. 4, pp. 1535–1556, 2005.
7. A. V. Gaponov-Grekhov and V. L. Granatstein (Ed.), *Applications of High Power Microwaves*, (Boston): Artech House, 1994.
8. J. Benford, J. A. Swegle, and E. Schamiloglu, *High Power Microwaves* 3rd edn (New York: CRC Press, 2015.
9. R. G. Carter, *Microwave and RF Vacuum Electronic Power Sources*. Cambridge, U.K.: Cambridge Univ. Press, 2018.
10. S. E. Tsimring, *Electron Beams and Microwave Vacuum Electronics*, Hoboken, NJ, USA: Wiley, 2007.
11. J. B. Pendry, A. J. Holden, W. J. Stewart, and I. Youngs, "Extremely low frequency plasmons in metallic mesostructures," *Phys. Rev. Lett.*, vol. 76, no. 25, pp. 4773–4776, 1996.
12. D. R. Smith, D. C. Vier, T. Koschny, and C. M. Soukoulis, "Electromagnetic parameter retrieval from inhomogeneous metamaterials," *Phys. Rev. E - Stat. Nonlinear, Soft Matter Phys.*, vol. 71, no. 3, pp.p. 036617, 2005.
13. J. B. Pendry, A. J. Holden, D. J. Robbins, and W. J. Stewart, "Magnetism from conductors and enhanced nonlinear phenomena," *IEEE Trans. Microw. Theory Tech.*, vol. 47, no. 11, pp. 2075–2084, 1999.
14. F. Bilotti, A. Toscano, L. Vegni, K. Aydin, K. B. Alici and E. Ozbay, "Equivalent-

circuit models for the design of metamaterials based on artificial magnetic inclusions," *IEEE Trans. Microw. Theory Techn.*, vol. 55, no. 12, pp. 2865–2873, 2007.
15. J. D. Baena, J. Bonache, F. Martín, R. M. Sillero, F. Falcone, T. Lopetegi, M. A. G. Laso, J. G. García, I. Gil, M. F. Portillo, and M. Sorolla, "Equivalent-circuit models for split-ring resonators and complementary split-ring resonators coupled to planar transmission lines," *IEEE Trans. Microw. Theory Techn.*, vol. 53, no. 4, pp. 1451–1461, 2005.
16. I. McGregor, and K. Hock, "Complementary split-ring resonator-based deflecting structure," *Phys. Rev. ST Accel. Beams*, vol. 16, 090101, 2013a.
17. M. Choi, S. H. Lee, Y. Kim, S. B. Kang, J. Shin, M. H. Kwak, K. Y. Kang, Y. H. Lee, N. Park, and B. Min, "A terahertz metamaterial with unnaturally high refractive index," *Nature*, vol. 470, pp. 369–374, 2011.
18. D. Schurig, J. J. Mock, and D. R. Smith, "Electric-field-coupled resonators for negative permittivity metamaterials," *Appl. Phys. Lett.*, vol. 88, no. 4, pp. 041109, 2006.
19. D. R. Smith, "Analytic expressions for the constitutive parameters of magnetoelectric metamaterials," *Phys. Rev. E*, vol. 81, no. 3, pp. 036605, 2010.
20. B. J. Arritt, D. R. Smith, and T. Khraishi, "Equivalent circuit analysis of metamaterial strain-dependent effective medium parameters," *Journal of Applied Physics*, vol. 109, no. 7, pp. 073512, 2011.
21. H. Chen, L. Ran, J. Huangfu, X. Zhang, and K. Chen, "Left-handed materials composed of only S-shaped resonators," *Phys. Rev. E*, vol. 70, no. 5, pp. 057605, 2004.
22. T. F. Gündoğdu, K. Güven, M. Gökkavas, C. M. Soukoulis, and E. Özbay, "A planar metamaterial with dual-band double-negative response at EHF," *IEEE J. Sel. Topics Quantum Electron.*, vol. 16, no. 2, pp. 376–379, 2010.
23. M. Kafesaki, I. Tsiapa, N. Katsarakis, Th. Koschny, C. M. Soukoulis, and E. N. Economou, "Left-handed metamaterials: The fishnet structure and its variations," *Phys. Rev. B*, vol. 75, no. 23, pp. 235114, 2007.
24. T. Roy, D. Banerjee, and S. Kar, "Studies on multiple-inclusion magnetic structures useful for millimeter-wave left-handed metamaterial applications," *IETE Journal of Research*, vol. 55, no. 2, pp. 83–89, 2009.
25. S. K. Datta, L. Kumar, and B. N. Basu, "Investigation into a metamaterial supported helix slow-wave structure,"in *Proc. IEEE Int. Vac. Electron. Conf. (IVEC)*, Bangalore, India, pp. 211–212, 2011.
26. A. K. Varshney, R. Guha, S. K. Datta, and B. N. Basu, "Dispersion control of helical slow-wave structure by double-negative metamaterial loading," *J. Electromagn. Waves Appl.*, vol. 30, no. 10, pp. 1308–1320, 2016.
27. N. Purushothaman and S.K. Ghosh, "Performance improvement of helix TWT using metamaterial," *Journal of Electromagnetic Waves and Applications*, vol. 27, no. 7, pp. 890–900, 2013.
28. A. K. Varshney, R. Guha, S. Biswas, P. P. Sarkarb, S. K. Datta and B. N. Basu, "Tape-helix model of analysis for the dispersion and interaction impedance characteristics of a helix loaded with a double-negative metamaterial for potential application in vacuum electron devices,," *J. Elect. Waves App.*, vol. 33, no. 2, pp. 138–150, 2019.
29. R. Guha, A. K. Bandyopadhyay, A. K. Varshney, S. K. Datta, and B. N. Basu, "Investigations into helix slow-wave structure assisted by double-negative metamaterial," *IEEE Tran. Electron Devices*, vol. 65, no. 11, pp. 5082–5088, 2018.
30. R. Guha and S. K. Ghosh, "Dispersion control of a helix slow-wave structure by I-shaped metamaterial loading for wideband traveling-wave tubes," URSI-RCRS, IIT-BHU, Varanasi, India, 2020.

31. A. V. Galdetskiy, "On the use of metamaterials for increasing of output power of multibeam klystrons," *Proc. IEEE Int. Vacuum Electron. Conf. (IVEC)*, Paris, France, 2013.
32. A. Rashidi and N. Behdad, "Metamaterial-enhanced travelling wave tubes," *Proc. IEEE Int. Vac. Electron. Conf. (IVEC)*, Monterey, CA, USA, pp. 199–200, 2014.
33. Y. S. Tan and R. Seviour, "Metamaterial mediated inverse Cherenkov acceleration," *Proceedings of* Proceedings of International Particle Accelerator Conference (IPAC'10), Kyoto, Japan, pp. 4378–4438, 2010.
34. Y. S. Tan and R. Seviour, "Wave energy amplification in a metamaterial-based traveling-wavestructure," *Europhys. Lett.*, vol. 87, no. 3, pp. 34005, 2009.
35. T. Rowe, J. H. Booske, and N. Behdad, "Metamaterial-enhanced resistive wall amplifiers: theory and particle-in-cell simulations," *IEEE Trans. Plasma Sci.*, vol. 43, no. 7, pp. 2123–2131, 2015.
36. Z. Duan, J. S. Hummelt, M. A. Shapiro, and R. J. Temkin, "Subwavelength waveguide loaded by a complementary electric metamaterial for vacuum electron devices," *Phys. Plasmas*, vol. 21, no. 10, pp. 103301, 2014.
37. S. A. Ramakrishna, "Physics of negative refractive index materials,"*Rep. Prog. Phys.*, vol. 68, pp. 449–521, 2005.
38. Y. Wang, Z. Duan, F. Wang, S. Li, Y. Nie, Y. Gong, and J. Feng, "S-band high-efficiency metamaterial microwave sources," *IEEE Trans. Elect. Dev.*, vol. 63, no. 9, pp. 3747–3752, 2016.
39. P. Narasimhan, S. Jain, N. Gurjar, N. Kumar and S. K. Ghosh, "Design of thin wire metamaterial-based interaction structure for backward wave generation," *IEEE Trans. Elect. Dev.*, vol. 67, no. 3, pp. 1227–1233, 2020.
40. D. M. French and D. Shiffler, "High power microwave source with a three dimensional printed metamaterial slow-wave structure," *Rev. Sci.Instrum.*, vol. 87, no. 5, pp. 053308, 2016.
41. G. Wu, Q. Li, X. Lei, C. Ding, X. Jiang, S. Fang, R. Yang, F. Wang, L. Yue, Y. Gong, and Y. Wei, "Design of a cascade backward-wave oscillator based on metamaterial slow-wave structure," *IEEE Trans. Elect.Dev.*, vol. 65, no. 3, pp. 1172–1178, 2018.
42. J. S. Hummelt, S. M. Lewis, M. A. Shapiro, and R. J. Temkin, "Design of a metamaterial-based backward-wave oscillator," *IEEETrans. Plasma Sci.*, vol. 42, no. 4, pp. 930–936, 2014.
43. X. Lu, J. C. Stephens, I. Mastovsky, M. A. Shapiro, and R. J. Temkin, "High power long pulse microwave generation from a metamaterial structure with reverse symmetry," *Phys. Plasmas*, vol. 25, no. 2, pp. 023102, 2018a.
44. M. Liu, E. Schamiloglu, S. C. Yurt, A. Elfrgani, M. I. Fuks, and C. Liu, "Coherent Cherenkov-cyclotron radiation excited by an electron beam in a two-spiral metamaterial waveguide," *AIP Adv.*, vol. 8, no. 11, pp. 115107, 2018.
45. H. Seidfaraji, A. Elfrgani, C. Christodoulou, and E. Schamiloglu, "A multibeam metamaterial backward wave oscillator," *Phys. Plasmas*, vol. 26, no. 7, pp. 073105, 2019.
46. M. A. Shapiro, S. Trendafilov, Y. Urzhumov, A. Alù, R. J. Temkin, and G. Shvets, "Active negative-index metamaterial powered by an electron beam," *Phys. Rev. B, Condens. Matter*, vol. 86, no. 8, pp. 085132, 2012.
47. D. R. Smith, S. Schultz, P. Markoš, and C. M. Soukoulis, "Determination of effective permittivity and permeability from Reflectionreflection and transmission coefficients," *Phys. Rev. B*, vol. 65, no. 19, pp. 195104, 2002.
48. X. Chen, T. M. Grzegorzyk, B.-I. Wu, J. Pacheco, Jr., and J. A. Kong, "Robust method to retrieve the constitutive effective parameters of metamaterials," *Phys. Rev. E*, vol. 70, no. 1, pp. 016608, 2004.

49. P. K. Jain and B. N. Basu, "Electromagnetic wave propagation through helical structures," in *Electromagnetic Fields in Unconventional Materials*, O. N. Singh and A. Lakhtakia, Eds.Hoboken, NJ, USA: Wiley, 2000.
50. B. N. Basu, *Electromagnetic Theory and Applications in Beam-Wave Electronics*, Singapore, New Jersey, London, Hong Kong: World Scientific Publishing Co. Inc., 1996.
51. V. Kesari and B. N. Basu, *High Power Microwave Tubes: Basics and Trends*, Volumes 1 and 2, Morgan and Claypool Publishers, San Rafael (California)/Bristol: IOP Publishing 2018.
52. C. K. Birdsall, G. R. Brewer, and A. V. Haeff, "The resistive-wall amplifier," *Proc. IRE*, vol. 41, no. 7, pp. 865–875, 1953.
53. C. K. Birdsall and J. R. Whinnery, "Waves in an electron stream with general admittance walls," *J. Appl. Phys.*, vol. 24, no. 3, pp. 314–323, 1953.
54. T. Rowe, N. Behdad, and J. H. Booske, "Metamaterial-enhanced resistive wall amplifier design using periodically spaced inductive meandered lines," *IEEE Trans. Plas. Sci.*, vol. 44, no. 10, pp. 2476–2484, 2016.
55. S. Tretyakov, *Analytical Modeling in Applied Electromagnetics*. Norwood, MA, USA: Artech House, 2003.
56. X. Wang, S. Li, X. Zhang, S. Jiang, Z. Wang, H. Gong, Y. Gong, B. N. Basu, and Z. Duan, "Novel S-band metamaterial extended interaction klystron," *IEEE Elect. Dev. Lett.*, vol. 41, no. 10, pp. 1580–1583, 2020.
57. X. Wang, Z. Duan, X. Zhan, F. Wang, S. Li, S. Jiang, Z. Wang, Y. Gong, and B. N. Basu, "Characterization of metamaterial slow-wave structure loaded with complementary electric split-ring resonators," *IEEE Trans. Microw. Theo. Tech.*, vol. 67, no. 6, pp. 2238–2246, 2019.
58. A. Weiss and A. Grbic, "Electron beam coupling to an NRI transmission-line metamaterial," *IEEE Trans. Plas. Science*, vol. 43, no. 3, pp. 796–803, 2015.
59. R. Marqués, J. Martel, F. Mesa, and F. Medina, "Left-handed-media simulation and transmission of EM waves in subwavelength split-ring resonator-loaded metallic waveguides," *Phys. Rev. Lett.*, vol. 89, no. 18, pp. 183901, 2002.
60. H. Xu, Z. Wang, J. Hao, J. Dai, L. Ran, J. A. Kong, and L. Zhou, "Effective-medium models and experiments for extraordinary transmission in metamaterial-loaded waveguides," *Appl. Phys. Lett.*, vol. 92, no. 4, pp. 041122, 2008.
61. S. Hrabar, J. Bartolic, and Z. Sipus, "Waveguide miniaturization using uniaxial negative permeability metamaterial," *IEEE Trans. Antennas Propag.*, vol. 53, no. 1, pp. 110–119, 2005.
62. J. Esteban, C. Camacho-Penalosa, J. Page, T. Martin-Guerrero, and E. Marquez-Segura, "Simulation of negative permittivity and negative permeability by means of evanescent waveguide modes—Theory and experiment," *IEEE Trans. Microw. Theory Techn.*, vol. 53, no. 4, pp. 1506–1514, 2005.
63. I. S. Nefedov, X. Dardenne, C. Craeye, and S. A. Tretyakov, "Backward waves in a waveguide, filled with wire media," *Microw. Opt. Technol.Lett.*, vol. 48, no. 12, pp. 2560–2564, 2006.
64. X. Tang, Z. Duan, X. Ma, S. Li, F. Wang, Y. Wang, Y. Gong, and J. Feng, "Dual band metamaterial Cherenkov oscillator with a waveguide coupler," *IEEE Trans. Electron Devices*, vol. 64, no. 5, pp. 2376–2382, 2017.
65. J. S. Hummelt, X. Lu, H. Xu, I. Mastovsky, M. A. Shapiro, and R. J. Temkin, "Coherent cherenkov-cyclotron radiation excited by an electron beam in a metamaterial waveguide," *Phys. Rev. Lett.*, vol. 117, no. 23, pp. 237701, 2016.
66. H. J. Curnow, "A general equivalent circuit for coupled-cavity slow-wave structures," *IEEE Trans. Microw. Theory Techn.*, vol. 13, no. 5, pp. 671–675, 1965.
67. S. C. Yurt, M. I. Fuks, S. Prasad, and E. Schamiloglu, "Design of a metamaterial

slow wave structure for an O-type high power microwave generator," *Phys. Plasmas*, vol. 23, no. 12, pp. 123115, 2016.
68. P. A. Cherenkov, "Visible emission of clean liquids by action of γ radiation," *Dokl. Akad. Nauk SSSR*, vol. 2, pp. 451–454, 1934.
69. I. M. Frank, and I. E. Tamm, "Coherent visible radiation of fast electrons passing through matter," *Dokl. Akad. Nauk SSSR*, vol. 14, pp. 109–114, 1937.
70. V. L. Ginzburg, "The quantum theory of light radiation of an electron uniformly moving in a medium," *Journal of Physics (Moscow)*, vol. 2, pp. 441–452, 1940.
71. V. G. Veselago, "The electrodynamics of substances with simultaneously negative values of ε and μ," *Soviet Physics Uspekhi*, vol. 10, no. 4, pp. 509–514, 1968.
72. Z.Duan, B.-I. Wu, J. Lu, J. A. Kong, and M. Chen, "Cherenkov radiation in anisotropic double-negative metamaterials," *Opt. Express*, vol. 16, no. 22, pp. 18479–18484, 2008a.
73. Z. Duan, B.-I. Wu, J. Lu, J. A. Kong, and M. Chen, "Reversed Cherenkov radiation in a waveguide filled with anisotropic double negative metamaterials," *J. Appl. Phys.*, vol. 104, no. 6, pp. 063303, 2008b.
74. Z. Duan, B.-I. Wu, J. Lu, J. A. Kong, and M. Chen, "Reversed Cherenkov radiation in unbounded anisotropic double-negative metamaterials," *J. Phys. D Appl. Phys.*, vol. 42, pp. 185102, 2009a.
75. Z. Duan, B.-I. Wu, S. Xi, H. Chen, and M. Chen, "Research progress in reversed Cherenkov radiation in double-negative metamaterials," *Prog. Electromagn. Res.*, vol. 90, pp. 75–87, 2009b.
76. Y. O. Averkov and V. M. Yakovenko, "Cherenkov radiation by an electron bunch that moves in a vacuum above a left-handed material," *Phys. Rev. B*, vol. 72, pp. 205110, 2005.
77. J. Hummelt, S. Lewis, H. Xu, M. Shapiro, I. Mastovsky, and R. Temkin, "Fabrication and test of a high power S-band metamaterial backward-wave oscillator," *Proc. IEEE Int. Vac. Electron. Conf. (IVEC)*, pp. 1–2, 2015.
78. Z. Duan, X. Tang, Z. Wang, Y. Zhang, X. Chen, M. Chen, and Y. Gong, "Observation of the reversed Cherenkov radiation," *Nature Commun.*, vol. 8, pp. 14901, 2017.
79. T. Rowe, P. Forbes, J. H. Booske, and N. Behdad, "Inductive meandered metal line metamaterial for rectangular waveguide linings," *IEEE Trans. Plas. Sci.*, vol. 45, no. 4, pp. 654–664, 2017.
80. N. N. Esfahani, K. Schünemann, N. Avtomonov and D. Vavriv, "Epsilon near zero loaded magnetrons, design and realization," European Microwave Conference (EuMC), pp. 454–457, 2015a.
81. N. N. Esfahani and K. Schünemann, "Application of metamaterials in spatial harmonic magnetrons," IEEE MTT-S International Microwave Symposium, pp. 1–4, 2015b.
82. N. N. Esfahani. "Low-current cathode spatial harmonic magnetrons: Analysis and realization based on metamaterial loaded slow wave structures,"*Int. J. Micro. Wire. Tech.*, vol. 10, no. 5-6, pp. 613–619, 2018.
83. A. D. Andreev and K. J. Hendricks, Metamaterial-like cathodes in multicavity magnetrons *IEEE Trans. Plas. Sci.*, vol. 40, no. 9, pp. 2267–2273, 2012.
84. M. Thumm, "High-power microwave transmission systems, external mode converters and antenna technology," in *Gyrotron Oscillators: Their Principles and Practice*,C. J. Edgombe, Ed. ch. 13, pp. 365–401, London, U.K.: Taylor & Francis, 1993.
85. W. Fu, S. Hu, C. Zhang, X. Guan and Y. Yan, "Study on a quasi-optical mode converter for gyrotron based on metamaterial," 44th Int. Conf. Infr., Millim., and Tera. Waves (IRMMW-THz), pp. 1–2, 2019.

86. S. Antipov, L. Spentzouris, W. Gai, W. Liu, and J. G. Power, "Double-negative metamaterial research for accelerator applications," *Nucl. Instrum. Methods Phys. Res. A, Accel. Spectrom. Detect. Assoc. Equip.*, vol. 579, no. 3, pp. 915–923, 2007.
87. I. McGregor and K. M. Hock, "Complementary split-ring resonator based accelerating structure," *J. Instrum.*, vol. 8, pp. 05009, 2013b.
88. E. Sharples and R. Letizia, "Investigation of CSRR loaded waveguide for accelerator applications," *J. Instrum.*, vol. 9, pp. 11017, 2014.
89. X. Lu, M. A. Shapiro, and R. J. Temkin, "Modeling of the interaction of a volumetric metallic metamaterial structure with a relativistic electron beam," *Phys. Rev. Accel. Beams*, vol. 18, no. 8, pp. 081303, 2015.
90. X. Lu, M. A. Shapiro, I. Mastovsky, R. J. Temkin, "A metamaterial wagon wheel structure for wakefield acceleration by reversed cherenkov radiation," *Proc. 9th Int. Part. Accel. Conf. (IPAC)*, Vancouver, BC, Canada, pp. 4681–4683, 2018b.
91. D. P. Starinshak and J. D. Wilson, "Investigating dielectric and metamaterial effects in a terahertz traveling wave tube amplifier," NASA, Cleveland, OH, USA, Tech. Rep. NASA/TM-2008-215059, 2008.
92. Z. Duan, C. Guo, J. Zhou, J. Lu, and M. Chen, "Novel electromagnetic radiation in a semi-infinite space filled with a double-negative metamaterial," *Phys. Plas.*, vol. 19, no. 1, pp. 013112-1–013112-5, 2012.
93. Z. Duan, C. Guo, X. Guo, and M. Chen, "Double negative-metamaterial based terahertz radiation excited by a sheet beam bunch," *Phys. Plas.*, vol. 20, no. 9, pp. 093301, 2013.
94. N. Bai, C. Feng, Y. Liu, H. Fan, C. Shen, and X. Sun, "Integrated microstrip meander line traveling wave tube based on metamaterial absorber," *IEEE Trans. Elect. Dev.*, vol. 64, no. 7, pp. 2949–2954, 2017.
95. N. Bai, W. Xiang, J. Shen, C. Shen and X. Sun, "A *Ka*-band folded waveguide traveling wave tube with lumped resistance metamaterial absorber," *IEEE Trans. Elect. Dev.*, vol. 67, no. 3, pp. 1248–1253, 2020.
96. X. Tang, Z. Duan, X. Shi, Y. Zhang, Z. Wang, Y. Gong, and J. Feng, "Sheet electron beam transport in a metamaterial-loaded waveguide under the uniform magnetic focusing," *IEEE Trans. Elect. Dev.*, vol. 63, no. 5, pp. 2132–2138, 2016.
97. N. Bai, M. Shen and X. Sun, "Investigation of microstrip meander-line traveling-wave tube using EBG ground plane," *IEEE Trans. Elect. Dev.*, vol. 62, no. 5, pp. 1622–1627, 2015.
98. T. Fu, Y. B. An, and Z. B. Ouyang, "A novel Cherenkov oscillator based on microcavity in photonic crystal waveguide," Prog. Elect. Res. Symp. (PIERS), pp. 1225–1228, 2016.
99. Y. Gong, H. Yin, Y. Wei, L. Yue, M. Deng, Z. Lu, X. Xu, W. Wang, P. Liu, and F. Liao, "Study of traveling wave tube with folded-waveguide circuit shielded by photonic crystals," *IEEE Trans. Elect. Dev.*, vol. 57, no. 5, pp. 1137–1145, 2010.
100. R. Letizia, M. Mineo, C. Paoloni, "Photonic crystal-structures for THz vacuum electron devices," *IEEE Trans. Elec. Dev.*, vol. 62, no. 1, pp. 178–183, 2015.
101. R. Letizia, M. Mineo and C. Paoloni, "Photonic crystal-coupler for sheet beam THz vacuum electron tubes," *IEEE Elec. Dev. Lett.*, vol. 37, no. 9, pp. 1227–1230, 2016.
102. S. Gong, M. Hu, R. Zhong, X. Chen, P. Zhang, T. Zhao, and S. Liu, "Electron beam excitation of surface plasmon polaritons," Opt. Expr. vol. 22, pp. 19252–19261, 2014.
103. W. X. Tang, H. C. Zhang, H. F. Ma, W. X. Jiang, T. J. Cui, "Concept, theory, design, and applications of spoof surface plasmon polaritons at microwave frequencies," *Adv. Opt. Mate.* vol. 7, pp. 1800421, 2019.

Index